IFIP Advances in Information and Communication Technology 381

IFIP – The International Federation for Information Processing

IFIP was founded in 1960 under the auspices of UNESCO, following the First World Computer Congress held in Paris the previous year. An umbrella organization for societies working in information processing, IFIP's aim is two-fold: to support information processing within its member countries and to encourage technology transfer to developing nations. As its mission statement clearly states,

> *IFIP's mission is to be the leading, truly international, apolitical organization which encourages and assists in the development, exploitation and application of information technology for the benefit of all people.*

IFIP is a non-profitmaking organization, run almost solely by 2500 volunteers. It operates through a number of technical committees, which organize events and publications. IFIP's events range from an international congress to local seminars, but the most important are:

- The IFIP World Computer Congress, held every second year;
- Open conferences;
- Working conferences.

The flagship event is the IFIP World Computer Congress, at which both invited and contributed papers are presented. Contributed papers are rigorously refereed and the rejection rate is high.

As with the Congress, participation in the open conferences is open to all and papers may be invited or submitted. Again, submitted papers are stringently refereed.

The working conferences are structured differently. They are usually run by a working group and attendance is small and by invitation only. Their purpose is to create an atmosphere conducive to innovation and development. Refereeing is also rigorous and papers are subjected to extensive group discussion.

Publications arising from IFIP events vary. The papers presented at the IFIP World Computer Congress and at open conferences are published as conference proceedings, while the results of the working conferences are often published as collections of selected and edited papers.

Any national society whose primary activity is about information processing may apply to become a full member of IFIP, although full membership is restricted to one society per country. Full members are entitled to vote at the annual General Assembly, National societies preferring a less committed involvement may apply for associate or corresponding membership. Associate members enjoy the same benefits as full members, but without voting rights. Corresponding members are not represented in IFIP bodies. Affiliated membership is open to non-national societies, and individual and honorary membership schemes are also offered.

Lazaros Iliadis   Ilias Maglogiannis
Harris Papadopoulos (Eds.)

Artificial Intelligence Applications and Innovations

8th IFIP WG 12.5 International Conference, AIAI 2012
Halkidiki, Greece, September 27-30, 2012
Proceedings, Part I

 Springer

Volume Editors

Lazaros Iliadis
Democritus University of Thrace
Department of Forestry and Management of the Environment
Pandazidou 193, 68200 Orestiada, Greece
E-mail: liliadis@fmenr.duth.gr

Ilias Maglogiannis
University of Central Greece
Department of Computer Science and Biomedical Informatics
Papasiopoulou 2-4, PC 35100 Lamia, Greece
E-mail: imaglo@ucg.gr

Harris Papadopoulos
Frederick University
Department of Computer Science and Engineering
7 Yianni Frederickou Str., Pallouriotissa, 1036 Nicosia, Cyprus
E-mail: h.papadopoulos@frederick.ac.cy

ISSN 1868-4238   e-ISSN 1868-422X
ISBN 978-3-642-43653-6   ISBN 978-3-642-33409-2 (eBook)
DOI 10.1007/978-3-642-33409-2
Springer Heidelberg Dordrecht London New York

CR Subject Classification (1998): I.2.6-8, I.5.1, I.5.3-4, I.2.1, I.2.3-4, I.2.10-11, H.2.8, I.4.3, I.4.8, K.3.1, H.3.4, F.1.1, F.2.1, H.4.2, J.3

Typesetting: Camera-ready by author, data conversion by Scientific Publishing Services, Chennai, India

Printed on acid-free paper

Springer is part of Springer Science+Business Media (www.springer.com)

Preface

After 50 years of research in Artificial Intelligence (AI), the dream of intelligent machines that use sophisticated and advanced approaches is becoming a reality. AI researchers have already created systems capable of tackling complicated and challenging problems. Scientists have developed analytical models and corresponding systems that can mimic human behavior and cognition: they can understand speech, beat expert human chess players, and perform countless other feats that can have a potential impact on our everyday lives. It is a fact that humans are a species that learn by training plus trial and error, so it can be considered rational to see AI more as a blessing and less as an inhibition. On the other hand the misuse of AI technology is always a potential.

The eighth AIAI conference was supported and sponsored by the International Federation for Information Processing (IFIP). AIAI is the official conference of the IFIP Working Group 12.5 "Artificial Intelligence Applications". IFIP was founded in 1960 under the auspices of UNESCO, following the first World Computer Congress, held in Paris the previous year. The first AIAI conference was held in Toulouse, France in 2004 and since then it has been held annually, offering scientists the chance to present the achievements of AI applications in various fields.

This Springer volume belongs to the IFIP AICT series. It contains the papers that were accepted to be presented orally at the main event of the eighth AIAI conference. A second volume contains the papers accepted for the eight workshops that were organized as parallel events, namely: the second AIAB, the first AIeIA, the second CISE, the first COPA, the first IIVC, the third ISQL, the first MHDW, and the first WADTMB. More details on the workshops will be given in the following paragraphs.

The eighth AIAI conference was held during September 27–30, 2012 on the Sithonia peninsula of Halkidiki, Greece. The diverse nature of the papers presented demonstrates the vitality of AI computing approaches and proves the very wide range of AI applications as well. On the other hand, this volume contains basic research papers, presenting variations and extensions of several existing methodologies.

The response to the call for papers was more than satisfactory with 98 papers initially submitted to the main event. All papers were peer reviewed by at least two independent academic referees. Where needed a third referee was consulted to resolve any conflicts.

A total of 43.9% of the submitted manuscripts (44 papers) have been published in these proceedings as full papers whereas 5.1% have been published as short ones. The authors of the accepted papers of the main event came from 17 different countries, namely: Belgium, Croatia, Cyprus, the Czech Republic, France, Germany, Greece, Iran, Italy, Pakistan, Poland, Portugal, Romania, Russia, Tunisia, the UK, and the USA.

Three keynote speakers were invited to lecture at AIAI 2012:

1. Dr. Danil Prokhorov from Toyota Research Institute NA, Ann Arbor, Michigan delivered a talk entitled "Computational Intelligence in Automotive Applications".
2. Prof. David Robertson from the University of Edinburgh talked on "Knowledge Engineering on a Social Scale".
3. Prof. Dr. Bernard De Baets from KERMIT, Ghent University talked on: "Monotonicity Issues in Fuzzy Modelling, Machine Learning and Decision Making".

Also, two tutorials were organized in the framework of the AIAI 2012:

1. Prof. Tatiana Tambouratzis from the University of Piraeus focused on "Identification of Key Music Symbols for Optical Music Recognition and On-Screen Presentation".
2. Prof. Costin Badica from the University of Craiova focused on "Negotiations in Multi-Agent Systems".

The accepted papers of AIAI 2012 are related to the following thematic topics:

- Artificial Neural Networks
- Bioinformatics
- Clustering
- Control Systems
- Data Mining
- Engineering Applications of AI
- Face Recognition - Pattern Recognition
- Filtering
- Fuzzy Logic
- Genetic Algorithms, Evolutionary Computing
- Hybrid Clustering Systems
- Image and Video Processing
- Multi Agent Systems
- Multi Attribute DSS
- Ontology - Intelligent Tutoring systems
- Optimization, Genetic Algorithms
- Recommendation Systems
- Support Vector Machines - Classification
- Text Mining

A total of eight workshops were organized as parallel events to AIAI 2012. Each one of these workshops was related to a specific AI topic, and was managed by internationally well-recognized colleagues, who put together the specific workshop programs mainly by invitation to prominent authors.

All workshops received a high response from scientists from all parts of the globe, from Europe to Australia, and we would like to thank all participants for this. More specifically, scientists from 13 countries (Australia, Belgium, Cyprus, Finland, France, Germany, Greece, Italy, Romania, South Korea, Spain, the UK, and the USA) submitted interesting and innovative research papers to the eight workshops.

- We are grateful to Profs. Harris Papadopoulos, Efthyvoulos Kyriacou (Frederick University, Cyprus) Prof. Ilias Maglogiannis (University of Central Greece) and Prof. George Anastassopoulos (Democritus University of Thrace, Greece) for their common effort towards the organization of the Second Artificial Intelligence Applications in Biomedicine Workshop (AIAB 2012).

- We wish to express our gratitude to Prof. Achilleas Kameas and Dr. Antonia Stefani (Hellenic Open University, Greece) for adding the First AI in Education Workshop: Innovations and Applications (AIeIA 2012) to the family of the AIAI workshops.

- We are very happy to see that AIAI workshops are repeated every year with the presentation of new and fresh research efforts. Many thanks to Prof. Andreas Andreou (Cyprus University of Technology) and Dr. Efi Papatheocharous (University of Cyprus) for the organization of the Second International Workshop on Computational Intelligence in Software Engineering (CISE 2012).

- We are also very happy about the organization of the First Conformal Prediction and its Applications Workshop (COPA 2012) by Prof. Harris Papadopoulos (Frederick University, Cyprus) and Profs. Alex Gammerman and Vladimir Vovk (Royal Holloway, University of London, UK).

- The First Intelligent Innovative Ways for Video-to-Video Communication in Modern Smart Cities Workshop (IIVC 2012) was an important part of the AIAI 2012 event and it was driven by the hard work of Drs. Ioannis P. Chochliouros and Ioannis M. Stephanakis (Hellenic Telecommunications Organization - OTE, Greece), and Profs. Vishanth Weerakkody (Brunel University, UK) and Nancy Alonistioti (National & Kapodistrian University of Athens, Greece).

- It was a pleasure to host the Third Intelligent Systems for Quality of Life Information Services Workshop (ISQL 2012) for one more time in the framework of the AIAI conference. We wish to sincerely thank Profs. Kostas Karatzas (Aristotle University of Thessaloniki, Greece), Lazaros Iliadis (Democritus University of Thrace, Greece), and Mihaela Oprea (University of Petroleum-Gas of Ploesti, Romania) for the presentation of AI applications in the crucial topics of sustainable development and quality of life.

- We would like thank Profs. Spyros Sioutas, Ioannis Karydis, and Katia Kermanidis (all with the Ionian University, Greece) for their hard work in organizing the First Mining Humanistic Data Workshop (MHDW 2012).

- Finally, we would like to thank Profs. Athanasios Tsakalidis and Christos Makris (all with the University of Patras, Greece) for the very successful organization of the First Workshop on Algorithms for Data and Text Mining in Bioinformatics (WADTMB 2012).

After eight years, the AIAI conference has become a mature well-established event with loyal followers and it has plenty of new and high-quality research results to offer to the International scientific community. We hope that these proceedings will be of major interest to scientists and researchers world wide and that they will stimulate further research in the domain of artificial neural networks and AI in general.

September 2012   AIAI 2012 Chairs

Organization

Executive Committee

General Chair

Tharam Dillon   Curtin University of Technology, Australia

Honorary Chairs

Max Bramer   University of Portsmouth, UK
Andreas Andreou   Cyprus University of Technology, Cyprus
Dominic Palmer Brown   Dean London Metropolitan University, UK

Program Committee Co-chairs

Lazaros Iliadis   Democritus University of Thrace, Greece
Ilias Maglogiannis   University of Central Greece
Haris Papadopoulos   Frederick University, Cyprus

Workshop Chair

Kostas Karatzas   Aristotle University of Thessaloniki, Greece
Spyros Sioutas   Ionian University, Greece

Advisory Chair

Chrisina Jayne   University of Coventry, UK

Organizing Chairs

Yannis Manolopoulos   Aristotle University of Thessaloniki, Greece
Elias Pimenidis   University of East London, UK

Web Chair

Ioannis Karydis   Ionian University, Greece

Program Committee

Members

Aldanondo, Michel
Alexandridis, Georgios
Anastassopoulos, George
Andreadis, Ioannis
Badica, Costin
Bankovic, Zorana
Bessis, Nick
Caridakis, Georgios
Charalambous, Christoforos
Chatzioannou, Aristotelis
Constantinides, Andreas
Donida Labati, Ruggero
Doukas, Charalampos
Fachantidis, Anestis
Fernandez de Canete, Javier
Flaounas, Ilias
Magda, Florea Adina
Fox, Charles
Gaggero, Mauro
Gammerman, Alex
Georgiadis, Christos
Georgopoulos, Efstratios
Hajek, Petr
Hatzilygeroudis, Ioannis
Kabzinski, Jacek
Kalampakas, Antonios
Kameas, Achilles
Karpouzis, Kostas
Karydis, Ioannis
Kefalas, Petros
Kermanidis, Katia
Kitikidou, Kyriakh
Kosmopoulos, Dimitrios
Koutroumbas, Kostantinos
Kurkova, Vera
Kyriacou, Efthyvoulos
Lazaro, Jorge Lopez
Lorentzos, Nikos

Lykothanasis, Spyridon
Malcangi, Mario
Maragkoudakis, Manolis
Marcelloni, Francesco
Margaritis, Kostantinos
Mouratidis, Harris
Nicolaou, Nicoletta
Onaindia, Eva
Oprea, Mihaela
Papatheocharous, Efi
Partalas, Ioannis
Pericleous, Savas
Plagianakos, Vassilis
Rao, Vijay
Roveri, Manuel
Sakelariou, Ilias
Samaras, Nikos
Schizas, Christos
Senatore, Sabrina
Sgarbas, Kyriakos
Sideridis, Alexandros
Spartalis, Stephanos
Stamelos, Ioannis
Stephanakis, Ioannis
Tambouratzis, Tatiana
Tsapatsoulis, Nikos
Tscherepanow, Marko
Tsiligkiridis, Theodoros
Tsitiridis, Aristeidis
Tsoumakas, Grigorios
Tzouramanis, Theodoros
Verykios, Vassilios
Voulgaris, Zacharias
Vouyioukas, Demosthenis
Vovk, Volodya
Yialouris, Kostas
Yuen, Peter

Table of Contents – Part I

Learning and Data Mining

Fuzzy Logic

Classification Pattern Recognition

Multi Agent Systems

Multi Attribute DSS

Clustering

Image-Video Classification and Processing

Engineering Applications of AI and Artificial Neural Networks

Table of Contents – Part II

Second Artificial Intelligence Applications in Biomedicine Workshop (AIAB 2012)

First AI in Education Workshop: Innovations and Applications (AIeIA 2012)

Second International Workshop on Computational Intelligence in Software Engineering (CISE 2012)

First Conformal Prediction and Its Applications Workshop (COPA 2012)

First Intelligent Innovative Ways for Video-to-Video Communication in Modern Smart Cities Workshop (IIVC 2012)

Third Intelligent Systems for Quality of Life Information Services Workshop (ISQL 2012)

First Mining Humanistic Data Workshop (MHDW 2012)

First Workshop on Algorithms for Data and Text Mining in Bioinformatics (WADTMB 2012)

A Probabilistic Approach to Organic Component Detection in Leishmania Infected Microscopy Images

Pedro Alves Nogueira and Luís Filipe Teófilo

LIACC – Artificial Intelligence and Computer Science Lab., University of Porto, Portugal
FEUP – Faculty of Engineering, University of Porto – DEI, Portugal
{pedro.alves.nogueira,luis.teofilo}@fe.up.pt

Abstract. This paper proposes a fully automated method for annotating confocal microscopy images, through organic component detection and segmentation. The organic component detection is performed through adaptive segmentation using a two-level Otsu's Method. Two probabilistic classifiers then analyze the detected regions, as to how many components may constitute each one. The first of these employs rule-based reasoning centered on the decreasing harmonic patterns observed in the region area density functions. The second one consists of a Support Vector Machine trained with features derived from the log likelihood ratios of incrementally Gaussian mixture modeling detected regions. The final step pairs the identified cellular and parasitic components, computing the standard infection ratios on biomedical research. Results indicate the proposed method is able to perform the identification and annotation processes on par with expert human subjects, constituting a viable alternative to the traditional manual approach.

Keywords: Leishmania; Gaussian mixture models; Computer Vision, SVM.

1   Introduction

Leishmania is the parasite responsible for Leishmaniasis, a disease currently affecting over 12 million people throughout 88 countries [1]. Leishmaniasis is treatable by chemotherapeutics, which, nevertheless, suffer from poor administration regimens and high host toxicity [2]. Although the disease is not generally deadly, it severely damages the immune system, leaving the body exposed to other deadly pathogens, which often prove fatal [2]. The inadequate means to treat Leishmaniasis render the research for new treatments an urgent task.

Research in microscopy images produces large amount of data, which require anywhere from days to weeks to classify and annotate. In a single laboratory the number can easily ascend to thousands of images with merely a dozen different experiments. Not only does this detract the researchers from exploring new alternatives, as it also introduces inter-person variance, as many images are extremely cluttered and contain several hundreds of cells and parasites. This results in a time consuming and mentally straining process, which expresses itself as a decaying function over time as the subject gets tired,

L. Iliadis et al. (Eds.): AIAI 2012, IFIP AICT 381, pp. 1–10, 2012.

frustrated or bored. These reasons justify the need for the development of automatic mechanisms to replace or aid researchers in the annotation task, for which and to the best of our knowledge no current solution exists. The proposed method provides a fully automatic pipeline for the identification of cells and parasites in Leishmania infected microscopy imaging, enabling more accurate annotations.

Pertaining this paper's organisation, it is structured as follows. Section 2 describes the main characteristics of fluorescence microscopy imaging, as well as the dataset used in this study. Section 3 discusses the state of the art in cell identification and segmentation in microscopy imaging. Section 4 briefly describes the proposed method, followed by the description of its steps. In section 5, the results for the segmentation and classifiers, as well as from the method's application to two real drug trials are presented. Finally, section 6, presents conclusions on the developed work, commenting on its performance and readiness for real-world applicability.

2 Fluorescence Microscopy Imaging

In contrast to the classical optical microscopy, the use of fluorescence microscopy allows simultaneous labeling of different cell components, which can be easily distinguished based on the fluorescence properties of their specific dyes [3]. The images collected for this study used three fluorophores [3], which emitted three distinct wavelengths. These corresponded to the cell nuclei DNA (in blue), cytoplasmic and nuclear DNA (in red) and the parasitic DNA (in green). This provided three separate sets of data per image (Figure 1), motivating the identification of cells, parasites and cytoplasm individually in the three channels as independent images.

Although very popular, fluorescence microscopy imaging (FMI) presents some well-known issues that also characterized our datasets. The most noticeable issues include: non-linear illumination (due to poor lighting conditions and sub-optimal experimental setup), photo bleaching, varying contrast, Gaussian noise, chromatic aberrations and overlapping cells and parasites (due to various focal planes).

In this study 794 fluorescence microscopy images from random drug trials with different experimental setups were collected and used. These images were collected through a light microscope and annotated manually by a Leishmania research team at the INEB/IBMC laboratory. Refer to section 5 for further details.

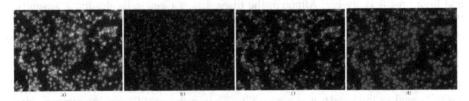

Fig. 1. Details of a fluorescence microscopy image. a) Original image; b) Cell nuclei channel; c) Parasite nuclei channel; d) Cytoplasmic channel.

3   Related Work

Microscopy image analysis has been an active field for several decades. In related work, Liao et al. [4] used a simple thresholding method, coupled with mathematical morphology and contextual shape detection to detect white blood cells. However, their approach does not tolerate cells outside the defined conditions (e.g.: poorly segmented regions, forming a cell cluster region). An automated method for cellular membrane segmentation is described in [5]. This method also allows the reconstruction of unstained tracts through the nuclear membranes as a spatial reference. Jiang et al have also proposed white blood cell segmentation using scale-space filtering and watershed clustering in HSV color space [6].

Park has proposed bone marrow cell segmentation through an iteratively relaxed watershed algorithm [7]. However, this work is sensitive to illumination and noise conditions, since it overly relies on the fixed mean color values of each patch for the relaxation procedure. Begelman [8] performs cell nuclei segmentation using color, shape features and a fuzzy logic engine. This work is more robust than the aforementioned one because the extracted shape features, serve as an auxiliary classification input. However, it is still not able to account for non-circular geometries or abnormally colored cells due to the implemented rules' simplicity.

Yu proposes using an adaptive thresholding technique to detect cell nuclei, which are then expanded via level sets to determine cell boundaries [9]. Yan proposes a similar approach [11]. Yan improves on Yu by replacing the adaptive thresholding with a distance map of the initial adaptive histogram-based thresholding step. This distance map is then used to create a watershed transform, serving as a region list representing the level-set seed points. The only drawback to this approach is that it is not able to deal with highly cluttered images, as the distance map would not provide enough information to accurately parameterize the watershed transform, thus leading to an erroneous number/location of seed points for the level-set step.

From this review, it is clear that most of the literature does not attempt or is unable to deal with highly cluttered or overlapped image regions. This is a major concern in microscopy image analysis as the great majority of real-world data is heavily cluttered and saturated. The proposed method in this paper aims at addressing this issue.

4   Proposed Method

The proposed method focuses on developing robust methods for identifying and segmenting cellular and parasitic agglomerations in confocal microscopy images. The method starts by splitting the original image f in three channels: f_c (blue); f_p (green); f_{cyt} (red). Each channel is then normalized and segmented through Otsu's Method. This yields three region vectors, corresponding each one to cellular DNA, parasitic DNA and weak[1] DNA signatures (cytoplasm). Low-level features are computed for

---

[1] In order to register a weak DNA signature, this fluorophore must be highly sensitive. Thus, it also registers the cell nuclei's DNA. Since the cell nuclei can be trivially subtracted through set operations involving the cell channel, we denominated this channel as the cytoplasmic channel.

the first two aforementioned region vectors, and then used to train a rule-based classifier and a Support Vector Machine, both of which attempt to determine how many cells or parasites each region contains. To resolve disputes between these two classifiers, a voting system taking into account both of the classifiers' error margins is employed. Each region is then further segmented into the predicted number of sub-regions by Gaussian unmixing. Figure 2 can be inspected bellow for a more structured and clear understanding of the described pipeline.

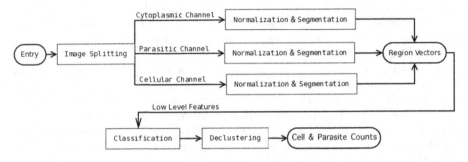

Fig. 2. Developed method's architecture

4.1 Pre-processing and Segmentation

The method first splits each of the target image's f color channels, as they are independently processed. Each image channel is then normalized and segmented into background and foreground components. An initial study on the general image characteristics was conducted in order to choose an appropriate segmentation technique. In this study, the intensity values of 120 randomly selected images presented clear bi-modal distributions for all color channels (Figure 3).

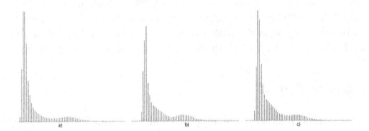

Fig. 3. Bimodal distributions observed in the RGB components of 120 images (averaged). a) Red color component; b) Green color component; c) Blue color component.

Thus, Otsu's Method presented itself as a fitting approach due to its, low temporal and spatial complexity, non-parameterisable characteristics and segmentation principle. Otsu's Method's principle assumes a bi-modal distribution in the target dataset, for which it attempts to determine an optimal threshold value by minimizing intra-class variance [12]:

$$\sigma_b^2(t) = \sigma^2 - \sigma_\omega^2(t) = \omega_1(t)\,\omega_2(t)\,[\mu_1(t) - \mu_2(t)]^2 \qquad (1)$$

Thus, each color channel was binarised using Otsu's Method and then proceeded to a connected component analysis, resulting in a region vector representing the cellular, parasitic and cytoplasmic regions present in the image. Note that the cytoplasmic regions are not used in this work, as they are intended for associating cell-parasite pairs. They were, however, computed and integrated into the method in hindsight of future work. Figure 4 depicts the result of the segmentation step for the cellular channel.

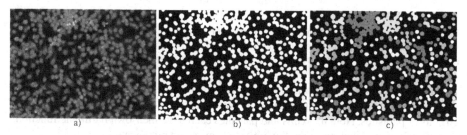

Fig. 4. Segmentation output of a moderately cluttered image. a) Original image; b) Segmentation output; c) Visual representation of the cellular region vector obtained from the connected component analysis (randomly color-coded).

Following the region extraction, a low-level feature vector $F_i=[a, ll_{1..N}, d^1_{1..N}(ll), d^2_{1..N}(ll)]$, is computed for each region. The features comprising the vector are: a, the area value (in pixels); ll, the log-likelihood ratios for modelling the region with 1 to N Gaussian mixtures; $d^1(ll)$, the first discrete derivate of ll and $d^2(ll)$, the second discrete derivate of ll. The area feature was used to define the rule-based classifier. The log-likelihood ratios and their derivates were used to train the SVM classifier.

4.2 Classification

The classification step is based on the assumption that the regions obtained from the segmentation process may not always correspond to a single cell or parasite. Based on this, it attempts to discern in how many sub-regions each region must be split. This is achieved by employing two separate classifiers: a rule-based classifier (RBC) and a support vector machine (SVM).

The RBC exploits the low area overlap percentage observed between cell and parasite pairs. This low overlap percentage is due to the high depth of field observed in the collected images, resulting in a near-perfect 2D cross-section of the 3D space within the tissue sample. Following this principle, and as single cells/parasites presented normal distributions, it was hypothesised that the area functions for cells and parasites could be approximated by a harmonic function. Since larger multi-nucleic regions are increasingly less frequent, the functions present a decreasing harmonic pattern. Although simple, this approach was found to be quite accurate (around 89.0% accuracy, hitting a maximum value of 98,9% when considering an error of ±2 regions as acceptable). Figure 5 illustrates a detail for one of the described decreasing harmonic functions.

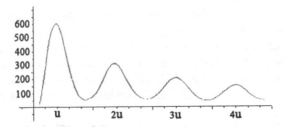

Fig. 5. Detail of a decreasing harmonic function described by the normalized (cellular) area values of 120 randomly sampled images (~870 regions per image: 106.318 region samples total). Horizontal axis: area values; Vertical axis: area value occurrences.

The classifier was programmed with rules reflecting this concept and taught to ignore values outside its knowledge space, as there was no data to accurately model regions over 10 cells or parasites. Note that cells and parasites each have their own harmonic function, since their area distributions do not exhibit the same standard deviation.

The SVM classifier relies on the concept that circles and ellipses can be described as Gaussian distributions and, as macrophages and parasites partake such geometry, clusters of these objects can be formulated as a mixture problem. Our conjectured hypothesis was that: as mixtures are added to the modeling process of each region, the improvement rate is described in the log-likelihood ratio evolution sequence. Thus, if an initial annotated dataset with the correct number of mixtures N is available, it should be possible to model a function that is able to predict this N for new, non-annotated observations. Following this hypothesis, the classifier was trained with a subset of features from the main feature vector, consisting of the log-likelihood ratios, it's first and second order discrete derivates. The training set was obtained from roughly 150 regions, modeled with $N = [1..15]$ Gaussian mixtures. Various machine-learning classifiers were tested. Ultimately, a SVM model was chosen as it achieved the highest (85,3%) sequential split and cross-validation classification rates.

In order to reconcile diverging predictions, a voting system was developed. As previously mentioned, the RBC exhibits an overall accuracy of nearly 98%, when considering an error margin of ±2. The SVM is generally much more accurate[2], but its error margins are, in counterpart, much wider. Making use of these characteristics, the voting system makes its choice based on the assumption that the correct decision does not deviate more than 2 from the RBC, so if the SVM's prediction is within this window, it is considered correct and incorrect if otherwise.

4.3   Declustering

Having obtained a prediction for the number of nuclei in each region, these were un-mixed using the Expectation-Maximization [13] method. The algorithm was parameterized with a minimum standard deviation of 10^{-6}, and a maximum of 200 iterations.

---

[2] Note that the 85% accuracy percentage referred for the SVM classifier refers to the classification of only multi-nucleic regions, as the accuracy ratings for the RBC refer to multi and uni-nucleic regions, giving it a considerable advantage.

To minimize runtime, the seeds for the centroids of each mixture were set by the averaged centroids of a 10 fold cross-validated K-Means. Figure 6, portrays the method's expert performance, even in the presence of large nucleic clusters.

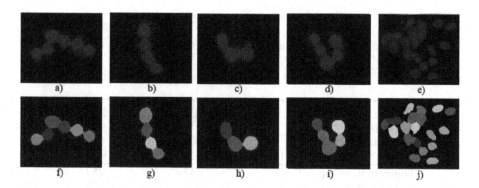

Fig. 6. Several *declustering* examples for 6, 4, 3, 5 and 11 region clusters. a) - e) Original region patches. f) - j) Respective *declustered* patches.

5   Results

In order to evaluate the practical applicability of our method, we choose to compare the method's classification output with the final annotations made by biomedical researchers in real drug trials. Upon surveying the current drug trials undergoing in the IBMC lab two specific drug trails were chosen. These were labeled trial 1 (T_1) and trial 2 (T_2). Trial one was chosen for presenting a low number of regions, most of which were difficultly differentiated from the background, thus straining the method's segmentation step. Trial two was chosen due to the sheer number of oversaturated and overlapping regions, hence straining mostly the classification step. These two trials were held as complementary, therefore constituting a complete test of the foreseeable experimental conditions in real-world applications. Our ground truth was taken as the individual annotations of three biomedical researchers. The researchers were asked to carefully perform the annotations in separate days and double-check them, so as to minimize human error.

5.1   Individual Component Results

In order to understand the method's general behavior across experimental conditions we measured its segmentation and classification accuracy in both trials. The following section details the summary ratings obtained.

Regarding the classification step, since the method assumes that the cell/parasite identification process is not completely performed in the segmentation step, multi-nucleic regions were considered as well segmented results. A region was considered ill segmented if: a) it was not detected or b) its geometry was not correctly identified. Figure 7 exemplifies these criteria.

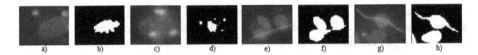

Fig. 7. Examples of ill-segmented regions

Table 1 presents a summary of the obtained segmentation accuracy on both trials.

Table 1. Accuracy ratings for the segmentation step of both datasets

	Macrophages	Parasites
Segmentation total (T_1)	4873	6113
Ground truth (T_1)	5007	6437
Accuracy percentage (T_1)	97,32	94,96
Segmentation total (T_2)	7633	2571
Ground truth (T_2)	8014	2783
Accuracy percentage (T_2)	95,24	92,38

Regarding the rule-based classifier, it showed itself capable of identifying regions with 98.34% accuracy, when considering a ±2 error margin acceptable. This result was obtained by manually inspecting 1500 distinct regions from both datasets. No distinction was made between cellular or parasitic regions.

To train the SVM classifier, the log-likelihood ratios and their first and second order derivates were computed from mixture modeling 150 random clustered regions. The SVM classifier was trained using John Plat's sequential minimal optimization algorithm and an RBF kernel [14]. Validation was performed through a 66% sequential split (SS), for which the classifier obtained an 85.3% classification accuracy.

5.2 Stress Testing

In order to assert our method's real-world applicability, the computed infection ratios were compared with each of the three manual annotations. To account for inter-person variation, annotation values were modeled as a normal distribution and the method considered accurate if its output did not deviate more than 2 standard deviations from the mean value. Intra person variance was eliminated from this test, as each subject was instructed to carefully perform the annotations in a single pass. Table 2 presents the acceptable error margins for both trials.

Table 2. Stress-test error margins for both trials

	Cells (total)	Parasites (total)	Infected Cells
Annotation Mean (T_1)	3020	4037	1873
Annotation Standard Deviation (T_1)	885	1110	546
Algorithm Error (T_1)	1353	1574	947
Annotation Mean (T_2)	5069	1967	1024
Annotation Standard Deviation (T_2)	294	34	38
Algorithm Error (T_2)	223	133	28

In trial one we observe a large standard deviation, both in the cellular and parasitic counts, which also has a bleeding effect into the infect cell count. Due to these large discrepancies, the method easily fits within the defined boundaries, actually being closer to one standard deviation in total parasites and infected cells. Hence, the method performs in a manner suitable for real-world application for images exhibiting similar experimental conditions as trial one.

Regarding trial two, the manual annotation discrepancies seen in trial one are no longer present, thus contributing to the low standard deviation verified. This further increased the error observed in the segmentation results. Although the algorithm error in total cells and parasites for trial two is not considerable in absolute numbers, it is in relative distance to the standard deviation, which translates to an error of over 3 standard deviations for the total cell count.

In sum, the method passed all tests, except the parasite detection category in trial two. Since this trial presented little to no multi-nucleic regions, no fault could be attributed to the classification step. Thus, the low parasitic detection (and ensuing error) falls upon the segmentation process, indicating future improvements should be directed at this step. The results indicate our method is robust to highly cluttered images, being able to expertly split region clusters and compute infection ratios.

6   Conclusions

In this work a robust, automatic analysis methodology for cell and parasite detection in fluorescence microscopy imaging was suggested. The proposed method has shown itself robust to poor lighting conditions and high cluttering indexes, falling well with-in the error margins observed in expert biomedical researcher annotations.

The obtained results demonstrate the method is capable of performing the image analysis task adequately and in less time than a human expert. Being a computer program, the method also boasts from being immune to traditional human errors related to distraction, fatigue or subjectivity. Since human errors are the major source of ambiguity in the traditional annotation process, we consider our alternative to be a more suitable choice. This claim is supported with the fact that the method has a fixed error margin; meaning, the error does not randomly vary through time, as human error does. The attained results in both stress tests further support our claim, proving the method's suitability for this specific task. As an added benefit, using the proposed method two or more drug trials can be safely compared as to their effectiveness, whereas if considering human error, the comparison would require validation via multiple annotations to attenuate the uncertainty generated by inter and intra-person variance.

Future work should focus on the methods used in the segmentation step, possibly employing mean shift or normalized cuts techniques, as well as increasing the training datasets of both classifiers. The built processing pipeline was made to be modular and applicable to other image types, thereby easily expandable to solve similar problems. This work has been successfully integrated with a pre-existing image annotation framework and is currently used in the INEB/IBMC laboratories in Portugal.

References

1. Ryan, K.J., Ray, C.G.: Sherris Med. Microbio., pp. 749–754. McGraw-Hill (2004)
2. Myler, P., Fasel, N.: Leishmania: After The Genome. Caister Academic Press (2008)
3. Spring, K.R.: MicroscopyU: Introduction to Fluorescence Microscopy (2010)
4. Liao, Q., Deng, Y.: An Accurate Segmentation Method For White Blood Cell Images. In: Proceedings IEEE International Symposium on Biomedical Imaging, pp. 245–248 (2002)
5. Ficarra, E., Cataldo, S.D., Acquaviva, A., Macii, E.: Automated Segmentation of Cells With IHC Membrane Staining. IEEE Transactions on Biomedical Engineering 58(5), 1421–1429 (2011)
6. Jiang, K., Liao, Q., Dai, S.: A Novel White Blood Cell Segmentation Scheme Using Scale-Space Filtering And Watershed Clustering. In: Proceedings Second International Conference on Machine Learning and Cybernetics (2003)
7. Park, J., Keller, J.M.: Fuzzy Patch Label Relaxation in Bone Marrow Cell Segmentation. In: IEEE International Conference on Computational Cybernetics and Simulation, pp. 1133–1138 (1997)
8. Begelman, G., Gur, E., Rivlin, E., Rudzsky, M., Zalevsky, Z.: Cell Nuclei Segmentation Using Fuzzy Logic Engine. In: Proceedings IEEE International Conference on Image Processing (2004)
9. Yu, W., Lee, H.K., Hariharan, S., Bu, W., Ahmed, S.: Level Set Segmentation of Cellular Images Based on Topological Dependence. In: Bebis, G., Boyle, R., Parvin, B., Koracin, D., Remagnino, P., Porikli, F., Peters, J., Klosowski, J., Arns, L., Chun, Y.K., Rhyne, T.-M., Monroe, L. (eds.) ISVC 2008, Part I. LNCS, vol. 5358, pp. 540–551. Springer, Heidelberg (2008)
10. Yan, P., Zhou, X., Shah, M., Wong, S.T.C.: Automatic Segmentation of High-Throughput RNAi Fluorescent Cellular Images. IEEE Transaction On Information Technology in Biomedicine 12(1) (2008)
11. Morse, B.S.: Brigham Young University, SH&B, Section 5 (2000)
12. Freeman, H.: On the encoding of arbitrary geometric configurations. IRE Transactions on Electronic Computers, 260–268 (1961)
13. Platt, J.: Fast Training of Support Vector Machines using Sequential Minimal Optimization. In: Schoelkopf, B., Burges, C., Smola, A. (eds.) Advances in Kernel Methods - Support Vector Learning. MIT Press (1998)
14. Reynolds, D.: Gaussian Mixture Models. MIT Lincoln Laboratory, 244 Wood St., Lexington, MA 02140, USA (2007)

Combination of M-Estimators and Neural Network Model to Analyze Inside/Outside Bark Tree Diameters

Kyriaki Kitikidou, Elias Milios, Lazaros Iliadis, and Minas Kaymakis

Democritus University of Thrace,
Department of Forestry and Management of the Environment and Natural Resources,
Pandazidou 193, 68200, Orestiada, Greece
kkitikid@fmenr.duth.gr

Abstract. One of the most important statistical tools is linear regression analysis for many fields such as medical sciences, social sciences, econometrics and more. Regression techniques are commonly used for modelling the relationship between response variables and explanatory variables. In this study, inside bark tree diameter was used as the dependent variable and outside bark diameter and site type as independents. While generally it is assumed that inside and outside bark diameters are linearly correlated, linear regression application is weak in the presence of outliers. The purpose of this study was to develop a Multi-Layer Perceptron neural network model which considered significant variables from an a priori developed robust regression model. The application of robust regression could be considered in selecting the input variables in a neural network model.

Keywords: Artificial Neural Networks; M-Estimation; Robust Regression.

1 Introduction

In experimental science measurements are typically repeated for a number of times, yielding a sample size n. Then, we summarize the measurements by a central value and we measure their variability, i.e. we estimate location and scale. These estimates should preferably be robust against outliers. The estimator's stylized empirical influence function should be smooth, monotone increasing for location, and decreasing–increasing for scale. When the values of the observations tend to depart from the main body of data, maximum likelihood estimators (M-estimators) can be considered.

According to Loetsch, Zöhrer and Haller (1973), the relation between inside and outside bark tree diameter is expressed by the simple linear regression model:

$$\hat{d}_{in} = b_0 + b_1 d$$

where $\hat{d}_{in}$ is the estimated breast height (1.3 m above the ground) diameter inside bark, d is the breast height diameter outside bark, and b_0, b_1 regression coefficients.

The aim of the current study is to incorporate variables that are found to be significant in a robust regression model, into a Multi-Layer Perceptron (MLP) Artificial Neural Network (ANN) model, so as to estimate the same dependent variable (diameter inside

L. Iliadis et al. (Eds.): AIAI 2012, IFIP AICT 381, pp. 11–18, 2012.

bark). We find an attempt to combine robust estimation and neural networks in several researches (Chen 1994, Lee 1999, Suykens et al. 2002, Kuana and Whiteb 2007, Aleng et al. 2012). In this study, we used inside and outside bark tree diameters to illustrate the pre-mentioned idea, assuming that the two variables are already linearly related but simple linear regression is challenging to apply, because of the presence of outliers.

2   Materials and Methods

Data used in this study were collected from 30 *Populus tremula* dominant trees on Athos peninsula (northern Greece), simply randomly selected (Kitikidou et al. 2012). Trees were selected from two site types. Site I was the most productive site, while site II was the less productive. For all the analyses the SPSS v.19 statistical package was used (IBM SPSS 2010). Descriptive statistics for inside and outside bark breast height (1.3 m above the ground) diameters are given in Table 1.

Table 1. Descriptive statistics for all variables used in the analysis

			Mean	Standard Deviation	Minimum	Maximum
Site	Good	diameter outside bark (cm)	25.697	3.649	19.500	31.600
		diameter inside bark (cm)	24.825	3.471	18.950	30.250
	Bad	diameter outside bark (cm)	14.195	3.121	10.350	18.900
		diameter inside bark (cm)	13.513	3.030	9.775	18.150

2.1   Robust Regression

Robust regression is a weighted average of independent parametric and nonparametric fits to the data. A statistical procedure is considered robust if it performs reasonably well, even when the assumptions of the statistical model are not met. If we assume that dependent and independent variables are linearly correlated, least squares estimates and tests perform quite well, but they are not robust in the presence of outliers in the data set (Rousseeuw and Leory 1987).

Robust regression analyses have been developed as an improvement to least squares estimation in the presence of outliers. This method does not exclude the outliers, which can be real observations, but tends to down-weight cases with large residuals (McKean 2004). The primary purpose of robust regression analysis is to fit a model which represents the majority of the data. Robust regression is an important tool for analyzing data that are contaminated with outliers. This method can be used to detect outliers and to give results resistant in the presence of outliers. Many methods have been developed to deal with such problems. In this case, we proposed the robust M-estimation to model response data, since linear regression between inside and outside bark diameters detected outliers in the data set.

Robust regression is an alternative of least squares estimation regression analysis designed to avoid limitations of traditional parametric analyses. The least squares estimation in a regression model is not sufficiently robust to extreme data. It is quite common for different nature of data to separate distinctly from the data distribution. In such cases, the influence of extreme data can be minimized using robust M-estimators.

Huber (1973) proposed an M-estimator that uses maximum likelihood formulations by deriving optimal weighting for the data set in non-normal conditions. In other words, the M-estimator minimizes the role of the residuals. As in the case of M-estimation of location, the robustness of the estimator is determined by the choice of weight function.

Consider the linear model

$$Y_i = a + \beta_1 x_{i1} + \beta_2 x_{i2} + \ldots + \beta_n x_{in} + \varepsilon_i = x_i' \beta + \varepsilon_i \tag{1}$$

for the ith of n observations. The fitted model is

$$Y_i = b_0 + b_1 x_{i1} + b_2 x_{i2} + \ldots + b_n x_{in} + e_i = x_i' b + e_i \tag{2}$$

The Huber M-estimators minimize the objective functions

$$\sum_{i=1}^{n} \rho(e_i) = \sum_{i=1}^{n} \rho\left(Y_i - x_i' b\right) \tag{3}$$

where the function $\rho(.)$ is usually chosen in a way that represents some weighting of the ith residual. This weighting means that outlying observations have their weights reduced and thus the estimates are affected less by such noise. A weighting of zero means that the observation is classified as an outlier.

Differentiating (1) with respect to the regression coefficients $\hat{\beta}_j$ yields

$$\sum_{i=1}^{n} \psi(e_i) x_i = 0 \tag{4}$$

where $\psi(.)$ is the derivative of $\rho(.)$ and the corresponding M-estimator is the maximum likelihood estimator. If we define the weight function $w(e_i) = \dfrac{\psi(e_i)}{e_i}$, then the estimating equations can be written as

$$\sum_{i=1}^{n} w(e_i) x_i = 0 \tag{5}$$

Solving the estimating equations is similar to a weighted least-squares estimation. The weights, however, depend upon the residuals, the residuals depend upon the estimated coefficients, and the estimated coefficients depend upon the weights. An iterative solution (called iteratively reweighted least-squares (IRLS)) is therefore required.

2.2 Multi-Layer Perceptron (MLP) Artificial Neural Network (ANN)

Consider a neural model, which consists of an input layer, one or several hidden layers and an output layer. The neurons in the neural network are generally grouped into layers. Signals flow in one direction from the input layer to the next, but not within the same layer. Success of the application of a neural network relies on the training network. Among the several learning algorithms available, back-propagation has been the most popular and most widely implemented. Basically, the back-propagation training algorithm with three-layer architecture means that, the network has an input layer, one hidden layer and an output layer (Bishop 1995; Ripley 1996; Haykin 1998; Fine 1999).

In this research the hidden layer is fixed at one in order to reduce the complexity of the network, and to increase its computational efficiency (Haykin 1998). For the network with N input nodes, H hidden nodes and one output node, the values $\hat{Y}$ are given by:

$$\hat{Y} = g_2 \left(\sum_{j=1}^{H} w_j h_j + w_0 \right) \tag{6}$$

where w_j is an output weight from hidden node j to output node, w_0 is the bias for output node, and g_2 is an activation function. The values of the hidden nodes h_j, $j=1,\dots,H$ are given by:

$$h_j = g_1 \left(\sum_{i=1}^{N} v_{ji} X_i + v_{j0} \right), j=1, \dots, H \tag{7}$$

where v_{ji} is the input weight from input node i to hidden node j, v_{j0} is the bias for hidden node j, X_i are the independent variables where $i=1,\dots, N$ and g_1 is an activation function. The schematic representation of the neural network used in this study is given in Fig. 1.

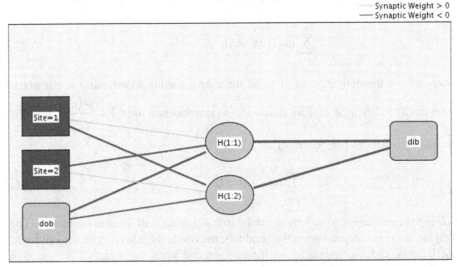

Hidden layer activation function: Hyperbolic tangent

Output layer activation function: Identity

Fig. 1. Multi-layer perceptron network structure with one hidden layer including two neurons (H(1:1), H(1:2)), two input variables (Sites 1 and 2, dob: diameter outside bark) and one output variable (dib: diameter inside bark)

From the initial 30 data records (30 *Populus tremula* trees), 70% were used for training (21 records) and 30% for testing (9 records).

3   Results-Discussion

3.1   Results of Robust Regression Analysis

We applied robust regression analysis as an alternative to the least squares regression model, because fundamental assumptions weren't met due to the nature of the data. R squared value was calculated at 0.9996, indicating a great ability of the model to predict a trend. Table 2 displays the final weighted least square estimates. The weighted estimates provide resistant results against outliers. As we see in this Table, both independent variables differ significantly from zero (sig. <0.05). That leads us in using both variables for the inside bark diameter estimation through the MLP neural network.

Table 2. Parameter estimates from robust regression model

| Parameter | B | Std. Error | 95% Wald Confidence Interval | | Hypothesis Test | | |
			Lower	Upper	Wald Chi-Square	df	Sig.
(Intercept)	-0.098	0.1080	0-.310	0.114	0.825	1	0.364
[Site=1]	0.283	0.0923	0.102	0.464	9.392	1	0.002
[Site=2]	0[a]	.	.	.	.	.	.
dob	0.959	0.0077	0.944	0.974	15558.112	1	0.000
(Scale)	0.018						

Dependent Variable: diameter inside bark (cm)
Model: (Intercept), Site, dob (diameter outside bark)
a. Set to zero because this parameter is redundant.

3.2   Results of MLP Neural Network

The input variables are selected based on the significant variables of the robust regression model (sites and diameter outside bark). Back propagation algorithm was applied and by following trial and error approach we ended up in two neurons included in the single hidden layer. Hence, the appropriate neural network architecture which results in the best multilayer neural network model is composed of two input variables, two hidden nodes and one output node. This can be represented by equations (6) and (7).

Training this architecture using the Levenberg-Marquardt back propagation training algorithm and using hyperbolic tangent transfer function in the hidden layer and identity function in the output layer, the R^2 for training and testing data were equal to 0.99964 and 0.99955, respectively, showing that the model gives very accurate estimations of the inside bark diameter (Table 3). This procedure was completed in just one step, as we can see in iteration history (Table 4).

Table 3. Model summary of the MLP ANN model

Training	Sum of Squares Error	0.012
	Root Mean Squared Error (RMSE)	0.032
	R square	0.99964
Testing	Sum of Squares Error	0.003
	Root Mean Squared Error (RMSE)	0.045
	R square	0.99955

Dependent Variable: diameter inside bark (cm)

Table 4. Iteration history of the MLP ANN model

		Parameter			
Iteration	Update Type	(Intercept)	[Site=1]	dob	(Scale)
0	Initial	-0.098241	0.282914	0.958909	0.016168
1	Scoring[a]	-0.098143	0.282807	0.958892	0.017964

Redundant parameters are not displayed. Their values are always zero in all iterations.
Dependent Variable: diameter inside bark (cm)
Model: (Intercept), Site, dob
a. All convergence criteria are satisfied.

Parameter estimates of the MLP neural network are given in Table 5. The observed and predicted values of the dependent variable in Fig. 2 show clearly that the MLP network does an exceptional job in predicting inside bark diameters.

Table 5. Parameter estimates from the MLP ANN model

		Predicted		
		Hidden Layer 1		Output Layer
Predictor		H(1:1)	H(1:2)	dib
Input Layer	(Bias)	-0.061	0.226	
	[Site=1]	0.187	-0.386	
	[Site=2]	-0.328	0.162	
	dob	-0.638	-0.306	
Hidden Layer 1	(Bias)			-0.046
	H(1:1)			-1.264
	H(1:2)			-1.108

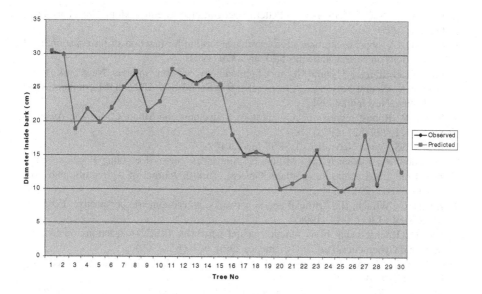

Fig. 2. Observed and predicted values of the MLP ANN

4   Conclusions

The MLP neural network model for inside bark diameter estimation, with selection of input variables from a robust regression model, discovered two input variables; diameter outside bark and site type. The R^2 for training and testing were 0.99964 and 0.99955, respectively, which are quite large. Hence, significant variables from a robust regression, when we prefer to avoid ordinary least squares estimation in the presence of outliers, could be considered in selecting the input variables of a neural network model.

Artificial neural networks have shown a great promise in predicting variables in many scientific fields, because of their fast learning capacity. However, when the training patterns incur larges errors caused by outliers, the networks interpolate the training patterns incorrectly. Using significant variables from robust procedures as input variables will result in better capability of approximation to underlying functions and faster learning speed (Lee 1999). In this way, a neural network can be robust against gross errors and have an improved rate of convergence, since the influence of incorrect samples is notably suppressed (Chen 1994).

References

1. Aleng, N., Mohamed, N., Ahmad, W., Naing, N.: A new strategy to analyze medical data using combination of M-estimator and Multilayer Feed-Forward Neural Network Model. European Journal of Scientific Research 1, 79–85 (2012)

2. Bishop, C.: Neural Networks for Pattern Recognition, 3rd edn. Oxford University Press, Oxford (1995)
3. Chen, D.: A robust backpropagation learning algorithm for function ap-proximation. IEEE Transactions on Neural Networks 5(3), 467–479 (1994)
4. Fine, T.: Feedforward Neural Network Methodology, 3rd edn. Springer, New York (1999)
5. Haykin, S.: Neural Networks: A Comprehensive Foundation, 2nd edn. Macmillan College Publishing, New York (1998)
6. Huber, P.: Robust regression: Asymptotics, conjectures and Monte Carlo. The Annals of Statistics 1, 799–821 (1973)
7. IBM SPSS Neural Networks 19. SPSS Inc. (2010)
8. Kitikidou, K., Kaymakis, M., Milios, E.: Site index curves for young Populus tremula stands on Athos peninsula (northern Greece). Turkish Journal of Agriculture and Forestry 36, 55–63 (2012)
9. Kuana, C., Whiteb, H.: Artificial neural networks: an econometric perspective. Econometric Reviews 13(1), 1–91 (1994)
10. Lee, C.: Robust radial basis function neural networks. IEEE Transactions on Systems, Man, and Cybernetics, Part B: Cybernetics 29(6), 674–685 (1999)
11. Loetsch, F., Zöhrer, F., Haller, K.: Forest Inventory, vol. II. BLV Verlagsge-sellschaft, Munchen (1973)
12. McKean, J.: Robust Analysis of Linear Models. Statistical Science 19(4), 562–570 (2004)
13. Ripley, B.: Pattern Recognition and Neural Networks. Cambridge University Press, Cambridge (1996)
14. Rousseeuw, P., Leory, A.: Robust regression and outlier detection. Wiley Series in Probability and Statistics (1987)
15. Suykens, J., De Brabanter, J., Lukas, L., Vandewalle, J.: Weighted least squares support vector machines: robustness and sparse approximation. Neurocomputing 48(1-4), 85–105 (2002)

Multi-classify Hybrid Multilayered Perceptron (HMLP) Network for Pattern Recognition Applications

Fakroul Ridzuan Bin Hashim, John J. Soraghan, and Lykourgos Petropoulakis

Centre for Excellent in Signal and Image Processing,
University of Strathclyde, Department of Electronic & Electrical Engineering,
204 George St, Glasgow G1 1XW, United Kingdom
fakroul.bin-hashim@strath.ac.uk

Abstract. This paper introduces a Hybrid Multilayered Perceptron (HMLP) based classifier known as the Multi-Classify HMLP network (MCHMLP). This network is shown to be able to enhance the performance accuracy when compared to the conventional HMLP network. The Multi-Classify HMLP network architecture is trained using a Modified Recursive Prediction Error (MRPE). This study uses three benchmark datasets in order to measure the capability of the network. The results show that the proposed Multi-Classify HMLP network provides a significant improvement over the conventional HMLP network for pattern recognition applications.

1 Introduction

Artificial neural networks (ANNs) can be used to solve problems without the need to create a model of a real biological system. ANNs pattern recognition and classification units are widely used in medicine, banking analysis, stock market prediction, telecommunication industry and many other technology and science areas. Mat-Isa *et al.*, [1] have shown that neural networks are able to identify the risk level in patients with pre-cancerous cervix. Neural networks are used to predict the stock market to ensure that profits can be enhanced or losses can be minimized [2]. Liu *et al.* use neural networks as the prediction agent in measuring the earthquake tremor levels that can occur based on the identified features. In [3] ANNs have been used in aiding earthquake predictions along with the related tremors analysis. In another study, Mat-Isa *et al.* showed that ANNs can provide good results in identifying the shape of aggregates. The aggregates are identified and separated into groups before being used in construction and those aggregates with good shape are selected to ensure the strength of the concrete at the highest level [4].

A Multilayered Perceptron (MLP) neural network is a conventional example of an ANN network. There are several modified versions of the MLP networks. The Hybrid MLP (HMLP) network is one such version, which has added connections to feed information from the input layer directly to the output layer. Mashor [5] has shown an increased accuracy of results provided by the HMLP network compared to the conventional MLP networks. The ability to make predictions of the neural network is dependent on network training. There are many types of training algorithms for

L. Iliadis et al. (Eds.): AIAI 2012, IFIP AICT 381, pp. 19–27, 2012.

training neural networks to increase prediction. Popular training algorithms include Back Propagation (BP), Lavenberg Marquardt (LM), Bayesian Regularization (BR), Recursive Least Squares (RLS), and Recursive Prediction Error (RPE). Mashor in [6] has shown that the accuracy could be improved if some modifications were made to the RPE training algorithms. The Modified Recursive Prediction Error (MRPE) is the modified version of the RPE training algorithm which can increase the accuracy of predictions of the conventional RPE algorithm when training the HMLP network. He also stated that the MRPE training algorithm is better suited for use with the HMLP network as it resulted in more accurate predictions than other approaches.

In this study, modifications to the HMLP network will be made to improve the performance of the resulting network. The presented method proposes that the classification process is done repetitively in multi-stages. The output outcomes from a HMLP network in the first stage will be presented as input to another HMLP network in the second stage. Courtney *et al.* [7] in their research used a multi-stage neural network to measure the colour constancy and the colour induction using 'NEXUS'. Nexus [8] is a large scale interactive neural simulator designed to simulate a huge number of networks with approximately 1.65 million connections. Lai *et al.* [9] used multi-stage neural networks to predict the foreign exchange rate (FOREX). Multiple training algorithms were used to train the network before the pruning phase took place in order to produce the selected base models. A multi-stage neural network was implemented to recognise the pattern of a mammogram screen [10]. Back propagation with a Kalman filter was applied as the training algorithm to perform the classification.

In the work presented in this paper three datasets, taken from the University of California – Irvine (UCI) repository [11] are used in order to investigate whether the proposed Multi-Classify HMLP network can provide more accurate results than conventional networks. These were: the Breast Cancer Wisconsin (BCW) dataset, the Liver Disorder (LD) dataset the Pima Indians Diabetes (PID) dataset. The remainder of the paper is organized as follows. In section 2 the proposed method is shown in detail followed by the results and discussion in section 3. Conclusions are presented in section 4.

2   Multi-classify Hybrid Multilayered Perceptron Network

It is well known [5] that modelling a linear framework using a standard nonlinear MLP network is not a good decision. For this reason a hybrid version of the MLP network was introduced known as the HMLP network. This involved adding linear connections linking the input layer directly to the output layer without going through the hidden layer, an approach which proved to significantly improve performance. Thus, the HMLP network was shown to be sufficiently competent for designing optimal structures both for linear and nonlinear systems. By directly linking the network input to the output node through a number of weighted connections, HMLP allows the formation of a linear system. Additional connections can be made as represented by the dashed line shown in Figure 1.

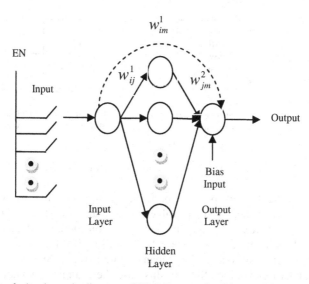

Fig. 1. A schematic diagram of HMLP network with one hidden layer

For the output node, the output of the network is given by:

$$\hat{y}(t) = \sum_{j=1}^{n_h} w_{jk}^2 F\left(\sum_{i=1}^{n_i} w_{ij}^1 x_i^0(t) + b_j^0 \right) + \sum_{i=1}^{n_i} w_{ik}^1 x_i^0(t)$$

$$\text{for } 1 \le j \le n_h \text{ and } 1 \le k \le m \tag{1}$$

where w_{ik}^1 denotes the weights of the additional linear connection between the input and output layers, n_h is the hidden node while m is the number of outputs of the network.

In this study we propose a modified version of the HMLP network that is formed by repeating the classification process in a second stage. The proposed method is affected by adding another HMLP network to the output layer of the conventional HMLP network.

Figure 2 shows the block diagram of the architecture of the proposed classification method using two HMLP networks. These networks are in cascade so that the output of the first network is the input of the second network. The first HMLP network is fed with training input data. Each data consists of several parameters representing the data features. These features correspond to the various attributes used for the classification in each case. Detailed explanation of these attributes is given in section 3 of the paper. Here the network was trained in order to find the best performance of the classification process.

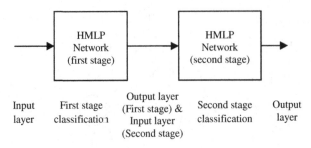

Fig. 2. Block diagram of the proposed network architecture

It is considered that the best possible performance has been attained when the error of the network has reached a small value and remains constant even though the number of nodes increases. At that moment it is considered that the network has obtained the highest level of convergence. Several statistical techniques are available to represent the error of system such as Mean Square Error (MSE), Sum Square Error (SSE), and Standard Deviation (SD). In this case the Standard Deviation is used to represent the error of the network.

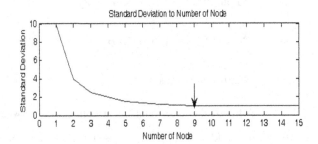

Fig. 3. A graph of Standard Deviation variation with increasing Number of Network Nodes

Figure 3 shows that the smallest SD occurred after using 9 nodes (arrow). After this point the value of SD remains unchanged although the number of nodes increases. With the highest convergence reached, a set of test data is provided to test the HMLP network. A good trained network is indicated by the ability to represent the whole dataset at the training phase. Once the best performance has been achieved for the first HMLP its outcome is then used as input to the second stage HMLP network. The same training and testing procedures applied to the first HMLP network are also used to train and test the second stage HMLP network. In this case only one feature is represented for each data input since the first stage HMLP network provides only one feature at the output layer as outcome.

Several aspects of the network training must be fulfilled in order to obtain results with the highest possible accuracy. The type of network structure must be matched by an appropriate training algorithm and an adequate size of data must exist to train the network, thus avoiding poor results. Neural networks cannot tolerate small size datasets because they are trained by the history of the dataset and try to adjust their weights to best fit the supplied data. The objective of applying the multi-stage classification is to redo the classification that happens at the first stage network and provide the second stage network an opportunity to improve the overall result. Each neural network has its own convergence limitations based on the structure of the network, type of training algorithm used and the size of the data available for training. If the accuracy of the network keeps giving the same results even when the number of iterations is increased, this means the network is near or at its highest convergence point. So, by using the output of the first stage as input for the second network we could improve on the convergence limitations of the first HMLP network. The second stage network will re-classify the wrong classification held at the first stage network. The function of the second HMLP network is to automatically readjust the classification thresholds thus improving the performance obtained compared to a single HMLP.

3   Benchmark Datasets

The performance of the proposed MCHMLP network using MRPE as the training algorithm is evaluated by comparing it with a multi-stage MLP network, and conventional HMLP, MLP, HRBF and RBF networks. The 3 benchmark datasets from the UCI machine learning repository [11], i.e. the Breast Cancer Wisconsin, the Liver Disorder, and the Pima Indians Diabetes, have been used to examine the performance of the MCHMLP network. The results are then compared to the results from the aforementioned networks.

The Breast Cancer Wisconsin dataset contains a set of subjects with breast cancer and characterized by 9 attributes. The subjects have been diagnosed either with benign tumors or malignant tumors. The dataset has 699 entries with 458 benign tumors and 241 malignant tumors. The 9 attributes are: clump thickness, uniformity of cell size, uniformity of cell shape, marginal adhesion, single epithelial cell size, bare nucleoli, bland chromatin, normal nucleoli and mitoses. The dataset is divided into two groups with 400 comprising the training dataset while 299 make up the testing dataset for the neural networks.

The Liver Disorder dataset classifies the existence of liver disorder in subjects with active alcoholic activity. Blood samples are tested for each subject and, according to the results the subject is placed in group 1 or in group 2. The dataset has 345 entries and is divided into two groups with 145 entries in group 1 and 200 entries in group 2. The 6 attributes are used to classify the subjects into these groups are: mean corpuscular volume (mcv), alkphos alkaline phosphotase, alamine aminotransferase (sgpt), aspartate aminotransferase (sgot), gamma-glutamyl transpeptidase (gammagt) and drinks number of half-pint equivalents of alcoholic beverages drunk per day. For the purpose of training and testing the neural networks, this dataset is divided into two subsets with 200 entries for the training dataset and 145 for the testing dataset.

The Pima Indians Diabetes dataset is taken from female subjects with at least 21 years old of Pima Indians heritage. The subjects have been diagnosed either as healthy or as diabetics. The dataset has 8 attributes: number of pregnancies, plasma glucose concentration, diastolic blood pressure, triceps skin fold thickness, 2-h serum insulin, body mass index, diabetes pedigree, and last but not least, the age of the subject. The database has 768 entries with 268 having tested positive for diabetes. For the purpose of training and testing the neural networks, this dataset is divided into a training set with 400 entries and a testing set with 368 entries.

4   Experimental Results

Several types of networks have been used and compared in this study using the 3 datasets mentioned earlier. The MCHMLP along with the multi-stage MLP, the HMLP, the Hybrid Radial Basis Function (HRBF), the MLP, and the Radial Basis Function (RBF) networks were used to classify all three datasets. Training was carried out using the following techniques: the proposed MCHMLP, the multi-stage MLP, the HMLP and the MLP networks were trained using the MRPE training algorithm to obtain the data classification. The HRBF and the RBF networks were trained using the Given's Least Square (GLS) for clustering the dataset.

Table 1. Neural Networks performance using the Breast Cancer Wisconsin dataset for the training and testing phases

Network	Accuracy, (%)		Number of Nodes	Standard Deviation, (SD)
	Training Phase	Testing Phase		
MCHMLP	97.25	97.03	15(fs) 6(ss)	2.32(fs) 0.93(ss)
Multi-stage MLP	94.50	92.30	14(fs) 2(ss)	2.48(fs) 0.95(ss)
HMLP	95.00	94.66	15	2.32
MLP	81.75	80.63	14	2.48
HRBF	82.00	76.58	1	5.20
RBF	75.25	73.58	1	5.92

*first stage (fs), second stage (ss).

Table 2. Neural Networks performance using the Liver Disorder dataset for the training and testing phases

Network	Accuracy, (%)		Number of Nodes	Standard Deviation, (SD)
	Training Phase	Testing Phase		
MCHMLP	77.00	75.86	9(fs) 2(ss)	4.69(fs) 3.76(ss)
Multi-stage MLP	71.00	70.34	5(fs) 3(ss)	6.30(fs) 3.96(ss)
HMLP	74.50	73.17	9	4.69
MLP	69.00	67.43	5	6.30
HRBF	68.00	60.00	2	6.50
RBF	44.50	40.00	2	6.70

*first stage (fs), second stage (ss).

Table 3. Neural Networks performance using the Pima Indians Diabetes dataset for the training and testing phases

Network	Accuracy, (%)		Number of Nodes	Standard Deviation, (SD)
	Training Phase	Testing Phase		
MCHMLP	77.50	77.05	12(fs)	3.68(fs)
			4(ss)	3.06(ss)
Multi-stage MLP	65.25	64.90	14(fs)	3.78(fs)
			6(ss)	3.74(ss)
HMLP	73.50	72.98	12	3.68
MLP	64.75	62.57	14	3.78
HRBF	58.00	51.52	2	5.92
RBF	56.00	51.52	2	4.85

*first stage (fs), second stage (ss).

Tables 1, 2 and 3 show the performance results of the neural networks using the three datasets. It is clear from these tables that the MCHMLP outperforms all the other networks, obtaining the highest overall accuracy for every dataset used. The performances of the multi-stage MLP, the HMLP and the MLP networks are quite good, however they are significantly lower that the performance of the MCHMLP. The results clearly show that when additional networks are added to the conventional HMLP and MLP structures the accuracy of the results increases considerably for a relatively small increase in the number of network nodes due to the introduced second stage. The RBF family of networks (RBF and HRBF) are able to classify the three datasets but no better than the MLP family.

The standard deviation in this study is used as the stopping criterion of network iteration to measure the convergence level. The MCHMLP shows the highest convergence and accuracy performance for the BCW dataset by using 15 and 6 nodes at the first and second stage networks with 2.32 and 0.93 of SD respectively. For the Liver Disorder dataset, the f network needed 9 nodes for the first and 2 nodes for the second stages. In the case of the PID dataset 12 and 4 nodes were required in each stage respectively. From the tables of the results it is obvious that the MLP family needs a larger number of nodes to reach the highest convergence when compared to the RBF family. Billings et al. in their study state the RBF family is able to perform faster than MLP networks, where the weight can be pre-fixed since they involve only linear optimization [12]. Although the RBF networks may be able to produce faster training sessions, are unable to produce as accurate results. This is well illustrated in our test cases. These results are also supported by Mat-Isa et al. who, in their research, found that the performance of the RBF family of networks cannot be better than that of the MLP family [1].

The results in Tables 2 and 3 also indicate that the MCHMLP offers superior performance however the difference in performance is not as clear as in Table 1. In the case of Liver Disorder (Table 2) and Pima Indian Diabetes (Table 2) problem occurs when there is a large overlap between two groups in the datasets. The datasets are divided into groups A and B but, parts of the data belong to either group A or group B. So, datasets with a high overlap lead to a difficult classification process compared to datasets with a small overlap. The MCHMLP network is shown to be able to obtain the best classification results under all tested conditions in this study compared to the other networks. Nonetheless, datasets with a high overlap reduce the accuracy of the classification. However, Tables 1, 2 and 3 show that the MCHMLP network is capable to generalize the datasets in the training phase and obtains good results in the testing phase.

5   Conclusion

In this study, a multi-classify HMLP network trained with MRPE as the training algorithm has been proposed for pattern recognition applications. The performance of the proposed network was measured by using three benchmark datasets. The results show that the multi-classify HMLP network is able to classify the Breast Cancer Wisconsin, Liver Disorder and Pima Indian Diabetes with significant accuracy (and in some cases with very high accuracy) and, in all cases, the performance was better than any of the other networks used in this study. Moreover, in all three cases examined, the introduction of the second stage network resulted in improved standard deviation results at the second stage, a direct indication of improved convergence of the network. The additional overhead of extra nodes and increased training time was not significant.

References

[1] Mat-Isa, N.A., Mashor, M.Y., Othman, N.H.: An automatic cervical pre-cancerous diagnostic system. Journal of Artificial Intelligent in Medical 42(1), 1–11 (2008)

[2] Naeini, M.P., Taremian, H., Hashemi, H.B.: Stock market value prediction using neural networks. In: International Conference on Computer Information Systems and Industrial Management Applications (CISIM), pp. 132–136 (2010)

[3] Liu, Y., Liu, H., Zhang, B., Wu, G.: Extraction of if-then rules from trained neural network and its application to earthquake prediction. In: Proceedings of the Third IEEE International Conference on Cognitive Informatics, pp. 109–115 (2004)

[4] Mat-Isa, N.A., Al-Batah, M.S., Zamli, K.Z., Azizi, K.A., Joret, A., Mat-Noor, N.R.: Suitable features selection for the HMLP and MLP networks to identify the shape aggregate. Journal of Construction and Building Materials 22(3), 402–410 (2008)

[5] Mashor, M.Y.: Hybrid multilayered perceptron networks. International Journal of System Sciences 31(6), 771–785 (2000)

[6] Mashor, M.Y.: Modified recursive prediction error algorithm for training layered neural network. International Journal of The Computer, The Internet and Management 11(2), 29–36 (2003)

[7] Courtney, S.M., Finkel, L.H., Buchsbaum, G.: A multistage neural network for color constancy and color induction. IEEE Transactions on Neural Networks 6(4), 972–985 (1995)

[8] Sajda, P., Finkel, L.H.: NEXUS: A simulation environment for large-scale neural systems. Simulation (1992) ISSN: 00375497/92.

[9] Lai, K.K., Yu, L., Huang, W., Wang, S.: Multistage Neural Network Metalearning with Application to Foreign Exchange Rates Forecasting. In: Gelbukh, A., Reyes-Garcia, C.A. (eds.) MICAI 2006. LNCS (LNAI), vol. 4293, pp. 338–347. Springer, Heidelberg (2006)

[10] Zheng, B., Qian, W., Clarke, L.P.: Multistage neural network for pattern recognition in mammogram screening. In: IEEE International Conference on Neural Networks, IEEE World Congress on Computational Intelligence, vol. 6, pp. 337–344 (1994)

[11] Newman, D.J., Hettich, S., Blake, C.L., Merz, C.J., Aha, D.W.: UCI Repository of machine learning database. Department of Information and Computer Science, University of Carlifonia, Irvine (1998), http://archieve.ics.uci.edu/ml/databases.html (last accessed on March 10, 2012)

[12] Billings, S.A., Wei, H.L., Balikhin, A.: Generalized multiscale radial basis function networks. Journal of Neural Network 20, 1081–1094 (2007)

Support Vector Machine Classification of Protein Sequences to Functional Families Based on Motif Selection

Danai Georgara[1,*], Katia L. Kermanidis[1], and Ioannis Mariolis[2]

[1] Department of Informatics, Ionian University,
7 Tsirigoti Square, 49100 Corfu, Greece
dgeorgara@yahoo.com, kerman@ionio.gr
[2] Information Technologies Institute, CERTH, Thessaloniki, Greece
ymariolis@iti.gr

Abstract. In this study protein sequences are assigned to functional families using machine learning techniques. The assignment is based on support vector machine classification of binary feature vectors denoting the presence or absence in the protein of highly conserved sequences of amino-acids called motifs. Since the input vectors of the classifier consist of a great number of motifs, feature selection algorithms are applied in order to select the most discriminative ones. Three selection algorithms, embedded within the support vector machine architecture, were considered. The embedded algorithms apart from presenting computational efficiency allowed for ranking the selected features. The experimental evaluation demonstrated the usefulness of the aforementioned approach, whereas the individual ranking for the three selection algorithms presented significant agreement.

Keywords: PROSITE database, protein classification, feature selection, machine learning.

1   Introduction

Assigning putative functions to protein sequences constitutes one of the most challenging tasks in functional genomics. Protein function is often correlated with highly conserved sequences of amino-acids called motifs. Hence, motif composition is often used to assign functional families to novel protein sequences. However, many proteins usually contain more than one motif and several motifs can belong to proteins assigned to different families. Therefore, in order to reliably assign a protein to a certain family it is often required to identify motif combinations that are present in that protein. Data mining or machine learning algorithms offer some of the most effective approaches to the discovery of such unknown relationships between collections of motifs and families.

---

* Corresponding author.

L. Iliadis et al. (Eds.): AIAI 2012, IFIP AICT 381, pp. 28–36, 2012.

Wang et al. use the decision tree method for assigning protein sequences to functional families based on their motif composition [1]. The datasets used in the experiments were extracted from the PROSITE database [2]. The experimental results showed that the obtained decision tree classifiers presented a good performance.

Hatzidamianos et al. present a preprocessing software tool, called GenMiner [3], which is capable of processing three important protein databases and transforming data into a suitable format for the Weka data mining suite and MS SQL Analysis Manager 2000. A decision tree technique was used for mining protein data and the experimental results confirmed the system's capability of efficiently discovering properties of novel proteins.

Psomopoulos et al. propose a finite state automata data mining approach, which is used to induce protein classification rules [4]. The form of the extracted rules is X→Y, where X is a set of motifs and Y a set of protein families. Results outperformed those obtained in [1] and [3].

Merschmann and Plastino propose a new data mining technique for protein classification based on Bayes' theorem, called highest subset probability (HiSP) [5]. To evaluate their proposal, same datasets as in [4] were used. The results have shown that the proposed method outperforms previous methods based on decision trees [3] and finite state automata [4].

Diplaris et al. present a comparative evaluation of several machine learning algorithms for the motif-based protein classification problem [6]. The results showed that a Support Vector Machine classifier provided the least mean error rate.

In the present study a Support Vector Machine classifier (SVM) is trained using a set of proteins with known function. Each protein in this set is represented by a binary input vector produced using a motif vocabulary. The classifier aims to assign novel protein sequences to one of the protein families that appear in the training set. Since the input vectors consist of a great number of motifs, Feature Selection Algorithms (FSAs) are applied in order to select the most discriminative motifs. Three FSAs, embedded within the SVM architecture, were considered. The first FSA, called Recursive Feature Elimination (RFE) [7], conducts feature selection in a sequential backward elimination manner using as a criterion the amplitude of the weights of the SVM. The second FSA is called discriminative function pruning analysis (DFPA) feature subset selection method [8]. The basic idea of the DFPA method is to learn the SVM discriminative function from training data using all input variables available first, and then to select feature subset through pruning analysis. The third, called prediction-risk-based feature selection (SBS) [9], evaluates the features by computing the change of training error when the features are replaced by their mean values. Moreover, Stepwise Discriminant Analysis (SDA) has been applied to the complete feature set, in order to compare the embedded techniques to a filter FSA.

This paper is organized as follows. In Section 2 the aforementioned feature selection methods are presented, whereas in Section 3 the experimental results are demonstrated. The paper concludes in Section 4.

2   Materials and Methods

2.1   Support Vector Machine Classification

Support Vector Machines (SVMs) [10]–[12] is a very popular choice for performing classification tasks. Since the structural risk minimization principle of SVMs chooses discriminative functions that have the minimal risk bound, SVMs are less likely to overtrain data than other classification algorithms. Because of their useful properties they were selected in this study for assigning functional families to protein sequences.

SVM is a linear machine performing binary classification. It is based on the large margin classification principle, according to which the discriminating hyperplane maximizes the margin between certain training data points of each class, called support vectors.

In some cases, using nonlinear SVMs can improve classification results. The key idea of nonlinear SVMs is mapping patterns non linearly from the input space to a transformed space, usually of higher dimensions and then perform classification in the transformed space using linear support vector machines. However, the nonlinear mapping is not explicitly performed; instead kernel functions are employed to compute the inner products between support vectors and the pattern vectors in the transformed space. The most popular kernels are Gaussian and Polynomial.

As mentioned above SVMs perform binary classification. In order to apply SVMs to multi-class problems a modification is required. In this study, the One Versus All (OVA) multi-class extension has been employed. This approach performs K binary classification between the instances of each class and the instances of all the remaining classes, where K denotes the number of different classes. OVA-SVM was preferred over other multi-class extensions not only because of its simplicity, but more importantly because it also allows for a straightforward extension of the aforementioned embedded feature selection methods.

2.2   Feature Selection

Despite the good generalization ability of SVMs, it is a good practice to reduce the feature space removing any redundant or noisy features. However, SVM feature selection based on wrapper methods [13] is inefficient. This is because it involves training a large number of SVM classifiers, with each training being computational expensive. In case of multi-class SVM the computational cost increases by a factor of K, where K is the number of different classes. On the other hand filter methods [14] that present low computational cost, do not take into account the applied classification scheme and are not very effective. A good compromise between efficiency and performance are the embedded techniques, which exploit the architecture of the classifier in order to derive the most important features. In this study, three embedded feature selection algorithms are considered.

Recursive Feature Elimination. Perhaps the most popular feature selection technique embedded to the SVMs is Recursive Feature Elimination (RFE) presented in [7]. It is based on the amplitude of the separating hyperplane's weights. In each step only the features with the highest weights are selected and the SVM classifier is retrained. In this study, the OVA multi-class extension proposed in [15] has been adopted. According to this approach in order to select the features the sum of the squared weights is calculated over the K binary classifications as shown in the following equation.

$$J_j = \frac{1}{K} \sum_{k=1}^{K} \left(w_j^{\,k} \right)^2 . \tag{1}$$

In (1) J_j denotes the cost for not selecting feature j, $w_j^{\,k}$ denotes the separating hyperplane's weight that corresponds to the jth feature and the binary classifier for the kth class, whereas K denotes the number of different classes.

Discriminative Function Pruning Analysis. The basic idea of the DFPA algorithm [8] is to learn the SVM discriminative function from training data using all input variables available first, and then perform pruning analysis in order to select feature subset. The pruning is implemented using a forward or backward selection procedure, combined with a linear least square estimation algorithm. The method takes advantage of the linear-in-the-parameter structure of the SVM discriminative function. In this study the backward selection procedure was preferred, since RFE also performs backward elimination of the features. Moreover, like in the RFE case, the method has been extended to apply to OVA SVM in a similar manner, i.e. averaging the selection criterion over the classes.

Prediction Risk Based Feature Selection Method. This method, originally proposed by Moody et al. [16], evaluates the features by computing the change of training error when the features are replaced by their mean values. As argued by Li et al. [9], it may be more attractive than the two previous methods, since it uses the multi-class classification SVMs directly, instead of averaging the results of the binary classifications. In that study the selection procedure was called Sequential Backward Search (SBS), and this name is also adopted for the rest of this paper.

3   Experimental Results

The dataset used for SVM training and evaluation, called genbase28, was extracted from the PROSITE database using GenMiner [3]. It contains 2934 proteins belonging to 28 classes and was also used in [4][5][6] for protein classification based on machine learning techniques. Every instance of this dataset corresponds to a certain protein and consists of the protein name, the subset of the 1185 database motifs that correspond to that protein and the name of the functional family assigned to that protein.

As discussed in [5], it is an extremely class-imbalanced dataset. Thus, pre-processing has been conducted removing proteins of the same class having identical input vectors. Then, proteins of poorly represented classes, containing less than 10 instances, were discarded. This resulted into a refined dataset of 878 proteins represented by 268 motifs and belonging to 13 classes. A vector with binary values denoting the presence (or absence) of each one of the 268 remaining motifs is assigned to each protein. These binary vectors constitute the input vectors of the SVM classifiers.

Various SVM kernels were tested on the above dataset using stratified 10-fold cross validation. More specifically, three kernel types were considered, linear, second degree polynomial, and Gaussian. Linear SVM produced the best results presenting a classification error rate of 21.07% ±3.88%, whereas analysis of variance [17] resulted to rejecting the null hypothesis that the classification error rate is the same for all three kernels at the 95% confidence level. Therefore, for the remaining experiments only the linear kernel was employed.

Linear SVM's performance was evaluated also in the case all instances were used for training, resulting to a classification error rate of 11.06%. Since the test error is about twice the training error it is safe to assume that despite the large margin property of the SVM some overfitting takes place. A common strategy to avoid overfitting is to employ feature selection techniques in order to reduce the number of features of the classifier.

A popular feature selection technique is Stepwise Discriminant Analysis [18]. It is a filter method that is easy to implement and is also computationally efficient. Using SDA, 50 motifs are selected out of the original 268, and the classification error rate drops to 16.97%, which is a significant improvement with respect to the original results. However, the classification error rate is still high compared to the training error.

As a next step, a very popular feature selection algorithm embedded to the SVM architecture, called Recursive Feature Elimination (RFE) was considered. During the experiment, in every step of the RFE algorithm only a single motif is eliminated, whereas the algorithm terminated when only one motif remained. This approach allows for ranking the motifs with respect to the order they are eliminated. The performance of SVM-RFE was evaluated for each subset of selected features using 10-fold cross-validation. The lowest classification error rate, 14.46% ±3.05%, has been achieved in case of 52 motifs. In that case the training error was 12.1% indicating that overfitting is significantly reduced.

A different feature selection algorithm embedded to the SVM architecture, called DFPA was also considered. Motif elimination and algorithm termination is similar to the RFE case, also allowing for motif ranking. The performance of SVM-DFPA was evaluated for each subset of selected features using 10-fold cross-validation. The lowest classification error rate, 14.81% ±3.46%, has been achieved in case of 45 motifs. In that case the training error was 12.22% indicating that overfitting is also significantly reduced. These results are very close to the ones achieved by RFE, while fewer motifs were selected.

Another feature selection algorithm embedded to the SVM architecture, called SBS was also considered. Like in the case of RFE and DFPA, the SBS allows for motif ranking using the same elimination and termination rules. The performance of SVM-SBS was evaluated for each subset of selected features using 10-fold cross-validation. The lowest classification error rate, 14.92% ±3.09%, has been achieved in case of 24 motifs. In that case the training error was 13.71% indicating that overfitting is also significantly reduced. These results are very close to the ones achieved by RFE and DFPA, while very few motifs were selected.

All three embedded feature selection algorithms presented similar results. However, these results were based only on the best subset of each algorithm. Further analysis is considered regarding the performance for all subsets and the agreement of the algorithms on the ranking of the motifs. In the diagram of Fig. 1 are illustrated the learning curves for all three methods. More specifically, the classification error rate % is plotted against the number of selected motifs of each method.

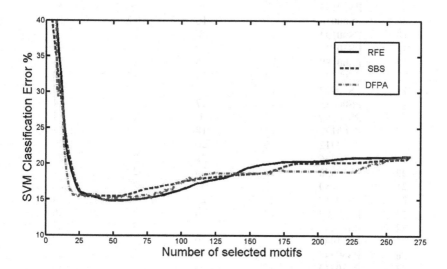

Fig. 1. Classification error rates for the three embedded FSAs, with respect to the number of selected motifs

The three plots are very similar, when the number of motifs is not extremely low. If there are less than 12 motifs SVM-SBS outperforms the other two methods, whereas SVM-RFE presents on average slightly better results than the other two methods.

All methods present their best performance in between 20 and 80 selected motifs. Therefore, the centre of this interval, namely 50 motifs, is selected to test for the agreement of the methods on motif ranking. In that case, all methods share 37 common motifs out of the total of 50. In particular, SVM-RFE shares 46 common motifs with SVM-DFPA, whereas SVM-SBS shares 39 common motifs both with SVM-RFE and SVM-DFPA.

In Table 1 the 37 common motifs are presented, where the individual ranking of each method is also given. In the last column the median of the three individual rankings is estimated. These results indicate that there is a significant agreement between the three methods as to which motifs of this dataset are important for protein classification.

Table 1. The 37 motifs common in all three methods' 50 motifs selections

	Motif	RFE rank	DFPA rank	SBS rank	Median rank
1	PS00022	3	2	2	2
2	PS01186	1	3	16	3
3	PS50114	4	13	3	4
4	PS50109	5	6	6	6
5	PS50071	6	8	1	6
6	PS00561	8	7	7	7
7	PS00562	7	15	8	8
8	PS00193	9	9	5	9
9	PS00010	2	10	17	10
10	PS00192	10	5	15	10
11	PS50322	15	4	10	10
12	PS00188	11	16	9	11
13	PS00187	12	14	12	12
14	PS50079	16	12	4	12
15	PS00025	14	17	11	14
16	PS50099	17	1	14	14
17	PS00177	13	18	19	18
18	PS50312	36	20	18	20
19	PS50326	20	32	21	21
20	PS50830	21	33	20	21
21	PS50280	22	37	23	23
22	PS50318	37	24	13	24
23	PS50089	19	26	27	26
24	PS50129	18	27	34	27
25	PS01040	25	29	37	29
26	PS50044	29	34	28	29
27	PS50313	47	30	22	30
28	PS50324	33	11	32	32
29	PS50316	45	22	33	33
30	PS50016	31	35	36	35
31	PS50325	50	23	35	35
32	PS50215	39	36	24	36
33	PS00402	26	38	40	38
34	PS00136	27	39	41	39
35	PS00875	32	40	39	39
36	PS00012	28	44	42	42
37	PS50303	49	43	30	43

It should be mentioned that the FSA methods produce lower classification error rates than those presented in [4], [5] and [6]. However, the focus of this study is on motif selection and ranking, and pre-processing was performed on that basis resulting to a smaller dataset with less than 28 classes. Therefore, it does not make much sense to perform direct comparisons of the classification results to those of the previous studies and this is why the experimental results do not include such comparisons.

All experiments were conducted using Matlab 7.8.0 programming environment. SVM classification was performed using the SVM-KM toolbox [18], whereas the embedded feature selection methods were implemented by the authors.

4   Conclusion

In this work three FSAs embedded to the SVM architecture, were employed and evaluated on a protein classification task. The classification scheme was used to assign protein sequences to functional families, based on binary features denoting the presence or absence of motifs in their sequences.

A real dataset extracted from the PROSITE database has been employed for the evaluation of the aforementioned schemes. The experimental results demonstrated that all three feature selection methods can greatly reduce the test error of the used data set. The reduction magnitude is about 7 % of the test error on the total feature set. This indicates that the data set has some redundant or even noisy motifs, which decrease the performance of the learning machine. Moreover, the three feature selection algorithms presented significant agreement on their rankings. Therefore, prior to protein sequence classification, even with robust classifiers like SVM, it is highly recommended that the motifs comprising the feature vectors should be carefully selected. In future work the fusion of the individual ranking results will be studied in order to derive even more robust selections, improving even more the generalization ability of the employed classifier.

Acknowledgments. The authors wish to thank Dr. Luiz Merschmann for providing the experimental dataset and the anonymous reviewers for their useful comments and suggestions.

References

1. Wang, D., Wang, X., Honavar, V., Dobbs, D.: Data-driven Generation of Decision Trees for Motif-based Assignment of Protein Sequences to Functional Families. In: Atlantic Symposium on Computational Biology, Genome Information Systems & Technology (2001)
2. PROSITE, http://prosite.expasy.org/
3. Hatzidamianos, G., Diplaris, S., Athanasiadis, I., Mitkas, P.A.: GenMiner: A Data Mining Tool for Protein Analysis. In: 9th Panhellenic Conference on Informatics, Thessaloniki, Greece, pp. 346–360 (2003)
4. Psomopoulos, F., Diplaris, S., Mitkas, P.A.: A Finite State Automata Based Technique for Protein Classification Rules Induction. In: 2nd European Workshop on Data Mining and Text Mining for Bioinformatics, Pisa, Italy, pp. 54–60 (2004)
5. Merschmann, L., Plastino, A.: A Lazy Data Mining Approach for Protein Classification. IEEE Transactions on Nanobioscience 6, 36–42 (2007)

6. Diplaris, S., Tsoumakas, G., Mitkas, P.A., Vlahavas, I.: Protein Classification with Multiple Algorithms. In: Bozanis, P., Houstis, E.N. (eds.) PCI 2005. LNCS, vol. 3746, pp. 448–456. Springer, Heidelberg (2005)
7. Guyon, I., Weston, J., Barnhill, S., Vapnik, V.: Gene Selection for Cancer Classification Using Support Vector Machines. Machine Learning 46, 389–422 (2002)
8. Mao, K.Z.: Feature Subset Selection for Support Vector Machines through Discriminative Function Pruning Analysis. IEEE Transactions on Systems, Man, and Cybernetics 34, 60–67 (2004)
9. Li, G.-Z., Yang, J., Liu, G.-P., Xue, L.: Feature Selection for Multi-class Problems Using Support Vector Machines. In: Zhang, C., Guesgen, H.W. (eds.) PRICAI 2004. LNCS (LNAI), vol. 3157, pp. 292–300. Springer, Heidelberg (2004)
10. Vapnik, V.: The Nature of Statistical Learning Theory, 2nd edn. Springer, New York (1999)
11. Scholkopf, B.: Support Vector Learning. Oldenburg-Verlag, Munich (1997)
12. Burges, C.: A Tutorial on Support Vector Machines for Pattern Recognition. In: Data Mining Knowledge Discovery 2, pp. 121–167 (1998)
13. Kohavi, R., John, G.H.: Wrappers for Feature Subset Selection. Artificial Intelligence 97, 273–324 (1997)
14. Kudo, M., Sklansky, J.: Comparison of Algorithms that Select Features for Pattern Classifiers. Pattern Recognition 33, 25–41 (2000)
15. Zhou, X., Tuck, D.P.: MSVM-RFE: Extensions of SVM RFE for Multiclass Gene Selection on DNA Microarray Data. Bioinformatics 23, 1106–1114 (2007)
16. Moody, J., Utans, J.: Principled Architecture Selection for Neural Networks: Application to Corporate Bond Rating Prediction. In: Moody, J.E., Hanson, S.J., Lippmann, R.P. (eds.) Advances in Neural Information Processing Systems 4, pp. 683–690. Morgan Kaufmann Publishers, Inc. (1992)
17. Hogg, R.V., Ledolter, J.: Engineering Statistics. MacMillan, New York (1987)
18. Einslein, K., Ralston, A., Wilf, H.S.: Statistical Methods for Digital Computers. John Wiley & Sons, New York (1977)
19. Canu, S., Grandvalet, Y., Guigue, V., Rakotomamonjy, A.: SVM and Kernel Methods Matlab Toolbox. Perception Systèmes et Information, INSA de Rouen, Rouen, France (2005)

A Multi-objective Genetic Algorithm for Software Development Team Staffing Based on Personality Types

Constantinos Stylianou[1] and Andreas S. Andreou[2]

[1] Department of Computer Science, University of Cyprus
cstylianou@cs.ucy.ac.cy
[2] Department of Computer Engineering and Informatics, Cyprus University of Technology
andreas.andreou@cut.ac.cy

Abstract. This paper proposes a multi-objective genetic algorithm for software project team staffing that focuses on optimizing human resource usage based on technical skills and personality traits of software developers. Human factors are recognized as critical aspects affecting the rate of success of software projects, as well as other properties, such as productivity, software quality, performance, and job satisfaction. However, managers often rely solely on technical criteria to staff their projects, which risks overlooking these important aspects of software development, such as the abilities and work styles of developers. The behaviour and scalability of the algorithm was validated against a series of hypothetical projects of varying size and complexity, and also through a real-world project of an SME in the local IT industry. The approach demonstrated a sufficient ability to generate both feasible and optimal staffing solutions by assigning developers most technically competent and suited personality-wise for each project task.

Keywords: Software Project Management, Team Staffing, Genetic Algorithms, Personality Type Matching, Five-Factor Model.

1   Introduction

For many years, researchers in the area of software engineering have argued that human factors should be taken into consideration when developing software since human resources are the most, if not only, crucial resource available for software development companies. Hence, most research targets software project managers since it is their responsibility to assign these resources to tasks and their selection directly influences the success of a software project, especially with respect to critical software development issues such as performance, productivity, quality, and job satisfaction.

Assigning software developers to tasks is not a simple process as each developer has their own strengths and weaknesses, which often extend beyond the "academic" knowledge acquired from a university degree, for example, or the experience gained from using a specific tool or technology over a number of years. They also concern traits and behaviours in the form of abilities and competencies that develop from each

L. Iliadis et al. (Eds.): AIAI 2012, IFIP AICT 381, pp. 37–47, 2012.

individual's personality and psychological processes. For this reason, it is argued that software project managers need to look beyond technical skill-based and experience-based methods when selecting developers, as it is equally important to deal with interpersonal relationships and social aspects present in software development processes and organizations [1]. The goal, therefore, of the proposed approach is to support software project managers of SMEs in the selection and allocation of the most suitable developers to tasks, by attempting to optimize the assignments based on technical skills and personality traits of developers using a multi-objective genetic algorithm.

The remainder of this paper is organized as follows: Section 2 provides a brief overview of the most recent research attempts proposed regarding the inclusion of human factors in software project team staffing. Section 3 describes the method used to identify the various occupations found in the software development industry together with their characteristics and requirements in terms of technical skills and personality traits. In section 4, a description of the methodology is provided, in which various aspects of the multi-objective optimization approach are described. Next, section 5 presents the evaluation of the proposed approach, describing the projects used in the experiments carried out followed by a discussion on the results obtained. Finally, section 6 concludes the paper with a synopsis and comments on future work.

2   Related Work

One of the most common human factors affecting software development addressed in literature is the area of personality, with many different investigations carried out over the years. Some attempts have looked into studying the various types of personality of software development professionals. The results from such studies can provide helpful insight about the type of personalities attracted to the software development industry, and also can be used by companies looking to recruit or release personnel, or by software project managers attempting to assign developers to tasks. Some of the questions asked in these studies focus on whether software professionals share a common personality type [2] and whether they differ from the rest of the general population [3]. Some studies concentrate on specific professions, such as systems analysts [4], while others examine occupations from all development phases [5]. More recently, Varona, et al. carried out a survey of existing studies that attempted to profile IT-related professions in order to identify trends and changes, and to form a better understanding of the software industry's human resources [6]. Another area of research has focused on assessing the effects of personality on various properties of software development. For instance, personality has been explored with respect to team effectiveness and performance [7, 8], cohesion [9], software quality [10], as well as job satisfaction [11]. More recent research has also concentrated on how personality influences pair programming in agile methodologies [12, 13].

Research work relating to team staffing and formation have also been carried out, whereby attempts to build teams for software projects take into account human factors such as developers' personality types [14] and capabilities [15]. Some of the team staffing approaches proposed employ computational intelligence techniques to aid with assigning developers to tasks. For instance, Martínez et al. [16] proposed RAMSET as a methodology for assigning roles in software engineering teams that adopted a learning approach based on an adaptive network-based fuzzy inference system to recommend the best resource allocation possible. Also, André, Baldoquín and Acuña [17] formalized a set of rules to match the personality types of developers to fixed project roles defined by a set of generic and technical competencies. They then transformed these rules into objective functions and constraints, and applied them in various heuristic algorithms (such as, random restart hill-climbing, simulated annealing and Tabu search) to carry out optimal assignment of developers to roles.

Despite much research being conducted concerning human factors, and in particular personality, most focuses on exploring and investigating their effects in various aspects of software development. There is still, however, a great need for tools to support software project managers incorporate these factors systematically in their staffing activities. Therefore, a major contribution of this work is providing such a tool, which carries out team staffing in an automated fashion and that, in addition, employs computational intelligence through the application of multi-objective optimization to handle the balancing of personality traits and technical skills and knowledge.

3   Personality Traits of Software Development Occupations

One aspect of the optimization approach implemented in this paper concerns the evaluation of selected developers based on the suitability of their personality traits with respect to the type of task they have been assigned to carry out. To do this, the software professions most commonly found in SMEs of today's software development industry were first identified using the 2010 Standard Occupational Classification (SOC), which serves as a systematized taxonomy of the majority of existing professions identified by the U.S. Bureau of Labor Statistics [18]. Then, detailed analysis of each profession was carried out using the Occupational Information Network (O*NET) Resource Center [19], which provides a content model and an online database defining standardized and occupation-specific descriptors of each profession using the SOC system coding. Each occupation's job-related and worker-related characteristics and requirements were retrieved containing information on: the abilities and work styles of workers, the skills required by workers, and the work activities of occupations. Once these key requirements and characteristics were identified, the most suitable personality traits required by developers to carry out activities of each profession were then were associated with corresponding personality traits. These are expressed using five basic domains of personality, which comprise the Five-Factor Model (FFM), originally identified in 1961 by Tupes and Cristal [20] and later operationalized by Costa and McCrae through the NEO-Personality Inventory (NEO-PI) [21]. The Five-Factor Model has been widely adopted in many academic and application disciplines where personality measures have been required, and is a common instrument in cases involving career and personnel assessment. Specifically, the five domains are described as follows:

- *Neuroticism* reflects the level to which an individual is predisposed to experiencing negative emotions, such as sadness, embarrassment, fear and anger.
- *Extraversion* refers to the level to which an individual engages with their external world through interpersonal interactions, as well as their energy and predisposition to experiencing positive emotions.
- *Openness* to experience concerns an individual's tendencies regarding intellectual curiosity, creativity and variety in interests and experiences.
- *Agreeableness* involves interpersonal orientation with regards issues, such as compassion, social harmony, cooperation, and trust.
- *Conscientiousness* relates to the degree of self-discipline and control of impulses, and also ambition and organization.

Finally, the desired level of each of the domain was determined so as to ascertain whether a profession requires either a {1:low, 2:medium or 3:high} level of that particular domain. The same was applied to personality traits possessed by developers so that comparisons between the two can take place using a simple distance measure.

An important issue here is that of the validation of each profession's associated personality traits that were selected as desirable. A large number of studies were used for this purpose, such as [2, 3, 4, 5, 6], as well as other related material, for example, career handbooks suggesting the best occupation based on personality types. However, the validation process is not currently in the scope of this paper, since the desired personality types can be easily modified and, furthermore, the study's focus is on how well the chosen encoding performs in optimizing developer assignments.

4   Description of Methodology

The goal of the proposed approach is to allow project managers of SMEs in the software industry to staff their projects with the most suitable teams taking into account the technical knowledge and skills of available developers as well as their personality traits and abilities. These, however, may be viewed as conflicting in some cases, where a developer may be technically capable but does not hold the appropriate traits required by activities of a task, or vice-versa. For example, if a programming task requires skills in a specific programming language, a developer possessing a high level of such skills may not necessarily possess a low level of extraversion, which is one of the desired traits of programming tasks. On this basis, a software project manager would encounter difficulties in trying to make the best selection and assignment of resources whilst trying to balance the two. Also, due to the fact that there are many different possible combinations to examine, a software project manager will be required to perform an exhaustive assessment of all possible permutations, which only becomes more difficult when the number of tasks to be performed and the number of available developers increases. Therefore, in order to decrease the search space and handle the NP-hard nature of such a problem [22], a multi-objective optimization approach was adopted. In particular, two objective functions were modelled in an implementation of the Non-dominated Sorting Genetic Algorithm II (NSGA-II), introduced by Deb in 2002 [23]. With this technique, a set of optimal solutions will be produced suggesting a collection of possible assignments of developers to project tasks aiming to satisfy the two aforementioned considerations. Furthermore, due to the involvement of constraints influencing the feasibility of solutions, the Constrained NSGA-II algorithm was applied. The algorithm promotes the solution diversity using a crowded comparison operator during its selection procedure and population reduction process. In addition, because parent and offspring populations are combined before non-dominated sorting takes place, elitism (i.e., preservation of the best solutions) is always ensured. Further details of the NSGA-II can be found in [23].

4.1   Encoding and Representation

Each software project consists of a number of tasks that need to be carried out. For each task, developers need to be assigned to perform the activities involved. Therefore, since it is the selection of developers that forms the basis of evaluation, each project task is denoted by a string of bits, and each bit represents one specific developer. If a bit in the string has a value of '1', then this signifies that the specific developer is assigned to work on the task, whereas a value of '0' indicates that the developer has not been selected for the task. Overall, if a software project consists of T tasks and there are E available developers, then each solution would be represented by an individual in the algorithm using (TxE) bits.

Since it is possible that a development company follows its own method to evaluate technical skills and knowledge, each developer is simply required to be rated based on each of the skills required by task activities in a normalized form in the range [0, 1], meaning that low possession of skill will be denoted by a value closer to zero, and high possession of a skill will be denoted by a value closer to one. Project managers can use different metrics, such as experience or IT-related aptitude tests, or can even use the experience requirements and occupation-specific information set suggested by the O*NET Content Model. As part of on-going research efforts in this area, a method to rate and match developers' skills and knowledge is currently in progress. On the other hand, regarding personality traits, developers are assessed specifically using the NEO-Five-Factor Inventory and their five domain scores are used.

4.2   Objective Functions and Constraints

A total of three objective functions were created for the evaluation of each solution along with two constraints to measure the degree of feasibility of each solution.

Technical Skills Objective Function (f_1). This maximization function is responsible for evaluating a solution based on the assigned developers' levels of technical skills and knowledge required by the activities of each task. The objective function's value is calculated for each task by adding the maximum skill level possessed by the developers assigned to the average skill level possessed by the developers assigned, as shown in Eq. (1).

$$f_1 = \max(skill\_level\_of\_assigned\_developers) + \\ + avg(skill\_level\_of\_assigned\_developers) \tag{1}$$

Personality Traits Objective Function (f_2). This maximization function is responsible for evaluating a solution based on developers' personality traits. Using Eq. (2), a distance is computed between the desired levels of the FFM domains for the profession and the selected developer's levels of the FFM domains. If more than one developer is assigned to work on a task, then the average of the distances is used.

$$f_2 = \sum_{i=1}^{5} |desired\_level\_in\_domain_i - developer\_level\_in\_domain_i| \tag{2}$$

Team Size Objective Function (f_3). This minimization function is responsible for evaluating a solution based on the number of developers assigned to each task. The inclusion of such a fitness measure is important in order to maximize resource utilization and avoid unnecessary assignments. For each task, the inverse of the number of team members is calculated, as shown in Eq. (3).

$$f_3 = \frac{1}{number\_of\_assigned\_developers} \tag{3}$$

Skills Satisfied Constraint (c_1). The purpose of this constraint function is to measure the feasibility of a solution with respect to the technical skills required by activities of a task. Using Eq. (4), a solution is feasible only if for all activities, there is at least one developer selected to carry it out who possesses the necessary technical skills.

$$c_1 = \frac{number\_of\_unsatisfied\_skills}{total\_number\_of\_project\_skills\_required} \tag{4}$$

Developer Availability Constraint (c_2). The second constraint implemented involves checking the feasibility of a solution regarding the availability of developers when assigned to tasks that are carried out simultaneously. An assumption made in this approach is that no developer can work on more than one task at any given point in time. Furthermore, it is also assumed that the project's schedule is fixed beforehand and, thus, each task has already been allocated a specific "time slot" to be carried out. Eq. (5) determines this constraint value by calculating the number of days a developer has been assigned to more than one task throughout the duration of the project.

$$c_2 = \frac{number\_of\_days\_assigned\_with\_conflicts}{total\_number\_of\_days\_assigned\_in\_the\_project} \tag{5}$$

4.3   Algorithm Parameters and Settings

For the execution of the multi-objective genetic algorithm, a population of 100 individuals were used. The fast non-dominated sorting procedure was applied to rank the individuals in terms of their fitness and feasibility, after which a tournament selection of size 4 was used to select which parents were to enter the mating pool. The best two parents were then set to produce offspring by the application of a recombination operator (one-point crossover) with a likelihood of 0.80, and a mutation operator (single bit-flip mutation) with a $(1/number\_of\_project\_tasks)$ probability. The population evolves by repeating the steps from the selection of individuals until the termination criteria are met or the maximum number of iterations have been reached (set at 2000).

5   Results of Experiments

5.1   Design of Experiments

Two experiments were carried out to evaluate the proposed methodology. First, in order to evaluate the validity and scalability of the multi-objective optimization algorithm, two hypothetical software projects consisting of 20 and 30 tasks each were created based on the input of several project managers of SME software development companies as to the basic structure, size and complexity of the type of software projects they usually undertake. For both projects, the skill levels and personality traits of the available software developers were selected so as to represent the best-case and worst-case scenarios for the proposed team staffing approach. For the best-case scenario, all available developers possessing the highest skill levels also possessed the most suitable personality traits. On the other hand, for the worst-case scenario, all the available developers possessing the highest skill levels possessed the least suitable personality traits, and vice-versa. Through these two extreme cases, both the behaviour and correctness of the optimization approach could be observed and analyzed, and also the competitive nature of the objective functions could be investigated.

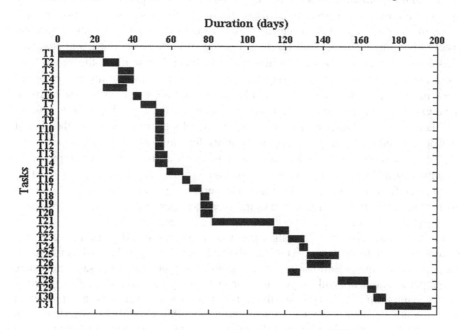

Fig. 1. Real-world software project schedule (vessel policies management system)

In the second experiment, a real-world software project developed by a local IT company was used that related to the implementation of a vessel policies management system for a large insurance brokers company. The project, whose schedule is shown in Fig. 1, consisted of 31 tasks involving project management, design, programming and testing activities. A total of four developers were available to carry out the project, each possessing a varied set of skills at different levels, as well as unique personality traits, which were determined by administering the NEO-FFI-3 [21].

5.2   Results and Discussion

In the first experiment, the algorithm was executed 10 times for both hypothetical projects using the best-case and worst-case scenarios. The results are shown in Table 1.

Table 1. Results obtained from the first experiment using hypothetical software projects

Experiment	Average Number of Unique Solutions	Execution Time (min.)
Best-case scenario (20 tasks)	1	22.56
Worst-case scenario (20 tasks)	94	22.86
Best-case scenario (30 tasks)	1	23.90
Worst-case scenario (30 tasks)	95	26.08

For both hypothetical projects, the algorithm managed to provide both feasible and optimal solutions when performing team staffing in the best-case scenario. Specifically, the algorithm always managed to assign the most suitable developer with respect to both the technical skill levels and personality traits possessed, and never assigned a developer who was less suited in either aspect. Furthermore, the Pareto front consisted of individuals representing the same optimal solution. This was expected since in the "optimistic" case there would always be only one possible ideal assignment existing for each task. In the worst-case scenario, the algorithm's job was to try to balance the two objective functions, since no developers possessed both the highest skill levels and most suitable personality traits for any task. For both hypothetical projects, it was observed that this time the individuals of the Pareto front represented a number of different solutions, as seen in Fig. 2. Such behaviour again was anticipated since for each task either skill levels or personality traits could be given preference – but not both due to the nature of the characteristics of the available developers. The general behaviour of the algorithm was, thus, validated as correct.

In the second experiment, the algorithm was also executed 10 times. However, the results obtained from these executions showed that the algorithm did not actually produce any optimal solutions but, in fact, consistently generated the same infeasible developer assignments with respect to their availability (constraint c_2). This observation is mainly attributed to the small number of available developers in combination with a relatively high number of concurrent tasks within the project. Specifically, in cases where several tasks requiring the same skills and personality traits were set to

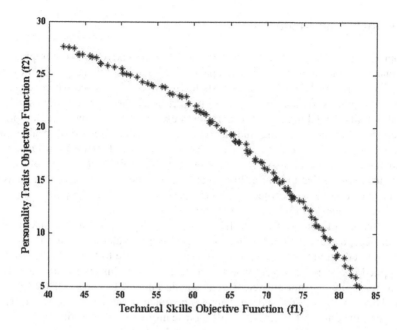

Fig. 2. Pareto front of hypothetical project (30 tasks) worst-case scenario (f_1 vs. f_2)

execute simultaneously (e.g., tasks T8-T14 in Fig. 1), the algorithm would encounter difficulties in finding an optimal assignment of developers possessing these to the tasks, simply because the available resources were insufficient for such concurrent scheduling of tasks.

What's interesting, however, is that whilst consulting with the project's manager to understand the method that was used to allocate human resources to tasks, it was established that the project suffered from schedule overruns due to the lack of available resources. This shows the potential of such approach as it can be used as means to pinpoint possible staffing caveats for project managers, who would then be able to seek a solution by either revising their original schedule of tasks or even consider hiring or buying the services of developers for tasks that could not be optimally allocated resources. To further investigate this, the experiment was repeated without inclusion of the developer availability constraint. The results this time showed that the algorithm was able to produce both feasible and optimal assignments, averaging around 95 unique Pareto front solutions over 10 executions. In some cases developers possessing high levels of skills required for tasks but less suitable personality traits were chosen and in other cases developers most suited with respect to personality traits were preferred even if they possessed lower levels of skills for the tasks they were assigned to. This further enhanced the initial belief that the two objectives can indeed be competing. In such a case, it is up to the project manager to decide which of the two options has higher priority and, therefore, should be followed.

6   Concluding Remarks

The approach described in this paper proposes the innovative use of computational intelligence as a means to help software project managers solve the problem of team staffing. In particular the approach suggests the use of a multi-objective genetic algorithm for assigning software developers to project tasks based on technical skills and personality traits. By taking into consideration these factors, it allows software project managers to view team staffing and human resource allocation from an alternative perspective, since software project success is considered to be largely influenced by the human factors present in software development. The results obtained from several experiments indicate that the algorithm is capable of generating adequate and feasible solutions, balancing the two objectives where necessary, and has the potential to constitute a decision support tool for software project team staffing.

One of the major contributions of this approach is that it can allow project managers to foresee possible resource issues arising during development. With respect to either or both constraint functions, if the algorithm is unable to find feasible solutions when applied to a specific project whose schedule is fixed, this could indicate that the available resources are not sufficient or adequate enough to carry out the software project. This is very useful for project managers since it would allow them to revise their project schedule and attempt to staff their team in a slackened timeframe. Alternatively, without modifying the project's schedule, a project manager may use the results to recruit extra resources (possessing either higher skill levels or more suitable personality traits or both). In a similar way, the approach can be used to examine whether the development company has the required capacity in terms of human resources before bidding for a software project.

As part of future work, improvements can be made to the way in which developer skill levels are measured and evaluated. Also, additional objective functions can be introduced, such as the minimization of a project's cost based on developers' salaries. Attempts have also been made to collect data from development companies for the purposes of further evaluation using real-world projects, in addition to comparison of the approach with other computational intelligence techniques. Another possible future enhancement can involve integrating the proposed approach with other techniques for software project management activities taking into account the solutions generated from the current optimization method. This could allow for a comparative analysis of the effectiveness of the approach when combined and used in parallel with other models. Possible applications include feeding the solutions generated by the algorithm into an intelligent scheduling mechanism or, alternatively, using the output of assigned developers to help predict the cost of a software project (per phase or as a whole).

References

1. Amrit, C.: Coordination in Software Development: The Problem of Task Allocation. In: 27th International Conference on Software Engineering, pp. 1–7. ACM, New York (2005)
2. Moore, J.E.: Personality Characteristics of Information Systems Professionals. In: 1991 Conference on Computer Personnel Research, pp. 140–155. ACM, New York (1991)
3. Wynekoop, J.L., Walz, D.B.: Revisiting the Perennial Question: Are IS People Different? ACM Database 29, 62–72 (1998)

4. Smith, D.C.: The Personality of the Systems Analysts: An Investigation. ACM SIGCPR Computer Personnel 12, 12–14 (1989)
5. Capretz, L.F., Ahmed, F.: Making Sense of Software Development and Personality Types. IT Prof. 12, 6–13 (2001)
6. Varona, D., Capretz, L.F., Pinero, Y., Raza, A.: Evolution of Software Engineers' Personality Profile. ACM SIGSOFT Soft. Eng. Notes 37, 1–5 (2012)
7. Peeters, M.A.G., van Tuijl, H.F.J.M., Rutte, C.G., Reymen, I.M.M.J.: Personality and Team Performance: A Meta-Analysis. Eur. J. Personality 20, 377–396 (2006)
8. Capretz, L.F., Ahmed, F.: Why do we Need Personality Diversity in Software Engineering? ACM SIGSOFT Soft. Eng. Notes 35, 1–11 (2010)
9. Karn, J.S., Syed-Abdullah, S., Cowling, A.J., Holcombe, M.: A Study into the Effects of Personality Type and Methodology on Cohesion in Software Engineering Teams. Behav. Inf. Technol. 26, 99–111 (2007)
10. Fernández-Sanz, L., Misra, S.: Influence of Human Factors in Software Quality and Productivity. In: Murgante, B., Gervasi, O., Iglesias, A., Taniar, D., Apduhan, B.O. (eds.) ICCSA 2011, Part V. LNCS, vol. 6786, pp. 257–269. Springer, Heidelberg (2011)
11. Acuña, S.T., Gómez, M., Juristo, N.: How do Personality, Team Processes and Task Characteristics Relate to Job Satisfaction and Software Quality? Inform. Software Tech. 51, 627–639 (2009)
12. Sfetsos, P., Stamelos, I., Angelis, L., Deligiannis, I.: An Experimental Investigation of Personality Types Impact on Pair Effectiveness in Pair Programming. Empir. Softw. Eng. 14, 187–226 (2009)
13. Salleh, N., Mendes, E., Grundy, J., Burch, G.S.J.: An Empirical Study of the Effects of Conscientiousness in Pair Programming using the Five-Factor Personality Model. In: 32nd ACM/IEEE International Conference on Software Engineering, pp. 577–586. ACM, New York (2010)
14. Rutherfoord, R.H.: Using Personality Inventories to Help Form Teams for Software Engineering Class Projects. ACM SIGCSE Bulletin 33, 73–76 (2001)
15. Acuña, S.T., Juristo, N.: Assigning People to Roles in Software Projects. Softw. Pract. Exper. 34, 675–696 (2004)
16. Martínez, L.G., Rodríguez-Díaz, A., Licea, G., Castro, J.R.: Big Five Patterns for Software Engineering Roles Using an ANFIS Learning Approach with RAMSET. In: Sidorov, G., Hernández Aguirre, A., Reyes García, C.A. (eds.) MICAI 2010, Part II. LNCS, vol. 6438, pp. 428–439. Springer, Heidelberg (2010)
17. André, M., Baldoquín, M.G., Acuña, S.T.: Formal Model for Assigning Human Resources to Teams in Software Projects. Inform. Software Tech. 53, 259–275 (2011)
18. Standard Occupation Classification, Bureau of Labor Statistics, U.S. Department of Labor, http://www.bls.gov/soc
19. Occupational Information Network, Employment and Training Administration, U.S. Department of Labour, http://www.onetcenter.org
20. Tupes, E.C., Christal, R.E.: Recurrent Personality Factors Based on Trait Ratings. Technical Report ASD-TR-61-97, Lackland Air Force Base, Personnel Laboratory, Air Force Systems Division (1961)
21. McCrae, R.R., Costa, P.T.: NEO Inventories Professional Manual. PAR Inc., Florida (1992)
22. Pan, N., Hsiao, P., Chen, K.: A Study of Project Scheduling Optimization using Tabu Search Algorithm. Eng. Appl. Artif. Intel. 21, 1101–1112 (2008)
23. Deb, K., Pratap, A., Agarwal, S., Meyarivan, T.: A Fast and Elitist Multi-Objective Genetic Algorithm: NSGA-II. IEEE Trans. on Evol. Comput. 6, 181–197 (2002)

An Empirical Comparison of Several Recent Multi-objective Evolutionary Algorithms

Thomas White[1] and Shan He[1,2,*]

[1] School of Computer Science
[2] Center for Systems Biology, School of Biological Sciences, The University of Birmingham,
Birmingham, B15 2TT, UK
s.he@cs.bham.ac.uk

Abstract. Many real-world problems can be formulated as multi-objective optimisation problems, in which many potentially conflicting objectives need to be optimized simultaneously. Multi-objective optimisation algorithms based on Evolutionary Algorithms (EAs) such as Genetic Algorithms (GAs) have been proven to be superior to other traditional algorithms such as goal programming. In the past years, several novel Multi-Objective Evolutionary Algorithms (MOEAs) have been proposed. Rather than based on traditional GAs, these algorithms extended other EAs including novel EAs such as Scatter Search and Particle Swarm Optimiser to handle multi-objective problems. However, to the best of our knowledge, there is no fair and systematic comparison of these novel MOEAs. This paper, for the first time, presents the results of an exhaustive performance comparison of an assortment of 5 new and popular algorithms on the DTLZ benchmark functions using a set of well-known performance measures. We also propose a novel performance measure called unique hypervolume, which measures the volume of objective space dominated only by one or more solutions, with respect to a set of solutions. Based on our results, we obtain some important observations on how to choose an appropriate MOA according to the preferences of the user.

Keywords: Multi-objective optimisation, Evolutionary Algorithms, Comparison, Genetic Algorithms.

1 Introduction

Many real-world optimisation problems involve multiple objectives, which are generally incommensurable and often conflicting. These so-called multi-objective optimisation problems are notoriously difficult to solve. In recent years, Multi-Objective Evolutionary Algorithms (MOEAs) have been attracting more attention due to their superior performance over traditional multi-objective optimisation algorithms, in terms of effectiveness and robustness. This general trend is reflected by an exponential increase of MOEA applications to many real-world problems. The most popular MOEAs are algorithms based on Genetic Algorithms (GAs), one notable example being NSGA-II [1]. Recently, several novel MOEAs have been proposed by extending traditional EAs

---

* Corresponding author.

L. Iliadis et al. (Eds.): AIAI 2012, IFIP AICT 381, pp. 48–57, 2012.

such as Simulated Annealing (SA) or novel EAs such as Scatter Search (SS) and Particle Swarm Optimiser (PSO) to handle multi-objective problems. These novel MOEAs were tested on different set of benchmark functions with varying number of function evaluations. There are some comparison studies that exist for these algorithms, such as [2], in which a set of six representative state-of-the-art multi-objective PSO algorithms were compared. However, to the best of our knowledge, there exists no such fair and systematic comparison of the novel MOEAs this work concerns.

In this paper, we tested several representative novel MOEAs due to their recency and reported success. They include AMOSA [3], a multi-objective SA algorithm; OMOPSO [4] and SMPSO, multi-objective PSO algorithms; and AbYSS [5], a multi-objective SS algorithm. For brevity, we roughly outline each algorithm in section 2.1, and refer the reader to the original papers for a more in-depth description. We also select NSGA-II [1] as our baseline algorithm for comparison.

Apart from a fair and comprehensive comparison of these four novel MOEAs, another main contribution of this paper is the proposal of a novel performance measure, unique hypervolume. This measures the volume of objective space dominated only by one solution, with respect to a set of solutions. We demonstrate the interesting insights this statistic can provide in the interpretation of algorithm performance results.

The remainder of this paper is organized as follows. Section 2 describes the MOEAs, the performance measures and the proposed unique hypervolume. In Section 3, we describe our experiments and discuss the the results. Section 4 concludes the paper.

2 Methods

2.1 Novel MOEAs

AMOSA [3] is a multi-objective adaptation of the original SA algorithm. In the single objective case, simulated annealing performs iterative perturbations upon a solution, accepting the change if it is beneficial and rejecting negative changes with a certain probability, where the probability is exponentially reduced as the algorithm progresses. Many significant changes are necessary to adapt this approach to multi-objective problems. The most important change made by this algorithm is the addition of an archive of non-dominated solutions, and the various probabilities regarding acceptance and rejection rely on the domination status of the new solution with respect to the archive.

OMOPSO [4] and SMPSO [6] are both multi-objective extensions of PSO, an algorithm inspired by the collective behaviour of social animals, like the flocking of birds. The PSO algorithm maintains a population, or so-called swarm, of particles, each a potential solution of the given problem. These particles move in the search space according to some simple rules, relative to each other, to converge upon a better solution. The multi-objective PSO algorithms OMOPSO and SMPSO both select leaders in the particles that are non-dominated with respect to the swarm. In OMOPSO, crowding distance [1] was used to filter out leader solutions. Two mutation operators are proposed to accelerate the convergence of the OMOPSO algorithm. In SMPSO, a constant σ is assigned to each particle of the swarm and of an external archive then select a particle as the leader of the particles with the closest σ value of the external archive.

AbYSS [5] is a MOA based on the well-known SS algorithm. SS algorithm differs from other EAs by using a small population, or so-called the reference set, in which the individuals are combined to construct new solutions systematically. AbYSS extended the original single-objective SS algorithm by incorporating concepts from the multi-objective field, such as Pareto dominance, density estimation, and an external archive to store the non-dominated solutions.

2.2   Performance Measures

Measuring and comparing the performance of multi-objective algorithms remains an unsolved problem in the field. The prevailing opinion in the literature is that performance should be measured not only according to how closely the resulting set of solutions converges on the global optimal Pareto front, but also how spread the solutions are across the breadth of the front. An excellent survey of performance measures can be found in [7]. In these paper, we adopt a large set of popular performance measures to evaluate the four novel MOEAs:

- **Epsilon** from [8], calculates the minimum size translation in objective values for one set of solutions to cover another reference set. In the case of these experiments, the reference set used was the known globally optimal Pareto front values.
- **The hypervolume indicator**, originally described in [9], defines the size of multi-dimensional space that a set of solutions cover, without counting the same area twice.
- **Spread**, as defined in [1], which calculates how uniformly spread across the objective space the solutions are within their set.
- **The cover function** from [10], which calculates the proportion of one solution set that is dominated by another set. Thus, if $C(X, Y) = 0.9$, 90% of solutions in Y are dominated by solutions in X. Note that the converse $C(Y, X)$ must also be considered separately; while most of X could be dominated by a few members Y, most of Y could also be dominated by a few members of X.
- **The Purity function** described in [11] unifies the solutions returned by a set of algorithms on a particular problem, and returns the proportion of each that remain nondominated in the set.
- **The unique hypervolume measure**, as described in detail below.

Unique Hypervolume (UH). UH is a property that can be calculated for one solution, with respect to another solution or a set of solutions. Essentially, it quantifies the volume of objective space dominated only by that solution, and not by any other. This idea is represented visually in Fig. 1.

UH has the following advantageous properties:

- It rewards domination proportional to the amount one solution dominates another.
- It implicitly punishes clustering of solutions.
- It rewards distinctly original and innovative solutions.
- It rewards diversity only when this diversity is objectively beneficial, in the context of all other solutions found.

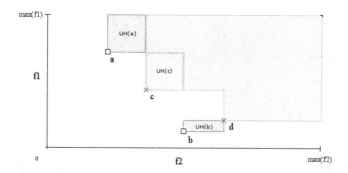

Fig. 1. A diagram showing the UH of 4 points on a two-dimensional minimisation problem. Points **a** and **b** belong to one set of solutions, denoted by a square and red lines, whereas **c** and **d** belong to a second set, denoted by a cross and blue lines. In this diagram, each point dominates the area directly above and to the right of it. The area (or in higher dimensional problems, hypervolume) it alone dominates is its UH. Note that point **b** dominates point **d**; therefore point **d** has no UH, but its presence restricts the UH of point **b**.

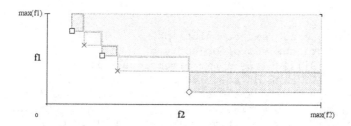

Fig. 2. In this diagram, we demonstrate one of the many situations where the results of UH contradict those of original hypervolume. Here we see three sets of results, colour coded as red, blue and purple. The red and blue sets undoubtedly contain more hypervolume than the purple set. However, the lone purple solution is noteworthy in that it covers a more original combination of objective space, making it a rarer and thus more valuable trade-off point.

UH bears conceptual similarities to the D metric [12]; a measure we have not actually observed in use. The key differences between the two lie in the behaviour in the case of one solution dominating another, and cases where solutions of the same set are neighbours. In Fig. 3 we demonstrate the differences by illustrating the area calculated by each measure on two sets of Pareto fronts. It can be seen from this diagram that UH is a more difficult metric to score well in. They both share the properties of rewarding useful and diverse solutions; however, UH punishes solutions within a set that are not diverse in comparison to each other. Therefore, we can learn more from this new metric because it also reflects the spread of solutions along the Pareto front.

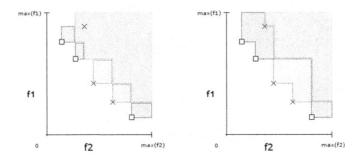

Fig. 3. A demonstration of the difference between the UH (left) and D metric (right), on two identical sets of Pareto fronts.

The UH measure concerns the same property of objective space as the original hypervolume measure, which is notoriously difficult to calculate in an efficient manner. However, our UH measure is comparatively simple to calculate, and does not require a calculation of hypervolume a priori. A simple and efficient method for calculating the UH of a solution is given in Fig. 4.

```
Definitions:
let S denote the set of solutions
let O denote the set of objectives
let side[|S|][|O|] be a two dimensional array
rank(s, o) : rank of solution s according to o
next(s, o) : next ranked solution according to objective o
score(s, o) : objective value of solution s in objective o
max(o) : maximum objective value of objective o
for each objective o in O
    sort the set S according to o
  for each solution s in S
      if ( rank( s, o ) == |S| )
          side[s][o] = max(o) - score(s, o)
        else
            side[s][o] = score(next(s, o), o) - score(s, o)
unique_hypervolume(s) = product( side[s] )
```

Fig. 4. Pseudocode for the calculation of UH

In prose, we simply sort the set of solutions according to each objective, and record the distance from each solution to the next worst in that objective. If the solution is the worst in the set in an objective, we record the distance to the worst possible objective score. The product of these values gives us the volume of a hyperrectangle in objective space, which only that solution has dominated. It should be noted that this method requires knowledge of each maximum objective value, assuming each objective is to be minimised. If this data is not available, an alternative is to substitute the worst value found in each objective.

This measure can be used in many ways, such as comparing sets of solutions found by two MOEAs, or finding the most novel solution within a set. In this paper, we can accumulate all the solutions found by every algorithm on a problem into one set, and total the UH of solutions found by each algorithm with respect to that set.

3   Experiments and Results

The experiments were performed using a customised version of the jMetal framework [13]. This software package already included the algorithms NSGA-II, OMOPSO, SMPSO and AbYSS, to which we added our implementation of AMOSA. jMetal also has a built in set of quality indicators to score the results of a single run of an algorithm on a problem, aggregating those scores to calculate the average for each algorithm, for comparison in that manner. This is very useful, however it does not facilitate comparison measures that require access to the result solutions themselves. For the purposes of this study, and to enable an implementation of UH, the software was modified accordingly.

Each algorithm was executed on each of the 7 problems from the Deb, Thiele, Laumanns and Zitzler (DTLZ) problem family [14]. We used the default configurations regarding the number of variables and objectives for each problem within jMetal, along with the default function evaluation budget of 25,000 for those problems. The algorithms are evaluated according to the final set of solutions returned at the end. To ensure reliable and statistically significant results, each such experiment was repeated 30 times.

We provide the empirical results of our experiments in the following tables. Note that the best performance in each measure on each problem is shaded a dark grey, and the second best is a lighter grey.

Table 1. Epsilon. Mean and standard deviation.

	AMOSA	NSGA2	OMOPSO	SMPSO	AbYSS
DTLZ1	$1.50e-01_{5.6e-02}$	$1.48e-01_{1.2e-01}$	$4.69e+01_{1.4e+01}$	$5.81e-02_{6.4e-03}$	$1.20e-01_{1.2e-01}$
DTLZ2	$1.70e-01_{3.4e-02}$	$1.25e-01_{1.8e-02}$	$6.07e-01_{1.0e-01}$	$1.39e-01_{1.2e-02}$	$1.29e-01_{1.6e-02}$
DTLZ3	$8.83e-01_{7.6e-01}$	$5.04e+00_{2.4e+00}$	$4.34e+02_{9.0e+01}$	$3.22e-01_{2.3e-01}$	$5.22e+00_{3.0e+00}$
DTLZ4	$4.39e-01_{1.7e-01}$	$1.13e-01_{1.7e-02}$	$1.11e+00_{2.2e-01}$	$1.25e-01_{2.9e-02}$	$1.64e-01_{1.6e-01}$
DTLZ5	$4.69e-02_{1.4e-02}$	$1.07e-02_{1.9e-03}$	$4.35e-01_{9.4e-02}$	$4.72e-03_{4.0e-04}$	$5.10e-03_{7.3e-04}$
DTLZ6	$1.00e-01_{6.7e-02}$	$8.47e-01_{7.3e-02}$	$5.83e+00_{1.1e+00}$	$2.67e-02_{1.2e-01}$	$6.55e-01_{2.0e-01}$
DTLZ7	$1.52e+00_{3.1e-01}$	$1.78e-01_{2.1e-01}$	$1.03e+01_{2.1e+00}$	$1.74e-01_{3.9e-02}$	$1.65e+00_{7.3e-01}$

From the results, it is difficult to select the best method that achieved the best performance across all the 6 performance measures. However, as a baseline algorithm proposed in 2002, NSGA-II performed moderately well on the 7 problems in terms all the measures. Of the 5 MOEAs, SMPSO appeared to be the best algorithm in terms of the distance to the Pareto-optimal front measured by Epsilon: it yields 5 of the best values and 1 second best on of the 7 problems. We can see that SMPSO obtained 3 of the best

Table 2. Hypervolume. Mean and standard deviation.

	AMOSA	NSGA2	OMOPSO	SMPSO	AbYSS
DTLZ1	$6.32e-01_{1.1e-01}$	$5.07e-01_{3.2e-01}$	$0.00e+00_{0.0e+00}$	$7.37e-01_{5.0e-03}$	$6.49e-01_{2.1e-01}$
DTLZ2	$3.45e-01_{1.4e-02}$	$3.74e-01_{5.6e-03}$	$1.89e-02_{6.5e-02}$	$3.51e-01_{4.9e-03}$	$3.82e-01_{6.3e-03}$
DTLZ3	$3.40e-02_{3.6e-02}$	$1.86e-03_{1.0e-02}$	$0.00e+00_{0.0e+00}$	$2.88e-01_{8.4e-02}$	$0.00e+00_{0.0e+00}$
DTLZ4	$7.48e-02_{5.4e-03}$	$3.75e-01_{4.6e-03}$	$1.18e-02_{6.4e-02}$	$3.57e-01_{1.3e-02}$	$3.67e-01_{5.3e-02}$
DTLZ5	$8.08e-02_{6.3e-03}$	$9.29e-02_{1.7e-04}$	$3.39e-03_{1.7e-02}$	$9.37e-02_{9.5e-05}$	$9.40e-02_{4.0e-05}$
DTLZ6	$7.51e-02_{9.6e-03}$	$0.00e+00_{0.0e+00}$	$3.16e-03_{1.7e-02}$	$9.17e-02_{1.7e-02}$	$4.89e-05_{2.6e-04}$
DTLZ7	$2.46e-01_{2.1e-02}$	$2.80e-01_{6.0e-03}$	$8.57e-03_{4.6e-02}$	$2.74e-01_{5.9e-03}$	$2.44e-01_{2.5e-02}$

Table 3. Spread. Mean and standard deviation.

	AMOSA	NSGA2	OMOPSO	SMPSO	AbYSS
DTLZ1	$1.20e+00_{1.7e-01}$	$8.47e-01_{1.9e-01}$	$6.09e-01_{8.7e-02}$	$6.78e-01_{5.0e-02}$	$9.51e-01_{1.1e-01}$
DTLZ2	$8.34e-01_{8.6e-02}$	$6.98e-01_{5.0e-02}$	$5.58e-01_{5.6e-02}$	$6.15e-01_{3.1e-02}$	$7.71e-01_{5.4e-02}$
DTLZ3	$1.30e+00_{1.6e-01}$	$1.02e+00_{1.3e-01}$	$6.09e-01_{5.6e-02}$	$8.74e-01_{2.9e-01}$	$9.38e-01_{9.7e-02}$
DTLZ4	$1.73e+00_{2.2e-01}$	$6.72e-01_{3.9e-02}$	$1.08e+00_{1.4e-01}$	$6.76e-01_{6.5e-02}$	$6.57e-01_{1.1e-01}$
DTLZ5	$1.04e+00_{7.8e-02}$	$4.40e-01_{4.0e-02}$	$5.79e-01_{8.9e-02}$	$1.72e-01_{5.9e-02}$	$1.35e-01_{1.8e-02}$
DTLZ6	$1.50e+00_{8.5e-02}$	$8.40e-01_{5.5e-02}$	$5.93e-01_{9.8e-02}$	$1.49e-01_{1.1e-01}$	$8.99e-01_{6.7e-02}$
DTLZ7	$1.32e+00_{6.3e-02}$	$7.40e-01_{4.3e-02}$	$6.65e-01_{6.3e-02}$	$6.99e-01_{3.8e-02}$	$7.03e-01_{5.4e-02}$

Table 4. Coverage. Relating this table to the conventional C(X,Y) notation, the algorithm named in the row heading represents X, and the algorithm named in the column heading replaces Y. For example, C(AMOSA, NSGA2) = 0.460 on DTLZ1.

		AMOSA	NSGA2	OMOPSO	SMPSO	AbYSS
	AMOSA	–	0.460	1.0	0.293	0.252
	NSGA2	0.698	–	1.0	0.515	0.317
DTLZ1	OMOPSO	0.003	0.120	–	0.0	0.0
	SMPSO	0.804	0.58	1.0	–	0.347
	AbYSS	0.889	0.691	1.0	0.663	–
	AMOSA	–	0.196	0.965	0.530	0.025
	NSGA2	0.417	–	0.968	0.651	0.017
DTLZ2	OMOPSO	0.102	0.040	–	0.125	0.008
	SMPSO	0.373	0.142	0.954	–	0.02
	AbYSS	0.538	0.480	0.976	0.809	–
	AMOSA	–	0.969	1.0	0.128	1.0
	NSGA2	0.533	–	1.0	0.1232	0.996
DTLZ3	OMOPSO	0.0	0.0	–	0.001	0.0
	SMPSO	1.0	1.0	1.0	–	1.0
	AbYSS	0.243	0.951	1.0	0.121	–
	AMOSA	–	0.037	0.831	0.043	0.020
	NSGA2	0.037	–	0.936	0.454	0.017
DTLZ4	OMOPSO	0.070	0.024	–	0.052	0.003
	SMPSO	0.100	0.220	0.930	–	0.029
	AbYSS	0.129	0.492	0.968	0.682	–
	AMOSA	–	0.196	0.916	0.168	0.015
	NSGA2	0.765	–	0.951	0.439	0.047
DTLZ5	OMOPSO	0.327	0.062	–	0.022	0.004
	SMPSO	0.780	0.388	0.951	–	0.075
	AbYSS	0.887	0.601	0.986	0.706	–
	AMOSA	–	1.0	0.957	0.109	1.0
	NSGA2	0.007	–	0.953	0.018	0.225
DTLZ6	OMOPSO	0.021	1.0	–	0.044	1.0
	SMPSO	0.021	1.0	0.959	–	1.0
	AbYSS	0.017	1.0	0.953	0.027	–
	AMOSA	–	0.262	0.911	0.296	0.125
	NSGA2	0.191	–	0.975	0.566	0.047
DTLZ7	OMOPSO	0.032	0.046	–	0.056	0.016
	SMPSO	0.276	0.425	0.972	–	0.078
	AbYSS	0.309	0.562	0.970	0.641	–

Table 5. Purity. We also report how many solutions each algorithm returned in total, and how many of those remained non-dominated with respect to solutions of other algorithms (thus (X/Y) means it returned Y solutions, of which X were not dominated by any other.)

	AMOSA	NSGA2	OMOPSO	SMPSO	AbYSS
DTLZ1	0.027 (81/2979)	0.271 (813/3000)	0.0 (0/720)	0.258 (775/3000)	0.546 (1638/2997)
DTLZ2	0.196 (590/3000)	0.466 (1399/3000)	0.012 (13/1074)	0.124 (374/3000)	0.924 (2773/3000)
DTLZ3	0.0 (0/2960)	0.0 (0/2334)	0.0 (0/686)	0.468 (1156/2467)	0.0 (0/2802)
DTLZ4	0.082 (197/2385)	0.477 (1433/3000)	0.019 (11/565)	0.248 (745/2999)	0.929 (2788/3000)
DTLZ5	0.083 (251/2998)	0.346 (1039/3000)	0.006 (5/806)	0.222 (668/3000)	0.851 (2552/2998)
DTLZ6	0.979 (2937/3000)	0.0 (0/3000)	0.039 (85/2138)	0.810 (2416/2982)	0.0 (0/2953)
DTLZ7	0.533 (1599/3000)	0.317 (953/3000)	0.009 (6/613)	0.230 (690/2999)	0.738 (2214/2999)

Table 6. Total UH

	AMOSA	NSGA2	OMOPSO	SMPSO	AbYSS
DTLZ1	$2.05e-12$	$7.79e-11$	0.0	$7.91e-11$	$1.14e-10$
DTLZ2	$1.25e-10$	$9.13e-10$	$4.13e-12$	$3.01e-10$	$1.69e-09$
DTLZ3	0.0	0.0	0.0	$2.11e-40$	0.0
DTLZ4	0.0	$2.62e-09$	$7.49e-12$	$8.98e-10$	$4.40e-09$
DTLZ5	$2.51e-11$	$4.17e-10$	$2.53e-12$	$2.63e-10$	$1.06e-09$
DTLZ6	$2.59e-09$	0.0	$3.97e-10$	$7.31e-09$	0.0
DTLZ7	$7.33e-11$	$1.59e-09$	$7.45e-12$	$1.66e-09$	$1.89e-09$

values and 2 second best in the Hypervolume measure, and it also obtained 1 of the best values and 5 second best values in the Spread measure.

The AbYSS algorithm scored the highest purity rating on 5 of the 7 test problems. The coverage results in these cases confirm that it is rare for the solutions of another algorithm to cover more than 2% of its solutions. Curiously, on the remaining 2 problems, all of its solutions were fully dominated by others. Despite having so many non-dominated solutions, this did not result in relatively high amounts of UH; information which is open to interpretation. Our opinion is that its frequently poor Spread results are suggestive of tightly clustered solutions on the Pareto front, which UH strongly punishes. In support of this, we can say that on problem 4, when it has the best spread and second best hypervolume, it has the best UH.

Interestingly, OMOPSO performed the poorest across all 7 problems in terms of Epsilon, despite regularly scoring the best according to the Spread metric. The purity results for OMOPSO reveal it is consistently yielding less solutions than the other algorithms. Our observation of the inferior performance of OMOPSO compared with other MOEAs is consistent with the results in [2], where the authors found that while OMOPSO outperformed other multi-objective PSO algorithms, second only to SMPSO.

The performance of AMOSA is also not satisfactory. In terms of Epsilon and Hypervolume, it only obtained 2 second best values. It is consistently outperformed by the other MOEAs in terms of both Spread and UH.

Problem 2 produced an example of the interesting information UH can provide. The best hypervolume score was awarded to AbYSS with $3.82e-01$, whereas NSGA-II was a very close second with $3.74e-1$. However, NSGA-II found nearly double the amount of UH as AbYSS, a surprising difference given the initial figures. This result also coincides with NSGA-II performing superior to AbYSS on problem 2 in both Spread and Epsilon.

Perhaps the most remarkable set of results emerged on problem 3. Some solutions of SMPSO dominated the solutions found by every other algorithm. The purity measure shows that 1156 of its 2467 total solutions were preserved on the non-dominated front, showing that this was a consistently superior performance, not a single fluke result. The total UH of SMPSO in problem 3 being incredibly low indicates that those 1156 solutions were closely clustered together.

Choosing the best performer depends upon the preferences of the user. On the one hand, an indicator that reflects poorly on SMPSO is purity; in this it is often outperformed by AbYSS, NSGA-II and AMOSA. Looking at the coverage results, we can see that it is common for AbYSS or NSGA-II to dominate over 50% of its solutions. Given that it frequently came best or second best in terms of Epsilon, Hypervolume and Spread, we must assume that the few remaining non-dominated solutions are responsible for the significant scores. Thus, if the priority is to find a smaller set of solutions closest to the global optimal Pareto front, SMPSO would be the correct algorithm to select. However, if finding a broader set of trade-off solutions is necessary, AbYSS may be preferred due to its impressive performance in purity.

4   Conclusion

In this paper, we conducted a fair and systematic comparison of four representative novel MOEAs proposed in recent years, adopting the most popular MOEA, NSGA-II as the baseline algorithm for comparison. We employed a range of well-known performance measures to evaluate the performance of the 5 MOEAs on the DTLZ problem family. We also proposed a novel performance metric, called unique hypervolume (UH), which can effectively quantify the volume of objective space dominated uniquely by the solution of an algorithm against those of other algorithms. Based on our results, we observed that for finding a smaller set of good solutions, SMPSO would be the best choice. If we are more interested in discovering the trade-off region in multi-objective problems, AbYSS is preferred. In terms of future work, due to the interesting insight into the results gained through the measurement of total unique hypervolume, especially its ability to quantify the novelty of solution within a set, we intend to explore this idea further. More specifically, we will be using the measure to observe performance of multi-objective optimisation algorithms on design problems, where innovation is particularly valued.

Acknowledgment. Mr Thomas White and Dr Shan He are supported by EPSRC (EP/J01446X/1).

References

1. Deb, K., Pratap, A., Agarwal, S., Meyarivan, T.: A fast and elitist multiobjective genetic algorithm: Nsga-ii. IEEE Transactions on Evolutionary Computation 6(2), 182–197 (2002)
2. Durillo, J.J., García-Nieto, J., Nebro, A.J., Coello, C.A., Luna, F., Alba, E.: Multi-Objective Particle Swarm Optimizers: An Experimental Comparison. In: Ehrgott, M., Fonseca, C.M., Gandibleux, X., Hao, J.-K., Sevaux, M. (eds.) EMO 2009. LNCS, vol. 5467, pp. 495–509. Springer, Heidelberg (2009)

3. Bandyopadhyay, S., Saha, S., Maulik, U., Deb, K.: A simulated annealing-based multiobjective optimization algorithm: Amosa. IEEE Transactions on Evolutionary Computation 12(3), 269–283 (2008)
4. Sierra, M.R., Coello Coello, C.A.: Improving PSO-Based Multi-objective Optimization Using Crowding, Mutation and ϵ-Dominance. In: Coello Coello, C.A., Hernández Aguirre, A., Zitzler, E. (eds.) EMO 2005. LNCS, vol. 3410, pp. 505–519. Springer, Heidelberg (2005)
5. Nebro, A.J., Luna, F., Alba, E., Dorronsoro, B., Durillo, J.J., Beham, A.: AbYSS: Adapting Scatter Search to Multiobjective Optimization. IEEE Transactions on Evolutionary Computation 12(4) (August 2008)
6. Nebro, A., Durillo, J., García-Nieto, J., Coello Coello, C., Luna, F., Alba, E.: Smpso: A new pso-based metaheuristic for multi-objective optimization. In: 2009 IEEE Symposium on Computational Intelligence in Multicriteria Decision-Making (MCDM 2009), pp. 66–73. IEEE Press (2009)
7. Okabe, T., Jin, Y., Sendhoff, B.: A critical survey of performance indices for multi-objective optimisation. In: The 2003 Congress on Evolutionary Computation, CEC 2003, vol. 2, pp. 878–885. IEEE (2003)
8. Fonseca, C., Knowles, J., Thiele, L., Zitzler, E.: A tutorial on the performance assessment of stochastic multiobjective optimizers. In: Third International Conference on Evolutionary Multi-Criterion Optimization (EMO 2005), vol. 216 (2005)
9. Zitzler, E., Thiele, L.: Multiobjective evolutionary algorithms: A comparative case study and the strength pareto approach. IEEE Transactions on Evolutionary Computation 3(4), 257–271 (1999)
10. Zitzler, E., Deb, K., Thiele, L.: Comparison of multiobjective evolutionary algorithms: Empirical results. Evolutionary Computation 8(2), 173–195 (2000)
11. Bandyopadhyay, S., Pal, S., Aruna, B.: Multiobjective gas, quantitative indices, and pattern classification. IEEE Transactions on Systems, Man, and Cybernetics, Part B: Cybernetics 34(5), 2088–2099 (2004)
12. Zitzler, E.: Evolutionary algorithms for multiobjective optimization: Methods and applications. Shaker (1999)
13. Durillo, J.J., Nebro, A.J.: jMetal: A java framework for multi-objective optimization. Advances in Engineering Software 42(10), 760–771 (2011)
14. Deb, K., Thiele, L., Laumanns, M., Zitzler, E.: Scalable multi-objective optimization test problems. In: Proceedings of the Congress on Evolutionary Computation (CEC 2002), Honolulu, USA, pp. 825–830 (2002)

Fine Tuning of a Wet Clutch Engagement
by Means of a Genetic Algorithm

Yu Zhong[1], Abhishek Dutta[1], Bart Wyns[1], Clara-Mihaela Ionescu[1], Gregory Pinte[2],
Wim Symens[2], Julian Stoev[2], and Robin De Keyser[1]

[1] Department of Electrical energy, Systems and Automation (EeSA),
Ghent University, 913 Technologiepark, Zwijnaarde, 9052, Belgium
{Yu.Zhong,Dutta.Abhishek,Bart.Wyns,
Claramihaela.Ionescu,Robain.DeKeyser}@ugent.be
[2] Flanders' Mechatronics Technology Center (FMTC),
Celestijnenlaan 300D, Heverlee, 3001, Belgium
{Gregory.Pinte,Wim.Symens,Julian.Stoev}@fmtc.be

Abstract. In many practical engineering applications, a feed-forward control is
often used to control the system with some parameterized signals, for example,
a wet clutch system. Usually these signals are designed empirically. In this paper, firstly, genetic algorithm (GA) will be used to optimize parameters. Then
by knowing the system response of the test bench in the frequency domain, GA
will be used again to fine tuning this parameterized signal. The result is then
compared to those performances of using signal without fine tuning step. It is
shown that after applying the fine tuning method, the resulted signal can
achieve a better performance.

Keywords: Genetic algorithm, parameterized signal, discrete Fourier transform, system identification, wet clutch.

1 Introduction

A clutch is a mechanical device which provides the transmission of power/torque
from one input component to an output component. This is a typical nonlinear system
which also is very hard to model since inside the clutch, there are hydraulic, electronic and mechanical parts. Now the control strategy is feeding forward a parametric
signal in which the parameters' values are set empirically. Usually with this method,
the engagement performance is not optimal. In this paper, as a model free optimizing
method, genetic algorithm (GA) will be implemented to tune the signal. After the
tuning process, we can achieve a fast smooth engagement.

The paper is organized as follows: section 2 will introduce the background including the wet clutch system, as well as current feed forward control strategy; section 3
will propose the fine tuning method based on the genetic algorithm; section 4 will
show the fine tuning results; section 5 will draw the concluding remarks.

L. Iliadis et al. (Eds.): AIAI 2012, IFIP AICT 381, pp. 58–67, 2012.

2   Background

2.1   Wet Clutch System and Test Bench

A wet clutch is a clutch that its friction plates are immersed in a cooling lubricating fluid which keeps the surfaces clean and gives smoother performance and longer life. Fig. 1 shows the design of a wet clutch. An electro-hydraulic pressure-regulated proportional valve regulates the pressure inside the clutch, such that the position of a piston which presses the multiple clutch disks together can be controlled. In the closing process of a wet clutch, the piston should be placed as far as possible from the clutch disks to avoid energy loss due to viscous friction of the oil between the disks. A returning spring inside of the clutch could keep the piston in this position when the clutch is not actuated. Once the command to engage the clutch is received, the left chamber of the clutch is filled with oil, so that the piston is pushing forward.

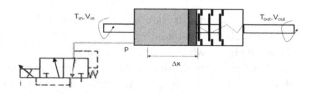

Fig. 1. The mechanical design of a wet clutch, i is the triggle signal for the valve, p is the flow pressure to push the piston, T_{in} and V_{in} are the input torque and velocity, T_{out} and V_{out} are the output torque and velocity, and Δx is the displacement of the piston

When evaluating the engagement performance of a wet clutch, it is necessary to consider both fast and smooth. One typical performance of the clutch is shown in Fig. 2. The unit for the torque measurement is $N{\cdot}m$, while the unit for the output speed is *rpm*. For confidential reasons, all results in this paper will be scaled. And thus the objectives to be optimized can be defined numerically in (1) and (2).

$$T_f = [number\ of\ samples(output\ speed < v_{out})] \qquad (1)$$

$$\Gamma = |mi\,n(toruqe)\,| \qquad (2)$$

The test bench used in this paper contains an electromotor and a flywheel. The electromotor drives a flywheel via two mechanical transmissions: one transmission is controlled in this project; the other transmission is used to vary the load and to adjust the braking torque. The sampling frequency on this setup is 1000Hz.

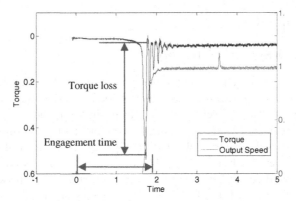

Fig. 2. Illustrative example of the performance of a wet clutch

2.2   Current Control Strategy

Currently, wet clutches used in industry are filled with a feed-forward parametric signal of the current to the electro-hydraulic valve. Fig. 3 shows a typical paramete-rized, feed-forward current signal, which is sent to the valve [1].

The shape of this signal perfectly illustrates the underlying idea behind the actual industrial control design. First, a step signal with height a and width b is sent to the valve to generate a high pressure level in the clutch. With this pressure, the piston will overcome the force from the return spring, and start to get closer to the clutch disks. After this pulse, the signal will give some lower current with height c and width d to decelerate the piston and trying to position it close to the clutch disks. Once the piston is close to the clutch disks and with very low velocity, a force is needed to push the piston forward so that the clutch disks are compressed together. This force will be generated by the pressure which is caused by the step current with height e and width f. Then a ramp current signal with slope *alpha* and the end height g is sent to the valve so that the pressure inside the clutch will increase again gradually. In order to secure the full closing of the clutch, the signal will be kept as a high current level afterwards. Many research efforts can be found in tuning such kind of signals in order to achieve a good engagement [2, 3].

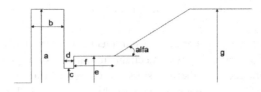

Fig. 3. Typical parameterized signal for controlling the wet clutch system

3   Fine Tuning the Signal

3.1   Brief Introduction of the Proposed Tuning Method

In this paper, a fine tuning method mainly based on Genetic Algorithm (GA) will be proposed to solve this problem. In this new method, the values of the parameters' will be first determined by using a non-dominate sorting genetic algorithm (NSGA) [4, 5]. Then the Fast Fourier transform (FFT) will be applied on the desired signal which is selected by using a user defined weight factor, and its nearest neighbors. Within the obtained Pareto front, one focus group is selected by using a user defined weight vector. A band limited signal composed by harmonics is used to approximate these signals. By using discrete Fourier transform (DFT) [6, 7], these signals/individuals within the focus group will give the ranges of the corresponding frequencies. These amplitudes of the signals will become the tuning parameters in the fine tuning stages, while the range is also determined by the upper and lower limit get from the focus group. GA will be employed again to tune these parameters. A weighted sum method [8, 9] will be used here to turn the multi-objective optimization into a single objective one. After the final optimized parameters are obtained, the inverse discrete Fourier transform (IDFT) [10] is used to convert back into the signal. The final signal is expected to give a better performance.

3.2   Forming the Pareto Front

The genetic algorithm (GA) is a search heuristic that mimics the process of natural evolution. Recently, GA has received considerable attention as a novel approach to multi-objectives optimization problems, resulting in a fresh body of research and applications. Typically, there are two ways in GA for handling the multi-objectives optimization problems; one is use the weighted sum method to transfer the multi-objectives problem to a single objective problem; the other one is based on the non-dominated sorting method to find out the Pareto front solution. The nondominated sorting Genetic Algorithm (NSGA) appears to achieve the most attention in the evolutionary algorithm literature and has been used in various studies. The key point of this method is the nondominated sorting and the fitness assignment. Before the selection is performed, the population is ranked on the basis of an individual's nondomination. Nondomination here is defined as: in a minimization problem, if a vector $X1$ is partially less than $X2$, we say $X1$ dominates $X2$, and any member of such vectors that is not dominated by any other member is said to be nondominated [4]. Then dummy fitness value is assigned according to the rank, individuals in the same rank share same fitness value.

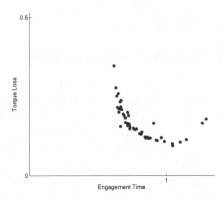

Fig. 4. Individuals of last generation

Table 1. GA parameters

Number of Decision Variables	4 (in Fig. 3, parameter b, c, d and f)
Number of Individuals	50
Maximum Number of Generations	50
Number of Objectives	2
Crossover Rate	0.9
Mutation Rate	0.1

After determined the objectives for optimization (in section 2.1), the NSGA is running. The parameters in running the GA are shown in Table 1. The GA is manually stopped at generation eight since over 90% of the individual within the generation converges at the so called Pareto front (Fig. 4).

3.3   Defining the Fine Tuning Objectives and Parameters

Along the Pareto front, the performance of the clutch shows different characteristics. The individuals gathering at the upper left area give an aggressive engagement which will experience the fast closure of the clutch disks and a large torque loss. On the other hand, the individuals laying on the lower right part could provide a smoother engagement in the cost of time. The Pareto front can offer the flexibility of in choosing the proper signals with different requirements. When in the practical usage, the desired signal is selected by a user defined weight vector $w=(w_1, w_2)$, w_1 and w_2 are the respected weight of two objectives and yields

$$w_1 + w_2 = 1 \tag{3}$$

The fine tuning targets are then defined as:

$$\{P||P - l| < \sigma, P \in Pareto\ front\} \tag{4}$$

In which σ is the distance and l is the line defined as:

$$l:\ y = (w_1/w_2)x \tag{5}$$

In this study, we treat the objective 1 (engagement speed) and objective 2 (torque loss) with same importance. Thus, the weight vector is defined as $4w_1=w_2=0.5$ (due to the difference scaling factors for both objectives). The distance is selected at 0.05 (scaled value). The selected focus group is shown in Fig. 5 with red color.

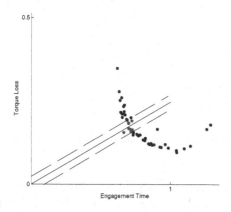

Fig. 5. Individuals as the fine tuning targets (focus group, in red)

The shape of the signals within the focus group is shown in Fig. 6. The rough shape of the signals belonging to the focus group share some similarities, the differences located on the length of first step and the height of the second step.

If we take the first second of the above signals as one period from a periodical signal, then it can be decomposed into a Fourier series [11]. This transform could bring the signal function from the time domain to the frequency domain. Using a Fourier series, the signal $x(t)$ in the time domain can be expressed as a sum of sines and cosines of different frequencies:

$$x(t) = \sum_{-\infty}^{\infty} c_k e^{j2\pi kft} \tag{6}$$

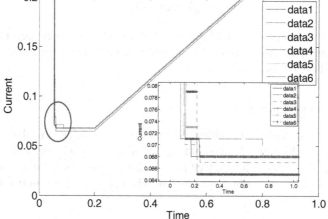

Fig. 6. Signals' shapes of focus group

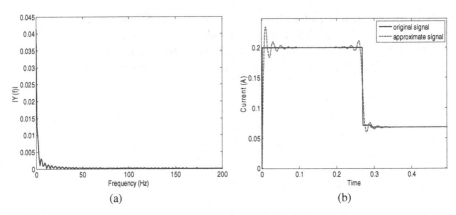

Fig. 7. (a) Spectrum analysis for target signal, single-sided; (b) Original signal and approximated signal

One way to determine the frequency range of the harmonics to form the approximated signals is using the DFT, we can analysis the individuals within the focus group in order to get the information about theirs spectrums. From the results of spectrum analysis (Fig. 7(a)), we can notice that it is not necessary to keep all the harmonics from the transform since in the high frequency range, the amplitudes are very small. Thus, instead of taking the whole frequency range, the first 50 harmonics are used to generate a band limited signal which is approximating the original signal by using the inverse discrete Fourier transform (IDFT). The original part of the signal has a length of $1s$, which gives a fundamental of $1/1=1Hz$. By taking the first 50 harmonics, the bandlimited for the approximated signal will be 50Hz. The approximated signal and the original signal are shown in Fig. 7(b).

In such presentation, the approximated signal has 50 harmonics; the amplitudes of these 50 harmonics offer the possibility for the fine tuning. The tunable parameters are the amplitudes and the range for every amplitude is defined as:

$$c_{ijk} \in [\min(c_k), \max(c_k)] \tag{9}$$

In which, c_{ijk} is the amplitude for kth harmonic of jth individual in ith generation. And the c_k is the amplitude vector for kth harmonics which is obtained by transforming the individuals within the focus group.

3.4   Fine Tuning Process

After determine the decision variables and the range of them, genetic algorithm again is used to find out the optimized combinations. Unlike the NSGA used in section 2.1, this time a weight vector will be added here so that this multi-objective problem will be transformed into a single objective problem. The new objective is defined as:

$$Obj = w_1 Obj_1 + w_2 Obj_2 \tag{10}$$

Now the optimization becomes minimize(Obj). In order to accelerate the convergence, penalty rules are added. For individual P_i, if either $Obj_{i1} > Obj_{1max}$ or $Obj_{i2} > Obj_{2max}$, the final objective value of such individual will be set very high. In which, Obj_{jmax} is defined as:

$$Obj_{jmax} = \{max\,(Obj_j)|P,\ j = 1,2\} \tag{11}$$

The individuals within the focus group are treated as the initial population.

4   Results

The GA used in section 3.4 stopped at 6th generation. The result is then compared with the performance of the central individual of the focus. Fig. 8 shown the signals, in which the blue line is the signal located in the center of the focus group; and the red line is the final fine tuned signal.

Fig. 9 showed the performance of the different signals. For the confidential requirement from the project partner, all the measurements are given in scaled measurement, thus the units are not provided in the figures. The output torques caused by the above two signals are illustrated. We can notice that with a further fine tuning, the performance could be optimized.

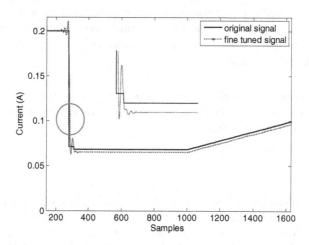

Fig. 8. The comparison of the signals

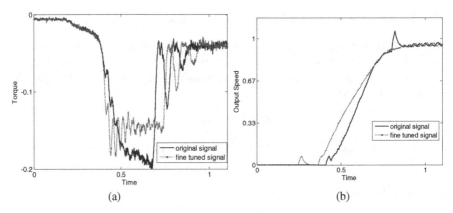

Fig. 9. (a) The comparison of the output torque; (b) The comparison of the engagement speed

5   Concluding Remarks and Future Work

In this paper, a new tuning method based on genetic algorithm is proposed to obtain the optimal signal profile in engaging a wet clutch system. The objective for this problem is to achieve a smooth engagement in the shortest time. Conventional, a parameterized feed forward signal which the values of those parameters are determined either empirically or through some other learning algorithms is used in this system.

This method could also be extended to optimize other feed forward parameterized signal. Besides the better performance, this tuning method can also transform the parameterized signal with sharp edges (discontinuity) into continuous signal. In some cases like the high speed motion with sudden direction changes, this property will beneficial a lot in practical usage.

Acknowledgements. This work has been carried out within the framework of the LeCoPro project (grant nr. 80032). C.M. Ionescu is financially supported by the Flanders Research Center (FWO).

References

[1] Hebbale, K.V., Kao, C.-K., McCulloch, D.E.: Adaptive electronic control of power-on upshifting in an autamatic transmission. US Patent No. 5,282,401 (1994)

[2] Depraetere, B., Pinte, G., Swevers, J.: Iterative optimization of the filling phase of wet clutches. In: The 11th International Workshop on Advanced Motion Control, Nagaoka, Japan, pp. 94–99 (2010)

[3] Lazar, C., Caruntu, C.-F., Balau, A.-E.: Modelling and predictive control of an electro-hydraulic actuated wet clutch for automatic transmission. In: IEEE International Symposium on Industrial Electronics, Bari, Indonesia, pp. 256–261 (2010)

[4] Srinivas, N., Deb, K.: Multiobjective function optimization using nondominated sorting genetic algorithms. Evolutionary Computation 3, 221–248 (1995)

[5] Fonseca, C.M., Fleming, P.J.: Multiobjective optimization and multiple constraint handling with evolutionary algorithms I: A unified formulation., Reasearch report 564, Department of Automatic Control and Systems Engineering, University of Sheffield (1995)

[6] Oppenheim, A.V., Schafer, R.W., Buck, J.R.: Discrete-time signal processing. Prentice Hall, Upper Saddle River (1999)

[7] Brigham, E.O.: The fast Fourier transform and its applications. Prentice Hall, Englewood Cliffs (1988) ISBN 0-13-307505-2

[8] Ishibuchi, H., Murata, T.: A multiobjective genetic local search algorithm and its application to flowshop scheduling. IEEE Transactions on System, Man, and Cybernetics 28(3), 392–403 (1998)

[9] Gen, M., Cheng, R.: Genetic algorithms and engineering optimization. John Wiley & Sons, New York (2000) ISBN 0-471-31531-1

[10] Duhamel, P., Piron, B., Etcheto, J.M.: On computing the inverse DFT. IEEE Transactions on Acoust., Speech and Signal 6(2), 285–286 (1988)

[11] Proakis, J.G., Manolakis, D.G.: Digital signal processing, principles, algorithms, and applications, 3rd edn. Prentice Hall (1996) ISBN 0-13-394338-9

A Representational MDL Framework for Improving Learning Power of Neural Network Formalisms

Alexey Potapov and Maxim Peterson

St. Petersburg National Research University of Information Technologies,
Mechanics and Optics, Kronverkskiy pr. 49,
197101 St. Petersburg, Russia
{pas.aicv,maxim.peterson}@gmail.com

Abstract. Minimum description length (MDL) principle is one of the well-known solutions for overlearning problem, specifically for artificial neural networks (ANNs). Its extension is called representational MDL (RMDL) principle and takes into account that models in machine learning are always constructed within some representation. In this paper, the optimization of ANNs formalisms as information representations using the RMDL principle is considered. A novel type of ANNs is proposed by extending linear recurrent ANNs with nonlinear "synapse to synapse" connections. Most of the elementary functions are representable with these networks (in contrast to classical ANNs) and that makes them easily learnable from training datasets according to a developed method of ANN architecture optimization. Methodology for comparing quality of different representations is illustrated by applying developed method in time series prediction and robot control.

Keywords: representational minimum description length, artificial neural networks, machine learning.

1 Introduction

Machine learning is one of the key paradigms in artificial intelligence, and ANNs pretend to be one of the universal approaches to learning. Such properties as self-adaptability, ability to generalize, and others are usually ascribed to ANNs as opposed to the traditional algorithms [1]. But the origin of these properties is rarely strictly explained. They are frequently grounded only by the presence of analogy between ANNs and real neural networks. At the same time, it can be said from bionic point of view that formal and biological neurons have almost nothing in common. Existing biophysically detailed neuronal models imitate their behavior much more precisely. However, learning is the very feature of neurons that has no plausible biophysical models, and thus cannot be really borrowed by ANNs.

As the result, there is no general theory of ANN learning. Particular training algorithms are proposed for all specific ANN architectures. Moreover, these

L. Iliadis et al. (Eds.): AIAI 2012, IFIP AICT 381, pp. 68–77, 2012.

algorithms are quite classical and external to ANNs. Neuroglial networks with learning algorithms encoded within astrocytic nets [2] are rather interesting, because different learning rules can be made internal for them. However, the origin and structure of these rules remain unclear. Thereupon, recurring declarations about self-learning capabilities of ANNs and their distinctions from the traditional algorithms seem to be paradoxical. Apparently, ANNs don't solve the problem of machine learning. In particular, one difficult issue for them is overlearning, which has no good explanation in the ANN theory. Indeed, overlearning is typically prevented by restricting the training time. Nevertheless, popularity of ANNs is not accidental, but one should strictly describe their benefits in order to improve them further.

In this paper, ANNs are analyzed in the inductive inference framework. Induction is interpreted as construction of a model optimally describing the given data. Such the inductive inference methods include a model quality criterion, a model space, and a search strategy as their components [3].

The Kolmogorov complexity [4] with the set of all algorithms as the model space is known to be the correct basis for universal inductive inference [5]. However, the search problem is unsolvable here. Even computable analogs such as the Minimum Description Length (MDL) principle [6] and its counterparts [7, 8] are non-strictly applied in practice. Even so, they help to solve overlearning problem [9]. Here, information-theoretic analysis of ANNs is deepened on the base of the recent Representational MDL (RDML) principle [10], which takes into account that models are always constructed within certain representations. ANNs can also be interpreted as a specific algorithmic representation. This interpretation helped us to explain their good properties, which have an effect on all the component of induction.

The main contribution of this work is an extension of the information-theoretic approach from construction of a single ANN with some architecture to optimizing some ANN formalism as a data representation for a given set of learning tasks. Application of the RMDL framework for comparing quality of different ANN-based representations is illustrated with the use of a developed particular ANN type tested on the tasks of time series prediction and robot control.

2   Previous Works

Let's consider ANN learning as the induction problem, which requires introduction of a model space, a model quality criterion, and a search procedure.

The most widely used model selection criterion is MAP (maximum a posterior probability) calculated on the base of the Bayes' rule. However, its usage leads to the fundamental problem of prior probabilities [11], which cannot be inferred within statistical methods. These probabilities are sometimes ignored resulting in the maximum likelihood approach that leads to overlearning.

Theoretical solution of this problem was given quite long ago by several authors [6, 7, 12] on the base of Kolmogorov complexity introduced in the algorithmic information theory. Kolmogorov complexity of the given string (data) D is defined as the length $l(H)$ of the shortest program (model) H for the Universal Turing Machine (UTM) U reproducing the given string $U(H)=D$:

$$K(D) = \min_{H}[l(H) \mid U(H) = D] \ . \tag{1}$$

Here, the model space contains all the algorithms. It appeared to be useful to divide each model into two parts: regular and random components. The best model can be defined using the conditional Kolmogorov complexity:

$$H^* = \arg\min_{H}[l(H) + K(D \mid H)] \ , \tag{2}$$

where $K(D \mid H) = \min_{R}[l(R) \mid U(HR) = D]$, and the string R is the input to the program H necessary to reconstruct the data D (or the data description encoded with the model H). This leads to the MDL principle, in which the best model is determined by minimizing the sum $l(H)+K(D|H)$. Using connection between information quantity and probability, one can write [13]: $-\log_2 P(H)=l(H)$ and $-\log_2 P(D|H)=K(D|H)$.

Prior probabilities of models are defined on the base of lengths of corresponding UTM programs leading to the universal priors, which are still under discussion [14]. Besides the search problem, these priors are inapplicable, because particular induction tasks require large amount of prior information influencing the model selection criterion. For this reason, the MDL principle is used in its inexact verbal form, and heuristic coding schemes are contrived in order to compute description lengths.

In particular, the MDL principle was applied in this heuristic way to solve the problem of ANN architecture optimization [15-17]. In these works, components of the description length are calculated within some ANN coding schemes. Partial solution of the overlearning problem is achieved here, because increase of the model precision at the cost of increase of its complexity (the number of neurons and connections) is allowed only in the case, when it decreases the total description length [9, 18].

However, heuristic coding schemes are ungrounded. They introduce non-optimal inductive bias into model selection criteria, and specify arbitrary model space containing regularities, which can be inadequate for the given learning task. For example, activation functions are rarely considered as model components that also should be optimized. Moreover, ANNs with restricted architecture (e.g., radial basis function networks [19] or multilayer perceptrons [9]) are typically considered. The possibility of inclusion of ANN formalisms themselves into the optimization process on the base of information-theoretic criteria has not been considered yet, despite the fact that such the optimization can be more significant than optimization of a single ANN within particular formalisms.

3   Methodology

Particular ANN formalisms define algorithmically incomplete model spaces, which adequacy is not analyzed. Consequently, regularities in the data can be inexpressible within the selected formalism leading to bad learning capabilities. This fact is typically ignored because of the well-known proof that ANNs are universal approximators [20]. This proof is cited even in the papers devoted to the MDL-based ANNs [17]. This contradiction arises from a lack of understanding that approximation of any function with preset accuracy is insufficient in machine learning. If some regularity in the data is not expressible in the given model space, overlearning cannot be completely avoided: Ptolemy's epicycles can approximate planetary orbits only with precision restricted by observation errors, because they don't capture underlying regularities and thus cannot generalize in contrast to Kepler's model. Arbitrary regularities are expressible only with the use of the algorithmically complete model space, but the search in this space is currently unachievable. Thus, narrowing the model space is unavoidable in applied tasks. But this just means that special attention should be paid to the problem of the model space selection (also, in the case of ANNs). Heuristic coding schemes not only specify restricted model spaces, but also introduce inductive bias assigning different complexities for models.

The formalized notion of representation replaces heuristic coding schemes within the RMDL principle that focuses attention on the selection of an optimal representation for any given class of inductive inference tasks. This principle was applied to construction of "essentially" learnable computer vision methods [21], but its significance can be extended on the fundamental issues of machine learning.

It can be noted that ANNs of the same type are trained on different data samples independently. Thus, we can set a mass problem of induction. Let a set $\mathbf{D}=\{D_i\}$ of data samples is given, and the best model for each sample D_i should be constructed independently. Data samples can contain mutual information (at least in the form of similar regularities). Consequently, the sum of their individual complexities $K(D_i)$ will be much higher than the complexity of their concatenation $K(D_1...D_n)$.

Thus, the universal criterion (1) is not applicable in this case, because individual models of D_i will be much worse than their common model. However, we can include some prior information S into the method developed specially for the mass problem $\mathbf{D}$, and each data sample D_i will be described conditionally with the given S.

Independent descriptions of D_i with the given S can be almost as efficient as their common description. Here, one should require that $(\forall D \in \mathbf{D})(\exists H,R)U(SHR)=D$. Such program S can be called *representation*, within which a model H can be constructed for each data D. Optimal representation can be chosen using

$$K(D_1 D_2...D_n) \approx \min_S \left(\sum_{i=1}^{n} K(D_i \mid S) + l(S) \right), \quad S^* = \arg\min_S \left(\sum_{i=1}^{n} K(D_i \mid S) + l(S) \right). \quad (3)$$

This gives the mentioned RMDL principle for choosing the best representation for the given set $\mathbf{D}$, which minimizes the summed description length of all data samples and the representation itself. Each ANN formalism or architecture has its own executing algorithm, which precisely corresponds to the definition of the representation. These algorithms aren't affected during ANN learning on the specific data D_i, but they can be optimized for each mass problem $\mathbf{D}$ using the criterion (3).

4   A Model Representation with Dynamic ANNs

The criterion (3) gives quantitative evaluation of representation quality, but it can also be estimated qualitatively by analysis of a set of expressible regularities. For example, outputs of a feedforward network with linear activation function are linear combinations of its inputs independent from the number of hidden neurons. Thus, such the ANN can represent only linear models. Because of this severe limitation, nonlinear activation functions are typically introduced. However, let's consider linear dynamic (recurrent with continuous time) ANN containing M neurons, which activities $x_i(t)$ follow the law:

$$x_i'(t) = \frac{dx_i(t)}{dt} = \sum_{j=1}^{M} w_{ji} x_j(t) , \tag{4}$$

where w_{ji} are connection weights constituting a matrix $\mathbf{W}$.

Starting from some initial values $x_i(0)$, activities $x_i(t)$ will evolve producing some functions as an output. Well-known general solution of the system of homogeneous linear differential equation has the form $\exp(\mathbf{W}t)$ corresponding to the mixture of harmonic, exponential, and polynomial functions, which appear to be representable by such ANNs. Consequently, even linear dynamic ANNs can be called "universal approximators" that can fit any regular function. One interesting application is time series forecasting, in which the data $D=\{\mathbf{y}(t_1),...,\mathbf{y}(t_n)\}$ is given, where the values $\mathbf{y}(t_i)=(y_1(t_i),..., y_N(t_i))$ of N-dimensional vector are observed at some moments of time $t_i \in [0,T_{max}]$. The task is to predict values $\mathbf{y}(t)$ for $t>T_{max}$.

Such the connection weights w_{ij} and such the initial activities $x_i(0)$ should be found that the activities $x_i(t)$ are most precisely correspond to the values $y_i(t)$. Naïve approach leads to minimization of the mean-square error:

$$E^2 = \frac{1}{n} \sum_{i=1}^{n} \sum_{j=1}^{N} \left(y_j(t_i) - x_j(t_i) \right)^2 . \tag{5}$$

The number of neurons M should be not less than the dimension N of the vector $\mathbf{y}$, but it can be larger. In this case, additional neurons can be treated as hidden dynamic variables. They are not included into the MSE criterion (5). Apparently, increase of the number of additional neurons will result in decrease of the MSE as well as in overfitting. In accordance with the MDL principle, the model complexity should also be taken into account in addition to the description length of the data encoded within the model that can be estimated as $nN\log_2 E$ (accurate to a constant).

ANN model description includes information about the number of neurons (roughly $\log_2 M$ bits), established connections (roughly $\log_2 K + \log_2 C(K,M^2)$ bits), their weights ($0.5K\log_2 n$ bits), and initial values of activity ($0.5M\log_2 n$ bits). Total MDL criterion for the ANN with M neurons and K connections can be roughly estimated as

$$L = nN \log_2 E + \log_2 M + \log_2 K + \log_2 C_{M^2}^{K} + 0.5(M + K)\log_2 n . \tag{6}$$

To find the best ANN, one should consider and optimize ANNs with different number of neurons and connections. In order to reduce computational complexity of this process, we utilized an iterative scheme, in which new neurons are consequently added and redundant connections are removed if these operations result in reduction of the description length criterion (6). We considered and implemented a combination of several optimization techniques (stochastic gradient descent, genetic algorithms, and simulated annealing) for optimizing ANNs with fixed architecture. Unfortunately, detailed analysis of this search problem goes beyond the scope of the paper.

Experimental validation of the developed algorithm showed that low-sized ANNs are automatically chosen if the data D is generated using combinations of harmonic, polynomial, and exponential functions. These ANNs extrapolated the given functions with relative errors less than 2% on interval $[T_{max}, 2T_{max}]$. Such precision is difficult to achieve with the use of conventional ANNs with nonlinear activation function, because all these elementary functions are not simultaneously representable by such ANNs. But they are representable by linear dynamic ANNs, which can extract these regularities from few data points and can make good predictions following from their high efficiency in terms of the RMDL principle.

Even linear dynamic ANNs can be rather useful, but still they define very restricted model space. Only extension of representable regularities can help to increase their learning power (and extrapolation capabilities) further. Thus, some type of nonlinearity should be introduced. However, typically used nonlinear activation functions violate the representability of the mentioned elementary functions.

In this context, it is not surprising that hybrid systems gain growing popularity. They include methods with different representable regularities, e.g., nonlinear ANNs and linear auto-regressive models [1]. However, the search problem in the heterogeneous model spaces is more difficult. Here, we propose a homogeneous representation, within which both linear and nonlinear models can be described.

It is natural to construct such the extension of the linear ANNs that will also incorporate models of nonlinear dynamic theory. These models are typically described with differential equations, which can be linear or can contain nonlinearity.

We propose to introduce optional nonlinearity by adding connections from neurons to other connections ("synapses on synapses"). The 2nd-order connections exert nonlinear influence on signals propagating through ordinary connections, but don't change the connection weights themselves. Neurophysiologic prototypes are the modulating neurons. These connections can be introduced in the following way. Consider the system containing 3 neurons shown on the figure 1a.

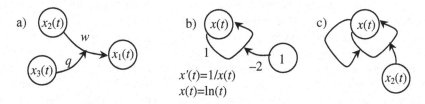

Fig. 1. General form of "connection on connection" (a); minimal (b) and automatically constructed (c) ANNs reproducing logarithmic function

Let the postsynaptic neuron activity be described by the following equation:

$$x_1'(t) = w x_2^{q x_3(t)+1}(t) \ . \tag{7}$$

ANNs with this structure can simply represent power functions as well as logarithmic functions (Fig. 1b). Chaotic modes of the dynamic systems are also representable. For example, a network reproducing the Lorenz attractor can be easily (manually) constructed. It can be seen that these ANNs define the wide model space containing perspicuous regularities. The mentioned above learning (search) procedure can be applied to these extended dynamic ANNs almost without modifications.

Our experiments showed that simple non-chaotic functions are automatically recovered rather reliably. However, even for the basic elementary functions the best network is not always constructed, because of complexity of the search problem. For example, the network on the figure 1c was reconstructed for the logarithmic function. This network contains two unnecessary connections. Nevertheless, its extrapolation error on the doubled time interval appeared to be less than 1%.

The search problem becomes very difficult in the case of chaotic time series. Although different chaotic sequences are representable, the necessary ANN can hardly be found by the direct approximation of the data points. Apparently, this difficulty is connected with instability of chaotic trajectories of the dynamic systems that results in very non-monotonic landscape of the quality criterion under optimization. Since the individual trajectories of chaotic dynamic systems are almost irreproducible, it is more reasonable to reconstruct their invariant measure. This also can be done within the RMDL framework applied to the dynamic ANNs, if one uses such the representation that encodes the values $\mathbf{y}(t_i)$ in accordance with the hypothesized invariant measure (defined by a particular ANN) instead of encoding the deviations of the output of this ANN from $\mathbf{y}(t_i)$. Unfortunately, discussion of this method goes beyond the scope of the paper and requires additional research.

5   Experiments

At first, the developed ANN type and the method for its optimization were tested on the well-known Wolf annual sunspot time series. Wolf numbers till 1979 were used as the training sample. The constructed ANN contained 4 neurons, 11 connections, and 2 second-order connections. Obtained prediction MSE value for 1980–1988 years equals to 220. The other methods mentioned in [18] show the MSE between 214 and 625. Thus, the proposed ANN type is usable. Judging by the prediction accuracy and the ANN size, it can be concluded that overlearning is avoided.

Then, we performed comparison of the ANN-based representations on mass problems. These representations included linear ANNs, ANNs with non-linear activation function, and ANNs with second-order connections. The data samples D_i were taken from a number of financial time series. The complexity of representations under comparison is similar, so we ignored $l(S)$ term in the criterion (3); however, this term can be crucial in more advanced cases in order to avoid overlearning on the level of representations. Table 1 shows the value of the RMDL criterion (divided on the number of data samples), and the relative prediction error (10 points ahead).

Table 1. The values of the RMDL criterion and the relative error for different types of ANNs

ANN type	RMDL, bits	error, %
Linear	651	15,8
Activation function	617	10,1
2nd–order connections	608	9,9

Although we obtained an agreement between the short-term prediction precision and the RMDL criterion in average, one can agree with the statement: "MSE and NMSE are not very good measures of how well the model captures the dynamics" [18]. One can hope that the MDL criterion is the better measure of how well the model captures the underlying regularities, and the RMDL principle helps to extend this criterion on representations.

Another considered mass problem for the ANN-based representations is the robotic control. Here, we used a wheel robot with two motors and sonar for measuring the distance to the obstacles. In this case, additional sensory neuron was included into the network. The training data samples D_i were obtained by recoding the sensory input and the motor commands from the robot under the manual control used for obstacle avoidance. It should be pointed out that the quality criterion included only the approximation precision of motor commands (not the sensory input), and such the network could be constructed that directly approximates commands ignoring the sensory data. An example of successful extrapolation of a sequence of motor commands is given on the figure 2. It can be seen that the prediction results are relatively successful (and the robot controlled by the trained network performs free roaming with adequate reaction to obstacles). This implies that such the ANN was constructed, in which the sensory neuron was connected to the rest network in such the way that it helped to increase both approximation quality (in terms of the MDL criterion) and prediction accuracy.

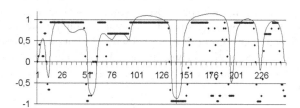

Fig. 2. Robot control commands reproduction (extrapolation is after vertical solid line)

The results of estimation of the RMDL criterion on a set of data samples were 1826 bits for linear ANNs, 1793 bits for ANNs with the nonlinear activation function, and 1798 bits for ANNs with 2nd-order connections.

In this case, the extrapolation precision is meaningless, because human chooses direction of movement during the obstacle avoidance randomly. Sometimes, the choice made by an ANN precisely corresponds to the human choice; but it cannot be guaranteed. So, one can rely only on the RMDL criterion that will hopefully reflect general adequacy of the robot movement.

The RMDL criterion values for both ANN types are compatible, but the ANN with the 2nd-order connections showed more interesting behavior. This result is understandable, because the control command sequences (as non-smooth functions) are not expressible within all the ANN-based representations under comparison.

6   Conclusions

The problem of comparison of learning power of ANN formalisms was considered as the optimization of model representations in the tasks of inductive inference. Model spaces defined by different ANN types are subsets of the set of all algorithms, so they can be optimized within the approach based on the algorithmic information theory. The simplicity of the descriptions of regularities, which presence is expected in the datasets, specifies the inductive bias defining prior probabilities of corresponding models and thus necessary amount of information for their reconstructions.

Such new modification of the ANN formalism was proposed, within which regularities corresponding to the elementary functions are representable as opposed to the ANNs with nonlinear activation functions. The method for optimization of such ANNs was developed. The number of neurons and connections between them is also controlled by the information-theoretic criterion in order to avoid overlearning. The methodology based on the RMDL principle for comparing quality of different representations was proposed and experimentally verified with the use of the developed method on tasks of time series prediction and robot control. Different representations appeared to be more efficient depending on the task.

Our research showed that the RMDL principle can be used to compare the quality of representations while solving inductive mass problems. However, efficiency of representations cannot be reduced only to their RMDL quality. Even in the case of very simple nonlinear representations, it is very difficult to find the best model even if it exists in the specified model space. Representations should give not only the optimal inductive bias for some mass problem, but also should make the model search process more efficient. The speed priors are known [22], but one can expect that they also depend on the representation. Thus, the speed priors can be extended with the notion of representation, or equivalently the RMDL principle should incorporate the model search speed.

References

1. Khashei, M., Bijari, M.: An Artificial Neural Network (p, d, q) Model for Timeseries Forecasting. Expert Systems with Applications 37, 479–489 (2010)
2. Pazos, A.B.P., Gonzalez, A.A., Pazos, F.M.: Artificial NeuroGlial Networks. In: Dopico, J.R.B., Calle, J.D., Sierra, A.P. (eds.) Encyclopedia of Artificial Intelligence, pp. 167–171. Hershey, New York (2009)
3. Baxter, R.A.: Minimum Message Length Inference: Theory and Applications. Ph.D. thesis. Department of Computer Science. Monash University. Clayton. Australia (1996)

4. Kolmogorov, A.N.: Logical Basis for Information Theory and Probability Theory. IEEE Trans. Inform. Theory. IT-14, 662–664 (1968)
5. Solomonoff, R.: Algorithmic Probability Solve the Problem of Induction. Oxbridge Research, P.O.B. 391887, Cambridge, Mass. (1997)
6. Rissanen, J.J.: Modeling by the Shortest Data Description. Automatica-J. IFAC. 14, 465–471 (1978)
7. Wallace, C.S., Boulton, D.M.: An Information Measure for Classification. Comput. J. 11, 185–195 (1968)
8. Vitanyi, P., Li, M.: Ideal MDL and Its Relation to Bayesianism. In: Proceedingof ISIS: Information, Statistics and Induction in Science, pp. 282–291 (1996)
9. Zhao, Y., Small, M.: Minimum Description Length Criterion for Modeling of Chaotic Attractors with Multilayer Perceptron Networks. IEEE Transactions on Circuits and Systems I 53(3), 722–732 (2006)
10. Potapov, A.S.: Comparative Analysis of Structural Representations of Images Based on the Principle of Representational Minimum Description Length. Journal of Optical Technology 75(11), 715–720 (2008)
11. Li, M., Vitanyi, P.M.B.: Philosophical Issues in Kolmogorov Complexity. In: Kuich, W. (ed.) ICALP 1992. LNCS, vol. 623, pp. 1–15. Springer, Heidelberg (1992)
12. Solomonoff, R.: A Formal Theory of Inductive Inference, parts 1-2. Information and Control 7, 1–22, 224–254 (1964)
13. Vitanyi, P., Li, M.: Minimum Description Length Induction, Bayesianism, and Kolmogorov Complexity. IEEE Trans. on Information Theory 2, 446–464 (2000)
14. Wood, I., Sunehag, P., Hutter, M.: (Non-)Equivalence of Universal Priors. Solomonoff 85th Memorial Conference. abs/1111.3854 (2011)
15. Lappalainen, H.: Using an MDL-Based Cost Function with Neural Networks. In: Proc. IJCNN, pp. 2384–2389 (1998)
16. Wang, J.-S., Hsu, Y.-L.: An MDL-Based Hammerstein Recurrent Neural Network for Control Applications. Neurocomputing 74, 315–327 (2010)
17. Molkov, Y.I., Mukhin, D.N., Loskutov, E.M., Feigin, A.M., Fidelin, G.A.: Using the Minimum Description Length Principle for Global Reconstruction of Dynamic Systems from Noisy Time Series. Physical Review E. 80, 046207(1-6) (2009)
18. Small, M., Tse, C.K.: Minimum Description Length Neural Networks for Time Series Prediction. Physical Review E. 66, 066701(1–12) (2002)
19. Leonardis, A., Bischof, H.: An Efficient MDL-Based Construction of RBF Networks. Neural Networks 11(5), 963–973 (1998)
20. Hornik, K., Stinchcombe, M., White, H.: Multilayer Feedforward Networks are Universal Approximators. Neural Networks 2(5), 359–366 (1989)
21. Potapov, A.S., Malyshev, I.A., Puysha, A.E., Averkin, A.N.: New Paradigm of Learnable Computer Vision Algorithms Based on the Representational MDL Principle. In: Proceeding of SPIE, vol. 7696, p. 769606 (2010)
22. Schmidhuber, J.: The Speed Prior: A New Simplicity Measure Yielding Near-Optimal Computable Predictions. In: Kivinen, J., Sloan, R.H. (eds.) COLT 2002. LNCS (LNAI), vol. 2375, pp. 216–228. Springer, Heidelberg (2002)

On the Design and Training of Bots
to Play Backgammon Variants

Nikolaos Papahristou and Ioannis Refanidis

University of Macedonia,
Department of Applied Informatics,
Egnatia 156, Thessaloniki, 54006, Greece
{nikpapa,yrefanid}@uom.gr

Abstract. Recently, a backgammon bot named *Palamedes* won the first prize in backgammon at the 16th Computer Olympiad. *Palamedes* is an ongoing work aimed at developing intelligent bots to play a variety of popular backgammon variants. Currently, the Greek variants *Portes*, *Plakoto* and *Fevga* are supported. A different neural network has been designed, trained and evaluated for each one of these variants. This paper presents the details of the architecture and the training procedure for each case. New expert features as inputs to the networks are also introduced, whereas experimental results demonstrate improvement over previous versions of *Palamedes*.

Keywords: TD(λ), Neural Networks, Self-Play, Backgammon, Plakoto, Fevga.

1 Introduction

Backgammon is one of the oldest board game of chance and skill that is very popular throughout the world with numerous tournaments and many popular variants. Variants of any game usually aren't as interesting as the standard version, but often offer a break in the monotony of playing the same game over and over again. In this paper we examine three popular variants of backgammon in Greece, namely *Portes*, *Plakoto* and *Fevga*, collectively called *Tavli*. In a traditional Tavli match these three games are played in turn, one after the other, until one of the players reaches a predefined number of points (usually seven). *Palamedes* [6] (Fig. 1) is an ongoing project dedicated to offer expert-level playing programs for Tavli and other backgammon variants.

The objective for each player of virtually all variants is to move all his checkers to the last quadrant (called the *home board*), so he can start removing them; a process called *bearing off*. The player that removes all his checkers first is the winner of the game. Players may also win a double game (2 points) when no checker of the opponent has been beared-off. Portes is essentially the same with standard backgammon; the main differences are: (1) the absence of the doubling cube and (2) the absence of triple wins (also called backgammons). The complete set of rules for standard backgammon, Plakoto, Fevga and other variants can be found in [2].

L. Iliadis et al. (Eds.): AIAI 2012, IFIP AICT 381, pp. 78–87, 2012.

In previous work [7,8], following the successful example of TD-Gammon [12,13,14,15] and other top playing backgammon programs, we trained neural networks (NN) using temporal difference learning to play *Portes*, *Plakoto* and *Fevga*, three variants very popular in Greece and neighboring countries. In this paper we present in detail the complete algorithm for the training of our NNs. We also present, for the first time, our Portes bot that won the gold medal in the Computer Olympiad in Tilburg, in November 2011. Furthermore, for the Plakoto and Fevga variants, we present new results that improve the performance upon our previous bots by adding new features, and we explain the logic behind our approach.

This paper is organized as follows. Section 2 reviews our training scheme for backgammon variants. Section 3 presents the expert features used in Portes and the additional expert features for Plakoto and Fevga, whereas Section 4 shows the experimental results. Finally Section 5 makes concluding remarks and discusses future work.

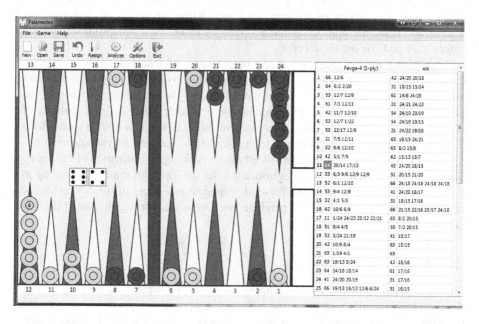

Fig. 1. *Palamedes*: A program for playing backgammon and variants

2   Training Procedure

This section presents the details of the procedure used to train the neural networks for the three backgammon variants examined in this paper.

2.1   Neural Network Architecture

The core function of the neural network is to score game positions. At any time when the program needs to decide which move to play from a set of legal moves, it scores all available states resulting from the current position/roll and selects the one with the highest estimated expected value. We use multilayer perceptrons (MLP) trained using the backpropagation algorithm [17].

The input layer of the NN is comprised of features capturing the position of the checkers on the board, also referred as "raw features", plus features that capture important concepts of the game, also called "expert features". The set of the features selected for each game is presented in Section 3.

We use one hidden layer in our backgammon NNs. The number of hidden neurons is 160 for backgammon, 100 for Fevga and 100 for Plakoto. These numbers were chosen based on preliminary experiments. A higher number of hidden neurons increases performance cost for evaluating each state. This results in increased thinking time for each move, especially when utilizing lookahead in greater depths (Section 4). Thus, the number of hidden neurons chosen is a compromise between performance and computational cost.

Three output neurons are used in the output layer, codenamed W, WD and LD. These correspond to the minimum probabilities needed by the bot in order to make an estimation of the game-theoretic value of a state: W is the probability of winning the game regardless of the number of points (single or double); WD is the probability of winning only a double game and LD is the probability of losing a double game. Both the hidden and the output layers use sigmoid activation functions for each neuron.

Using the above architecture, the procedure of obtaining an estimation of the game-theoretic value of each state is straightforward: set the inputs of the NN according to the board positioning, execute the forward-propagate procedure of the NN to update the outputs, and finally linearly combine the outputs according to the following formula: $V = 2 * W - 1 + WD - LD$.

2.2   Training the NN Using TDL

Training a neural network requires training examples in a supervised learning setting. We use TD(λ) algorithm [10] and the NN's backpropagation algorithm to update the TD error. The exact training procedure is summarized in Algorithm 1. This training scheme, named *reverse offline learning with target recalculation*, was selected among several similar self-play methods [8].

In the adopted training procedure, the updates are applied (Lines 5-15), after a self-play game (Line 4) is ended, starting from the last position of the game and ending at the first (Line 5). At each time step, we recalculate the target for each update (Lines 9-11) in order to get as much accuracy for the estimation of the example label as possible. The function *encoding* (Lines 9, 13), encodes the raw and expert features in their predefined positions at the input layer. Note that the value of the next state is inverted (Line 11). This is necessary because the NN plays the game for both sides always as the first player. When all the moves up to the first are updated, the algorithm starts a new self-game producing the moves according to the updated NN. The procedure is repeated until the selected stopping criterion is satisfied. Possible stopping criteria are: (1) a predefined number of self-play games is reached or (2) no more performance improvement according to a predefined benchmark is found after a prespecified number of self-play games.

Algorithm 1. Training a backgammon NN using TD(0)

```
        // nn:  the neural network that we want to train
        // nn.inputs: a vector representing the input layer
        // nn.outputs: a vector representing the output layer (W, WD, LD)
        // nn.target: a vector representing the target of the update
        // states: a vector holding the all the positions of a game
  1.    nn.initialize(input layer size, hidden layer size, output layer size = 3, learning rate α)
  2.    randomize(nn)      // randomize all weights to [-0.5, 0.5]
  3.    while (stopping condition) do
  4.        states = selfplaygame(nn)
  5.        for (t=T to 1 step -1) do
  6.            if(states(t) is terminal)
  7.                nn.targets = reward(states(t))
  8.            else
  9.                nn.inputs = encoding(states(t+1))
 10.                nn.forwardpropagate()       // calculate outputs
 11.                nn.targets = invert(nn.outputs)
 12.            endif
 13.            nn.inputs = encoding(states(t))
 14.            nn.forwardpropagate()  // calculate outputs
 15.            nn.backpropagate()      // apply backpropagation algorithm
 16.        endfor
 17.    end while
```

Algorithm 1 uses TD(λ) with λ=0, that is the current state is updated only according to the estimation of the next state (Lines 9-11). Thus the target of the update is $V_{target}(s_t)$ = $V(s_{t+1})$. If we want the target of the update to be based on future move estimates of the gameas well ($0<\lambda\leq1$), we can use the forward view of TD(λ) [11] and the target of the update becomes

$$V_{target}(s_t) = (1 - \lambda) \sum_{n=1}^{\infty} \lambda^{n-1} V(s_{t+n}) + \lambda^{T-t-1} V(s_T)$$

In case of $\lambda >0$, lines 8-10 of Algorithm 1 must be changed accordingly. Similarly to $V(s_{t+1})$, all values $V(s_{t+n})$ for n being any odd number must be inverted.

The updates of the network weights are done incrementally and not in a batch setting. This procedure is similar to stochastic or "online" training [17]. The main difference is that there are no fixed labels in the training examples; the labels are given by TD(λ). We prefer incremental training because it has been shown to perform at least equally to the standard batch training using fewer computational resources [17].

2.3   Choosing a Learning Rate α and a λ Parameter

One of the advantages of incremental training is that one can use a larger learning rate than in a batch setting. We also made some experiments with different values of λ with mixed results. In the Plakoto variant, values of λ>0.6 resulted in divergence, whereas lower values sometimes became unstable. So it was decided to keep λ=0 for this variant. For Portes and Fevga variants it was possible to increase the λ value without problems and this always resulted in faster learning, but unlike other reported results [16], final performance did not exceed experiments with λ=0.

Table 1. Selected values of α and λ parameters

Games Trained	Portes	Plakoto	Fevga
0-10000	λ=0.7 α=1	λ=0 α=0.3	λ=0.7 α=1
10000-100000	λ=0.7 α=0.3	λ=0 α=0.3	λ=0.7 α=0.3
100000-250000	λ=0.7 α=0.1	λ=0 α=0.1	λ=0.7 α=0.1
250000-500000	λ=0 α=0.3	λ=0 α=0.1	λ=0 α=0.3
500000-1500000	λ=0 α=0.1	λ=0 α=0.1	λ=0 α=0.1
1500000-5000000	λ=0 α=0.1	λ=0 α=0.01	λ=0 α=0.01
5000000-	λ=0 α=0.01	-	-

Previous experiments were conducted with constant λ and α=0.1. Following the above preliminary experiments we use a decreasing value for λ and α for the experiments in this paper (with the exception of Plakoto where λ is kept constant to zero). Starting with high values of λ=0.7 and α=1 we gradually decrease these values when performance starts to flatten. The exact values of these parameters are shown in Table 1. Using this setup the performance of Plakoto and Fevga variants maxes out at 5 million games and Portes at around 15 million games.

3   Expert Features

The features included in the input layer of each NN are divided to "raw" and "expert" features. Raw features present to the network the placement of each checker in the board while expert features are important game concepts that would otherwise be very difficult for the NN to infer from the raw encoding alone. The raw features of Plakoto and Fevga are presented in [7], while the raw features of our Portes NN are exactly the same as used in [14]. The remaining of this section presents the selected expert features for the Portes game as well as the new expert features that we used in Plakoto and Fevga. The remaining expert features of Plakoto and Fevga are described in [7].

3.1   Expert Features for Portes/Backgammon

All the expert features of our Portes/Backgammon bot are shown in Table 2. The features capture important game playing concepts according to the current literature from expert backgammon players. For example EnterFromBar_1 and

EnterFromBar_2 capture the concept of home board strength. This feature however is useless when the position has no contact (race feature). The NN takes care of combining the features in the correct way taking the current position into account. Additionally, the hidden neurons can create features not existent in the expert list if necessary. For example, we found that the prime formation (six consecutive made points) was handled correctly by the program so we did not include it in the list of expert features even if it is an important concept. The features PipDiff_1, PipDiff_2, PipBearoff_1, PipBearoff_2 were normalized to the [0, 1] interval by a dividing with 60.

Table 2. Expert features for the Portes/backgammon variant

Feature name	Description
HitProb_1	Probability of one player checker being hit on the next roll
HitProb_2	Probability of two player checkers being hit on the next roll
Race	Boolean feature showing the position is a no contact position
PipDiff_1	Pipcount difference when the player is behind (when ahead = 0)
PipDiff_2	Pipcount difference when the player is ahead (when behind = 0)
PipBearoff_1	Pipcount to bearoff for player on roll
PipBearoff_2	Pipcount to bearoff for opponent
EnterFromBar_1	Probability of player entering from bar
EnterFromBar_2	Probability of opponent entering from bar
OppContain_1	Probability of opponent's last checker escaping from player's home board
OppContain_2	Probability of opponent's second to last checker escaping from player's home board
UsContain_1	Probability of player's last checker escaping from opponent's home board
UsContain_2	Probability of player's second to last checker escaping from opponent's home board

Table 3. Expert features for the Plakoto variant

Feature name	Description
Race	Boolean feature showing the position is a no contact position
PipDiff_1	Pipcount difference when the player is behind (when ahead = 0)
PipDiff_2	Pipcount difference when the player is ahead (when behind = 0)
PipBearoff_1	Pipcount to bearoff for player on roll
PipBearoff_2	Pipcount to bearoff for opponent
ChFrontOfPin_1	Number of player checkers in front of last pin when the player has the opponent pinned in the player's homeboard
ChFrontOfPin_2	Number of opponent checkers in front of last pin when the opponent has the player pinned in the opponent's homeboard
Esc_Prob1	Escape probability of player's last made point
Esc_Prob2	Escape probability of opponent's last made point

3.2   New Expert Features for Plakoto

After manual examination and with the help of comments from users that downloaded *Palamedes*, we identified two key problems of our Plakoto bots. The first one presented itself in positions when the bot has pinned the opponent inside the bot's home board. In such positions it is advisable for the bot to "stack" its checkers in the pinned point whenever possible so as to prolong the duration of the pin even in the bearoff situation. Such a strategy most often leads to a double game. However our bots were positioning their checkers as if it was a normal bearoff, greatly reducing their chances for a double game. This problem was addressed by adding the ChFrontOfPin_1 and the ChFrontOfPin_2 features. These two features were scaled to [0, 1] interval by dividing each by 14. We also added the Esc_Prob1 and EscProb2 features hoping that the bot can advance its made points more fluidly, not leaving behind made points that cannot escape easily. Finally we added five features from Portes that are relevant to Plakoto as well. The complete set of features is shown in Table 3.

3.3   New Expert Features for Fevga

The most important concept in the Fevga variant is the existence of a prime formation. In previous work we addressed this by adding one binary feature for every type of prime when it was encountered in the game. While this resulted in the desired effect of the NN learning the concept of making primes when necessary, it did not always understand when it was important to prevent the opponent from making primes of its own. The bot could not understand by this feature alone when the opponent was close to making a prime so as to take immediate measures to disrupt his plan. The inclusion of 2-ply look-ahead improved the situation as now the bot had access to the next moves of the opponent but it would be desirable to have this knowledge without reverting to the computational expensive procedure of looking ahead at greater depths.

To address this problem we changed the binary features of making primes in the following way: When a prime is made the feature is set to one as before. When there is no prime present, instead of setting the feature to zero, we replaced it by a heuristic that computes the probability of making the prime. This was done both for the primes of the bot as well as for the primes of the opponent. Computing accurately this heuristic is very complex and takes much time especially for middle game positions. In order to keep the computational requirements low, we compute the heuristic only for the most common scenario: when there is only one checker left to make the prime. Positions where the prime needs two or more checkers to be achieved are less frequent and usually have smaller probability of success. Thus, the resulting heuristic is a compromise between accuracy and executing time.

These updated features resemble the way we added the pinning probabilities in the Plakoto variant [7]. It has the advantage of putting knowledge in the NN while at the same time keeping low the size of the inputs. We also added the features PipDiff_1,

PipDiff_2, PipBearoff_1, PipBearoff_2 of Portes and Plakoto, because they are relevant to Fevga as well.

We also experimented by combining the above new features with the intermediate reward procedure during the training of Fevga3 and Fevga5 bots [8]. Such a procedure results in a strategy that tries to build primes and maintain them at all cost. While the resulting performance was higher than previous bots, it was lower than Fevga6, i.e. without the intermediate reward. One possible explanation is that without the intermediate reward the bot can identify situations where a prime is not the best course of action. It seems that finding exceptions to the rule of building primes even with an incomplete heuristic is more fruitful than a "dogmatic" behavior regarding primes.

4   Experimental Results

Being consistent with our previous naming scheme, we name the new bots Plakoto-5 and Fevga-6. We compare them by taking the best set of trained weights and make them playing a tournament against a benchmark opponent without look-ahead (1-ply). For Plakoto and Fevga this benchmark is our best previous bot, namely Plakoto-4, and Fevga-4 respectively. For the Portes/Backgammon we chose the pubeval benchmark because we can indirectly compare the performance with others backgammon bots that published results against it. We also report on the performance when applying a simple look-ahead procedure using the expectimax algorithm [5] at 2-ply depth. The bot is awarded a +1 point for a single win, +2 points for a double win, -1 for a single loss, -2 for a double loss. The result of the tested games sum up to the form of estimated *points per game* (ppg) and is calculated as the mean of the points won and lost. The number of games played are 100000 for 1-ply and 10000 for 2-ply. In order to speed up the testing time of 2-ply, the expansion of depth-2 was performed only for the best 15 candidate moves (forward pruning). Table 4 presents the results.

Table 4. Performance of the new bots against benchmark opponents

Bot	Opponent	ppg
Portes-1(1-ply)	Pubeval (1-ply)	0.603
Plakoto-5(1-ply)	Plakoto-4(1-ply)	0.356
Plakoto-5(2-ply)	Plakoto-4(1-ply)	0.422
Fevga-6(1-ply)	Fevga-4(1-ply)	0.215
Fevga-6(2-ply)	Fevga-4(1-ply)	0.323

The performance of the Portes/Backgammon bot is comparable to most top playing bots. TD-Gammon [13] reported a 0.596 performance against pubeval [14] while another backgammon program, GNUBG[1], frequent participant to backgammon Computer Olympiads, recently reported in its mailing list similar performance

---

[1] http://www.gnubg.org

(0.6046 ppg) while using a more complex training scheme and three different NNs for three different stages of the game [3].

Since the training procedure and the NN architecture is the same for the old and new bots for the Fevga and Plakoto variants, it is safe to assume that the gain was due to the addition/alteration of the expert features. We believe that the common features of Portes that were added to Plakoto and Fevga played a minor role to the improved performance. More important for Fevga, was the alteration of the prime features, and for Plakoto, the addition ChFrontOfPin_1 andChFrontOfPin_2.

5   Conclusion and Future Work

We have presented the complete algorithm of our training scheme for backgammon variants that are included in Palamedes. The computer backgammon winner of the 2011 backgammon Computer Olympiad was also presented in full for the first time. Finally, we have managed to increase the performance of the Plakoto and Fevga variants by adding new expert features based on manual examination and user feedback.

We will continue the search for new features that could improve the playing strength of *Palamedes*. The heuristic for calculating the probability of making a prime formation on the next roll can be improved by including cases with two or more missing checkers, and by making it faster to compute.

Our experiments with different values for the learning rate α and the λ parameter show that the best choice for either of them is domain specific. Using our setup, it is possible to start the training with high values and gradually decrease them. As we did not exhaust all possible combinations, it may possible that an even more aggressive approach could yield faster learning. An algorithm that automatically decreases these parameters during training would be interesting to investigate as it would free the human designer of the otherwise cumbersome trial and error approach.

A difficult part of the work so far is the manual examination of the playing style of the trained bots by human experts. This is necessary because NNs cannot easily describe the concepts learned by examining the weights alone. We plan to improve our understanding of the playing style of our NNs by visualizing the weights and by extracting rules [1].

We also plan to increase the number of backgammon variants that can be handled by *Palamedes*. Interesting candidates towards this direction are the acey-deucey, gioul and gul-bara variants. Finally we plan to improve the look-ahead procedure by searching in greater depths and by utilizing cutoff algorithms as in [4].

References

1. Andrews, R., Diederich, J., Tickle, A.: Survey and critique of techniques for extracting rules from trained artificial neural networks. Knowledge-Based Systems 8(6), 373–389 (1995)
2. BackGammon Variants, http://www.bkgm.com/variants

3. GnuBg Mailing list post, http://lists.gnu.org/archive/html/bug-gnubg/2012-01/msg00034.html
4. Hauk, T., Buro, M., Schaeffer, J.: *-MINIMAX Performance in Backgammon. In: van den Herik, H.J., Björnsson, Y., Netanyahu, N.S. (eds.) CG 2004. LNCS, vol. 3846, pp. 51–66. Springer, Heidelberg (2006)
5. Michie, D.: Game-playing and game-learning automata. In: Fox, L. (ed.) Advances in Programming and Non-Numerical Computation, pp. 183–200 (1966)
6. Palamedes, http://ai.uom.gr/nikpapa/software.html
7. Papahristou, N., Refanidis, I.: Training Neural Networks to Play Backgammon Variants Using Reinforcement Learning. In: Di Chio, C., Cagnoni, S., Cotta, C., Ebner, M., Ekárt, A., Esparcia-Alcázar, A.I., Merelo, J.J., Neri, F., Preuss, M., Richter, H., Togelius, J., Yannakakis, G.N. (eds.) EvoApplications 2011, Part I. LNCS, vol. 6624, pp. 113–122. Springer, Heidelberg (2011)
8. Papahristou, N., Refanidis, I.: Improving Temporal Difference Learning Performance in Backgammon Variants. In: van den Herik, H.J., Plaat, A. (eds.) ACG 2011. LNCS, vol. 7168, pp. 134–145. Springer, Heidelberg (2012)
9. Pubeval source code backgammon benchmark player, http://www.bkgm.com/rgb/rgb.cgi?view+610
10. Sutton, R.S.: Learning to predict by the methods of temporal differences. Machine Learning 3(1), 9–44 (1988)
11. Sutton, R.S., Barto, A.G.: Reinforcement Learning: An Indroduction. MIT Press (1998)
12. Tesauro, G.: Practical issues in temporal differnce learning. Machine Learning 4, 257–277 (1992)
13. Tesauro, G.: Programming backgammon using self-teching neural nets. Artificial Intelligence 134, 181–199 (2002)
14. Tesauro, G.: Td-gammon, http://www.scholarpedia.org/article/Td-gammon
15. Tesauro, G.: Temporal Difference Learning and TD-Gammon. Communications of the ACM 38(3), 58–68 (1995)
16. Wiering, M.A.: Self-Play and Using an Expert to Learn to Play Backgammon with Temporal Difference Learning. Journal of Intelligent Learning Systems and Applications 2, 57–68 (2010)
17. Wilson, D.R., Martinez, T.R.: The general inefficiency of batch training for gradient descent learning. Neural Networks 16(10), 1429–1451 (2003)

Surrogate Modelling of Solutions
of Integral Equations by Neural Networks

Věra Kůrková

Institute of Computer Science, Academy of Sciences of the Czech Republic
Pod Vodárenskou věží 2, Prague 8, Czech Republic
vera@cs.cas.cz

Abstract. Surrogate modelling of solutions of integral equations by neural networks is investigated theoretically. Estimates of speed of convergence of suboptimal surrogate solutions to solutions described by Fredholm theorem are derived.

Keywords: surrogate modelling by neural networks, Fredholm integral equations, rates of approximation, model complexity.

1 Introduction

Surrogate modelling of functional relationships by neural networks has been successfully applied to empirical functions for which experimental evaluations are too expensive or to functions, which are described by known analytic formulas but numerical calculations of such formulas are too time consuming. For construction of surrogate models, results of such experiments or calculations merely at some selected points of the domains of the functions are needed. Values obtained by such expensive or time consuming procedures are then used as training samples of data for neural networks. The networks trained on such samples are called surrogate models. For example, neural networks have been successfully used in chemistry as surrogate models of empirical functions assigning to compositions of chemicals measures of quality of catalyzers produced by reactions of these chemicals or in biology as surrogate models of empirical functions classifying structures of RNA [1,2].

In the case of empirical functions, results obtained by surrogate modelling can only be used as suggestions to be confirmed by additional experiments as no other than empirical knowledge of the functions is available. In the case of functions whose mathematical expressions are known, although complicated and difficult to calculate numerically, these expressions might enable to estimate how close to these functions their surrogate models are. In particular, theoretical analysis can provide estimates of model complexity of neural networks needed for a sufficient accuracy of approximation of the functions which are replaced by these networks.

Recently, possibilities of neural networks in surrogate modelling of solutions of Fredholm integral equations were explored. Fredholm equations play an important role in many problems in applied science and engineering. They arise

L. Iliadis et al. (Eds.): AIAI 2012, IFIP AICT 381, pp. 88–96, 2012.

in image restoration, differential problems with auxiliary boundary conditions, potential theory and elasticity, etc. (see, e.g., [3,4,5]). Numerical calculations of values of solutions of Fredholm equations are time consuming as they involve computations of complicated expressions in terms of infinite Liouville-Neumann series with integrals coefficients. Thus various methods of approximate solutions of Fredholm equations have been developed. Traditional approach employed polynomial or wavelet interpolations of numerically calculated values at certain collocation points [6, Chapter 11]. As an alternative to the traditional approach, surrogate models formed by neural networks with perceptrons [7] and Gaussian radial-basis units [8] were proposed and explored experimentally. These experiments were motivated by higher flexibility of neural networks than flexibility of polynomials. Neural networks have more adjustable parameters than linear models as in addition to coefficients of linear combinations of basis functions, also inner coefficients of computational units are optimized during learning. Thus they are sometimes called variable-basis approximation schemas in contrast to traditional linear approximators which are called fixed-basis approximation schemas. In some cases, especially in the case of functions of large numbers of variables, neural networks achieve better approximation rates than linear models with much smaller model complexity [9,10]. Motivated by the experimental exploration [8,7] of surrogate models of solutions of Fredholm equations by neural networks, Gnecco et al. [11] initiated a theoretical analysis of this modelling. In [11], they estimated approximation errors in supremum norm for surrogate solutions by networks with kernel units induced by the kernels of the equations.

In this paper, we investigate surrogate solutions of Fredholm integral equation by networks with general computational units. Taking advantage of results from nonlinear approximation theory and integral representations of functions, we estimate how well surrogate solutions computable by neural networks can approximate exact solutions of Fredholm equations described by the Fredholm alternative theorem. We show that such surrogate solutions converge to the exact solutions and we give upper bounds on decrease of approximation errors measured in $\mathcal{L}^2$-norm with increasing model complexity of approximating networks.

The paper is organized as follows. In section 2, we introduce Fredholm integral equations and their surrogate solutions by neural networks. In section 3, using results from nonlinear approximation theory, we derive estimates of rates of approximation of solutions of Fredholm equations by neural networks. Section 4 is a brief discussion.

2   Solutions of Fredholm Equations

Solving an *inhomogeneous Fredholm integral equation of the second kind* on a domain $X \subseteq \mathbb{R}^d$ for a given $\lambda \in \mathbb{R} \setminus \{0\}$, $K : X \times X \to \mathbb{R}$, and $f : X \to \mathbb{R}$ is a task of finding a function $\phi : X \to \mathbb{R}$ such that for all $x \in X$

$$\phi(x) - \lambda \int_X \phi(y) K(x,y)\, dy = f(x). \tag{1}$$

The function ϕ is called a *solution, f data, K* a *kernel*, and a λ *parameter* of the equation (1).

Recall that tasks of finding *unknown causes* of *known consequences* have been studied in applied science under the name *inverse problems*. Formally, an inverse problem is defined by a linear operator $A : \mathcal{X} \to \mathcal{Y}$ between two function spaces. It is a task of finding for $g \in \mathcal{Y}$ (called *data*) some $f \in \mathcal{X}$ (called *solution*) such that $A(f) = g$.

Let T_K denotes the integral operator with a kernel $K : X \times X \to \mathbb{R}$ defined for every ϕ in a suitable function space as

$$T_K(\phi)(x) := \int_X \phi(y)\, K(x, y)\, dy . \tag{2}$$

Thus solving the Fredholm equation (1) is an inverse problem described by the linear operator $I_{\mathcal{X}} - \lambda T_K$, where $I_{\mathcal{X}}$ is the identity operator. Existence and uniqueness of solutions of this inverse problem follows from the Fredholm alternative theorem (see, e.g., [12, p.499]).

Theorem 1. *Let $\mathcal{X}$ be a Banach space, $T : \mathcal{X} \to \mathcal{X}$ be a compact operator, and $I_{\mathcal{X}}$ be the identity operator. Then the operator $I_{\mathcal{X}} + T : \mathcal{X} \to \mathcal{X}$ is one-to-one if and only if it is onto.*

Theorem 1 implies that when T is a compact operator and $1/\lambda$ is not its eigenvalue (i.e., there is no $\phi \in \mathcal{X}$ for which $T(\phi) = \frac{\phi}{\lambda}$), then the operator $I_{\mathcal{X}} - \lambda T$ is invertible (one-to-one and onto) [13, p. 112], [14, Section 1.3].

Fredholm theorem can be applied to spaces $(\mathcal{C}(X), \|.\|_{\sup})$ of bounded continuous functions on $X \subseteq \mathbb{R}^d$ with the supremum norm $\|f\|_{\sup} = \sup_{x \in X} |f(x)|$ and to spaces $(\mathcal{L}^2(X), \|.\|_{\mathcal{L}^2})$ of square integrable functions with the norm $\|f\|_{\mathcal{L}^2} = \left(\int_X f(x)^2\, dx \right)^{1/2}$. The following proposition is well-known and easy to check (see, e.g., [13, p. 112]).

Proposition 1. *(i) If $X \subset \mathbb{R}^d$ is compact and $K : X \times X \to \mathbb{R}$ is continuous, then $T_K : \mathcal{C}(X) \to \mathcal{C}(X)$ is a compact operator.*
(ii) If $X \subset \mathbb{R}^d$ and $K \in \mathcal{L}^2(X \times X)$, then $T_K : \mathcal{L}^2(X) \to \mathcal{L}^2(X)$ is a compact operator.

So by the Fredholm theorem, when $1/\lambda$ is not an eigenvalue of T_K and the assumptions of the Proposition 1 (i) or (ii) are satisfied, then for every f in $\mathcal{C}(X)$ or $\mathcal{L}^2(X)$, resp., there exists unique solution ϕ of the equation (1). Moreover,

$$\phi(x) = f(x) - \lambda \int_X f(y)\, R_K^\lambda(x, y)\, dy , \tag{3}$$

where $R_K^\lambda : X \times X \to \mathbb{R}$ is called a *resolvent kernel* [14]. Although a formula expressing the resolvent kernel is known, it is not suitable for efficient computation as it is expressed as an infinite Neumann series in powers of λ with coefficients in the form of iterated kernels [15, p.140]. Numerical calculations of values of solutions of Fredholm equations based on (3) are quite computationally demanding and so various methods of finding surrogate solutions of (1) have been explored.

A traditional approach employed linear methods such as polynomial inter-polation of approximations $\bar{\phi}(x_1), \ldots, \bar{\phi}(x_m)$ (computed by numerical methods) of the values $\phi(x_1), \ldots, \phi(x_m)$ of the solution ϕ at suitable points $x_1, \ldots, x_m$ from the domain X. In linear methods, approximating sets of functions are n-dimensional subspaces of the form

$$\text{span}\{g_1, \ldots, g_n\} := \left\{ \sum_{i=1}^{n} w_i g_i \mid w_i \in \mathbb{R} \right\}, \tag{4}$$

where $G = \{g_n \mid n \in \mathbb{N}_+\}$ is a set of functions with a *fixed linear ordering*. Nested linear subspaces of the form (4) can be generated, e.g., by sets of powers of increasing degrees or trigonometric functions with increasing frequencies.

A recent alternative to linear methods employs neural networks. Recall that one-hidden-layer networks with one linear output unit compute input-output functions from sets of the form

$$\text{span}_n G := \left\{ \sum_{i=1}^{n} w_i g_i \mid w_i \in \mathbb{R}, g_i \in G \right\}, \tag{5}$$

where the set G is sometimes called a *dictionary* [16] and n is the *number of hidden computational units*. This number can be interpreted as a measure of *model complexity* of the network. In contrast to linear approximation, the approximation scheme formed by sets $\text{span}_n G$ is called *variable-basis approximation* (the set G has no fixed ordering).

Often, dictionaries are parameterized families of functions modelling computational units, i.e., they are of the form

$$G_K(X, Y) := \{K(\cdot, y) : X \to \mathbb{R} \mid y \in Y\}, \tag{6}$$

where $K : X \times Y \to \mathbb{R}$ is a function of two variables, an input vector $x \in X \subseteq \mathbb{R}^d$ and a parameter $y \in Y \subseteq \mathbb{R}^s$. When $X = Y$, we write briefly $G_K(X)$.

In some contexts, K is called a *kernel*. However, the above-described computational scheme includes fairly general computational models, such as functions computable by perceptrons, radial or kernel units, Hermite functions, trigonometric polynomials, and splines. For example, with

$$K(x, y) = K(x, (v, b)) := \sigma(\langle v, x \rangle + b)$$

and $\sigma : \mathbb{R} \to \mathbb{R}$ a sigmoidal function, the computational scheme (5) describes one-hidden-layer *perceptron networks*. *RBF units* with an activation function $\beta : \mathbb{R} \to \mathbb{R}$ are modeled by kernel

$$K(x, y) = K(x, (v, b)) := \beta(v\|x - b\|).$$

Typical choice of β is the Gaussian function. *Kernel units* used in support vector machine have the form $K(x, y)$ where $K : X \times X \to \mathbb{R}$ is a symmetric positive semidefinite function [12].

Various learning algorithms optimize parameters $y_1, \ldots, y_n$ of the computational units as well as coefficients of their linear combinations $w_1, \ldots, w_n$ so that network input-output functions $\sum_{i=1}^{n} w_i K(., y_i)$ from the set $\mathrm{span}_n G_K$ fit well to training samples. In the case of surrogate solutions of Fredholm equations, training samples are the above described sets $\{(x_1, \bar{\phi}(x_1)), \ldots, (x_m, \bar{\phi}(x_m))\}$ of input-output pairs formed by selected points of the domain together with numerical approximations of values of the solution ϕ at these points.

3   Rates of Convergence of Surrogate Solutions

Estimates of model complexity of one-hidden-layer networks approximating solutions of Fredholm equations follow from inspection of upper bounds on rates of variable-basis approximation. Such rates have been studied in mathematical theory of neurocomputing for various types of dictionaries G and norms measuring approximation errors such as Hilbert-space norms and the supremum norm (see, e.g., [17,18]). Typically, such bounds have the form $\frac{\xi(h)}{\sqrt{n}}$, where n is the number of network units and $\xi(h)$ depends on a certain variational norm of the function h to be approximated, which is tailored to the dictionary of the network units. Variational norm is defined quite generally for any bounded nonempty subset G of a normed linear space $(\mathcal{X}, \|.\|_{\mathcal{X}})$. It is called *G-variation*, denoted $\|.\|_G$, and defined for all $f \in \mathcal{X}$ as

$$\|f\|_{G,\mathcal{X}} := \inf \{c > 0 \mid f/c \in \mathrm{cl}_{\mathcal{X}} \, \mathrm{conv}\,(G \cup -G)\},$$

where the closure $\mathrm{cl}_{\mathcal{X}}$ is taken with respect to the topology generated by the norm $\|.\|_{\mathcal{X}}$ and conv denotes the *convex hull*. So G-variation depends on the ambient space norm, but when it is clear from the context, we write merely $\|f\|_G$ instead of $\|f\|_{G,\mathcal{X}}$.

The concept of variational norm was introduced by Barron [19] for sets of characteristic functions. In particular, for the set of characteristic functions of half-spaces, the variational norm is tailored to the dictionary of functions computable by Heaviside perceptrons. Barron's concept was generalized in [20,21] to the concept of variation with respect to an arbitrary bounded set of functions and applied to various dictionaries of computational units such as Gaussian RBF or kernel units [22].

The following theorem from [21] is a reformulation of results by Maurey [23], Jones [24], Barron [17] in terms of G-variation. It gives an upper bound on rates of approximation by sets of the form $\mathrm{span}_n G$. For a normed space $(\mathcal{X}, \|.\|_{\mathcal{X}})$, $g \in \mathcal{X}$ and $A \subset \mathcal{X}$, we denote by $\|g - A\|_{\mathcal{X}} := \inf_{f \in A} \|g - f\|_{\mathcal{X}}$ the *distance* of g from A.

Theorem 2. *Let $(\mathcal{X}, \|.\|_{\mathcal{X}})$ be a Hilbert space, G its bounded nonempty subset, $s_G = \sup_{g \in G} \|g\|_{\mathcal{X}}$, $f \in \mathcal{X}$, and n be a positive integer. Then*

$$\|h - \mathrm{span}_n G\|_{\mathcal{X}}^2 \leq \frac{s_G^2 \|h\|_G^2 - \|h\|_{\mathcal{X}}^2}{n}.$$

Theorem 2 implies that for every $\varepsilon > 0$ and n satisfying

$$n \geq \left(\frac{s_G \, \|h\|_G}{\varepsilon} \right)^2 ,$$

a network with n units computing functions from the dictionary G approximates the function h within ε. So the size of the G-variation of the function h is a critical factor influencing model complexity of networks approximating h within an accuracy ε. Generally, it is not easy to estimate G-variation. However, for the special case of functions with integral representations in the form of "infinite networks", variational norms are bounded from above by the $\mathcal{L}^1$-norms of "output-weight" functions of these networks.

Theorem 3. *Let* $X \subseteq \mathbb{R}^d$, $Y \subseteq \mathbb{R}^s$, $w \in \mathcal{L}^1(Y)$, $K : X \times Y \to \mathbb{R}$ *be such that* $G_K(Y) = \{K(.,y) \,|\, y \in Y\}$ *is a bounded subset of* $(\mathcal{L}^2(X), \|.\|_{\mathcal{L}^2})$, *and* $h \in \mathcal{L}^2(X)$ *be such that for all* $x \in X$, $h(x) = \int_Y w(y) \, K(x,y) \, dy$. *Then*

$$\|f\|_{G_K(Y)} \leq \|w\|_{\mathcal{L}^1(Y)}.$$

Theorem 3 gives an upper bound on variation with respect to the dictionary G_K induced by the kernel K of the integral operator T_K which maps an "output-weight" function w of the "infinite network" $\int_Y w(y) \, K(.,y) \, dy$ to the function h. However, to apply Theorem 2 to approximation of solutions of Fredholm equations by surrogate models formed by networks with units from a general dictionary G, we need upper bounds on G-variation for functions expressed by an integral formula with the kernel K. The next proposition describes a relationship between variations with respect to two sets, G and F. Its application to G and G_K gives an upper bound on G. The proof of the next proposition follows easily from the definition of variational norm.

Proposition 2. *Let* $(\mathcal{X}, \|.\|_{\mathcal{X}})$ *be a normed linear space*, F *and* G *its bounded subsets such that* $c_{G,F} := \sup_{g \in G} \|g\|_F < \infty$. *Then for all* $h \in \mathcal{X}$, $\|h\|_G \leq c_{G,F} \|h\|_F$.

Combining Theorems 2, 3, and Proposition 2, we obtain the next theorem on rates of approximation of functions which can be expressed as $h = T_K(w)$ by networks with units from a dictionary G.

Theorem 4. *Let* $X \subseteq \mathbb{R}^d$, $K : X \times Y \to \mathbb{R}$ *be a bounded kernel, and* $h \in \mathcal{L}^2(X)$ *such that* $h = T_K(w) = \int_Y w(y) K(.,y) \, dy$ *for some* $w \in \mathcal{L}^1(Y)$, *where* $G_K(X,Y)$ *is a bounded subset of* $\mathcal{L}^2(X)$. *Let* G *be a bounded subset of* $\mathcal{L}^2(X)$ *with* $s_G = \sup_{g \in G} \|g\|_{\mathcal{L}^2}$ *such that* $c_{G,K} = \sup_{g \in G} \|g\|_{G_K}$ *is finite. Then for all* $n > 0$,

$$\|h - \mathrm{span}_n G\|_{\mathcal{L}^2} \leq \frac{s_G \, c_{G,K} \, \|w\|_{\mathcal{L}^1(Y)}}{\sqrt{n}}.$$

A critical factor in the upper bound from Theorem 4 is the $\mathcal{L}^1$-norm of the "output-weight function" w in the representation

$$h(x) = T_K(w)(x) = \int_Y w(y)\, K(x,y)\, dy$$

of h as an "infinite network" with units computing $K(.,y)$.

The solution ϕ of the Fredholm equation minus the function f representing the data, $\phi - f$, is the image of $\lambda\,\phi$ mapped by the integral operator T_K, i.e.,

$$\phi - f = \lambda \int_X f(y)\, R_K^\lambda(x,y)\, dy = \lambda \int_X \phi(y) K(x,y)\, dy \; = T_K(\lambda\,\phi).$$

Thus to apply Theorem 4 to approximation of $\phi - f$, we need to estimate the $\mathcal{L}^1$-norm of the solution ϕ itself as $\lambda\phi$ plays the role of the "output-weight" function in the infinite network $\lambda \int_X \phi(y) K(x,y)\, dy$.

Theorem 5. *Let $X \subset \mathbb{R}^d$ be compact, $K : X \times X \to \mathbb{R}$ be a bounded kernel such that $K \in \mathcal{L}^2(X \times X)$, $s_K := \sup_{y\in Y} \|K(.,y)\|_{\mathcal{L}^2}$, $\rho_K := \int_X \sup_{y\in X} |K(x,y)| dx$ be finite, G be a bounded subset of $\mathcal{L}^2(X)$ with $s_G = \sup_{g\in G} \|g\|_{\mathcal{L}^2}$ such that $c_{G,K} = \sup_{g\in G} \|g\|_{G_K}$ is finite, and $\lambda \neq 0$ be such that $\frac{1}{\lambda}$ is not an eigenvalue of T_K and $|\lambda|\, \rho_K < 1$. Then the solution ϕ of the equation (1) satisfies for all $n > 0$,*

$$\|\phi - f - \operatorname{span}_n G\|_{\mathcal{L}^2} \leq \frac{s_G\, c_{G,K}\, |\lambda|\, \|f\|_{\mathcal{L}^1}}{(1 - |\lambda|\, \rho_K)\, \sqrt{n}}.$$

Proof. By (1), we have for every $x \in X$,

$$|\phi(x)| \leq |\lambda|\, \|\phi\|_{\mathcal{L}^1} \sup_{y\in X} |K(x,y)| + |f(x)|.$$

Hence $\|\phi\|_{\mathcal{L}^1} \leq |\lambda|\, \rho_K\, \|\phi\|_{\mathcal{L}^1} + \|f\|_{\mathcal{L}^1}$ and so $\|\phi\|_{\mathcal{L}^1} (1 - |\lambda|\, \rho_K) \leq \|f\|_{\mathcal{L}^1}$. This inequality is non trivial only when $|\lambda| < \frac{1}{\rho_K}$. Thus we get $\|w\|_{\mathcal{L}^1} = |\lambda|\|\phi\|_{\mathcal{L}^1} \leq \frac{|\lambda|\, \|f\|_{\mathcal{L}^1}}{1 - |\lambda|\, \rho_K}$. The statement then follows from Theorem 4. $\qquad\square$

Theorem 5 estimates rates of approximation of the function $\phi - f$ by functions computable by networks with units from a dictionary G. The values of the function $\lambda \int_X f(y)\, R_K^\lambda(x,y)\, dy$ are difficult to be compute numerically. However, $\phi - f$ can be replaces with its approximation computable by a neural network. With increasing number n of units, input-output functions of such networks converge to the function $\phi - f$. When for a reasonable size of the network measured by the number n of units, the upper bound from Theorem 5 is sufficiently small, the network can serve as a good surrogate model of the solution of Fredholm equation.

The next corollary illustrates our results by an example of approximation of solutions of Fredholm equations with Gaussian kernels by Gaussian kernel and perceptron networks. Fredholm equations with Gaussian kernels arise, e.g., in image restoration problems [5]. By μ is denoted the Lebesgue measure on $\mathbb{R}^d$ and by $P_d^\sigma(X)$ the dictionary of functions on $X \subseteq \mathbb{R}^d$ computable by sigmoidal perceptrons.

Corollary 1. *Let $X \subset \mathbb{R}^d$ be compact, $b > 0$, $K_b(x,y) = e^{-b\|x-y\|^2}$, $\lambda \neq 0$ be such that $\frac{1}{\lambda}$ is not an eigenvalue of T_{K_b} and $|\lambda| < 1$. Then the solution ϕ of the equation $\phi - f = \lambda \int_X \phi(y) K(x,y)\, dy$ with f continuous satisfies for all $n > 0$*

$$\|\phi - f - \mathrm{span}_n\, G_{K_b}(X)\|_{\mathcal{L}^2} \leq \frac{\mu(X)\,|\lambda|\,\|f\|_{\mathcal{L}^1}}{(1 - |\lambda|\,\mu(X)\,\sqrt{n}}$$

and

$$\|\phi - f - \mathrm{span}_n\, P_d^\sigma(X)\|_{\mathcal{L}^2} \leq \frac{\mu(X)\,2d\,|\lambda|\,\|f\|_{\mathcal{L}^1}}{(1 - |\lambda|\,\mu(X)\,\sqrt{n}}.$$

Proof. It was shown in [25] that P_d^σ-variation (i.e., variation with respect to the dictionary formed by sigmoidal perceptrons) of the d-dimensional Gaussian is bounded from above by $2d$. The statement then follows by Theorem 5, Proposition 2, equalities $\tau_{K_b} = 1$, $\rho_{K_b} = \mu(X)$, and the estimate $s_{K_b} \leq \mu(X)$.   □

4   Discussion

Taking advantage of results from mathematical theory of neurocomputing holding for functions representable as "infinite neural networks" we derived estimates of rates of convergence of surrogate solutions of Fredholm equations to exact solutions guaranteed by Fredholm theory. The bounds are in the form $\frac{c}{\sqrt{n}}$ where n is the number of network units and c depends on $\mathcal{L}^1$-norm of the function f corresponding to data of the inverse problem given by the operator $I - \lambda T_K$, the parameter λ, the Lebesgue measure $\mu(X)$ of the domain $X \subset \mathbb{R}^d$, and a bound on $\|K(.,y)\|_{\mathcal{L}^2}$. The results hold under certain restrictions on the size of λ similarly as in the case of traditional linear approximation of solutions of Fredholm equations.

Acknowledgments. This work was partially supported by MŠMT grant COST INTELLI OC10047 and RVO 67985807.

References

1. Forrester, A., Sobester, A., Keane, A.: Engineering Design via Surrogate Modelling: A Practical Guide. Wiley (2008)
2. Baerns, M., Holeňa, M.: Combinatorial Development of Solid Catalytic Materials. Imperial College Press, London (2009)
3. Lovitt, W.V.: Linear Integral Equations. Dover, New York (1950)
4. Lonseth, A.T.: Sources and applications of integral equations. SIAM Review 19, 241–278 (1977)
5. Lu, Y., Shen, L., Xu, Y.: Integral equation models for image restoration: high accuracy methods and fast algorithms. Inverse Problems 26 (2010), doi: 10.1088/0266-5611/26/4/045006

6. Atkinson, K., Han, W.: Theoretical Numerical Analysis: A Functional Analysis Framework. Springer, Heidelberg (2005)
7. Effati, S., Buzhabadi, R.: A neural network approach for solving Fredholm integral equations of the second kind. Neural Computing and Applications (2010), doi:10.1007/s00521-010-0489-y
8. Golbabai, A., Seifollahi, S.: Numerical solution of the second kind integral equations using radial basis function networks. Applied Mathematics and Computation 174, 877–883 (2006)
9. Gnecco, G., Kůrková, V., Sanguineti, M.: Some comparisons of complexity in dictionary-based and linear computational models. Neural Networks 24, 171–182 (2011)
10. Gnecco, G., Kůrková, V., Sanguineti, M.: Can dictionary-based computational models outperform the best linear ones? Neural Networks 24, 881–887 (2011)
11. Gnecco, G., Kůrková, V., Sanguineti, M.: Accuracy of approximations of solutions to Fredholm equations by kernel methods. Applied Mathematics and Computation 218, 7481–7497 (2012)
12. Steinwart, I., Christmann, A.: Support Vector Machines. Springer, New York (2008)
13. Rudin, W.: Functional Analysis. McGraw-Hill, Boston (1991)
14. Atkinson, K.: The Numerical Solution of Integral Equations of the Second Kind. Cambridge University Press (1997)
15. Courant, R., Hilbert, D.: Methods of Mathematical Physic, vol. I. Wiley, New York (1989)
16. Gribonval, R., Vandergheynst, P.: On the exponential convergence of matching pursuits in quasi-incoherent dictionaries. IEEE Transactions on Information Theory 52, 255–261 (2006)
17. Barron, A.R.: Universal approximation bounds for superpositions of a sigmoidal function. IEEE Transactions on Information Theory 39, 930–945 (1993)
18. Girosi, F.: Approximation error bounds that use VC-bounds. In: Proceedings of ICANN 1995, Paris, pp. 295–302 (1995)
19. Barron, A.R.: Neural net approximation. In: Narendra, K. (ed.) Proc. 7th Yale Workshop on Adaptive and Learning Systems, Yale University Press (1992)
20. Kůrková, V.: Dimension-independent rates of approximation by neural networks. In: Warwick, K., Kárný, M. (eds.) Computer-Intensive Methods in Control and Signal Processing. The Curse of Dimensionality, pp. 261–270. Birkhäuser, Basel (1997)
21. Kůrková, V.: High-dimensional approximation and optimization by neural networks. In: Suykens, J., Horváth, G., Basu, S., Micchelli, C., Vandewalle, J. (eds.) Advances in Learning Theory: Methods, Models and Applications, pp. 69–88. IOS Press, Amsterdam (2003) (Chapter 4)
22. Kainen, P.C., Kůrková, V., Sanguineti, M.: Complexity of Gaussian radial-basis networks approximating smooth functions. Journal of Complexity 25, 63–74 (2009)
23. Pisier, G.: Remarques sur un résultat non publié de B. Maurey. In: Séminaire d'Analyse Fonctionnelle 1980-1981, École Polytechnique, Centre de Mathématiques, Palaiseau, France, vol. I(12) (1981)
24. Jones, L.K.: A simple lemma on greedy approximation in Hilbert space and convergence rates for projection pursuit regression and neural network training. Annals of Statistics 20, 608–613 (1992)
25. Kainen, P.C., Kůrková, V., Vogt, A.: A Sobolev-type upper bound for rates of approximation by linear combinations of Heaviside plane waves. Journal of Approximation Theory 147, 1–10 (2007)

A Regularization Network Committee Machine of Isolated Regularization Networks for Distributed Privacy Preserving Data Mining

Yiannis Kokkinos and Konstantinos G. Margaritis[*]

Parallel and Distributed Processing Laboratory, Department of Applied Informatics,
University of Macedonia, 156 Egnatia str., P.O. Box 1591, 54006, Thessaloniki, Greece

Abstract. In this paper we consider large scale distributed committee machines where no local data exchange is possible between neural network modules. Regularization neural networks are used for both the modules as well as the combiner committee in an embedded architecture. After the committee training no module will know anything else except its own local data. This privacy preserving obligation is a challenging problem for trainable combiners but crucial in real world applications. Only classifiers in the form of binaries or agents can be sent to others to validate their local data and sent back average classification rates. From this fully distributed and privacy preserving mutual validation a coarse-grained matrix can be formed to map all members. We demonstrate that it is possible to fully exploit this mutual validation matrix to efficiently train another regularization network as a meta learner combiner for the committee.

Keywords: Distributed processing, Regularization Networks, Committee Machines, privacy preserving, data mining.

1   Introduction

A committee machine [1] exhibits an intrinsically parallel and distributed architecture [2] in which multiple modules of independently trained neural networks are combined for the same task. An ensemble learner of this kind is an ideal candidate for data mining large scale physically distributed data repositories in institutions/organizations or Peer-to-Peer networks. However privacy preserving and scalability are crucial issues for these real life applications. The Regularization Networks [3][4][5] are kernel based classifiers known to use as hidden neurons the real training data points, to form the kernel functions and capture the data closeness approximate of the underlined problem distribution. Using real points is valuable when data features have discrete values, e.g., in cases of image processing, computer vision [1] and data mining [8], a fact that elevates such type of kernel based ridge regression methods [9] [10] to state of the art. For these distributed Regularization Network (RN) modules if another Regularization Network can be trained to act as a high level combiner then can serve as

---

[*] Corresponding author.

L. Iliadis et al. (Eds.): AIAI 2012, IFIP AICT 381, pp. 97–106, 2012.

a meta-learner in a distributed data mining system [11]. Such a meta-learner is looked at in detail here for the distributed privacy preserving case.

A committee of neural networks has excellent generalization capabilities, [6] since typically the committee error is reduced considerably by taking the average error of the combined networks. All neural networks classifiers are first trained in parallel based on local data to construct local data models. Then the committee combines [7] all individual decisions through proper weights to form the global data model used for collective decisions. Committees can be used in data mining physically distributed data repositories as well as peer-to-peer systems. Gather large volumes of distributed data to a single location for centralized data mining is usually unfeasible. The causes that prevent this lay in technical issues like limited network bandwidth and enormous main memory demands, practical issues like huge required training times, algorithmic issues in where mining algorithms operate only on data in main memory, and especially privacy preserving concerns that restrict the transferring of sensitive data.

To this context the task to build a global model from data distributed over workstations, without moving or sharing local data itself, and with little centralized coordination, is challenging. The trainable combiner requires a separate test set to find proper weights for the neural network modules. Thus it requires the use of extra information from their input-output mappings. At least two-by-two the classifiers must share either input vectors, or output vector results with respect to instances of an independent test set. Without exposing data between modules or without aggregating a portion of data this is tricky. A regularization network as a meta-learner committee of regularization networks trained by a simple distributed privacy preserving mutual validation matrix is presented here, as an effort to the solution of the problem.

2   Distributed Privacy-Preserving Data Mining

Distributed privacy-preserving data mining is the study of how to extract globally interesting models, associations, classifiers, clusters, and other useful aggregate statistics from distributed data without disclosing private information within the different participants. Data exchange and free flow of information is frequently prohibited by legal obligations or by commercial and personal concerns, since the participants may wish to collaborate, but might not fully trust each other. The basic idea of a secure multiparty computation is that a distributed computation is secure if at the end no party knows anything except its own input and the aggregate results. For example, secure sum protocol [12] computes the sum of a collection of numbers without revealing anything but the output sum. Classifiers which need total sums like Naive Bayes can be worked in this fashion. Data sets are usually distributed in horizontal partitions where different sites contain different sets of records with the same attributes. Classifier examples that have been generalized to this distributed privacy preserving data mining problem are the Naïve Bayes Classifier [13][14], the SVM Classifier with nonlinear kernels [15] and the k-nearest neighbour classifier [16]. Since for multiclass problems classical neural networks are proven the best over the years, here we present an embedded architecture Regularization Network committee machine of Regularization Networks distributed over workstations, of which the training leave the processor nodes with no extra knowledge for the other participant inputs.

3   A Regularization Networks Committee Machine

The committee training [17][18][19], like in neural network training has to find a proper weight for each individual neural network. A Regularization Network (RN) committee of embedded Regularization Networks classifiers illustrated in fig.1, is analysed in this section. An individual RN [3] [4] [5] has one input layer, one hidden layer, and one output layer. All real data points are loaded to the hidden neurons to form the kernel functions. For a training set $\{\bar{x}_i, y_i\}_{i=1}^N$, a kernel function $k(\cdot,\cdot)$, usually a Gaussian, and a Kernel matrix K with $K_{i,j} = k(\bar{x}_i, \bar{x}_j)$, the RN training phase finds optimum weights w for the output f() by solving in Reproducing Kernel Hilbert Space H_K a minimization problem for a regularized functional which consists of a usual data term plus a second regularization term that plays the role of the stabilizer [3] [4] [5]

$$\arg\min_{f \in H_K}\left\{\frac{1}{N}\sum_{i=1}^N (y_i - f(\bar{x}_i))^2 + \gamma\|f\|_K^2\right\} \tag{1}$$

For a class C the weights w_C is the solution of a linear system $(K + N\gamma I)w_C = y_C$, where I is identity matrix, K is the kernel matrix, $\gamma > 0$ is the regularization parameter and $y_C = (y_1, \ldots, y_N)$ are the desired output labels, 1 for class C and 0 for the others.

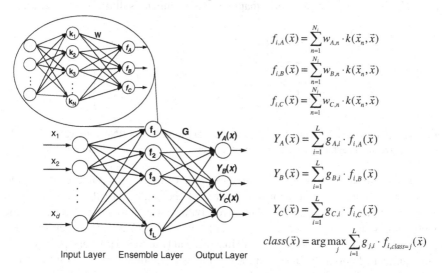

$$f_{i,A}(\bar{x}) = \sum_{n=1}^{N_i} w_{A,n} \cdot k(\bar{x}_n, \bar{x})$$

$$f_{i,B}(\bar{x}) = \sum_{n=1}^{N_i} w_{B,n} \cdot k(\bar{x}_n, \bar{x})$$

$$f_{i,C}(\bar{x}) = \sum_{n=1}^{N_i} w_{C,n} \cdot k(\bar{x}_n, \bar{x})$$

$$Y_A(\bar{x}) = \sum_{i=1}^{L} g_{A,i} \cdot f_{i,A}(\bar{x})$$

$$Y_B(\bar{x}) = \sum_{i=1}^{L} g_{B,i} \cdot f_{i,B}(\bar{x})$$

$$Y_C(\bar{x}) = \sum_{i=1}^{L} g_{C,i} \cdot f_{i,C}(\bar{x})$$

$$class(\bar{x}) = \arg\max_j \sum_{i=1}^{L} g_{j,i} \cdot f_{i,class=j}(\bar{x})$$

Input Layer   Ensemble Layer   Output Layer

Fig. 1. Architecture of a Regularization Network committee machine of embedded regularization networks for the three class problem

For the three class problem in fig.1 a Regularization Network module consists of a hidden layer with N neurons of kernel units and three linear outputs. $f_{i,class=j}(\bar{x})$ is the output of the RN module i for class j. The RN committee machine of L embedded RNs has also three outputs. While the training of local RNs can provide the weight vectors W_A, W_B and W_C, one for each class output, the high level weight vectors G_A, G_B and G_C for the outer regularization network committee are still unknown.

4   The Distributed Mutual Validation Matrix

To find the weight vectors G_A, G_B and G_C in fig.1 without reveal the training data vectors between modules we must first introduce in this section the mutual validation matrix S. Assume an ensemble of three hidden Regularization Networks namely RN(1), RN(2) and RN(3) that are trained independently from each other based on their local datasets. For privacy reasons the different locations cannot contribute or share any even smallest part of their data to other modules. Only RN networks in the form of binaries or agents can be sent to other modules to validate their data. Then schematically one can work with measures of classification rates between RN(1), RN(2) and RN(3) train patterns. The RN(1) classifies its own train data points to produce 1-1 measure. Likewise RN(2) classifies its own data to produce 2-2 measure and RN(3) produce 3-3 measure. These three are internal (or intra) measures, as they controlled by internal characteristics. In consequence RN(1) classifies train data of RN(2) to produce 1-2 measure and RN(2) classifies RN(1) data to produce 2-1 measure. In the same manner the measures 1-3, 3-1, 2-3 and 3-2 are produced. These six asymmetric measures are local (or inter) as they are based on the performance of neighbours data. A coarse-grained mutual validation matrix can be filled with these average rates. The validation set for one classifier is the train set of the other and vice versa. This mutual validation matrix maps the RN members is illustrated in fig. 2.

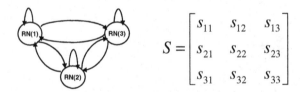

$$S = \begin{bmatrix} S_{11} & S_{12} & S_{13} \\ S_{21} & S_{22} & S_{23} \\ S_{31} & S_{32} & S_{33} \end{bmatrix}$$

Fig. 2. (A) An ensemble of three different Regularization Networks RN(1), RN(2) and RN(3), inter-connected with each other via accuracy measures, (B) the mutual validation matrix

The paradigm in fig. 2 is an illustrative example of the simple point-to-point communications involved. The diagonal measures of the matrix are the self-validation average positive hits of each RN. Upon receiving a Regularization Network i the module j apply it to classify its own local data and send back a simple average learning rate equal to $S(i, j) = \text{PositiveLocalHits}(j)/\text{LocalTrainSize}(j)$, where positive hits are the number of correctly classified local samples of module j and local train size is their number N_j of training points. In this way privacy preserving is achieved.

Distributed computations are required for the i-j, and j-i asymmetric measures between different RNs across the communication network. An asynchronous cycle is continually executed composed of commands, like sent local classifier, check for received classifier, compute local positive hits, and sent average. In terms of parallel and distributed computing this RN committee training approach is hybrid task as well as data parallel. Tasks are the RN classifiers which travel across the communication network and when arrive in a processing node are applied to the node's local data.

In this way one manages for the RN committee training to be transformed to a fully asynchronous embarrassingly parallel programming paradigm like the iterative decomposition [20]. Iterative decomposition occurs when a loop parallel execution can be done in some independent and unconnected manner. The work is statically decomposed, but the work assignments are dynamically distributed to processes. Each node may operate independently and communicate its own results to another node, making it an appropriate choice for various types of asynchronous cycles.

5   Training the Committee via the Mutual Validation Matrix

From the distributed and data privacy preserving mutual validation the coarse-grained validation matrix formed can map all modules. Although so far such a matrix was ignored, we demonstrate here that it is possible to fully exploit it to efficiently train another regularization network as a meta learner combiner for the committee.

Recall now that the weights per class output of each local Regularization network module i of the ensemble are given by solving a linear system of the form $(K + \lambda I)w = y$, where $\lambda = N_i\gamma$. While the weight vectors G_A, G_B and G_C for the outer regularization network committee machine are unknown, the simple mutual validation matrix S can now enter into the training procedure. The weights G_A, G_B and G_C can be found by considering the matrix S as the outer kernel and solving the linear equations, one for each class, in terms of the vectors Y_A, Y_B and Y_C.

$$G_A = (S + \lambda I)^{-1}Y_A$$
$$G_B = (S + \lambda I)^{-1}Y_B \tag{2}$$
$$G_C = (S + \lambda I)^{-1}Y_C$$

The vectors Y_A, Y_B and Y_C correspond to classes A, B and C have all size equal to the number L of ensemble modules. A value $Y_A(i)$ is positive local hits of RN classifier i per overall train size, produced by applying the ith RN module to the A class portion of its own i^{th} dataset (if not any set 0). Respectively $Y_B(i)$ is the overall classification rate produced by the ith RN module for the B class portion (if not any set 0) of its data, and $Y_C(i)$ similarly for the C class portion. The regularization parameter is again λ, and I is again the identity matrix.

Besides the simplicity of this method another interesting observation is that if the local neural network classifiers all are reduced to have only a single train example then, the mutual validation matrix reduces to a usual kernel matrix and the proposed committee machine in fig.1 switches to the single conventional Regularization Network. For systems where data movement across local sites is hindered like those studied at this point the above observation can serve only as a proof of correctness. However simply means that the more fine-grained the modules are, the more accurate the committee could become. Fine-grained modules can be accomplished by globally finding compact dense clusters of data points and training all the RN modules based on these data clusters. Then this RN committee of RNs method can be possibly extended for large kernel ridge regression approximation in open systems.

6   Experimental Results

The set of experiments aims at discovering the classification performance of the RN distributed privacy preserving committee tested on a separate test set of points. To this end several benchmarks are used, taken from the UCI machine learning data repository. We compare our method against majority voting. To show the efficiency of the method we must create highly unevenly and without stratification data partitions, otherwise accurate estimations may emerge simply from the fact that we use an ensemble. For the same reason a comparison with the majority voting rule is also done.

The experimental design is as follows:

1. A dataset is randomly split into a train set (70%) and a test set (30%) with stratification.
2. The train set is distributed unevenly, randomly and without stratification across a number of processors.
3. Every processor trains a local Regularization Neural Network classifier.
4. An asynchronous computing cycle is executed to find all entries of the proposed mutual validation matrix.
5. The high level RN committee is trained using the mutual validation matrix.
6. The final RN committee is tested on the test set.
7. This procedure is repeated 10 times for each benchmark dataset and each corresponding processor number, and the error rate results are averaged.
8. A single Regularization Neural Network is trained again on the same initial train set and tested on the same test set for comparison

In step 2 the uneven as well as random and without stratification choice of a particular processor's data is important for the experiment to simulate a real situation and to show the power of the method. To this end we allow a quarter of processors to randomly peek a population size between 5 and 300 train points. Likewise another quarter randomly peek a population size between 5 and 100. Similarly the remainder half of processors are allowed to have a size between 5 and 30. Then according to the total number of training points these population sizes are normalized, in favour of the smaller ones, for their sum to fit the total. This method produces a fairly uneven unstratified distribution, with half of processors populations being small. Many of them end up with no samples from some class. In addition small local populations are likely to produce singularities to the mutual validation matrix inversion, in order to make harder the proposed training method and show the benefits of the RN stabilizer. Other uneven and irregular distributions we try have worked as well as the former one.

On all tested datasets the RN committee outperforms majority voting. The Iris dataset has 150 examples, 4 input features and 3 classes. The Diabetes dataset has 768 examples, 8 input features and 2 classes. The Wisconsin breast cancer dataset has 683 examples, 9 input features and 2 classes. The Vehicle dataset has 846 examples, 18 features and 4 classes. The Glass dataset has 214 examples, 9 input features and 6 classes.The Wine dataset has 178 examples, 13 input features and 3 classes. Although the uneven un-stratified splitting produces highly irregular data distributions, the RN committee was found to perform better not only than majority voting but also slightly better than the single RN on datasets like the Iris, Wine and Wisconsin.

Table 1. Iris dataset results

RN modules	Majority Voting error	RN committee error	single RN error
10	5,2%	3,1%	3,3%
11	8,3%	2,7%	3,1%
12	6,3%	2,7%	2,7%
13	5,8%	2,5%	2,7%
14	4,2%	3,8%	2,5%
15	4,2%	3,5%	2,9%

In table 1, RN modules column indicates the number of RNs in the ensemble layer, and is used to show the range of the applicability. All error rates are measured by the ratio (falsely classified samples)/(total) on the test set. In the second column all module networks in the ensemble layer perform simple majority voting to produce error rate. The third column shows the proposed distributed privacy preserving RN committee error. The fourth column shows the single RN error when trained on the whole train set. The RN committee outperforms the majority voting and unexpectedly recovers the single RN error rate in most of the experimental cases. A marginally better performance of RN committee over the single RN is also present in some cases.

Table 2. Diabetes dataset results

RN modules	Majority Voting error	RN committee error	single RN error
50	30,2%	26,7%	25,9%
55	29,3%	25,4%	25,4%
60	29,7%	26,7%	24,6%
65	31,5%	25,9%	25,0%
70	32,3%	25,0%	24,1%
75	31,0%	26,7%	26,3%

In table 2, for the Diabetes dataset, the error rate of a single RN was found in last column to be about 25% on the same test set. The RN committee outperforms majority voting and manages to be as accurate as the single RN which was unexpected.

Table 3. Wisconsin dataset results

RN modules	Majority Voting error	RN committee error	single RN error
10	3,7%	3,5%	3,7%
20	3,6%	2,9%	3,2%
30	4,3%	3,6%	3,3%
40	4,4%	3,3%	3,5%
50	4,1%	3,5%	3,5%
60	4,9%	3,5%	3,8%

In table 3, for the Wisconsin dataset, the RN committee again performs better than majority voting and achieves the same error as the single RN, which also marginally overrun in some cases.

Table 4. Vehicle dataset results

RN modules	Majority Voting error	RN committee error	single RN error
5	27,3%	23,4%	21,5%
10	31,3%	25,4%	20,7%
15	33,2%	27,7%	21,1%
20	32,4%	27,3%	20,7%
25	33,6%	28,5%	20,7%
30	37,5%	30,1%	21,1%

In table 4, for the Vehicle dataset the single RN error rate in last column was found to be about 21%. The RN committee performs much better than majority voting and as expected in all cases the error produce where in between majority voting and the single RN error.

Table 5. Glass dataset results

RN modules	Majority Voting error	RN committee error	single RN error
10	38,8%	34,3%	32,8%
12	40,3%	38,8%	33,6%
14	38,8%	37,3%	32,1%
16	39,6%	36,6%	32,8%
18	43,3%	35,8%	34,3%
20	44,8%	37,3%	32,8%

In table 5, for the Glass dataset, again the RN committee performs much better than majority voting. As expected from the highly uneven and un-stratified data distribution across processors in all cases the RN committee error produce where in between majority voting and the single RN error.

Table 6. Wine dataset results

RN modules	Majority Voting error	RN committee error	single RN error
11	3,6%	2,7%	2,4%
12	2,5%	2,5%	2,5%
13	3,1%	3,1%	2,5%
14	3,8%	2,7%	2,9%
15	2,7%	1,8%	1,8%
16	3,6%	2,0%	2,2%
17	3,8%	2,7%	2,7%

In table 6, for the Wine dataset, again the RN committee error results are better than majority voting and once more are comparable to the single RN case. While we run the method 60 times for each dataset, more experiments are needed, and are therefore planed for future research in an extensive collection of benchmark datasets different in record size and feature complexity, together with another training method.

7   Conclusions and Future Work

For large scale distributed committee machines we consider the challenging case where no local data exchange is possible among the neural network classifiers. Regularization neural networks are used for both the classifiers as well as the combiner committee in an embedded architecture. After the RN committee training finished no RN module will know anything else except its own input local data vectors. The present study proposes a simple method to accomplish such a task. Using the distributed system a mutual validation matrix among them is computed asynchronously. The mapping is done based on classification rates between them. The train set of one becomes the validation set of the other. Then it is possible to exploit this mutual validation matrix to train another high level regularization network as a RN committee combiner for the individual RN modules. Experimental results were supportive, and the proposed privacy preserving RN committee outperforms the majority voting rule in all of the cases.

It must be noted here that as the mutual validation method improves accuracy, the gaining speed is also remarkable, producing a highly scalable system. The complexity of a single RN is about $O(N^3)$. For $N>1.000.000$, this algorithm is difficult to implement. It is possible for the RN committee machine presented here to assist in splitting the work without significant loss of accuracy. Training with fine-grained modules can be done by globally finding compact dense clusters of data. In the future we will try using a different mutual validation matrix for each class. Since these are asymmetric matrices and might have zeros in the diagonal we plan to resolve this issue by using regularized alternating least squares for the weights training.

Let a general loss function denoted as $V(f(x), y, u)$ where $f()$ the classifier, x the feature vector, y the label and u the parameters vector (weights etc.) of the classifier. The minimization of V with respect to u like in eq. 1 gives the solution of parameter vector u [4]. When two classifiers are compared versus a common separate test set, the comparison is made on their outputs, so their distance measure $d(i,j)$ is usually dependent on their pair of parameter vectors u_1 and u_2, meaning the outcome $d(i,j)$ is biased from their joint biases. The proposed mutual validation matrix method is independent of the joint parameter vectors u_1 and u_2. In our case an asymmetric measure $s(i,j)$ depends only on parameter vector u_1 of classifier i and thus is independent from the bias and variance of the classifier j. So it can be used as uncorrelated distance measure estimation for conventional ensemble training.

Unlike the on-line neural network training, or gradient descent methods, the training phase of a regularization network is always off-line, using kernel methods, and thus the stable solution is restricted to solving a linear system of the form $(K + \lambda I)w = y$. Thus with or without the privacy preserving constrain, the static training of a meta-learner RN committee which consists of any other type of classifier modules (SVM, RBF, MLP etc.) requires a coarse-grained high level kernel matrix. For example constrained regression training uses a covariance matrix usually computed from the average errors of classifiers to find module weights. In the future we plan to present extensive experiments that directly compare the proposed method performance with ensembles of RNs trained via bagging, constrained regression and stacking.

References

1. Tresp, V.: Committee Machines. In: Hu, Y.H., Hwang, J.N. (eds.) Handbook of Neural Network Signal Processing. CRC Press LLC (2002)
2. Drucker, H.: Fast Committee Machines for Regression and Classification. In: KDD 1997 Proceedings (1997)
3. Girosi, F., Jones, M., Poggio, T.: Regularization theory and neural networks architectures. Neural Computation 7, 219–269 (1995)
4. Evgeniou, T., Pontil, M., Poggio, T.: Regularization Networks and Support Vector Machines. Advances in Computational Mathematics 13, 1–50 (2000)
5. Poggio, T., Smale, S.: The mathematics of learning: Dealing with data. Notices of the American Mathematical Society 50(5), 537–544 (2003)
6. Bishop, C.M.: Neural Networks for Pattern Recognition. Oxford University Press, Oxford (1995)
7. Jain, A.K., Duin, R.P.W., Mao, J.: Statistical Pattern Recognition: A Review. IEEE Transactions on Pattern Analysis and Machine Intelligence 22(1), 4–37 (2000)
8. Wang, L., Fu, X.: Data Mining with Computational Intelligence. Springer (2005)
9. Bottou, L., Chapelle, O., DeCoste, D., Weston, J.: Large Scale Kernel Machines. Neural Information Processing Series. MIT Press, Cambridge (2007)
10. Kashima, H., Ide, T., Kato, T., Sugiyama, M.: Recent Advances and Trends in Large-scale Kernel Methods. IEICE Transactions on Information and systems E92-D (7), 1338–1353 (2009)
11. Prodromidis, A., Chan, P.: Meta-learning in a distributed data mining system: Issues and approaches. In: Proceedings of the Fourteenth International Conference on Machine Learning, pp. 211–218 (1998)
12. Clifton, C., Kantarcioglu, M., Vaidya, J., Lin, X., Zhu, M.: Tools for Privacy Preserving Distributed Data Mining. ACM SIGKDD Explorations 4(2), 1–7 (2003)
13. Kantarcioglu, M., Vaidya, J.: Privacy-Preserving Naive Bayes Classifier for Horizontally Partitioned Data. In: IEEE Workshop on Privacy-Preserving Data Mining (2003)
14. Yi, X., Zhang, Y.: Privacy-preserving naïve Bayes classification on distributed data via semi-trusted mixers. Information Systems 34(3), 371–380 (2009)
15. Yu, H., Jiang, X., Vaidya, J.: Privacy-Preserving SVM using nonlinear Kernels on Horizontally Partitioned Data. In: SAC Conference (2006)
16. Xiong, L., Chitti, S., Liu, L.: k nearest neighbour classification across multiple private databases. In: Proceedings of the ACM Fifteenth Conference on Information and Knowledge Management, November 5-11 (2006)
17. Hansen, L.K., Salamon, P.: Neural network ensembles. IEEE Transactions on Pattern Analysis and Machine Intelligence 12, 993–1001 (1990)
18. Perrone, M.P., Cooper, L.N.: When networks disagree: ensemble method for neural networks. In: Mammone, R.J. (ed.) Neural Networks for Speech and Image Processing. Chapman & Hall, Boca Raton (1993)
19. Krogh, A., Vedelsby, J.: Neural networks ensembles, cross validation and active learning. In: Advances in Neural Information Processing Systems 7. MIT Press, Cambridge (1995)
20. Wilson, G.: Parallel Programming for Scientists and Engineers. MIT Press, Cambridge (1995)

Improved POS-Tagging
for Arabic by Combining Diverse Taggers

Maytham Alabbas and Allan Ramsay

School of Computer Science,
Manchester University,
Manchester, UK
{alabbasm,ramsay}@cs.man.ac.uk

Abstract. A number of POS-taggers for Arabic have been presented in
the literature. These taggers are not in general 100% accurate, and any
errors in tagging are likely to lead to errors in the next step of natural
language processing. The current work shows an investigation of how
the best taggers available today can be improved by combining them.
Experimental results show that a very simple approach to combining
taggers can lead to significant improvements over the best individual
tagger.

Keywords: Combining systems, Arabic tagging.

1 Introduction

The process of assigning a correct POS tag (i.e. noun, verb, adverb or others) to
each word of a sentence is called part-of-speech(POS) tagging. This process is
considered an essential step for most natural language applications. In general,
however, POS-taggers make mistakes, and since tagging is the first step in most
natural language processing (NLP) systems these mistakes will lead to problems
in all subsequent stages of analysis. It is thus important to obtain the highest possible accuracy at this initial stage of processing. One popular technique
for improving tagging accuracy is *tagger combination*. This approach involves
combining different taggers to exploit the unique properties of each tagger and
reduce some of the random errors. This technique has been applied for different
languages such as English [1], Swedish [2],Telugu [3], Italian [4] and Polish [5]
and the results were encouraging, but has not to our knowledge been applied for
Arabic. The current work is step forward in this regard. We evaluate different
techniques for combining POS-taggers for Arabic.

The key problem for Arabic is that it is more ambiguous than many other
languages (e.g. English) for many reasons. Firstly, Arabic is written without
diacritics (short vowels and a range of other phonological effects), often leading
to multiple ambiguities [6]. This is particularly problematic because the diacritics
are often the only difference between different words (especially derived forms)
and between inflected forms of the same word. This matter makes analysis of
the language morphologically very challenging. This is because a certain lemma

L. Iliadis et al. (Eds.): AIAI 2012, IFIP AICT 381, pp. 107–116, 2012.

(or lexeme) in Arabic can be interpreted in various ways. Hence, a single word can have various senses, where determining the sense is based on the context in which the word is used. Furthermore, a noun in Arabic can be diacritised in three different ways for the nominative, accusative and genitive cases, which can be even more ambiguous at the structural or grammatical level.

In addition, Arabic is highly syntactically flexible [7]:

- It has a comparatively free word order, where sentence components can be exchanged without affecting the core meaning. This results in structural ambiguity, with each morphological analysis having more than a single meaning. So, besides the regular sentence of verb, subject and object (VSO), Arabic allows other potential surface forms such as SVO and VOS constructions. The potential of allowing variations on the canonical order leads to a large amount of ambiguity.
- Furthermore, Arabic (like Spanish, Italian and Japanese) is a pro-drop language [8]. The pro-drop can lead to structural ambiguity by leaving any syntactic parser with the challenge to determine if there is a dropped pronoun or not in the subject position, which is made worse by the fact that lots of Arabic verbs can have both transitive and intransitive forms, and further that it is generally impossible to tell the difference between active and passive forms by inspecting the surface form. In case that just one noun phrase (NP) follows one of these verbs, the ambiguity appears. In contrast with English because its canonical order is SVO, losing the subject (S) does not cause confusion about the object (O).
- In addition the copula is omitted in simple positive equational sentences (the sentences that did not contain verb), so that a sequence of a noun and a predicative item (i.e. another noun, an adjective or a prepositional phrase (PP)) may make a sentence.
- Finally, Arabic nouns can be linked together without any overt marker, whereas two English nouns are joined together by different markers, such as the suffix "-'s" on possessing noun, a possessive phrase "of" or a pronoun as "his/her". In the construct phrase the first noun must be indefinite solely (which can be in any case: nominative, genitive or accusative). The second noun may be either definite or indefinite (which is always in genitive case).

While these are strictly speaking syntactic issues, they have serious consequences for tagging, since they mean that the local context often fails to supply constraints on tags–to take a simple example, if you do not know whether the item following a verb is its subject or its object then you cannot use the proximity of the verb to determine the case marker.

2   The Taggers

We are interested in improving tagging by combine different taggers. We have carried out a number of experiments with state-of-the-art taggers (AMIRA 2.0 [9], MADA 3.1 [10] and a home-grown tagger, MXL [11], with comparable accuracy), using the Penn Arabic Treebank (PATB) [12] as a *gold-standard* corpus.

Gold-Standard Tags

We used PATB Part 1 v3.0 as a resource for our experiments. The words in the PATB are already tagged. This provides us with a benchmark for tagger evaluation. Even PATB tagging is not guaranteed to be 100% accurate, but it nonetheless provides as good a reference set as can be found.

The PATB uses a very fine-grained set of tags, which carry a great deal of syntactically relevant information (particularly case-marking). This tagset contains 305 tags, with for instance 47 tags for different kinds of verb and 44 for different kinds of noun. We carried out our experiments with a variant of this original tagset, and also with a coarser-grained version obtained by deleting details such as case- and agreement-markers.

AMIRA

The first tagger we use is AMIRA. This tagger is reported to achieve around 97% accuracy on its target tagset, which is about as good as any reported system.

Using AMIRA, however, highlights one of the problems that arise when we try to compare its retagged corpus with other corpora. The PATB is tagged, but with different tags from the ones used by AMIRA.[1] In order to compare AMIRA tags with other corpora tags, we will have to translate between the two tagsets.

This is a non-trivial task. The two tagsets have different numbers of tags (e.g. AMIRA has 130 fine-grained tags compared with 305 in PATB), and more importantly they make different kinds of distinctions. The AMIRA tagset, for instance, uses one tag (RP) to cover a range of particles which are subdivided into eight subclasses (EMPH_PART, EXCEPT_PART, FOCUS_PART, INTERROG_PART, RC_PART, NEG_PART, PART, VERB_PART) in the PATB; and it uses several tags to describe different kinds of verbs (VB, VBG, VBD, VBN, VBP) where the PATB just uses three (IV, PV, CV).

In order to overcome these problems, we use transformation-based retagging (TBR) [13,14] to recover from the mismatches between the two tagsets. TBR collects statistics about the local context in which erroneous tags have been assigned, and attempts to find rules based on this information to apply *after* the original tagger has been run. This technique will produce a small improvement in the performance of almost any tagger. Typically, taggers that achieve scores in the mid 90s are boosted by 2-3%–the lower the original accuracy, the greater the typical improvement. When we used it for comparing the original tags produced by AMIRA and the gold tags the score improved from around 89% to just over 95%. Some of this improvement arises simply from rules that spot that the two tagsets use different names for the same things (e.g. that things that are called JJ are called ADJ in the PATB) but some of it comes from learning how to split coarse-grained AMIRA tags into fine-grained PATB ones.

---

[1] We used the Extended Reduced Tag Set (ERTS) setting for AMIRA and then removed inflectional markers. This produced a set of tags that is very similar to the 25 tags in the Bies/RTS tagset, but with distinctions between nouns, adjectives and cardinal numbers.

There is a further problem with using AMIRA. The fact that Arabic allows a range of items to be cliticised (conjunctions, prepositions, pronouns) makes it difficult even to tokenise text reliably. This means that not only does AMIRA sometimes assign different tags from the PATB, it sometimes even splits the text into different numbers of tokens (i.e. AMIRA's tokeniser segments the text differently from the way that it is segmented in the PATB–AMIRA's tokeniser gives us around 97% agreement with gold-standard tokenisation).

In order to see whether this was the cause of the difference, we constructed a version of the corpus where we replaced PATB tags by a coarse copy of the AMIRA tags, using hand-coded substitution rules, and then replaced these by fine-grained AMIRA tags where the substituted tags were compatible with tags actually assigned by AMIRA. Thus if the PATB assigned a word the tag `ADJ` we replaced it by the AMIRA tag `JJ`. We then inspected the tags assigned by AMIRA itself: if the tag assigned to this word was one of AMIRA's fine-grained adjective tags, e.g. `JJR`, then we used this instead. If, on the other hand, the hand-coded replacement for the PATB tag was incompatible with the one assigned by AMIRA then we retained the hand-coded one. This gave us a version of the treebank that had the same number of tokens as the original PATB, with as many items as possible given the tags assigned by AMIRA and the others given hand-coded AMIRA equivalents of the original PATB tags.

MADA

MADA [15] uses a slightly extended version of the PATB Part 1 v3.0 tagset, with some extra classification of nouns. The fine-grained MADA retagged version contains 352 tags compared to 305 tags in the PATB corpus. Fortunately the MADA tags are a strict superset of the standard PATB set, and hence can be reduced to either the standard fine-grained version or our coarse version by omitting the extra classification of nouns, so we do not have the same problems using MADA with the PATB as we have with AMIRA.

We also applied TBR to the output of MADA, because although we were not faced with mapping incompatible tagsets in the same way as with AMIRA, using TBR nearly always provides a small improvement, amounting in this case to an increase from 94% to 96.7%. We applied the same technique that we used in AMIRA to get a MADA version of the corpus that has the same size as the gold-standard one because MADA's tokeniser also has small (1%) variance from the gold-standard.

Maximum-Likelihood Tagger

We also use an in-house maximum-likelihood tagger, which we will refer to as MXL. We have described this tagger in detail in [11]: we will simply outline the basic principles that it is based on and note its accuracy here.

MXL operates in two stages, as follows:

– In the first stage we use two simple kinds of statistic: (i) the conditional likelihood that a word which starts with the same three letters or ends with the same three letters as the one we are trying to tag has a given tag, and (ii) the transition probabilities between tags. We use a weighted combination of these to produce a maximum-likelihood guess at the current tag. This process produces about 95.2% accuracy.

– We then use TBR to refine the original set of hypotheses, leading to a final accuracy of 95.6%.

The advantage of this tagger is that because it was trained on the PATB, the tags it uses are exactly the PATB tags and the tokenisation is exactly the PATB tokenisation. We therefore do not need to overcome problems associated with mismatches between tag sets.

3   Improving POS-Tagging Accuracy

In the current work, we are interested in both fine-grained and coarse-grained tagsets. We therefore collapsed the original fine-grained set, by deleting inflectional markers (case, number, gender), to the coarse-grained set show in Table 1 (e.g. PATB has 305 fine-grained tags which become 39 coarse-grained tags, for instance, the fine-grained tags NOUN+CASE_DEF_ACC, NOUN+CASE_DEF_GEN, NOUN+CASE _DEF_GEN+POSS _PRON_3MS, NOUN+CASE_DEF_NOM are grouped to NOUN).

Table 1. Coarse-grained tagset

ABBREV	DET+NOUN_PROP	LATIN	PUNC
ADJ	DET+NUM	NEG_PART	PV
ADV	EMPH_PART	NOUN	PVSUFF_DO
CONJ	EXCEPT_PART	NOUN_PROP	RC_PART
CV	FOCUS_PART	NO_FUNC	REL_ADV
CVSUFF_DO	FUT+IV	NUM	REL_PRON
DEM_PRON	INTERJ	PART	SUB
DET	INTERROG_PART	POSS_PRON	SUB_CONJ
DET+ADJ	IV	PREP	VERB_PART
DET+NOUN	IVSUFF_DO	PRON	

Table 2 summarises the fine-grained and coarse-grained tags for PATB corpus and the three retagged corpora by the taggers: AMIRA, MADA and MXL. Table 3 summarises the accuracy of the three taggers on our gold standard set using the built-in tagsets and coarse versions of each, and shows the improvements that are obtained in each case by applying TBR. This provides a reference point: the best of the three taggers is MADA, which achieves 96.7% on the coarse-grained version of its built-in tagset if we also apply TBR and 93.6% on the fine-grained version, again after applying TBR. The goal of the current paper is to see whether we can improve on this by utilising the other taggers, despite the fact that their individual performance is worse.

Table 2. Coarse-grained and Fine-grained tags numbers, single tagger

POS	PATB/MXL	AMIRA	MADA
Coarse-grained	39	29	56
Fine-grained	305	130	351

Table 3. Tagger accuracies in isolation, with and without TBR

POS	TBR	AMIRA	MXL	MADA
Coarse-grained	×	89.6%	95.2%	94.1%
	√	95.3%	95.6%	**96.7%**
Fine-grained	×	84.3%	89.7%	91.7%
	√	88.8%	91.2%	**93.6%**

The first observation is that when the taggers agree on how to tag a given item they are more likely to be right than when they disagree. This is fairly obvious–if you have a set of taggers which assign different tags to an item then at least one of them must be wrong! Table 4 substantiates this observation–each column shows the precision (P) and recall (R) for a particular pair of taggers simply taking cases where they agree and leaving words on which they disagree untagged. Thus the combination of MXL and MADA achieves a precision of 99.5% on the coarse-grained tagset and 99% on the fine-grained one. Table 5 shows what happens when we combine all three taggers, either taking the majority view when at least two of them agree or demanding that all three agree. In the latter case we obtain a precision of 99.9% for the coarse-grained tagset and 99.2% for the fine-grained one.

Table 4. Precision (P) and Recall (R) and F-score for combinations of pairs of taggers, with and without TBR

Metrics	POS	TBR	MXL-AMIRA	MXL-MADA	MADA-AMIRA
P	Coarse-grained	×	97.7%	99.3%	96.3%
		√	98.8%	**99.5%**	**99.5%**
	Fine-grained	×	92.9%	98.9%	94.7%
		√	94.3%	**99%**	95.8%
R	Coarse-grained	×	84.3%	90.1%	86.1%
		√	94.1%	94.6%	91.9%
	Fine-grained	×	79.8%	83.3%	84.9%
		√	81.6%	86.4%	85.6%
F-SCORE	Coarse-grained	×	0.905	0.945	0.909
		√	0.964	0.97	0.955
	Fine-grained	×	0.858	0.904	0.895
		√	0.875	0.923	0.904

Table 5. Precision (P) and Recall (R) and F-score for combinations of three taggers, with and without TBR

Condition	POS	TBR	P	R	F-SCORE
At least two agree	Coarse-grained	×	95.1%	93.5%	0.943
		√	98.4%	97.8%	0.981
	Fine-grained	×	90.7%	87.8%	0.892
		√	92.8%	90.5%	0.916
All three agree	Coarse-grained	×	99.5%	83%	0.905
		√	**99.9%**	90.9%	0.952
	Fine-grained	×	99.1%	77.8%	0.871
		√	**99.2%**	79.8%	0.885

The cost, of course, is that the recall goes down, because there are places where they disagree, and in these cases no tag is assigned. What should we do in such cases?

3.1 Backoff Strategies

Our first proposal was to take the majority view when at least two of the taggers agreed, and to backoff to one or other when they there was no common view [16]. The results of this are shown in Table 6 where column one shows the result of backing off to MXL when there is no majority view and column two shows the result of backing off to MADA. The results for the coarse-grained tagset are markedly better than for any of the taggers individually, though there is only a very slight improvement over the original MADA scores for the fine-grained set. Interestingly, the best results are obtained by backing off to MXL rather than to MADA (98.2% vs. 97.9%, despite the fact that MADAs individual performance is better than MADA's (96.7% vs. 95.6%) (the results in this table and in Table 8 are for accuracy rather than precision or recall).

Table 6. Backoff to MXL or MADA when there is no agreement

POS	TBR	Otherwise, MXL tag	Otherwise, MADA tag
Coarse-grained	×	94.9%	93.8%
	√	**98.2%**	97.9%
Fine-grained	×	89.6%	88.8%
	√	91.8%	91.5%

The fact that backing-off to either MADA or MXL improved the performance suggested that it was worth investigating other backoff strategies. If majority vote + backoff to an arbitrary tagger is better than any single tagger in isolation, then perhaps there is something we can backoff to which will do even better.

We return to the observation that in cases where all three taggers disagree, at most one of them can be right. Given that they each employ different information about the material being tagged, it is likely that they are systematically prone to different kinds of errors.

We therefore collected statistics about the *kinds* of things they each get right. How likely, for instance, is MXL to be right when it assigns the tag NEG_PART? Table 7 shows an extract of this data, showing the likelihood that each tagger is right for a given assigned tag, e.g. for the given instance of the word غير *gyr*[2] "other than" the correct tag was NEG_PART (as in gold standard): MADA suggested NOUN, MXL suggested NEG_PART, AMIRA suggested RP. Because MXL is right 98.2% of the time when it suggests NEG_PART, whereas MADA is right 97.9% of the time when it suggests NOUN and AMIRA is right only 8.1% of the time when it suggests RP, NEG_PART was chosen. For إلى <*lA* "to", MADA's suggestion of EXCEPT_PART was accepted because MADA is right 100% of the time when it suggests EXCEPT_PART, which is better than the reliability of either of the other suggestions.

Table 7. Confidence levels for individual tags

Word	Gold standard	MADA	MXL	AMIRA	TAG
gyr	NEG_PART	NOUN (97.9)	NEG_PART (98.2)	RP (8.1)	NEG_PART
<*lA*	EXCEPT_PART	EXCEPT_PART (100.0)	SUB_CONJ (96.5)	RP (7.9)	EXCEPT_PART
...	...	...	...	...	...

Using this strategy for choosing what to do when all three taggers make different suggestions produces the results in Table 8.[3] The 'default' column reports the results when we simply chose the most confident proposal, whereas for 'backoff unless two agree' we took the majority verdict if two of the taggers agreed and the most confident if all three gave different results.

Table 8. Backoff to most confident tagger

POS	TBR	default	backoff unless two agree
Coarse-grained	×	97.3%	95.7%
	√	**99.5%**	99.2%
Fine-grained	×	95.6%	93.2%
	√	**96.0%**	94.5%

The results in Table 8 are both surprising and compelling. Simply taking the most confident of the three taggers produces 99.5% accuracy for the coarse-grained set, which is going to be hard to beat by much, and even for the fine-grained set it produces 96%. This improves over taking the majority verdict

---

[2] The transcription of Arabic examples follows Buckwalter's system for transcribing Arabic symbols. Available at: http://www.qamus.org/transliteration.htm.

[3] These results were obtained by 10-fold cross validation.

when at least two of the contributing taggers agree and backing off to the most confident one where there is no agreement,[4] and it beats each of the individual taggers by a fairly wide margin–the original error by MADA of 3.3% for the coarse-grained set has been reduced to 0.5%, a nearly sevenfold reduction.

4 Conclusions

We have shown a rather simple mechanism for combining taggers which can provide considerable improvements in accuracy. If you measure the likelihood that a tagger is right when it suggests a particular tag, and then take the suggestion with highest score in each case then you can decrease the error rate substantially. The key is that the different taggers tend to make different systematic mistakes. The accuracy statistics capture these systematic mistakes, so a low score is likely to reflect a case where the tagger is making one of its characteristic errors, and in such cases we take the output of one of the others. This simple approach outperforms strategies involving more subtle ways of combining the individual taggers, e.g. by taking the majority preference in cases where one exists. This is likely to be because two of the taggers (MADA and AMIRA) use very similar information, and hence where they make systematic mistakes they are likely to make the *same* systematic mistakes. In such cases they will tend to agree, and hence would outvote MXL if allowed the majority view to win, as suggested by the precision results in Table 5. The simple mechanism we have used provides a built-in resilience against this tendency.

In every case, including TBR makes a useful contribution. Again, TBR is most effective when the base tagger makes systematic errors. The version of TBR that we are using includes extra templates looking at the first three and last three letters of words in addition to the standard word-based templates. These extra templates pay attention to prefixes and suffixes, which carry much more information than is the case for English (where TBR has been most extensively applied).

The bottom line is that for the tagset in Table 1 we can obtain 99.5% accuracy when tagging freely occurring Arabic. It is going to be hard to improve substantially on this score. Given that Marton et al. [17] have argued that coarse-grained tagsets are actually more useful than fine-grained ones for parsing, which is the usual next step in the chain, we are fairly satisfied with this result.

Acknowledgments. We would like to thank Dr. Yasser Sabtan (Al-Azhar University, Eygpt) for important suggestions and for helpful discussions. Maytham Alabbas owes his deepest gratitude to Iraqi Ministry of Higher Education and Scientific Research for financial support in his PhD study. Allan Ramsay's contribution to this work was supported in part by Qatar National Research Foundation grant NPRP 09 - 046 - 6 - 001.

---

[4] Note that taking the majority verdict when all three agree and backing off to the most confident when there is not complete unanimity is **exactly** the same as simply taking the most confident one from the outset.

References

1. Brill, E., Wu, J.: Classifier combination for improved lexical disambiguation. In: Proceedings of the 17th International Conference on Computational Linguistics, vol. 1, pp. 191–195. Association for Computational Linguistics (1998)
2. Sjöbergh, J.: Combining pos-taggers for improved accuracy on swedish text. In: Proceedings of NoDaLiDa 2003 (2003)
3. RamaSree, R., Kusuma Kumari, P.: Combining pos taggers for improved accuracy to create telugu annotated texts for information retrieval. Dept. of Telugu Studies, Tirupathi, India (2007)
4. Søgaard, A.: Ensemble-based pos tagging of italian. In: IAAI-EVALITA, Reggio Emilia, Italy (2009)
5. Śniatowski, T., Piasecki, M.: Combining Polish Morphosyntactic Taggers. In: Bouvry, P., Kłopotek, M.A., Leprévost, F., Marciniak, M., Mykowiecka, A., Rybiński, H. (eds.) SIIS 2011. LNCS, vol. 7053, pp. 359–369. Springer, Heidelberg (2012)
6. Nelken, R., Shieber, S.: Arabic diacritization using weighted finite-state transducers. In: Proceedings of the Workshop on Computational Approaches to Semitic Languages at 43rd Meeting of the Association for Computational Linguistics (ACL 2005), pp. 79–86 (2005)
7. Daimi, K.: Identifying syntactic ambiguities in single-parse arabic sentence. Computers and the Humanities 35(3), 333–349 (2001)
8. Farghaly, A.: Subject pronoun deletion rule. In: Proceedings of the 2nd English Language Symposium on Discourse Analysis (LSDA 1982), pp. 110–117 (1982)
9. Diab, M.: Second Generation Tools (AMIRA 2.0): Fast and Robust Tokenization, POS Tagging, and Base Phrase Chunking. In: Proceedings of the Second International Conference on Arabic Language Resources and Tools, Cairo, Eygpt, The MEDAR Consortium, pp. 285–288 (April 2009)
10. Habash, N.: Introduction to Arabic Natural Language Processing. Synthesis Lectures on Human Language Technologies. Morgan & Claypool Publishers (2010)
11. Ramsay, A., Sabtan, Y.: Bootstrapping a lexicon-free tagger for arabic. In: Proceedings of the 9th Conference on Language Engineering ESOLEC 2009, Cairo, Egypt, pp. 202–215 (December 2009)
12. Maamouri, M., Bies, A.: Developing an Arabic treebank: Methods, guidelines, procedures, and tools. In: Proceedings of COLING, pp. 2–9 (2004)
13. Brill, E.: Transformation-based error-driven learning and natural language processing: a case study in part of speech tagging. Computational Linguistics 23(4), 543–565 (1995)
14. Lager, T.: μ-tbl lite: a small, extendible transformation-based learner. In: Proceedings of the 9th European Conference on Computational Linguistics (EACL 1999), pp. 279–280. Association for Computational Linguistics, Bergen (1999)
15. Habash, N., Rambow, O., Roth, R.: MADA+TOKAN: A Toolkit for Arabic Tokenization, Diacritization, Morphological Disambiguation, POS Tagging, Stemming and Lemmatization. In: Proceedings of the Second International Conference on Arabic Language Resources and Tools, Cairo, The MEDAR Consortium (2009)
16. Zeman, D., Žabokrtský, Z.: Improving parsing accuracy by combining diverse dependency parsers. In: Proceedings of the Ninth International Workshop on Parsing Technology, pp. 171–178. Association for Computational Linguistics (2005)
17. Marton, Y., Habash, N., Rambow, O.: Improving arabic dependency parsing with lexical and inflectional morphological features. In: Proceedings of the NAACL HLT 2010 First Workshop on Statistical Parsing of Morphologically-Rich Languages, pp. 13–21. Association for Computational Linguistics (2010)

Multithreaded Implementation of the Slope One Algorithm for Collaborative Filtering

Efthalia Karydi and Konstantinos G. Margaritis

University of Macedonia, Department of Applied Informatics
Parallel and Distributed Processing Laboratory
156 Egnatia str., P.O. Box 1591, 54006 Thessaloniki, Greece
Karydithalia@gmail.com, kmarg@uom.gr

Abstract. Recommender systems are mechanisms that filter information and predict a user's preference to an item. Parallel implementations of recommender systems improve scalability issues and can be applied to internet-based companies having considerable impact on their profits. This paper implements a parallel version of the collaborative filtering algorithm Slope One, which has advantages such as its efficiency and the ability to update data dynamically. The presented version is parallelly implemented with the use of the OpenMP API and its performance is evaluated on a multi-core system.

Keywords: Collaborative Filtering, Recommender Systems, Slope One Algorithm, Parallel Programming, OpenMP.

1 Introduction

Collaborative filtering based recommender systems introduce the users' opinion to the procedure of the recommendations generation. Collaborative filtering recommender systems have gained wide popularity. Thus, as the number of users and items of such systems increases, so inevitably does the amount of data. One of the most challenging factors in recommender systems, which is caused due to the data abundance, is the way to achieve high quality recommendations in the shortest time possible. Consequently, a great need is emerging. The achievement of quick data processing, in order to accomplish high quality recommender systems of an increased performance.

In this paper the Slope One algorithm [1] was chosen for parallelization due to the presence of advantages such as its speed and efficacy and the dynamically updatable data. A parallel implementation is introduced using OpenMP and the experimental results are evaluated.

The rest of this paper is organized as follows: In section 2 related work is discussed. Section 3 presents an overview of the different versions of the Slope One algorithm. Section 4 presents the proposed parallel implementation of the Slope One algorithm. The experimental results are analyzed in section 5.

L. Iliadis et al. (Eds.): AIAI 2012, IFIP AICT 381, pp. 117–125, 2012.

2   Related Work

Recently, the main versions of Slope One are being used together with other algorithms, such as data mining techniques, in order to accomplish faster and more effective recommender systems.

D. Zhang in [2] presented in 2009 a method that used the Slope One algorithm to produce the predictions of a user, and continued by using the Pearson Correlation metric to calculate the neighborhood of similar items and produce recommendations. The quality of the predictions was affected by the size of the set of similar items. This variation was more accurate than the traditional Collaborative Filtering algorithm.

Another method first divides the set of items into subsets, taking into account the kind of items requested by the user. In this manner, the dimension of the set of items, the number of ratings and some times even the number of users, is being reduced. The co-clustering of the data is accomplished by using the K-Means algorithm, taking as parameters the demographic data, that the user must have previously determined. After the dimensionality reduction, Slope One is being used to the smaller dataset to produce the predictions. This approach reduces the time needed for calculations and augments the predictions' accuracy [3].

Between other approaches that use the Slope One algorithm is the one presented in [4], which combines Slope One with Userrank. Userrank is based on Pagerank algorithm and attaches weights to each user, depending on how many related items he has rated. These weights are being used to calculate the differences between the items' ratings. Another algorithm that combines item based with user based collaborative filtering, was proposed in 2009. This algorithm uses Slope One to fill in the empty spaces of the array containing the ratings and on the new dense array applies user based collaborative filtering techniques [5]. Recently, another recommender system based on Slope One has been designed [6]. This approach selects Slope One for being more efficient in the item similarity calculation than other item based algorithms.

The need to accomplish fast real-time recommender systems that are able to handle enormous data, has led the research trends to the study and implementation of parallel collaborative filtering algorithms. [11] presents a parallel algorithm for collaborative filtering, whose purpose is to be scalable to very large datasets. The Alternating Least Squares with Weighted λ Regularization algorithm is implemented using parallel Matlab. [12] presents a parallel collaborative filtering algorithm based on the Concept Decomposition technique, which uses Posix Threads NPTL API on 32 cores and takes 3,2 μs to compute a prediction on Netflix dataset [8].

A distributed algorithm based on co-clustering, which is implemented with MPI and OpenMP, is presented in [13]. The Netflix Prize dataset is used on a 1024-node Blue Gene/P architecture and achieve training time of only 6 seconds and 1,78 μs prediction time per rating. Other parallel co-clustering algorithms exist, as the one described in [14], which simultaneously creates user and item neighborhoods by dividing among the processors the rows and colums of the matrix averages, and the dataflow implementation of a parallel co-clustering algorithm, presented in [15], which uses the Netflix dataset and achieves prediction runtime of 9,7 μs per rating.

Most recent attempts in the field of parallel collaborative filtering algorithms embrace the use of frameworks. To attain better scalability, frameworks such as Hadoop [9] and GraphLab [10] are extensively used. In [16] is implemented a user-based collaborative filtering algorithm on Hadoop, using the Netflix dataset on 9 dual-core processors. Item-based collaborative filtering algorithm is implemented using Hadoop in[17]. This approach seperates the three most excessive computations into four Map-Reduce phases, which are executed in parallel on a three node Hadoop cluster. An open source collaborative filtering library is implemented in [18], using the GraphLab parallel machine learning framework, and two approaches of SGD on Hadoop are presented in [19] and [20].

[21] implements the Weighted Slope One algorithm using Hadoop. This approach clusters users and assigns weights to each cluster. Then, the ratings are predicted using Weighted Slope One. Lately, the research community has drawn further attention to the Slope One algorithm and many approaches have been published [22,23,24,25].

3   Background and Notation

In this section is given an overview of the Slope One algorithm. Slope One defines in a pairwise mode, how much better is one item prefered than another by calculating the difference between the items' ratings. One of the main characteristics of the algorithm is that only the ratings of users who have evaluated some common items with the user for whom the prediction is being produced and this user's predictions are introduced in the predictions calculation.

Given a set χ, consisting of all the evaluations in the training set, and two items i and j with ratings u_i and u_j respectively, in a user's u evaluation ($u \in S_{j,i}(\chi)$), the average deviation of u_i regarding u_j is given by

$$dev_{j,i} = \sum_{i \in S_{j,i}(\chi)} \frac{u_j - u_i}{card(S_{j,i}(\chi))} \cdot \tag{1}$$

The average deviation of the items is used for the prediction of the rating that the user u would give to item j,

$$pred(u,j) = \bar{u} + \frac{1}{card(R_j)} \sum_{i \in R_j} dev_{j,i} \,, \tag{2}$$

where $R_j = \{i | j \in S(u), i \neq j, card(S_{j,i}(\chi)) > 0\}$ is the set of all relevant items and $card(S_{j,i}(\chi))$ is the number of all the evaluations in the set S that contain ratings for both items i and j.

Two additional versions of the Slope One algorithm exist. The Weighted Slope One, in which the number of observed ratings for each item, $c_{j,i} = card(S_{j,i}(\chi))$, is taken into account and the predictions are calculated according to

$$pred(u,j) = \frac{\sum_{i \in S(u)-\{j\}} (dev_{j,i} + u_i)c_{j,i}}{\sum_{i \in S(u)-\{j\}} c_{j,i}}. \tag{3}$$

In the Weighted Slope One version, if a pair of items has been rated by more users than another, this fact affect the predictions.

Another version is the Bi-Polar Slope One, which predicts only if an item will be liked by a user or not. Thus, a two value scale is used instead of a multivalued, which was used in the basic Slope One scheme to predict the exact rating a user would give to an item.

4   Multithreaded Implementation of Slope One

Through the constant increase of data, a need for better processing speed acquisition is emerging. To acomplish better speed, a parallel version of Slope One is implemented using OpenMP, which is described in this section.

The presented version can be applied to shared memory systems. In order for the predictions to be produced, the values of four arrays have to be calculated. The values of the array that contains the ratings that each user has inserted to the system, the values of the arrays which contain the differences and the frequences of the items' appearance in pairs, and finally the values of the deviation matrix. Since these calculations are the most time-consuming part of the code, they are computed in parallel, using the Data Parallel Model. According to this model data is being shared to the threads, and the computations are shared between all threads. The Task Parallel Model has been deliberately avoided, because the calculations needed to the formation of some of the above matrices involve the use of the rest of the matrices. Thus, the use of Task Parallel Model would delay the overall performance. After these calculations, the program calls two functions. One function computes the predictions and the other, the weighted predictions. A parallel region is defined in each of these functions, and each thread produces the predictions for the items that have not been rated.

Pseudocode.

```
Main procedure
main()
{
    1. Initialize OpenMP routines;
    2. All threads compute ratings, differences, frequencies
       and deviation matrices;
    3. For i=0 to users x items, if (ratings[i]==0) then call
       predictions and weighted predictions function;
}
Predictions Function
Predictions()
{
    1. Calculate prediction of a given user's rating for a given
```

```
item;
    2. Return prediction;
}
Weighted Predictions Function
Weighted predictions()
{
    1. Calculate weighted prediction of a given user's rating for a
    given item;
    2. Return weighted prediction;
}
```

5   Results and Analysis

5.1   Experimental Methodology

The MovieLens dataset, available from GroupLens Research [7], was used for the performance and scalability evaluation of the implementation discused above. MovieLens 100k was used for performance evaluation, and MovieLens 1M was divided into sub-datasets, augmenting the number of users in each one of them and was used for scalability analysis.

The results were compared to those of the sequential algorithm. To achieve this, the OMP_NUM_THREADS environment variable was used to the multithreaded implementation.

The experiments were performed on a system consisted of two CPU's, AMD opteron(tm) Processor 6128 HE, with eight cores each, 800MHz clock speed and 16GB RAM, under Ubuntu Linux 10.04 operating system. The OpenMp version 3.0 was used and time was measured by its omp_get_wtime() function.

5.2   Performance Analysis

As can be seen by the figure 1 the total execution time reduces as the number of used threads increases. With the use of 16 threads, the total execution time is reduced by 9 times over the sequential time. The total execution time measured, refers to the computation time and to both predictions' and weighted predictions' time and their storage to text files. Thus, the total time needed for the predictions' production is less than the total execution time measured in this implementation, because only one of the predictions' functions would be necessary in a recommender system.

The reduction of the total execution time for the MovieLens 100k dataset, in relation to the number of threads used, can be seen in figure 1. About 134 seconds are needed for the sequential execution, and 16 seconds for the multithreaded. Thus, the multithreaded implementation taking advantage of only 16 threads is 8 times faster than the sequential.

Both preprocess time and the time needed to calculate the predictions are reduced. Using 16 threads, the preprocess time in all datasets is reduced in

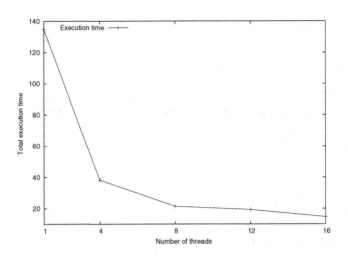

Fig. 1. Total execution time

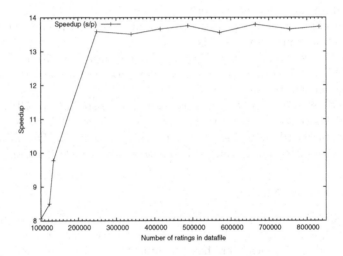

Fig. 2. Speedup on different datasets

a range from 13.9 to 15.33 times and predictions are generated faster. Both predictions' and weighted predictions' calculations are about 14.5 times faster in the multithreaded implementation. Regarding the number of predictions and weighted predictions produced per second, only 2.2 μs are needed per rating and 1.31 μs per rating using weights.

In figure 2 can be seen the ratio of the sequential implementation's total execution time to the parallel implementation's total execution time on the different datasets. As the dataset size increases, the multithreaded implementation achieves improvement 13.7 times over the sequential implementation.

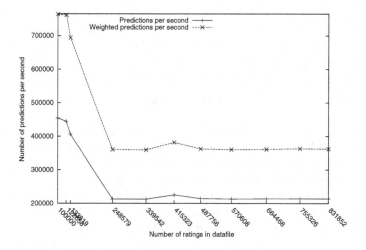

Fig. 3. Predictions and weighted predictions per second

Table 1. Density of the datasets and predictions per second

Ratings per dataset	Density %	Predictions per second	Ratings per dataset	Density %	Predictions per second
100000	6.30	454196	487756	4.11	214239
122658	4.54	444371	570608	4.12	213135
133819	4.46	406291	664468	4.20	213732
248579	4.19	212796	755326	4.24	214168
339542	4.29	212621	831852	4.20	214010
415323	4.20	225335			

Table 2. Performance on MovieLens 100k

Results		
	Sequential	**OpenMP**
Preprocess time	56.88 sec	3.87 sec
Total prediction time	47 sec	3.20 sec
Total weighted prediction time	27 sec	1.90 sec
Predictions per second	31847	454196
Weighted predictions per second	55630	763647
Total time	134.75 sec	16.71 sec
Speedup	—	8.06

The number of predictions and weighted predictions per second tends to stabilize to a certain number, as the density of the different datasets remains the same. Figure 3 shows the number of predictions and weighted predictions per second for each dataset. In all datasets after the one containing 248579 ratings, the density ranges between 4.11% and 4.29%. In table 1 can be seen the density of the different datasets and the number of predictions per second on each dataset. In table 2 can be seen some numerical results of the sequential and the multithreaded implementation.

6   Conclusions and Future Work

This paper describes a parallel implementation of the Slope One algorithm and evaluates its performance. Improvement in the execution time, up to 8 times over the sequential execution time, has been achieved. This fact proves that further optimization of the presented approach will not be in vain.

In future, optimization techniques will be performed and hybrid approaches will be implemented, in order to improve both execution time and scalability of the multithreaded implementation of Slope One.

References

1. Lemire, D., Maclachlan, A.: Slope One Predictors for Online Rating-Based Collaborative Filtering. In: SIAM Data Mining (SDM 2005), Newport Beach, California, pp. 471–476 (2005)
2. Zhang, D.: An Item-Based Collaborative Filtering Recommendation Algorithm Using Slope One Scheme Smoothing. In: Second International Symposium on Electronic Commerce and Security ISECS 2009 (2009)
3. Mittal, N., Govil, M., Nayak, R., Jain, K.: Recommender System Framework using Clustering and Collaborative Filtering. In: Third International Conference on Emerging Trends in Engineering and Technology ICETET, pp. 555–558 (2010)
4. Gao, M., Wu, Z., Jiang, F.: Userrank for item-based collaborative filtering recommendation. Information Processing Letters 111, 440–446 (2011)
5. Wang, P., Wu Ye, H.: A Personalized Recommendation Algorithm Combining Slope One Scheme and User Based Collaborative Filtering. In: International Conference on Industrial and Information Systems IIS 2009, pp. 152–154 (2009)
6. Sun, Z., Luo, N., Kuang, W.: One Real-time Personalized Recommendation Systems Based On Slope One Algorithm. In: 2011 Eighth International Conference on Fuzzy Systems and Knowledge Discovery (FSKD), pp. 1826–1830 (2011)
7. GroupLens Research, MovieLens Data Sets, http://www.grouplens.org/node/73
8. Netflix Prize, http://www.netflixprize.com/
9. Apache Hadoop, http://hadoop.apache.org/
10. GraphLab: A New Parallel Framework for Machine Learning, http://graphlab.org/
11. Zhou, Y., Wilkinson, D., Schreiber, R., Pan, R.: Large-Scale Parallel Collaborative Filtering for the Netflix Prize. In: Fleischer, R., Xu, J. (eds.) AAIM 2008. LNCS, vol. 5034, pp. 337–348. Springer, Heidelberg (2008)
12. Narang, A., Gupta, R., Joshi, A., Garg, V.K.: Highly scalable parallel collaborative filtering algorithm. In: 2010 International Conference on High Performance Computing (HiPC), pp. 1–10 (2010)
13. Narang, A., Srivastava, A., Katta, N.P.K.: Distributed Scalable Collaborative Filtering Algorithm. In: Jeannot, E., Namyst, R., Roman, J. (eds.) Euro-Par 2011, Part I. LNCS, vol. 6852, pp. 353–365. Springer, Heidelberg (2011)

14. George, T., Merugu, S.: A Scalable Collaborative Filtering Framework Based on Co-Clustering. In: Fifth IEEE International Conference on Data Mining ICD 2005, pp. 625–628 (2005)
15. Daruru, S., Marín, N., Walker, M., Ghosh, J.: Pervasive Parallelism in Data Mining: Dataflow solution to Co-clustering Large and Sparse Netflix Data. In: KDD 2009 Proceedings of the 15th ACM SIGKDD International Conference on Knowledge Discovery and Data Mining, pp. 1115–1123. ACM Press (2009)
16. Zhao, Z., Shang, M.: User-Based Collaborative-Filtering Recommendation Algorithms on Hadoop. In: 2010 Third International Conference on Knowledge Discovery and Data Mining, pp. 478–481 (2010)
17. Jiang, J., Lu, J., Zhang, G., Long, G.: Scaling-Up Item-Based Collaborative Filtering Recommendation Algorithm Based on Hadoop. In: 2011 IEEE World Congress Services (SERVICES), pp. 490–497 (2011)
18. Wu, Y., Yan, Q., Bickson, D., Low, Y., Yang, Q.: Efficient Multicore Collaborative Filtering. Matrix (2011)
19. Gemulla, R., Nijkamp, E., Haas, P., Sismanis, Y.: Large-scale matrix factorization with distributed stochastic gradient descent. In: Proceedings of the 17th ACM SIGKDD International Conference on Knowledge Discovery and Data Mining KDD 2011, pp. 69–77 (2011)
20. Ali, M., Johnson, C., Tang, A.: Parallel Collaborative Filtering for Streaming Data (2011), http://www.cs.utexas.edu/~cjohnson/
21. Chen, X., Hongfa, W.: Clustering Weighted Slope One for distributed parallel computing. In: Computer Science and Network Technology (ICCSNT), vol. 3, pp. 1595–1598 (2011)
22. Mi, Z., Xu, C.: A Recommendation Algorithm Combining Clustering Method and Slope One Scheme. In: Huang, D.-S., Gan, Y., Premaratne, P., Han, K. (eds.) ICIC 2011. LNCS (LNBI), vol. 6840, pp. 160–167. Springer, Heidelberg (2012)
23. Wang, Y., Yin, L., Cheng, B., Yu, Y.: Learning to Recommend Based on Slope One Strategy. In: Sheng, Q.Z., Wang, G., Jensen, C.S., Xu, G. (eds.) APWeb 2012. LNCS, vol. 7235, pp. 537–544. Springer, Heidelberg (2012)
24. Li, J., Sun, L., Wang, J.: A Slope One Collaborative Filtering Recommendation Algorithm Using Uncertain Neighbors Optimizing. In: Wang, L., Jiang, J., Lu, J., Hong, L., Liu, B. (eds.) WAIM 2011. LNCS, vol. 7142, pp. 160–166. Springer, Heidelberg (2012)
25. Gao, M., Wu, Z.: Personalized Context-Aware Collaborative Filtering Based on Neural Network and Slope One. In: Luo, Y. (ed.) CDVE 2009. LNCS, vol. 5738, pp. 109–116. Springer, Heidelberg (2009)

Experimental Identification of Pilot Response Using Measured Data from a Flight Simulator

Jan Boril and Rudolf Jalovecky

Faculty of Military Technology, University of Defence,
Kounicova 65, 662 10 Brno, Czech Republic
{jan.boril,rudolf.jalovecky}@unob.cz

Abstract. This paper describes the measuring of pilot response time to a sudden change in a controlled parameter whilst flying an aircraft. The authors of this paper created an analytical model of human behavior from the basic data of an automated regulation. The measurements have been done on a Cessna 152 simulator at the University of Hertfordshire, Hatfield. The tested pilots were pilot students with several tens of flight hours in real planes. The pilot's response to a sudden aircraft altitude change was measured. For analysis of the measured results a mathematical identification model in MATLAB® environment was used. The results obtained from MATLAB® confirm that the experimental measurements were successful.

Keywords: MATLAB®, Aircraft Control, Parameter Identification, Human Behavior Model, X-Plane, Pilot Response.

1 Introduction

In today's automated and digitalized world the stress is put on the development of both computers and artificial intelligence. However, a pilot or an operator is an indispensable part of any aircraft flying. Only time will tell if a pilot (operator) could be fully replaced by a machine and if so it will take a very long time. That's why the aircraft´s manufactures started to do research on the influence of the human factor. The human factor influences many processes of aircraft flying from the very beginning of entering the cockpit, through taking off and landing procedure to stopping the engines. Taking the human efficiency into account the emphasis is put on the ergonomics of the controls in the cockpit, the manipulation space of the pilot, the method of entering the pre-flight data into the Flight Management System, autopilot controls, etc. These factors are supposed to make a pilot's work easy and eliminate his psychological and physical workload while flying a plane.

How will the pilot react in an unpredictable flight situation [1] if one of the automated systems were to cut off or if a sudden change of position angles would accrue due to weather conditions? The authors of this paper focused on weather conditions causing a sudden change of altitude or other flight parameters. Using experimental measurements from the flight simulator a model situation was created where the pilot's task was to react as fast as he could and put the aircraft back to the

L. Iliadis et al. (Eds.): AIAI 2012, IFIP AICT 381, pp. 126–135, 2012.

same altitude using only an elevator. The data from this flight simulator was analyzed. Only the most believable and the most interesting data was input into the MATLAB® environment with use an algorithm to identify parameters of a transfer function.

To determine the optimal mathematical model of a pilot's behavior when flying an aircraft is, from an automated control point of view, a very difficult and complex task. The reason is that the parameters and time constants of the pilot (as a human) are time variables and are influenced by many unpredictable factors such as the pilot's experience, tiredness, stress, surrounding noise etc. To determine a human's behavior, within a control loop, in a given flight mode is possible only after obtaining a correct pilot response in a given mode in the correct time. The authors identified, modeled and simulated these responses by measuring the pilot's responses in a flight simulator. From this data the best realistic time constants representing the pilot's behavior were found. The future vision of the authors is to set limits to all the pilot behavior time constants depending on the level of their experience and psychological and physical condition.

2   Mathematical and Theoretical Background of the Experiment

2.1   Mathematical Model of a Pilot Behavior

A human-pilot character in the control system can be represented by a variety of complex block diagrams which more or less describe most of the possible factors affecting human behavior [2, 3]. Generally, it is not possible to create one universal model fully describing the human dynamic character in various situations during a flying process.

One possible model of human behavior dynamics is shown in the block diagram in Fig 1. It is very simplified, but very concise. There are 3 mutually interfering "blocks". The input – sensors are the pilot's sensory organs, from where the detected information goes into the central nervous system. The average speed of emotion transmission is in the range of 5 to 125 ms-1. In an automated control system this transmission feature can be represented by a transport delay. The response time mainly depends on the level of the pilot's internal stress, the actual pilot's condition and perhaps also on some other factors. Sensory organ features are in real life represented by a sensitivity level, adaptation ability and the ability to mutually cooperate. After processing the received signal a command to hand or leg muscles is sent to adjust the elevator, aileron and rudder deflections. For maintaining the requested flight parameters the pilot uses three different types of regulators [3]:

- Predictive regulator, keeping the required flight mode based on the pilot's received visual and sensory perception of the flight.
- Feedback regulator, created by correct visual and sensory perception of the required flight mode.
- Precognitive regulator, recalling the learnt maneuver from memory, i.e. a clear sequence of elevator, aileron and rudder deflections making the required aircraft movement.

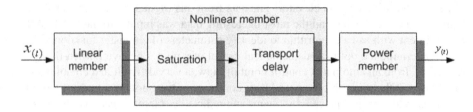

Fig. 1. One of the possible human behavior models in the machine

When analyzing any aircraft control with human behavior it is essential to take into account that all the human features are time variables and dependent on the actual pilot condition, psychological state, tiredness and ability to adapt to a new situation. This is all affected by long-term habits, education, training, etc. To create a mathematical model of a human in such a moment is not easy. For modeling human behavior a linear model is often being used (which is not quite correct for example regarding output value limitations) with a transport delay defined by a transfer function as follows [4, 5, 6, 7]:

$$F_{(s)} = \frac{Y_{(s)}}{X_{(s)}} = K \frac{(T_3 s + 1)}{(T_1 s + 1)(T_2 s + 1)} e^{-\tau s}, \tag{1}$$

where:

K – Pilot Gain – representing the pilot's ability to respond to an error in the magnitude of a controlled variable. Increasing of force on the steers in relation to their deflection (from 0.1 to 100).

T_1 – Lag Time Constant – describes the ease with which the pilot generates the required input i.e. reaction ability to rate of change of input signal (0.1 to 0.4s)

T_2 – Neuromuscular Lag Time Constant – represents the time constant associated with contraction of the muscles through which the control input is applied by the pilot. The dynamics properties of the pilot power member's components (0.05 to 0.2s).

T_3 – Lead Time Constant – reflecting the pilot's ability to predict a control input (0 to 2s).

τ – Represents a pure time delay describing the period between the decision to change a control input and the change starting to occur (0.1 to 0.3s)

This shape of the transfer function is based on the assumption and can be applied in cases where the pilot behaves as a linear member. In the real control loop to a certain extent non-linear elements are always take into account, as in the system pilot - aircraft. In the literature [2], for cases where the nonlinearity of actuator is take into account can be found the extended shape of above mentioned transfer function in the shape:

$$F_{(s)} = \frac{Y_{(s)}}{X_{(s)}} = K \frac{(T_3 s + 1)}{(T_1 s + 1)(T_2 s + 1)} e^{-\tau s} + remnant \ function, \qquad (2)$$

The design of the remnant function is complicated procedure because it attempts to represent the non-linear component of pilot behavior. It is primary source is the pilot´s ability to learn and adapt which results in non-linearity and non-steady behavior. The secondary contribution comes from such things as the experimental setup and experimentally injected noise that affect pilot response to other inputs. However, careful selection of the pilot model and task can help minimize remnant effect [2].

In fact the human operator does not perform controlling activities according to a linear model, but his control efforts are always loaded by negative effects of nonlinear elements such a hysteresis, dead zone, saturation or nonlinear variable gain. It is difficult to identify not only those elements but also include or placed elements into the regulation circuit which has multiple feedback.

2.2   Algorithm for Experimental Identification of Transfer Function Parameters

According to the selected type of transfer functions for the pilot model, it is possible for the time constants determination in human behavior models to use mathematical methods of the experimental identification of real systems.

If input and output signal and the approximate form of the transfer functions are known, is possible to use these methods to specify the parameters of the transfer function. A key advantage is the use of the simulation program MATLAB®, which already contains some functions for realization of necessary calculations.

Function "fminsearch" is looking for a minimum of scalar function of several variables. With its help the algorithm for experimental identification of transfer function parameters was assembled in the form:

$$F_{ei} = \frac{a_1 s + 1}{b_2 s^2 + b_1 s + 1}, \qquad (3)$$

with defined criterion condition

$$f_{min} = \sum (y_{id} - y)^2. \qquad (4)$$

The above algorithm cannot manage calculation of the transport delay. Therefore, the program was completed with a simple subprogram to search the beginning of the pilot response (output value was not zero respectively bigger than the entered low value). After the transport delay evaluation for the identification algorithm the input pulse was moved to the response beginning. The results of the transport delay calculations at all types of pilots practically agree (with an accuracy of calculation step 0.01s) with input value.

3   Description of the Experimental Workplace and the Measuring Methods of Pilot Responses

3.1   Experimental Workplace

The flight parameters and the generally measured values for transfer function parameter identification were measured during a three-month exchange program at the University of Hertfordshire, Hatfield. The university has a laboratory with flight simulators used for pilot training as well as for research purposes. The mentioned flight simulator is primarily intended for pilot's preparation especially for training flight procedures before flight, during and after the flight. The lab is under the auspices of a specialist in automated aircraft control. Our flight tests have been allowed only with good will of Dr. Rashid Ali. Based on his expert advice a Cockpit Simulator Cessna 152 was selected for our testing, see Fig. 2.

The flight simulator Cessna 152 consists of a Cessna 152 aircraft fuselage with two seats for crew. This fuselage is anchored to a static base fixed to the floor. The flight simulation was done by three projectors, projecting images onto a parabolic wall. Based on the research needs software X-Plane 9 from Laminar Research Company was used. The main advantage of this software is its precise and detailed simulation of flight physics for all individual aircrafts. The simulator as a whole is controlled by a PC - also called an Instructor Station. An instructor sitting at this station can change any flight parameters during the flight simulation. All control elements, flight instruments and control stick inside the cockpit are connected to the instructor station. The pilot can fully focus on flying the plane while the instructor can see all the real time parameters on his monitor.

Fig. 2. Cessna 152 Cockpit Simulator (University of Hertfordshire)

3.2   Measurement Methodology of the Pilot Response

The tested pilots were around the age of 23 and all holding Commercial Pilot Licenses (CPL). They were American university students on an exchange program at Hatfield, studying their final year of pilot studies. They all had several hours of flight experience on the Cessna 152 and Cessna 172 aircrafts. All tests were conducted in one day.

As the earlier created algorithm for transfer function parameter identification was made to process the input signal as a unit step function, the authors of this paper chose a unit step function (from a constant flight level) as an input signal. The test was conducted as follows. After an initial induction and simulator training the pilot was explained the test procedure and his task. The pilot's task was to take the plane into a straight horizontal flight. The instructor suddenly changed the aircraft altitude by 100 feet. In real situations such a decrease or increase in altitude can be caused by strong weather conditions or turbulences. The pilot's task was to put the aircraft back to the original altitude as fast as possible and stay there. He could do this by using only the aircraft elevator controls. The engine thrust was constant. The test was conducted with the same pilot several times in the same manner. Also the other pilots were tested in the same manner and under the same conditions. All of the data was recorded and stored in the instructor station.

3.3   Factors Affecting the Measurements

Some limiting factors, occurring during testing, affected the measured results. Firstly, in real situation the pilot senses any aircraft change by his senses organs. This cannot be ensured when using a simulator fixed to the floor. The tested pilots only sensed the altitude change visually by watching the altimeter in the cockpit and by expecting a sudden change. This fact largely influenced (increased) the time constant of the pilot transport delay between sensory perception of the change and a brain response.

After result evaluation and consultation with the pilots about the flight process the pilots talked about greater control sensitivity of the simulator compared to a real aircraft. Another factor lowering the realistic feel of the flight was a small observation angle as seen in Fig.2. Due to the distance and curvature of the screen used for image projecting the pilots didn't have 100% the same feeling as they would in a real aircraft cockpit.

4   Simulation Result Analysis Using an Algorithm for Experimental Identification of Transfer Function Parameters

The measurements from the simulator were analyzed using an algorithm for experimental identification of transfer function parameters. The authors have already created and tested such an algorithm. However, this was the first time realistically measured data from a simulator was applied. Four pilots were tested and each of them had to deal with four to six different changes of a flight altitude. The two cases below are the two best pilot's maneuvers, one going back up to the original altitude and one going back down to the original altitude. The last case demonstrates a badly conducted maneuver and the imperfection of the identification algorithm.

Fig.3 shows an almost perfect pilot maneuver when returning to the original altitude. This was the pilot's fourth trial which proves that the more trials the better the pilot gets. The pilot was able to recover the original altitude in 14 seconds only by using an elevator control. Undoubtedly, the time in which the pilot is able to recover the original altitude also depends on the type of aircraft. The pilot's response is copying the graph of PID regulator to which the pilot can be compared. Taking in account a standard deviation, the pilot's response chart is almost perfect. To a person's naked eye, there is almost no difference between the pilot response curve and the curve created by the mathematical model for identification of transfer function parameters.

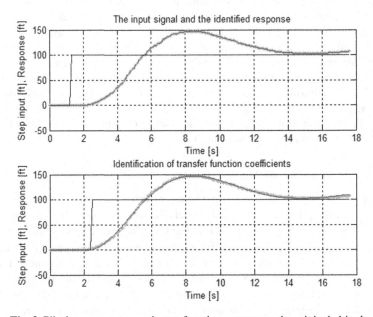

Fig. 3. Pilot's response to a unit step function – ascent to the original altitude

Fig.4 shows the best pilot's maneuver when descending to the original altitude. This result is absolutely unique. In the other tests none of the other pilots matched even slightly such a response curve. The identification algorithm approximated the pilot's response to the unit step function reasonably well. In Tab.1 important identification algorithm parameters are shown after several hundreds of iterations and also the pilot behavior time constants are shown, i.e. their product and their sum.

In Fig.5 there is clear evidence of the pilot's effort to come back to the original altitude. In this case the altitude recovery took longer and two aircraft oscillations occurred. Similar aircraft oscillations were found at least once for each pilot in their attempt to quickly descend back to the original altitude. The reason is that when descending the speed is naturally increasing and therefore the recovery maneuver is more difficult and the controls are more sensitive. It is also important to note that applying a 2nd order transfer function for this case of pilot response was inadequate. All the simulation parameters and the pilot time constants from this analysis were disproportionally higher than those in the two cases mentioned above. An improvement could be reached by applying a higher order transfer function, but those results would not be comparable with the other measured results.

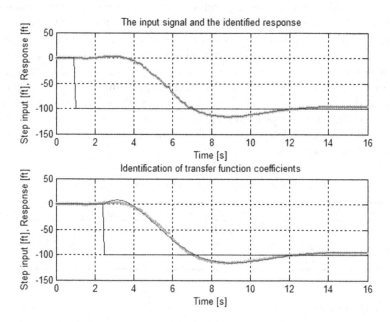

Fig. 4. Pilot's response to a unit step function – descent to the original altitude

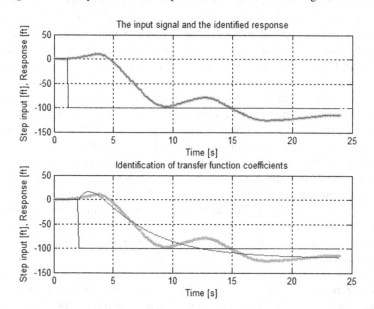

Fig. 5. Pilot's response to a unit step function – descent to the original altitude (oscillation)

All the conducted tests show fairly higher time delays than assumed in theory [3]. That is caused by a wide range of factors affecting both the method of testing and assessment and the identification algorithm itself. There are two main reasons for these higher delays. Firstly, that the pilot was detecting the altitude change only visually. Secondly, the pilots were not informed when the altitude would suddenly drop or increase by the 100 feet and thus taking longer to analyze the situation and react appropriately. The authors also discovered, from the identified data, that aircraft dynamics also play an important role in getting a higher time delay. It is clear from the measured control stick responses that pilots started the returning maneuver about 0.2-0.3 sec earlier than the plane started to ascent or descent. The last but not least factor affecting the time delay is the sampling frequency set at 0.1 sec disallowing more accurate time delay analyses.

Table 1. Parameters of the identified transfer functions

Figure	Standard deviation	Proportional gain	Transport delay	T_3	T_1T_2	T_1+T_2
3.	1.56	1.12	1.1	0.28	3.46	1.32
4.	2.12	0.97	1.4	-0.78	2.72	1.60
5.	10.33	1.20	0.8	-1.58	4.11	5.30

5   Conclusion

The authors of the paper conducted about twenty changes of a flight altitude on the Cessna 152 flight simulator from which they obtained a lot of data for parameter identification of a pilot behavior model transfer function. Only the most important and the most interesting pilot responses were chosen. When the measured time constants shown in Tab.1 were compared to the theoretical boundaries of the time constants it was clear that it is possible to simulate a realistic model of a pilot's behavior.

In the near future, the authors are planning to set up an aircraft simulator experimental laboratory in which testing and data assessment would continue to be improved. The algorithm for experimental identification will be refined so that the algorithm clearly separates the important pilot time constants needed for determining its limits.

The authors of this paper are aware that the last factor affecting their measurements is the pilot's simulator flight hours. This was also shown in the measurements where the pilots were getting better with each conducted trial. Their response time, in which they had to put the aircraft back into its original horizontal altitude, was shortening.

The above mentioned factor negatively affected both the pilot's lead time constant depending on his experience and lag time constant connected to his accustomed stereotype routine.

Acknowledgments. The paper was written under the umbrella of a project development department at the University of Defence – project K206 titled "Complex Electronic System for UAS" and supported by the association UDeMAG (University of Defence MATLAB Group).

References

1. Boril, J., Jalovecky, R.: Response of the Mechatronic System, Pilot - Aircraft on Incurred Step Disturbance. In: 53rd International Symposium ELMAR 2011, pp. 261–264. ITG, Zagreb (2011)
2. McRuer, D.T., Krendel, E.S.: Mathematical Models of Human Pilot Behavior. AGARD-AG-188 (1974)
3. Havlikova, M.: Diagnostic of Systems with a Human Operator, Doctoral Thesis, Brno University of Technology (2008) (in Czech)
4. Jalovecky, R., Janu, P.: Human – Pilot's Features During Aircraft Flight Control from Automatic Regulation Viewpoint. In: 4th International Symposium on Measurement, Analysis and Modeling of Human Functions, pp. 119–123. Czech Technical University in Prague, Czech Republic, Prague (2010)
5. Jalovecky, R.: Man in the Aircraft's Flight Control System. Advance in Military Technology – Journal of Science 4(1), 49–57 (2009)
6. Cameron, N., Thomson, D.G., Murray-Smith, D.J.: Pilot Modelling and Inverse Simulation for Initial Handling Qualities Assessment. The Aeronautical Journal 107(1744), 511–520 (2003)
7. Boril, J., Jalovecky, R.: Simulation of Mechatronic System Pilot - Aircraft - Oscillation Damper. In: ICMT 2011 - International Conference on Military Technologies, pp. 591–597. University of Defence, Brno (2011)

A Probabilistic Knowledge-Based Information System for Environmental Policy Modeling and Decision Making

Hamid Jahankhani[*], Elias Pimenidis, and Amin Hosseinian-Far

School of Architecture, Computing & Engineering, Docklands Campus,
University of East London- London E16 2RD, UK
{Amin,E.Pimenidis,Hamid.Jahankhani}@uel.ac.uk

Abstract. Decision making for setting new policies is a challenging process as the current policy making system is utterly flawed. A policy is introduced by the decision maker when the problem domain was fully consulted by experts in the field. Not always all the consultants and advisers agree on details or even basics of such a course of action. The need for an intelligent predictive system is emerging. Policy making on environmental issues are even shoddier as the environmental systems are habitually complex, and adaptive; and introduction of new technologies can easily affect the guiding strategies already taken. This paper outlines the principles of Knowledge Management Systems. It then reflects on Influence Diagrams' suitability for construction of such an information system through the use of the London Plan case study. An application of such a system is outlined by means of a probabilistic knowledge based IS which is developed by Influence Diagrams and can be utilized as an Environmental policy modeler and/or DSS.

Keywords: Knowledge Management System, Decision Support System, London Plan, Influence Diagram, Tacit and Explicit Knowledge.

1 Introduction to KMS

There are various applications for Knowledge Management Systems (KMS) theories including but not limited to Distributed Databases, Ontology, and Artificial Intelligence (AI). Knowledge Based systems which are one of the software tools in Intelligent Systems are the instruments to manage the knowledge [1]. The main focus would be on the Artificial Intelligence in this research.

Knowledge Management Systems can be classified into two main categories; Distributive and Collaborative aplications. Distributive knowledge management system is where database plays a vital role in shaping the knowledge base, hence there is a structured formal database involve. On the other hand the collborative knowledge management system partakes an informal internal knowledge[2].

2 Types of Knowledge and Knowledge Management Systems

Some experts still argue on the defintion of knowledge itself. knowledge is summation of information, skills, experience, and personal capabilities [3]. Other

---

[*] Corresponding author.

L. Iliadis et al. (Eds.): AIAI 2012, IFIP AICT 381, pp. 136–145, 2012.

scholars also attempted to categorize and formulated knowledge and its types. Another categorization of knowledge is outlined in Fig. 1:

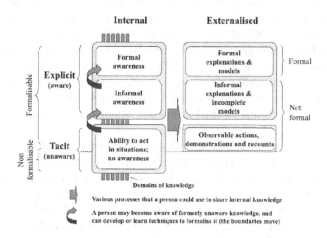

Fig. 1. Categories of Knowledge [4]

Defining knowledge is not enough as the knowledge should be retained and gathered. There are various frameworks for knowledge acquisition. The common knowledge acquisition framework is illustrated in Fig. 2:

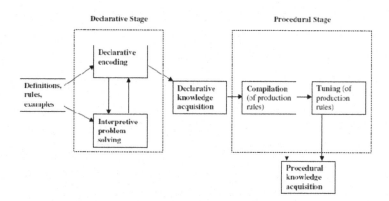

Fig. 2. Knowledge Acquisition Framework and Stages [5]

According to this general framework, the first step is the declarative stage where the rules, standards, examples and definitions should be clearly defined. Once the standards are declared, the problem solving and encoding stage begins, still within the declarative stage. Up to this point the knowledge is still in the declarative knowledge acquisition mode. In the start of the procedural stage, compilation should be performed on the declarative acquitted knowledge. Once the errors are fixed and the compilation is tuned, then the procedural knowledge acquisition is completed [5].

Furthermore, knowledge can be transformed while in use. The general categorization for knowledge transformation is internal, external and socialized knowledge (Fig. 2). The knowledge is initially in the internal form in tacit format. The tacit format indicates that the knowledge still contains areas where the use is not fully identified. The tacit knowledge within the internal domain will be transformed into explicit knowledge where the standards and usages will be clarified. By having the formal explanations and models the knowledge can be externalized. The externalization on its own does not have to be formal at all circumstances. Informal explanations and observations in the external domain based on the explicit internal knowledge can still provide a means for formalization. The set formal external knowledge which is explicit can be widely used and become a socialised knowledge format [6].

A dynamic system model which has the learning characteristic, complexity elaboration and non-linear features will be able to correspond to the research questions within the problem domain. The complexity, non-linearity, and adaptive characteristics are the main factors that should be considered for selection of proper tool. Fig. 3 represents different software tools for Artificial Intelligence. As it is preferred to use a predictive design one of the following should be selected for the design. Generally aritificial intelligent systems are divided into two main categories which are Knowledge Based systems and Computational Intelligence systems.

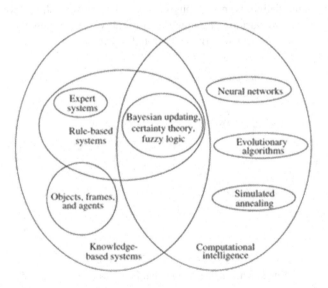

Fig. 3. Artificial Intelligence areas [7]

Some characteristics of the sustainability can be modeled using other types of system modeling techniques; for instance using cybernetic system approach through state space method for socio- ecological system. The latter has been modeled for the climate change by Schellnhuber [8].

3   Problem Area

Policitcians are having a demanding era in facing the sustainability challenges. Policy making as a rule happens after long periods of consultancies. Not always, all the consultants and polictical advisers come to a common ground in different scenrios. The use of predictive information Systems can be valuable in assisting the decision maker. These ISs have be predictive in a way that can approximate the outcomes of different decisions in a given scenrio. Modeling environmental states is quite a challenging process. The tradeoff between complexity, efficiency and accuracy of the Information System used for development of a DSS is crucial. Use of probabilistic inference is often overlooked. Influence Diagrams practices as the primary means for developing the knowledge base for complex environmental scenarios should be tested. Appraisal of Influence Diagrams and ID evaluation techniques can be tested by means of a case study research design type. The London Plan as a case study has had various discussions since 2004[9]. Using microeconometric theories, the required investment for policy 4A.2 has been calculated between 40 to 80 £Billion. Since the case study has these estimates using other quantitative modeling techniques, using Influence Diagrams and probablistic inference can be validated for these kind of modelings. Bayesian networks can be utilised for development of such a knowledge base, but chance nodes are required for development of a DSS. Therefore, Influence Diagram is a fine choice for the knowledge base construction. The objectives of this research would be:

1. Knowledge management representation of the knowledge body behind the London Plan sustainability domain. The London Plan is the case study used for validation of the proposed model.
2. Identifying an appropriate quantitative approach for modeling the financial domain of The London Plan.

4   Probabilistic Networks for Knowledge Types

Real-life state of affairs mostly modelled as group of entities demonstrating random variables in a "probabilistic network". Clever graphical illustration of dependence and independence relations between accidental variables is a "probabilistic network". Area of random variables could, for instance, help decision makers to identify the most beneficial decision in a given situation from the basis of a decision support system.

Probabilistic networks processes and symbolize probabilistic data. Representative elements of a probabilistic network are a quantitative and a qualitative element. The qualitative component sets (conditional) belief and independence assertions along with a set of chance variables, informational fondness, and preferred relations [10]. Graphical language visually encoded the statements of (conditional) dependence and independence, information preference, and favourite relations. On the other hand, the quantitative component identifies the potency of dependence relations by means of probability and utility theories [11].

The illustrative depiction of a probabilistic network, explains knowledge of a problem area in a clear-cut manner [12]. The illustrative depiction is perceptive and easy to understand, making it an ideal tool for passing statement of domain knowledge among experts, users, and systems. Therefore, the formalism of probabilistic networks is becoming an ever trendier domain knowledge representation for interpretation and decision making under uncertainty [11]. In this model, tacit and explicit knowledge are extracted using narrative extraction from plan policy reports and the discussions. The decision node is derived from the policy objectives. Therefore in this scenario, 60% CO_2 reduction as the aim objective makes the decision node. The suitability of Influence Diagram is assessed as follows.

5   Suitability of Influence Diagrams for Environmental Policy Modeling

Despite having various modeling techniques in the field of sustainability, probabilistic inference and Bayesian networks models should be given high priority. The complexity of sustainability scenarios and understanding the systems' resilience and boundary paradox are essential when analyzing and designing the case studies [12]. Among the quantitative techniques, agents and multi agent systems are quite useful as they can bring together the components of the system using agents' characteristics. Furthermore the artificial intelligence techniques used in multi agent systems would enable the system to adapt itself with the changes faced from outside the boundary. But there is a major drawback with multi-agent systems as they are too expensive for building large Decision Support System applications. Although they can fairly fit fine in a CAS (Complex Adaptive System) methodology in theory [13], but when it comes to practice, they are not responsive to all of the components. Use of neural networks as the tool for modeling the inputs and the outputs based on a network of linked components sound very functional. But the major drawback with the neural networks is when it comes to its training. Training of the network in a sustainability scenario is almost impossible. The longitudinal study may help for the training, but long intervals and wait time is required for the testing. Hence practically, they are not usable in large complex systems where training is nearly impossible. The system's dynamic models are also other quantitative techniques for sustainability modeling. The difficult nature of developing a system using systems' dynamic, has made it again impossible for the developer to develop the model. System dynamic might be useful for modeling some small trade-offs in the environment where increase in one component would lead to a drop in another. Stella might be a good IDE for these small projects; but again, if the system becomes large with various elements in the knowledge base, analysis, design and development of the model would be unattainable [14].

On the other hand, probabilistic inference, e.g. influence diagrams are fine tools for modeling sustainability. The adaptively of the model is an issue with probabilistic networks as the base for the model, but that can be easily fixed by incorporating an agent based engine. The use of goal oriented knowledge management systems theories would be a good further work for this study. This research study by means of the London Plan case study validates that despite having a complex scenario; modeling with ID would be feasible. The model would be a formal model where replication, testing and validation are possible using the available algorithms. The predictive nature of the Influence Diagrams, very good graphical interface for non-expert users, and also accurate mathematical and probability layer would be one of the finest approaches to model sustainability systems with complex nature. It still fit into the Complex Adaptive

System structure if the agents are incorporated to give the adaptively feature to the system. The use of knowledge management theories would help to build the knowledge base for the model [15]. Selection of a specific probabilistic network on its own should be rationale. The reason for selection of Influence Diagrams over Binominal Trees and Decision trees is explained using more details on the other two techniques; although they can be converted to each other in many scenarios.

6   KMS for the London Plan Case Study

The final model can be implemented as decision support system for the stakeholders. It also has the capability of being disseminated as a web based DSS tool via web services techniques. There is an overlap is between probabilistic networks and fuzzy systems. The ID uses the same theories, but incorporates the set graphical interface for better understanding and modeling the problem domain. Financial analysis of the London Plan policy 4A.x (X=1, 2, 3) is an appropriate case study for this validation, although other case studies might be used. The reason for selection of such a case study lies within the gap in discussions and analysis of the GLA plan. The London Plan Policy 4A.X proposes 60% reduction in London CO_2 compared to 1990 base, using CHP (Combined Heat & Power), Energy Efficiency, and Renewable Energy Systems expending solar and wind [9]. The data extraction is done using a theoretical narrative extraction, and the design base of this research would be quantitative practical base. Fig. 4 outlines the high level ID for this scenario.

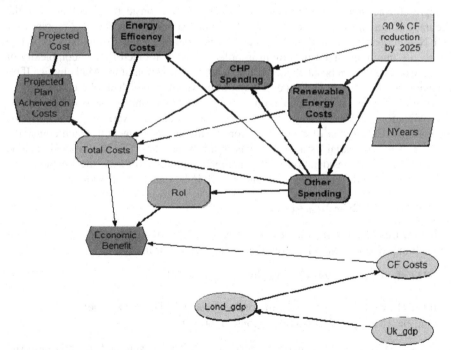

Fig. 4. High Level Influence Diagram as the Knowledge Base for Policy 4A.x Financial Assessment

Reasoning in such a model is dependent on the type of scenario under analysis. In this scenario, there is no 'Explaining Away' reasoning case between the nodes. Although, there are converging connections in the node orders, but explaining away cannot be inferred in the Extended ID developed. The converging inference can be assessed when the node orders are reversed using the reversal technique. There is no successor for this decision node and therefore the decision node removal cannot be considered for evaluating this ID. The reversal and node removal including barren node removal for this influence diagram can be further works of this piece of research. Casual reasoning in the model is where the UK-GDP has statistical dependence on the London GDP and that continues to all other proceeded nodes. There is various inductive reasoning in the ID where the parent nodes are not dependent to each other, i.e. 'Total Costs'.

$\varepsilon = \hat{A}^\alpha . \hat{C}^{1-\alpha}$ is the efficiency function for an ID. The parameters involved are Accuracy (A), and Complexity (C) [16]. In addition, α is the value set by the decision maker. This formula should be looked at with caution, as the complexity and accuracy of the ID should be normalized first [17]. Cobb's formula concerning the accuracy and complexity functions are as follows:

$$\hat{A} = N_{min} \frac{(N_{max} - N_{min}).(A - \underline{A})}{\overline{A} - \underline{A}}$$
$$\hat{C} = N_{min} \frac{(N_{max} - N_{min}).(C - \underline{C})}{\overline{C} - \underline{C}}$$

A represents accuracy and C represents complexity. The consideration is that when the accuracy is maximized, the complexity of minimized and vice versa. N_{min} represents 1 and N_{max} represents 2 in Cobb's research. The Accuracy and Complexity is scaled between a minimum and a maximum in order to assess a trade-off. The complexity of the ID expressions can be calculated using LeafCount function of the Mathematica IDE. Mathematics is software developed by Wolfram which can evaluate the complexity of functions and graphs. The calculation is simple and is counting the variables of any approximation within the ID. It does not only count the variables, but also the expressions defining the function. Therefore LeafCount is a function which counts the number of words, variables and constants in an expression [17]. The rationale behind this calculation is consideration of the memory required to process the function. The Complexity of an ID then is the summation of all individual complexities:

$$C = \sum_{i=0}^{n} C_i = \sum_{i=0}^{n} \sum_{j=0}^{m_i} \mathcal{L}\{\theta_{ij}\}$$

Using the LeafCount function we can simply determine the complexity of expressions and sub-expressions of the London Plan Policy 4A.x. For instance:

```
In[1]:=LeafCount [399943*Num_london_propertie]
Out[1]=3
In[2]:=Level [399943 + * + Num_london_propertie, Heads → True]
Out[2]= {Multiply, 399943, Num_london_propertie}
```

The complexity of the RoI node expression is 3. That is summed up to complexity outputs from the LeafCount function in Mathematica on other nodes. If there are combinatory expressions in the ID, then the NestList function of Mathematica, lists all the expression in combinations and then the LeafCount can be run. The complexity of the whole Extended Influence Diagram for considering the Leafcount function is done manually and the number 50 is reached.

For finding Accuracy of Influence Diagram accuracy, the mean squared error between the analytical decision rule and the ID decision rule should be calculated.

The found Accuracy and Complexity values are then traded off using the stated formulas. Cobb's minimum and Maximum (N) and be scaled using other values, but the technique remains the same. The efficiency of an ID can also be affected by the decision makers' amendments using α variable [17].

This model also validates the approximations on predicting the financial costs of the policy which gives practical values to this research study. The result of the research indicates the estimated costs of the London plan policy concerning $CO2$ reduction has been within the discussed outcomes. The output of the simulation states that the London plan would cost £Billion 68.84. This number is derived after subtraction of £Billion 12.06 Billion pounds return on investment from the total costs which was £Billion 80.9. The simulations are performed within Analytica IDE, but other environments can also be utilized for simulation.

However, Analytica is a simulation-based tool, and currently cannot prescribe an optimal solution for the detection network. Although unavailable in software tools, influence diagrams with continuous variables can be found in the literature, Again, none of the works is applicable to the parallel detection network structure. Now consider a variation where we remove the continuous variables, i.e., the measurements from the sensor, and transform the local decision makers to be chance nodes.

This yields an alternative model, which is equivalent a similar structure was introduced by Heckerman to analyze the value of information for diagnosis. In this configuration, the subordinate decision makers provide evidence or local decisions [18].

7   Conclusion, Limitation and Further Works

There are number of limitations to this model. One the limitations of this research study would be the all-embracing complexities involved. The complexities do not only arise from the financial domains of the policy, but also the social and environmental facets as well.

Another constraint of this research project has been the consideration of only the financial facets of the policy plan. The social limitations and all environmental restraints reside outside the boundary of this study. Knowing that this project involves an open system analysis, therefore, not considering those two facts would restrain the overall result of the research work.

On the case study, there are some zones where the plan policy has not clarified fully. For instance the choice of 60% reduction by 2050 within policy has not been explained. In addition to that a comparative financial analysis has not been provided in the policy. This could be used as a framework for comparing the results of this project and other similar projects and the policy plan results.

There are some limitations in Knowledge Management Systems.These limitations in KMS are defined as:

1. Tacit knowledge cannot be easily monitored and managed,
2. Involvement of the stakeholders in a dynamic and up to date management of their knowledge,
3. Interactions between stakeholders might be limited,
4. System is not necessarily adaptive [20].

Recent research on goal oriented models suggests that Knowledge Management Systems can be developed in a goal oriented friendly format. The initial proposal for goal oriented KMS was introduced in 2004, but still various research is taking place concerting this topic. In theory the goal oriented KMS would have the adaptively, innovation and replication characteristics [21]. Although this is out of scope of this study, but a future work on ID model developed in this research work might be implementation of its goal oriented KMS which has the mentioned here characteristics.

References

1. Valente, G.: Artificial Intelligent Methods in Operational Knowledge Management, Torino (2004)
2. Zack, M., Serino, M.: Knowledge management and Collaboration Technologies. The Lotus Institute, Lotus Development Corporation (1996)
3. Baker, M., Baker, M., Thorne, J., Dutnell, M.: Leveraging human capital. Journal of Knowledge management 1(1), 63–74 (1997)
4. Kalpic, B., Bernus, P.: Buesiness Process Modeling through Knowledge Management Perspective. Journal of Knowledge Management 10(03), 40–56 (2006)
5. McCall, H., Arnold, V., Sutton, S.: Use of knowledge management Systems and the Impact on the Acquisition of Explicit Knowledge. Journal of Information Systems 22(02), 77–101 (2008)
6. Kernstock, P.G.: A Web-Based knowledge Management System for information technology Education. Texas State university, Texas (2006)
7. Hopgood, A.: Intelligent Systems for Engineers and Scientists, 2nd edn. CRC press, New York (2001)
8. Schellnhuber, H.J.: Earth system' analysis and the second Copernican revolution. Nature (1999)
9. Greater London Authority: The London Plan (2011), from London. Gov., http://www.london.gov.uk/priorities/planning/vision/london-plan/replacement-process (retrieved 2011)
10. Uffe, B., Kjaerulff, A.L.: Bayesian Networks and influence Diagrams: a guide to construction and analysis. Springer (2007)
11. Lerz, J.L.: The Case for Using Probabilistic Knowledge in a Computer Chess Program, from Verizon.net (2012), http://mysite.verizon.net/vzesz4a6/current/id309.html (retrieved February 2011)

12. Hosseinian-Far, A., Jahankhani, H., Pimenidis, E., Wijeyesekera, D.C.: Reflections on Modeling Systems' Resilience. In: 5th SASTech Intl. Symposium, Mashhad, Iran (2011)
13. Hosseinian-Far, A., Jahankhani, H., Pimenidis, E., Wijeyesekera, D.C.: Reflections on the Need for an Improved Quantitative Approach. In: Advances in Computing and Technology Conference, London (2011)
14. Hosseinian-Far, A.: A Systemic Approach to an Enhanced Model for Sustainability. PhD Thesis (2012)
15. Hosseinian-Far, A., Pimenidis, E., Jahankhani, H., Wijeyesekera, D.C.: Financial Assessment of London Plan Policy 4A.2 by Probabilistic Inference and Influence Diagrams. In: Iliadis, L., Maglogiannis, I., Papadopoulos, H. (eds.) EANN/AIAI, Part II. IFIP AICT, vol. 364, pp. 51–60. Springer, Heidelberg (2011)
16. Cobb, B.R.: Measuring Efficiency in Influence Diagram Models (2008)
17. Wolfram: The Mathematica Book. Wolfram (2003)
18. Baye, M.R.: Managerial Economics and Business Strategy. Mc-Graw Hill, New York (2006)
19. Heckerman, D., Shachter, R.: A Decision Based View of Casualty. Microsoft Research (1995)
20. Nabeth, T., Angehrn, A.A., Roda, C.: Enhancing Knowledge Management Systems with Cognitive Agents. Journal of information Systems in Management 8(2) (2003)
21. Gray, P.H., Meister, D.B.: Knowledge sourcing effectiveness. Journal of Management Science 50(06), 821–834 (2004)

Conceptualization and Significance Study
of a New Appliation CS-MIR

Kaichun K. Chang[1], Carl Barton[1], Costas S. Iliopoulos[1,3], and Jyh-Shing Roger Jang[2]

[1]Dept. of Informatics, King's College London, Strand, London WC2R 2LS, England
[2] Dept. of Computer Science, Tsing Hua University, Taiwan
{ken.chang,carl.barton,c.iliopoulos}@kcl.ac.uk,
jang@cs.nthu.edu.tw
[3] Digital Ecosystems & Business Intelligence Institute, Curtin, Perth, Australia

Abstract. Numerous researches on Music Information Retrieval (MIR) have been estimated and linked with sparse representation method, few has paid enough attention on the application of compressive sensing and how it affects the reconstruction of MIR. This paper provides solid theoretical and various empirical evidence on the conceptualization, theoretical development, and implication of Compressive Sensing (CS), which to great extent, contributes to the application of the Music Information Retrieval.

Keywords: compressive sampling, music information retrieval, music identification.

1   Introduction

This paper describes the theoretical development of Compressive Sensing (CS) study suggested in the past literature and identifies the need for investigating deficiencies of current audio fingerprinting research. By reviewing researches of audio fingerprint and music information retrieval (MIR) that is affected by Compressive Sensing, we can see a transparent linkage between Compressive Sensing theory and Music Information Retrieval system. Therefore, we create a new application CS-MIR to explore more sophisticated issues related so as to generate more practical results for future music information researchers, within which the theoretical perspectives and empirical evidences derived from compressive sensing technologies as a new application related to MIR will be embedded in the discussion of this new theory. In this paper, we conclude that some evidence prove that the new CS-MIR application could be the feasible solution for music classification and music recognition technologies.

2   Deficiencies Of Existing Work In Audio Fingerprinting

2.1   Feature Extraction

When the sampling signals of audio items are obtained, the next step is to extract features from the audio item. The goal of feature extraction is to simplify a large

L. Iliadis et al. (Eds.): AIAI 2012, IFIP AICT 381, pp. 146–156, 2012.
© IFIP International Federation for Information Processing 2012

amount of data required to a small size for describing accurately. For the audio fingerprinting system, the extraction of features, which provide direct access to differentiating different items, is extremely crucial. As to the fingerprint extraction, some studies have explored various kinds of features that are robust to distortions, while the others mainly make an analysis on establishing a precise statistical fingerprint model and more appropriate distance measurements to improve the robustness of the system. Both approaches can achieve robust recognition.

Based on the outcome of these theoretical investigations, some practical audio fingerprinting systems are established, such as the Foosic Algorithm. It comes from Foosic, which is a free and open content project and employs its own free fingerprinting technology named libFooID. As to its feature extractor, firstly, all test music data converted to 32-bit float format and then down mixed to mono, and subsequently the next 100 seconds of audio data feed into a res-ampler to obtain 90 seconds of 8000Hz sampled output. The output is then converted into a frequency spectrum by using a Hann-windowed DFT (here a custom Split-Radix FFT routine used) applied to 8192 sample blocks. By the above processing, a 90 seconds audio turns into 87 frames with frequency spectra. Another example is the "Audio Spectrum Flatness" LLD (Low Level Descriptor) which uses a time-to-frequency mapping and some further computation on a block by block basis to generate an MPEG-7 compliant fingerprint. The fingerprint is based on the calculation of the Spectral Flatness Measure (SFM).

2.2   Fingerprint Modelling

Many works on audio fingerprint have discussed the modelling of audio fingerprints, and in most of the present fingerprint models, fingerprints are modelled as a series of binary numbers [1]. For example, Haitsma et al. moulded the fingerprint database as binary vectors. When testing an unknown audio, the fingerprints of the test audio are compared with the fingerprints of the other items in the database, and then the item which presents a low bit-error rate (BER) (commonly a certain threshold is set beforehand) is chosen. As mentioned above, the sign of the energy difference between the frame and block energies are taken as the fingerprint by Venkatachalam et al. This is the most fundamental approach for the fingerprint modelling.

Recently the researchers are aware of the importance of establishing a statistic model for the extracted features of audio clips or audio clips themselves, because the fingerprints should perform a dimensionality reduction of the original data significantly, provide accurate discrimination for different audio clips, and be invariant to distorted versions. For example, Kulesh et al. [3]. have used Mel-frequency spectral coefficients (MFCC) as the features of fingerprints, and the features are then modelled by a Gaussian mixture models (GMM). According to Cano et al. and Batlle et al. [2], the fingerprints are modelled as a sequence of hidden Markov models (HMM). Firstly, an alphabet of sounds describing an audio song (or item, clip) is extracted, and then these audio units are modelled with hidden Markov model. The unknown audio songs and the set of songs are divided into these audio units; these audio units are ended up with some symbols for the unlabelled song and a database of sequences representing the original songs. The unlabelled song is recognized using the Viterbi decoding of the unlabelled item fingerprint against the repository of fingerprints. In paper [4], Arunan Ramalingam et al. designed fingerprints by modelling the audio clip as a Gaussian mixture models (GMM). In paper [5], Hui Lin, et al. modeled the fingerprint with precise common component Gaussian mixture models (CCGMMs). By stabling these statistical models for audio clips, more accurate recognition of audio items can be obtained.

2.3   Search Method for Efficient Matching

Recognition of fingerprints relies on the fingerprint matching technology. Item matching is another important aspect in audio fingerprints, which is based on the distance measurement to decide which item in the database is mostly matching to the item. So the distance measurement for comparing fingerprints is also an important problem to be addressed in the fingerprint matching. In the available audio fingerprinting system, fingerprint matching is often performed using the square of the Euclidean distance for a fast computation and mathematical tractability for analysis. The Euclidean distance between vectors f and g is defined as:

$$d = \sqrt{\sum_{i=1}^{N}(f_i - g_i)^2} \tag{1}$$

Some other distance metric can also be considered in the system such as L1, L2 and KL metric. Once the measurement is determined, we will use the distance measurement to perform an efficient searching. It is very important for the fingerprinting algorithm that the algorithm needs both few bit errors and efficient searching [6]. In the well-known audio fingerprints systems, some searching algorithms are employed.

Philip Database search: a two-phase search algorithm that is based on only performing full fingerprint comparisons at candidate positions pre-selected by a sub-fingerprint search.

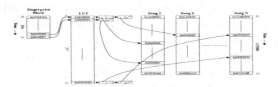

Fig. 1. Database layout

As shown in fig.1, the fingerprint database contains a lookup table (LUT) with all possible 32 bit sub-fingerprints as an entry. A hash table is often used instead of a lookup table in practice. In order to obtain an efficient searching algorithm, the sub-fingerprints that extracted from the database are registered in hash tables. Coarse search in the database is performed firstly, and output N songs as results. Distances between each song of these N songs and the query song are top N smallest. Then the exact search algorithm is used, as shown in fig.2.

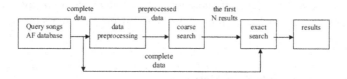

Fig. 2. Philip Database search

3   Theoretical Development of Compressive Sensing Theory

3.1   Rationalization of Compressive Sensing Model

Just as we have discussed above, the dimension reduction of musical signals are very important in audio fingerprinting and musical classification. Now many dimension reduction approaches have been proposed, such as LCA, which is a non-linear dimensionality reduction method, and NMF (Non-Negative Matrix Factorization) method. Compressive sensing theory provides a random measurement of signals, and proves to be able to not lose the information of the signals under the condition of enough number of measurement and incoherence between the measurement matrix and the transform matrix. Consequently, it is a natural compressive process of signals, which can also be regarded as the process of dimension reduction. Different with LCA and NMF, it is a linear method, which is characteristic of easy realization. In this section, firstly we present a detailed introduction on the compressive sensing theory.

Considering a real valued signal x (with length N), given that it is K-sparse in the sparse basis matrix ψ. Then considering an M×N measurement (random or determined) matrix ϕ (here M<<N, M is far less than N), where the rows of ψ are incoherent with the columns of ψ. In terms of matrix notation, we have $x = \psi \alpha$, in which α can be approximately using only K<<N (K is far less than N) nonzero entries. The compression sensing theory states that such a K-sparse signal x can be reconstructed by taking only $M = O(K \log N)$ number of liner, non-adaptive measurements, which can be described as follows:

$$y = \phi x = \phi \psi \alpha = A \alpha \qquad (2)$$

where y represents an M×1 sample vector, $A = \phi \psi$ is an M×N matrix. From it we can see that the implementation of compression sensing of a signal can be divided into three steps: first, transform the signal x to some sparse domain to make it K sparse under certain dictionary; second, make an observation of the sparse transformation coefficients to obtain the measurements; third, apply an optimization algorithm to retrieve the sparse coefficients, and then recover the original signal x.

From the model we can see that there are three important things in realizing a compressive sensing of signals, that is, finding a representation space in which the signal is sparse, determining a measurement matrix that is incoherent with the transform matrix, and using an efficient optimization algorithm to recover the sample from the limited number of measurement. In the following we discuss the measurement and optimization.

3.2   The Measurements Basis Matrix

The measurement basis matrix ϕ must allow the reconstruction of the length N signal from M<N measurement (the vector y). From the model we can see that M<N, so this problem appears to be an ill-conditioned problem. If, however, x is K-sparse in a given space and the K locations of the nonzero coefficients in are known, then the problem can be solved provided that M•K, a necessary and sufficient condition for this poorly defined problem to be well conditioned is that, for any vector v sharing the same K nonzero entries as and for some $\varepsilon > 0$, the following equation has to be satisfied,

$$1-\varepsilon \le \frac{\|Av\|_2}{\|v\|_2} \le 1+\varepsilon \tag{3}$$

That is, the matrix A must preserve the lengths of these particular K-sparse vectors. Of course, in general the locations of the K nonzero entries in are not known. However, a sufficient condition for a stable solution for both K-sparse and compressible signals is that the matrix A satisfies the above formula for an arbitrary 3K-sparse vector v. This condition refers to as the restricted isometry property (RIP condition) [7]. Paper [8] said that if the measurement matrix is incoherence with the sparse basis matrix ψ, it means that the rows $\{\phi_i\}$ of ϕ cannot sparsely represent the columns $\{\psi_i\}$ of ψ, so the matrix A will have the restricted isometry property.

3.3   The Optimization Based Reconstruction

To recover the signals from the measurement of signals, one should use some optimization algorithms to reconstruct the signal. Reconstruction aims to get the recovered signal from measurements y making use of the sparse prior or compressibility of signal. Although the sampling process is simply a linear projection, the reconstruction algorithm is highly non-linear. The step of reconstruction is equivalent to find the signal's sparse coefficient vectors α, which can be written as ℓ_0 optimization:

$$\min \|\alpha\|_0 \quad s.t. \ y = \phi\psi\alpha, \ x = \psi\alpha \tag{4}$$

where $\|\alpha\|_0$ represents the number of nonzero elements in vector α. But unfortunately the above equation is in general NP-hard. So the basis pursuit (BP) [16] optimization aims to minimize ℓ_1:

$$\min \|\alpha\|_1 \quad s.t. \ y = \phi \psi \alpha, \ x = \psi \alpha \tag{5}$$

where $\|\alpha\|_1 = \sum_i |\alpha_i|$.

There are many optimization algorithms to solve the above problem, and some fundamental algorithm includes the greedy pursuit based algorithm, orthogonal greedy pursuit based algorithm, convex optimization algorithm and some variation of them. For two-dimensional images, another popular reconstruction algorithm is through the minimization of total variation (often called as the min-TV method), which offer the reconstructed images with better visual quality at much higher computational cost. Several fast greedy algorithms have also been proposed, such as the matching pursuit algorithm and orthogonal matching pursuit (OMP), the stage-wise orthogonal matching pursuit (StOMP) and iterative thresholding.

In this dissertation, the two compressive sensing based technologies are both based on the above model. Taking the iterative hard thresholding (IHT), a remarkably straightforward and rapid nonlinear signal reconstruction algorithm, as an example, it involves the application of the operators ϕ and ϕ^T once in each iteration, the iteration operation of updating the estimation of x can be described as:

$$x' = x + \phi^T (y - \phi x) \tag{6}$$

After iteration, the image is first transformed through ψ to yield α . Then, the largest K coefficients of were kept, while the rest were set to zeros [11]. After that, the inverse transform of the linear sparse transformation matrix ψ is applied to yield the reconstructed image. In this optimization algorithm, the initial solution influences the final result to a large scale. We can refer to [10] for other crucial factor about the sensing method and recovery optimization algorithm.

3.4   Empirical Results of Implications of Compressive Sensing on MIR

As to the applications of audio fingerprints, several practical requirements have been discussed to realize a successful audio fingerprinting system. Firstly, reliable audio fingerprinting is desirable in automatic music recognition. The audio fingerprinting system should be capable of identifying the distorted or corrupted audio clips in the case of degradations, that is, the system should be very robust. Robust fingerprint extraction method should be capable of dealing with severe degradations such as audio compression and large signal-to-noise ratios should be employed to represent the musical signal. Secondly, it should identify the clips of the item in the database in a few seconds, that is, the system should be time efficient or an audio fingerprinting system should be compact, finally, the system should be of low computational complexity in forming the fingerprints and matching algorithm for finding the best match item in the database. [10] proposed a general framework for what we call compressive signal processing (CSP), an alternative approach in which signal processing problems are solved directly in the compressive measurement domain without first resorting to a full scale signal reconstruction. They concluded from their empirical results that detection, classification, and estimation enable the extraction of information from the samples, while the filtering enables the removal of irrelevant information and separation of signals into distinct components. While these choices do not exhaust the set of canonical signal processing operations, they believe that they provide a strong initial foundation. In practical applications of compressive sensing, an alternative approach has been made by [11]. as a new spectral compressive sensing (SCS) theory for general frequency-sparse signals, by which their new SCS algorithms significantly outperform the current state-of-the-art CS algorithms while providing provable bounds on the number of measurements required for stable recovery. Compared with this method, other researchers also attempted to introduce a fast and efficient framework for practical compressive sensing, on which the framework is mainly based a novel design of Structurally Random Matrix (SRM). This framework showcases that the number of measurements for exact signal reconstruction is almost minimal. Simulation results with several interesting SRM under various practical settings are also presented to verify the validity of the theory as well as to illustrate the promising potentials of the proposed framework.

After summarizing recent results of how random Toeplitz and Circulant matrices can be easily (or even naturally) realized in various applications, [12] have introduced fast algorithms for reconstructing signals from incomplete Toeplitz and circulant measurements; and exhibited computational results showing that Toeplitz and circulantmatrices are not only as effective as random matrices for signal encoding, but also permit much faster signal decoding. [13] then conducted a high frequency analysis of (probabilistic) recoverability by the L1-based minimization/regularization principles, in which the absence of noise has shown that the L1-based solution can recover exactly the target of sparsity up to the dimension of the data either with the multiple-input-multiple-output (MIMO) measurement for the Born scattering or with the single-input-multiple-output (SIMO)/ multiple-input-single-output (MISO) measurement for the exact scattering. Meantime, a numerical exploration of compressed sampling theory has come into being, driven by a new greedy pursuit algorithm that computes sparse vectors that are difficult to recover; although it allows us to challenge theoretical identifiability criteria based on poly-topes analysis and on restricted isometry conditions, the theoretical analysis without resorting to Monte-Carlo sampling tends to avoid worst case scenarios. Based on intentional aliasing of the frequency components of the periodic signal while the reconstruction algorithm exploits recent advances in sparse representations and compressive sensing, [14] address the problem of sub-Nyquist sampling of periodic signals and show designs to capture and reconstruct such signals, concluding that for such signals the Nyquist rate constraint can be imposed on strobe-rate rather than the sensor-rate. Although numerous works have emphasized on measuring inequalities of randomized compressive operators, Wakin, et al. took endeavour to derive a concentration of measure bound for block diagonal matrices where the nonzero entries along the main diagonal blocks are sub-Gaussian random variables, concluding that the concentration exponent, in the best case, scales as that for a fully dense matrix and that the energy distribution of the signal plays in distinguishing the best case from the worst. In the paper [15] authors discussed a streaming CS framework and greedy reconstruction algorithm, the Streaming Greedy Pursuit (SGP), to reconstruct signals with sparse frequency content, in which their experimental results on very long signals demonstrate the good performance of the SGP for validation. To further probe the compressed sensing issue, the paper [16] considered the problem of estimating a sparse signal from a set of quantized, Gaussian noise corrupted measurements by employing two methods and finding that compressed sensing can be carried out when the quantization is very coarse. The recent theory of Compressed Sensing states that a signal, e.g. a sound record or an astronomical image, can be sampled at a rate much smaller than what is commonly prescribed by Shannon-Nyquist. The sampling of a signal can indeed be performed as a function of its intrinsic dimension" rather than according to its cut-off frequency. Probabilistic matching pursuit for compressive sensing has recently caught some attention by certain researchers, who overturned a previous held assumption in compressive sensing research. A novel matching pursuit algorithm has been presented that uses the measurements to probabilistically select a subset of bases that is likely to contain the true bases constituting the signal. The algorithm is successful in recovering the original signal in cases where deterministic matching pursuit algorithms fail. It is also known that exact recovery is possible where the number of nonzero coefficients is up to one less than the number of measurements. Recent results in compressed sensing show that a sparse or compressible signal can be reconstructed from a few incoherent measurements. Since noise is always present in practical data acquisition systems, sensing and reconstruction methods are developed assuming a Gaussian (light-tailed) model for the corrupting noise. However, some researchers have found that when the underlying signal and measurements are corrupted by impulsive noise, commonly employed linear sampling operators fail to recover a close approximation of the signal, in this case, they proposed robust methods for sampling and reconstructing sparse signals in the presence of impulsive noise by employing nonlinear measurement operator based on the weighted myriad estimator and a geometric optimization method. Their simulation results demonstrate that the proposed robust methods significantly outperform commonly used compressed sensing sampling and reconstruction methods in impulsive environments, at the same time, providing comparable performance in less demanding environments.

As the development of the sparse signal recovery research, some researchers [17] have focused on an optimal decentralized algorithm that demonstrates its application in monitoring localized phenomena exploiting energy-constrained large-scale wireless sensor networks. In their proposed algorithm, simulation results just corroborate with previous research outcomes that the sensing performance is globally optimal and attains a high spatial resolution commensurate with the node density of the original network containing both active and inactive sensors. Other researchers [18] studied the number of measurements required to recover a sparse signal in Cm with L nonzero coefficients from compressed samples in the presence of noise and proved that O (L) (an asymptotically linear multiple of L) measurements are necessary and sufficient if L grows linearly as a function of M. This result improves on the existing literature that is more focused on variants of a specific recovery algorithm based on convex programming and that O (L log (M-L)) measurements are required in the sub-linear regime (L = o (M)). Another group of researchers [16] introduced a model-based Compressive Sensing theory that parallels the conventional theory and provides concrete guidelines on how to create model-based recovery algorithms with provable performance guarantees. This paper highlights a new class of structured compressible signals along with a new sufficient condition for robust structured compressible signal recovery that they dub the restricted amplification property, which is the natural counterpart to the restricted isometry property of conventional CS.

Recovering or estimating the initial state of a high-dimensional system can require a potentially large number of measurements. Those researchers Error! Reference source not found. extrapolated how this controversial issue can be significantly brought down as to certain linear systems when randomized measurement operators are utilised, upon which their work builds resent results from field of Compressive Sensing with a high-dimensional signal containing few nonzero entries that can be efficiently recovered from a small number of random measurements, and they also illustrate their results with simple case study of a diffusion system. Aside from permitting recovery of sparse initial states, their analysis has potential applications in solving inference problems such as detection and classification of more general initial state. There is one aspect pertaining to compressive sensing principles that mainly concerns iterative sparse recovery for inverse and ill-posed problems, researchers in this field have developed their results by providing compressed measurement models for ill-posed problems and recovery accuracy estimates for sparse approximations of the solution of the underlying inverse problem. The main ingredients are formulations that allow the treatment of ill-posed operator equations in the context of compressively sampled data, on which Tikhonov variational and constrained optimization formulations are relied. One important breakthrough to the classical compressive sensing framework lies in the area which the incorporation of joint sparsity measures allow the treatment of infinite dimensional reconstruction spaces, thereby to reassert that theoretical results are furnished with a number of numerical experiments.

Compressive sensing is an emerging field based on the revelation that a small collection of linear projections of a sparse signal contains enough information for stable, sub-Nyquist signal acquisition. When a statistical characterization of the signal is available, Bayesian inference can complement conventional CS methods based on linear programming or greedy algorithms. In recently years, belief propagation (BP) decoding has been utilized to perform approximate Bayesian inference, which implies the CS encoding matrix as a graphical model and that fast computation can be obtained by reducing the size of the graphical model with sparse encoding matrices. And simulation results show that focusing on a two-state mixture Gaussian model will not prevent CS-BP from being easily adapted to other signal models.

4   Conclusion

In conclusion, this paper has explored theoretical foundation and development of compressive sensing model, and has elaborated on how it affects music information retrieval (MIR) system. This paper also reviewed the requirements for compressive sensing by illustrating their natural fit to MIR, and by illustrating and critically analysing four applications of Compressive Sensing in MIR. The paper placed great emphasize on an intuitive understanding of compressive sensing by describing the compressive sensing reconstruction as a process of interference cancellation. There is also a focus on the extrapolation of the driving factors in its applications, ranging from limitations imposed by feature extraction, the characteristics of fingerprint modeling, effective matching concerns in MIR, to practical fingerprint systems and its feasibility. In the end, this paper summed up implications of compressive sensing as a fundamental and crucial theoretical basis and a syndicated branch of compressive sensing for further developing and deepening the further work in the Music Information Retrieval field. Besides, along with revealing the fact that the concepts and approaches, points discussed potentially allow entirely new applications of Music Information Retrieval, we can conclude that CS-MIR is still in its infancy. Many crucial issues remain unsettled. These include: optimizing sampling trajectories, developing improved sparse transforms that are incoherent to the sampling operator, studying reconstruction quality in terms of clinical significance, and improving the speed of reconstruction algorithms. The important point is that complex tasks can be addressed easily in the sparse domain for large datasets and future work will explore other information retrieval tasks such as similarity search. Music informatics researchers may recognise the majority of the signal representations and machine learning algorithms applied; however the source material has important differences from musical signals (e.g. its temporal structure) which necessitate differences in approach.

References

1. Haitsma, J., Kalker, T.: A highly robust audio fingerprinting system. In: Proc. 3rd Int. Conf. Music Information Retrieval, pp. 107–115 (October 2002)
2. Cano, P., Batlle, E., Kalker, T., Haitsma, J.: A review of algorithms for audio fingerprinting. In: Proc. Int. Workshop Multimedia Signal Processing (2002)
3. Kulesh, V., Sethi, I., Petrushin, V.: Indexing and retrieval of music via gaussian mixture models. In: Proc. 3rd Int. Workshop on Content Based Multimedia Indexing, Rennes, France, pp. 201–205 (September 2003)
4. Ramalingam, A., Krishnan, S.: Gaussian Mixture Modeling of Short-Time Fourier Transform Features for Audio Fingerprinting. IEEE Transactions on Information Forensics and Security 1(4), 457–463 (2006)
5. Lin, H., Ou, Z., Xiao, X.: Generalized Time-Series Active Search With Kullback–Leibler Distance for Audio Fingerprinting. IEEE Signal Processing Letters 13(8), 465–468 (2006)
6. Haitsma, J., Kalker, T.: Speed-change resistant audio fingerprinting using auto-correlation. In: Proc. Int. Conf. Acoustics, Speech, and Signal Processing, vol. 4, pp. 728–731 (April 2003)
7. Candès, E., Romberg, J., Tao, T.: Robust uncertainty principles: Exact signal reconstruction from highly incomplete frequency information. IEEE Trans. Information Theory 52, 489–509 (2006)
8. Tropp, J.A., Gilbert, A.C.: Signal recovery from random measurements via orthogonal matching pursuit. IEEE Trans. Information Theory 53(12), 4655–4666 (2007)

9. Gan, L.: Block compressed sensing of natural images. In: Proc. Int. Conf. on Digital Signal Processing (DSP), Cardiff, UK (2007)
10. Davenport, M.A., et al.: Signal Processing with Compressive Measurements. IEEE Journal of Selected Topics in Signal Processing, 1 (February 2010) ISSN: 1932-4533
11. Duarte, M.F., Baraniuk, R.G.: Spectral Compressive Sensing, Rice University Technical Report TREE-1005 (February 2010)
12. Yin, W., et al.: Practical Compressive Sensing with Toeplitz and Circulant Matrices, Tech. rep., CAAM, Rice University (2010)
13. Fannjiang, A.C.: Compressive Inverse Scattering I. High Frequency SIMO/MISO and MIMO Measurements. Inverse Problem 26(3) (2010)
14. Veeraraghavan, A., et al.: Coded Strobing Photography: Compressive Sensing of High-speed Periodic Eve. IEEE Transactions of Pattern Analysis and Machine Intellegence (PAMI) 33(4), 671–686 (2011)
15. Boufounos, P.T., Asif, M.S.: Compressive Sampling for Streaming Signals with Sparse Frequency Conten. In: CISS, [10] (March 2010)
16. Zymnis, A., Boyd, S., Candes, E.: Compressed Sensing with Quantized Measurements. IEEE Signal Processing Letters 17(2) (February 2010)
17. Ling, Q., Tian, Z.: Decentralized Sparse Signal Recovery for Compressive Sleeping Wireless Sensor Network. IEEE Transactions on Signal Processing 58(7) (July 2010)
18. Akcakaya, M., Tarokh, V.: Shannon Theoretic Limits on Noisy Compressive Sampling. IEEE Transactions on Information Theory 56(1), 492–504 (2010)

Physical Bongard Problems

Erik Weitnauer and Helge Ritter

CoR-Lab, CITEC, Bielefeld University,
Universitätsstr. 21-23, 33615 Bielefeld, Germany
{eweitnau,helge}@techfak.uni-bielefeld.de

Abstract. In this paper, we introduce Physical Bongard Problems (PBPs) as a novel and potentially rich approach to study the impact the constraints of a physical world have on mechanisms of concept learning and scene categorization. Each PBP consists of a set of 2D physical scenes which are positive or negative examples of a concept that must be identified. We discuss the properties that make PBPs challenging, analyze computational and representational requirements for a computational solver, and describe a first implementation of such a system. It can solve a subset of non-trivial PBPs using a version space approach for achieving its scene categorizations. The key element is a physics engine that is used both for the construction of information-rich physical features and for the prediction of how a given situation might evolve.

Keywords: concept learning, scene categorization, physical understanding, physics simulation, analogy making.

1   Introduction

Despite the complex and dynamic nature of the world we live in, we are able to make sense of what happens around us. Already in early childhood, we build a sophisticated conceptual knowledge of our physical reality and are able to predict and visualize the outcome of many dynamic situations. Attempts to understand these striking abilities and their underlying processes need to take into account the important role of our physical embodiment [1]. Being embodied in a physical world requires the ability to rapidly capture the 'essence' of situations with respect to their physical interaction properties. This includes recognizing configurations that can provide physical support for an intended action, judging the feasibility of moving pieces, or 'perceiving' the imminent instability in a particular arrangement.

Having studied aspects of this challenge using advanced robot platforms in the context of grasping and manipulation [2], we here wish to introduce a complementary approach whose aim is to provide a maximally parsimonious, yet very rich framework to study mechanisms of *physics-based categorization*. To this end we introduce a novel class of problems inspired by and extending earlier work on pattern categorization [3]. Their essential characteristic is the embedding of an analogy detection task in the domain of physical situations. We call these

L. Iliadis et al. (Eds.): AIAI 2012, IFIP AICT 381, pp. 157–163, 2012.

problems *Physical Bongard Problems* (PBPs). Each PBP consists of a set of 2D physical scenes, containing four positive and four negative examples of the concept to be learned.

We argue that this problem class is well suited for research on our ability to conceptualize physical situations and make appropriate decisions in dynamic and interactive settings. Insights can be gained both by analyzing how humans solve PBPs, e.g., using questionnaires or eye tracking techniques, and by combining this empirical work with the development and analysis of computational solvers. In this paper, we introduce the domain of PBPs and give an overview of their properties and what makes them intricate to solve. We discuss the role of physical knowledge for PBPs, how it can be modeled using a standard physics engine and how a particular version space based solver implementation performs.

2   Physical Bongard Problems

In the design of Physical Bongard Problems we took inspiration from the class of Bongard problems (BPs), which are a set of 100 visual pattern recognition and categorization tasks, originally created by M. Bongard and extended by Douglas Hofstadter and Harry Foundalis [3–5]. Each BP consists of twelve images, six of them on the left and six on the right, all with an arbitrary pattern in black and white. The task is to identify the conceptual distinction between both sides. While many of the BPs are solved by humans rather intuitively, their computational solution is still an outstanding challenge.

In Physical Bongard Problems, while the task is the same, the images are taken from a physical domain, shifting the focus away from low-level visual processing towards dynamics and interaction. Instead of arbitrary static patterns, the images contain snapshots of 2D physical scenes depicted from a side perspective. PBPs can be considered as BPs which are more constrained by being 'embedded in physics', on the one hand, but can, on the other hand, as a consequence represent concepts not within the reach of the non-physical BPs. The scenes in PBPs may contain arbitrary-shaped non-overlapping rigid objects which do not move at the time $t = t_0$ of the snapshot. The solution of PBPs can be based on descriptions of the whole scene or parts of the scene at any point in time or on the reaction of objects to simple kinds of interaction, e.g., pushing. We have so far designed 34 PBPs, which can be viewed online [6]. Figure 1 depicts four of them.

2.1   Challenging Aspects

There are several challenging aspects of PBPs that make them both intricate to solve and an interesting object of research. In the following list, the first three aspects are unique to PBPs, while the further ones are shared by PBPs and classical BPs.

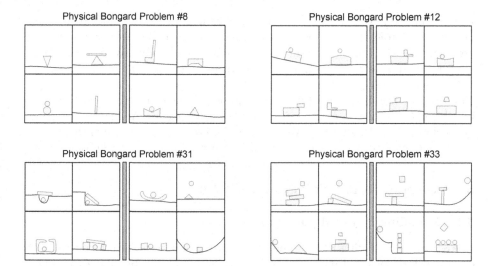

Fig. 1. Four Physical Bongard Problems. Solutions: (try yourself, first!) #8: configuration unstable vs. stable. #12: small object falls down vs. stays on large one. #31: circle is blocked vs. can be lifted. #33: construction stays intact vs. gets destroyed. See [6].

Physics. The need to invoke implicit physical knowledge of how the depicted object configuration will evolve (or respond to imagined physical interventions) for solving a problem is the main distinguishing characteristic of a PBP. This involves 'natural' assumptions, such as the association of some mass with each object and the presence of a downward directed gravity force. Using these assumptions, we can make physical judgments, e.g., about the stability of a configuration or predict likely states of motion (e.g., a 'ball' accelerating on a ramp).

Interaction. Physical understanding includes judgments about how objects might respond to imagined interventions. This is important in many situations in life, e.g., to judge whether some location can support my body, or how objects can be moved in a scene without causing unwanted inference to others.

Time. To see a scene as physical allows us to see it as a snapshot of a dynamical process. This connection generates a rich set of additional features arising from forward and backward predictions of the expected changes and can augment the scene with events that themselves are not depicted, like the collision of objects.

Grouping. Based on common features, relations or roles, several objects of one scene might have to be interpreted as a group to find a solution. There can be relations between groups or even groups of groups.

Focusing. In scenes with many objects, it is inefficient to consider the relations between all object pairs. Instead, a few important objects might have to be picked out while the others can be considered as 'background'.

Correspondence. When scenes contain groups of objects or relations between objects play a role, the mapping of two scenes requires to identify correspondence between two structured representations, which is highly non-trivial. This task is often referred to as analogy-making, an exciting research topic in itself.

Context. A suitable representation of a physical scene cannot be given a-priori, but depends on the context that is set by the other scenes. A single scene could be used in several PBPs and have a different interpretation in each of them, e.g., a different choice of what is the main object and what is the 'background'.

3   Computational Solvers for PBPs

3.1   Modeling Physical Knowledge

It is essential for solving PBPs to be able to predict and visualize the outcome of dynamic situations and interactions. We model this ability by giving the solver access to a physics engine[1] (PE). It is used in two ways: First, for the prediction of the unfolding of actions in the scenes. By constructing and simulating the scenes in the PE, the solver can inspect them at any time between the initial snapshot t_0 and the time all motion has stopped. Second, the engine is used to estimate physical object features. This includes features like object speed, acceleration and collision events, as well as concepts depending on interactions with objects in the scenes like pushing and pulling. We construct a basic notion of object stability by pushing the object briefly and observing its reaction. Its stability is judged by the distance it moves, where less movement correlates with more stability. A notion of the 'motion potential' or 'movability' of objects can be constructed by measuring the distance the objects can be pulled using a small force. A last feature derived from interaction is the role of an object as supporter of other objects in the scene. By imagining the scene without the object, i.e., by removing the object, it can be observed how the stability of the other objects depends on the removed one.

3.2   Implementation of a Basic Solver

An important and non-trivial decision for implementing any solver is the choice of a suitable input representation. Since PBPs are embedded in a physical world and only contain closed objects above some ground, the outlines and positions of these objects can be used as input representation without restricting the problem domain or making the problems significantly easier to solve. Using this input representation, we implemented a basic solver based on the version space algorithm for concept learning [7]. The hypothesis space contains all vectors `<side, numbers, distances, sizes, shapes, stabilities>`, where `side` is the side of the PBP ('left' or 'right'), `numbers` is a range of object count and the

---

[1] A physics engine is a piece of software that can perform physical simulations. We used the free Box2D physics engine in our experiments. See `http://box2d.org/`

remaining elements are disjunctions of feature values ('small', 'medium', 'large' or 'rectangle', 'circle', 'triangle', 'other' or 'stable', 'unstable', 'moving' or 'near', 'far', respectively). All elements except `side` can also take the value '?', in which case they match any scene. For example, the meaning of the hypothesis `<left, 1-3, ?, small or large, ?, stable>` is "all left scenes (and none of the right scenes) contain one to three objects that are small or large-sized and stable". The algorithm starts with a set of all possible hypotheses and then removes the incompatible ones for each scene. Finally, among the remaining hypotheses, the one with the shortest length is chosen as solution. If no solution could be found at $t = t_0$, the algorithms is applied to the scenes at $t = t_{end}$.

Results. The presented algorithm can solve the PBPs 1 to 5, 8, 11 and 18. It demonstrates the successful application of a physics engine in concept learning of dynamic physical scenes and constitutes a baseline for PBP solvers. Yet, due to its simplicity, the subset of PBPs that it solves still is small. It could be extended by adding more object and scene features. However, there are some principal limitations that cannot be overcome this way. Of the challenging aspects listed in Section 2.1, the present algorithm addresses *physics*, *interaction* and *time*, where the handling of time is only rudimentary and not sensitive to changes, durations or events. The other complexity sources of *grouping*, *focusing*, *correspondence* and *context sensitivity* cannot be adequately addressed by the present algorithm because 'flat' feature vectors are used to describe the scenes. Therefore, e.g., the selection of key objects in the scene and relations between objects cannot be captured (See PBP 12, 31 and 33 in Figure 1). The step to a more powerful algorithm will involve the use of structured scene representations. Building and mapping these representations is a task of analogy-making, and we will report in a subsequent paper on extensions along this line.

4   Related Work

The interpretation of and reasoning about physical scenes has a long tradition in artificial intelligence in the field of qualitative physics, where physical knowledge is represented as high-level logical rule systems [8, 9]. We chose to provide physical knowledge in another, less rigid and more analog form: with a physics engine. This way, we are not committed to a certain level of abstraction and can start building representations at a low level.

Traditionally, much research on concept learning has been done in the context of unstructured domains where a concept can be represented as a set of attribute values [10]. Recently, more attention was paid to learning structured concepts in the domain of description logics [11]. The learning of concepts from dynamic examples as presented in this work, has not been in the focus of concept learning research, so far.

There have only been few attempts to develop computational solvers for classical Bongard problems. The only solver that is able to come up with solutions of some BPs without using hand-crafted input representations is the 'Phaeaco' system by H. Foundalis [5]. It builds and maps representations in a process of analogy-making performed by a complex adaptive system. See [12] for a summary of computational approaches to analogy-making.

5   Conclusion

In this paper, we made two main contributions. First, we introduced Physical Bongard Problems as a novel research tool for concept learning and scene categorization by agents situated in a physical world. We discussed the aspects of Physical Bongard Problems that make them a challenge for computational solvers, which are *physics, interaction, time, grouping, focusing, correspondence* and *context sensitivity*. As a second contribution, we demonstrated how a physics engine can be effectively used to equip an algorithm with the physical understanding necessary to solve PBPs. The engine is used for both scene prediction and construction of information-rich physical features through simulated object interactions. We showed the feasibility of this approach with a basic PBP solver implementation and discussed its limitations, which are mainly a result of using unstructured collections of features as scene representations.

The step to a more powerful solver will require the use of structured representations and the extension of the basic solver with dynamic scene-encoding and structure-mapping capabilities. These two abilities are central topics in analogy-making and we are currently exploring how existing analogy-making algorithms can be adapted for the use in a PBP solver.

References

1. Pfeifer, R., Bongard, J., Grand, S.: How the body shapes the way we think: a new view of intelligence. The MIT Press (2007)
2. Ritter, H., Haschke, R., Röthling, F., Steil, J.J.: Manual Intelligence as a Rosetta Stone for Robot Cognition. In: Kaneko, M., Nakamura, Y. (eds.) Robotics Research. STAR, vol. 66, pp. 135–146. Springer, Heidelberg (2010)
3. Bongard, M.M.: Pattern Recognition. Hayden Book Co., Spartan Books, Rochelle Park (1970)
4. Hofstadter, D.R.: Gödel, Escher, Bach: an eternal golden braid. Harvester Press (1979)
5. Foundalis, H.E.: Phaeaco: A cognitive architecture inspired by Bongard's problems. PhD thesis, Indiana University (2006)
6. Weitnauer, E.: Physical bongard problems (2012), http://naive-physics.com/pbp/
7. Mitchell, T.M.: Generalization as search. Artificial Intelligence 18(2), 203–226 (1982)
8. Forbus, K.D.: Qualitative process theory: Twelve years after. Artificial Intelligence in Perspective 59(1), 115 (1994)
9. Kurtoglu, T., Stahovich, T.F.: Interpreting schematic sketches using physical reasoning. In: AAAI Spring Symposium on Sketch Understanding, pp. 78–85 (2002)

10. Goodman, N.D., Tenenbaum, J.B., Feldman, J., Griffiths, T.L.: A rational analysis of Rule-Based concept learning. Cognitive Science 32(1), 108–154 (2008)
11. Lehmann, J.: DL-Learner: learning concepts in description logics. The Journal of Machine Learning Research 10, 2639–2642 (2009)
12. Gentner, D., Forbus, K.D.: Computational models of analogy. Wiley Interdisciplinary Reviews: Cognitive Science 2(3), 266–276 (2011)

Taxonomy Development and Its Impact on a Self-learning e-Recruitment System

Evanthia Faliagka[1], Ioannis Karydis[3], Maria Rigou[1], Spyros Sioutas[3], Athanasios Tsakalidis[1], and Giannis Tzimas[2]

[1] Computer Engineering and Informatics Dept., University of Patras, Patras, Greece
{faliagka,rigou}@ceid.upatras.gr
[2] Dept. of Applied Informatics in Management & Economy,
Technological Educational Institute of Messolonghi, Messolonghi, Greece
{tsak,tzimas}@cti.gr
[3] Dept. of Informatics, Ionian University, 49100, Kerkyra, Greece
{karydis,sioutas}@ionio.gr

Abstract. In this work we present a novel approach for evaluating job applicants in online recruitment systems, using machine learning algorithms to solve the candidate ranking problem and performing semantic matching techniques. An application of our approach is implemented in the form of a prototype system, whose functionality is showcased and evaluated in a real-world recruitment scenario. The proposed system extracts a set of objective criteria from the applicants' LinkedIn profile, and compares them semantically to the job's prerequisites. It also infers their personality characteristics using linguistic analysis on their blog posts. Our system was found to perform consistently compared to human recruiters, thus it can be trusted for the automation of applicant ranking and personality mining.

Keywords: e-recruitment, personality mining, recommendation systems, data mining.

1 Introduction

In the recent years an increasing number of people turn to the web for job seeking and career development while a lot of companies use online knowledge management systems to hire employees, exploiting the advantages of the World Wide Web [1]. The information systems used to support these tasks are termed e-recruitment systems and automate the process of publishing position openings and receiving applicant CVs, thus allowing Human Resource (HR) agencies to target a very wide audience at a small cost. At the same time this situation may as well prove overwhelming to HR agencies that need to allocate human resources for manually assessing the candidate resumes and evaluating the applicants' suitability for the positions at hand. Automating the process of analyzing the applicant profiles to determine the ones that best fit the specifications of a given job position could lead to a significant gain in terms of efficiency. For example,

L. Iliadis et al. (Eds.): AIAI 2012, IFIP AICT 381, pp. 164–174, 2012.

it is indicative that SAT Telecom India reported 44% cost savings and a drop in average time needed to fill a vacancy from 70 to 37 days [2] after deploying an e-recruitment system.

Several e-recruitment systems have been proposed with an objective to speed-up the recruitment process, leading to a better overall user experience. *E-Gen* system [3] performs analysis and categorization of unstructured job offers (i.e. in the form of unstructured text documents), as well as analysis and relevance ranking of candidates. In contrast to a free text description, the usage of a common "language" in the form of a set of controlled vocabularies for describing the details of a job posting would facilitate communication between all parties involved and would open up the potential of the automation of various tasks within the process [4]. Another benefit from having postings annotated with terms from a controlled vocabulary is that the terms can be combined with background knowledge about an industrial domain. Job portals could offer semantic matching services which would calculate the semantic similarity between job postings and applicants' profiles based on background knowledge about how different terms are related. For example, if *Java* programming skills are required for a certain job and an applicant is experienced in *Delphi*, the matching algorithm would consider this person's profile a better match than someone else's who has the skill *SQL*, since *Delphi* and *Java* are more closely related than *SQL* and *Java*. This approach allows for comparison of job position postings and applicants' profiles using background knowledge instead of merely relying on the containment of keywords, like traditional search engines do.

CommOn framework [5] applies Semantic Web technologies in the field of HR Management, while *HR-XML* can partly support the "standardized" representation of competency profiles [6]. In this framework the candidate's personality traits, determined through an online questionnaire which is filled-in by the candidate, are considered for recruitment. In order to match applicants with job positions these systems typically combine techniques from classical IR and recommender systems, such as relevance feedback [3], semantic matching in job seeking and procurement tasks [7], Analytic Hierarchy Process [8,9] and NLP technology used to automatically represent CVs in a standard modeling language [10]. These methods, although useful, suffer from the discrepancies associated with inconsistent CV formats, structure and contextual information. In addition approaches that incorporate ontological information for determining the degree of position-to-applicant matching face significant complexity problems concerning the development of the required ontological structure and associations. This problem appears even when trying to reuse available ontologies (ontology discovery through evaluation to ontology integration and merging), a task that requires considerable manual work [11]. What's more, these methods are unable to evaluate some secondary characteristics associated with CVs, such as style and coherence, which are very important in CV evaluation.

Such approaches attempt to match terms found in CV descriptions to job position descriptions. In this work a different approach is adapted in the sense that the semantic matching primarily concerns applicant skills as denoted in the respective LinkedIn profile descriptions. Applicant skills are then semantically associated with equivalent concepts from job descriptions as specified by the recruiter, who constructs a list of required job position skills using a predefined IT skills hierarchy. Hierarchy skills are contained in the LinkedIn skills but also the hierarchy integrates even broader skills ending up to the root of "IT skills".

The system described in this work, attempts to solve the candidate ranking problem by applying a set of supervised learning algorithms in combination with a semantic skills matching mechanism, for automated e-recruitment. It is an integrated company oriented e-recruitment system that automates the candidate pre-screening and ranking process. Applicant evaluation is based on a predefined set of objective criteria, which are directly extracted from the applicant's LinkedIn profile. What's more, the candidate's personality characteristics, which are automatically extracted from his social presence [12], are taken into account in his evaluation. Our objective is to limit interviewing and background investigation of applicants solely to the top candidates identified from the system, so as to increase the efficiency of the recruitment process. The system is designed with the aim of being integrated with the companies' Human Resource Management infrastructure, assisting and not replacing the recruiters in their decision-making process.

2   System Overview

In this work, we have implemented an integrated company oriented e-recruitment system that automates the candidate evaluation and pre-screening process. Its objective is to calculate the applicants' relevance scores, which reflect how well their profile fits the position's specifications. In this Section we present an overview of the proposed system's architecture and candidate ranking scheme.

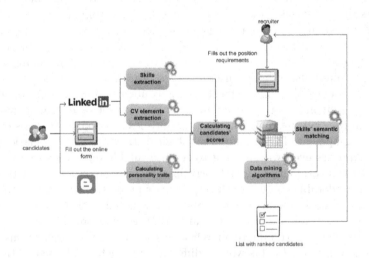

Fig. 1. System's Architecture

2.1   Architecture and Implementation

The proposed e-recruitment system implements automated candidate ranking based on a set of credible criteria, which will be easy for companies to integrate with their existing Human Resources Management infrastructure. In this study we focus on 5 complementary selection criteria, namely: Education (in years of formal academic training), Work Experience (in months of related experience), Loyalty (average number of years spent per job), Extraversion and skills. The system's architecture, which is shown in Figure 1, consists of the following components:

- *Job Application Module*: Implements the input forms that allow the candidates to apply for a job position.
- *Personality Mining Module*: If the candidate's blog URL is provided, applies linguistic analysis to the blog posts deriving features reflecting the author's personality.
- *Semantic Matching*: Calculates the semantic distance between candidate skills and prior experience, as extracted from the respective LinkedIn profile and job position requirements.
- *Applicant Grading Module*: Combines the candidate's selection criteria to derive the candidate's relevance score for the applied position. The grading function is derived through supervised learning algorithms.

The proposed e-recruitment system was fully implemented as a web application, in the Microsoft .Net development environment. Job applicants are given the option to authenticate using their LinkedIn account credentials to apply for one or more of the available job positions. This allows the system to automatically extract the selection criteria required for candidate pre-screening from the applicants' LinkedIn profile, so the user experience is streamlined. As part of the job application process, candidates are asked to fill-in the feed URI of their personal blog. This allows our system to syndicate the blog content and calculate the extraversion score with the personality mining technique presented in Section 2.3.

On the recruiter's side, the system after authentication provides access rights to post new job positions and evaluate job applicants. In the "rank candidates" menu, the recruiter is presented with a list of all available job positions and the candidates that have applied for each one of them. Upon the recruiter's request, the system estimates applicants' relevance scores and ranks them accordingly. This is achieved by calling the corresponding Weka classifier, via calls to the API provided by Weka software [13]. The recruiter can modify the candidate ranking, by assigning new relevance scores to the candidates. This will improve the future performance of the system, as the recruiter's suggestions are incorporated in the system's training set and the ranking model is updated. It must be noted here that the ranking model is initialized as a simple linear combination of the selection criteria, until sufficient input is provided from the recruiters to build a training set.

2.2 Semantic Matching

In the previous version of the system [14] it was found that except from senior positions that required domain experience and specific qualifications, our system performed consistently with a Pearson's correlation of up to 0.85. The present expanded version of the system tackles the problem of specific qualifications and experience in senior positions and demonstrates improved accuracy (as will be presented in Section 3) by deploying semantic matching technologies.

The data exchange between employers, applicants and job portals in a Semantic Web-based recruitment scenario is based on a set of vocabularies which provide shared terms to describe occupations, industrial sectors and job skills [15]. Semantic matching is a technique which combines annotations using controlled vocabularies with background knowledge about a certain application domain. In our case, the domain specific knowledge is represented by a taxonomy of IT skills (Figure 2). A taxonomy is defined as a set of categories or terms organized into a hierarchy with parent-child relationships and implied inheritance, meaning that a child term (ie, C) has all of the characteristics of its parent term (ie, Structured). A taxonomy only contains broader and narrower relationships.

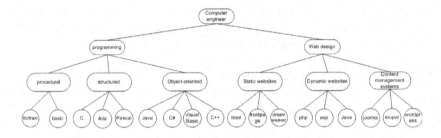

Fig. 2. Part of the implemented IT skills taxonomy

The implemented taxonomy serves a dual role:

1. Matches the applicants' skills as stated in the respective LinkedIn profile and the job position requirements as specified in the job description and rejects all candidates that don't fulfill the requirements.
2. Searches the text of job title and job description of the job experience section in the applicant's LinkedIn profile and identifies terms corresponding to skills required by the recruiter. Thus, in the current system version, the calculation of the job experience criterion takes into account only the job experience that concerns relative competencies.

It is important to clarify that in both cases we do not use a simple keyword search but a concept search. First, for the specific job position a skills search is applied to the candidate skills, as specified in the respective LinkedIn profile (Figure 3). In most cases a recruiter does not ask for specialized competencies but resorts to more general qualifications, such as object-oriented programming (as opposed to Java or C#). In this case the proposed algorithm searches the hierarchy tree and identifies the leaves with the node of the skill required by the recruiter as their lowest (nearest) common ancestor (for instance, object-oriented programming). Next, the identified leaves are examined to determine if there is a match with the skills stated by the candidate. In the case that these is no match then the candidate is excluded from the ranking process.

For those candidates that were found to have the necessary skills a second search is conducted to determine whether one or more of the candidate's past work experience belongs to the same domain of expertise as the job position of interest. The algorithm applied for this purpose can be briefly described as follows: Let $S1$ be the skills corresponding to a past job position $E1$ as stated in the work experience section (title or description text) of the respective LinkedIn profile. Also, let S be the skills required by the current job position, corresponding to the job position domain and may be found at any level of the hierarchy. If there is an overlap between S and $S1$ ($S \cap S1 \neq \emptyset$), the past job position $E1$ is regarded as relevant and thus is taken into account in the relevant job experience calculations.

Fig. 3. LinkedIn skills example

2.3 Personality Mining

Previous works have shown that by applying linguistic analysis to blog posts, the author's personality traits can be derived [16] as well as his mood and emotions [17]. The text analysis in these works is performed with LIWC (Linguistic Inquiry and Word Count) system, which analyzes written text samples and extracts linguistic features that act as markers of the author's personality. Pennebaker and King [18] have found significant correlations between these frequency counts and the author's personality traits, as measured by the Big-Five personality dimensions.

In this work, we focus on the extraversion personality trait, due to its importance in candidate selection. Extraversion is a crucial personality characteristic in positions that interact with customers, while social skills are important for team work. Specifically, the emotional positivity and social orientation of candidates, both directly extracted from LIWC frequencies, can act as predictors of extroversion trait [12]. We measured extraversion using the candidate's blog posts, which are input to the TreeTagger tool [19] for lexical analysis and lemmatization. Then, using the LIWC dictionary, our system classifies the canonical form of words output from TreeTagger in one of the word categories of interest (i.e. positive emotion, negative emotion and social words) and calculates the LIWC scores. Finally, the system estimates the applicant's extraversion score.

An expert recruiter has assigned extraversion scores to each of 100 job applicants with personal blogs, which were part of a large-scale recruitment scenario. The recruiter's scores were used to train a regression model, which predicts the candidates' extraversion from their LIWC scores in the posemo, negemo, social categories. In what follows, a linear regression model was selected as a predictor of the extraversion score E, as proposed in [20], due to its increased accuracy and low complexity. Equation 1 corresponds to the linear model that minimizes the Mean Square Error between actual values assigned by the recruiter and predicted scores output by the model:

$$E = S + 1.335 * P - 2.250 * N \tag{1}$$

where S is the frequency of social words (such as friend, buddy, coworker) returned from LIWC, P is the frequency of positive emotion works and N is the frequency of negative emotion words.

2.4   Candidate Ranking

The proposed system leverages machine learning algorithms to automatically build the applicant ranking models. This approach requires sufficient training data as an input, which consist of previous candidate selection decisions. Methods that learn how to combine predefined features for ranking by means of supervised learning algorithms are called "learning-to-rank" methods.

In the typical "learning to rank" process a training set is used that consists of past candidate applications represented by feature vectors, denoted as $x_i^{(k)}$, along with an expert recruiter's judgment of the candidates' relevance score, denoted as y_i. The training set is fed to a learning algorithm which constructs the ranking model, such that its output predicts the recruiter's judgment when given the candidates' feature vector as an input. In the test phase the learned model is applied to sort a set of candidate applications, and return the final ranked list of candidates.

In our problem, a scoring function $h(x)$ outputs the candidate relevance score, which reflects how well a candidate profile fits the requirements of a given job position. Then the system outputs the final ranked list by applying the learned function to sort the candidates. The true scoring function is usually unknown and an approximation is learned from the training set D. In the proposed system the training set consists of a set of N previous candidate selection examples, given as an input to the system (Equation 2):

$$D = \{(x_i, y_i) | x_i \in R^m, y_i \in R)\}_{i=1}^{N} \tag{2}$$

3 Experimental Evaluation

The proposed system was tested in a real-world recruitment scenario, to evaluate its effectiveness in ranking job applicants. The system's performance evaluation is based on how effective it is in assigning consistent relevance scores to the candidates, compared to the ones assigned by human recruiters.

In the recruitment scenario used in our tests, we compiled a corpus of 100 applicants with a LinkedIn account and a personal blog, as these are key requirements of the proposed system. The same corpus was used in a previous version of the system [14] for comparison reasons. The applicants were selected randomly via Google blog search API with the sole requirement of having a technical background, as indicated by the blog metadata (list of interests), as well as a LinkedIn profile. Our corpus of job applicants was formed by choosing the first 100 blogs returned from the profile search API that fulfilled our preconditions. We also collected three representative technical positions announced by an unnamed IT company with different requirements, i.e. a sales engineering position, a junior programmer position and a senior programmer position.

He sales engineering position favors a high degree of extraversion, while experience is the most important feature for senior programmers. Junior programmers are mainly judged by loyalty (as a company would not invest in training an individual prone to changing positions frequently) as well as education. What's more, each position has its own desired set of skills, which are semantically matched with the skill-set reported by each user at their LinkedIn profile. Specifically, the junior position requires programming skills in C++ or Java development languages, while the senior position requires a 5-year experience in J2EE technologies. The use of different requirements per position is expected to test the ability of our system to match candidates' profiles with the appropriate job position.

In our experiments, we assume that each applicant in the corpus has applied for all three available job positions. For each job position, applicants were ranked according to their suitability for the job position both by the system (automated ranking) and by an expert recruiter. Human recruiters had access to the same information as the system, i.e. the candidate's blog and LinkedIn profile. It must be noted though that despite the fact that the selection criteria are known to the system, the recruiter's interpretation of the data and the exact decision-making process is unknown and must be learned.

Table 1. Correlation coefficients for applicants' relevance scores vs. different machine learning models

Correlation coefficient	LR		M5' Tree		REP Tree		SVR, poly		SVR, PUK	
	TE	RE	TE	RE	TE	RE	TE	RE	TE	RE
Sales engineer	0.74	0.74	0.81	0.81	0.81	0.81	0.61	0.61	0.81	0.81
Junior programmer	0.79	0.81	0.85	0.85	0.84	0.86	0.81	0.81	0.84	0.86
Senior programmer	0.64	0.73	0.63	0.71	0.68	0.80	0.62	0.68	0.73	0.82

In our first experiment, we use Weka to evaluate the learning-to-rank models. Specifically, we test the correlation of the scores output from the system (i.e. model predictions) with the actual scores assigned by the recruiters, using the Pearson's correlation coefficient metric. Table 1 shows the correlation coefficients for 4 different machine learning models, namely: *Linear Regression* (LR), *M5'* model tree (M5'), *REP Tree* decision tree (REP), and *Support Vector Regression* (SVR) with two non-linear kernels (i.e. polynomial kernel and PUK universal kernel). For each machine learning model we show the results derived using the Total Experience for a candidate (TE) and those that derived using only the Relevant Experience (RE).

As it can be seen, the Tree models and the SVR model with a PUK kernel produce the best results. On the other hand Linear Regression performs poorly, suggesting that the selection criteria are not linearly separable. It must be noted here that all values are averages, obtained with the 10-fold cross validation technique. For the sales position, the recruiter's judgment is dominated by the highly subjective extraversion score, thus increasing the uncertainty of the overall relevance score. Still, the system was able to achieve a correlation coefficient of up to 0.81, depending on the regression model used. On the other hand, the selection of junior programmer candidates is based on more objective criteria such as loyalty and education, thus resulting in a slightly higher correlation coefficient, up to 0.86. Finally, the senior programmer's position exhibited high consistency, with a Pearson's correlation of up to 0.82.

Concerning the first job position (i.e. sales engineer), there was no difference in the results of the two approaches as the relevant experience has no effect on the score calculations. For this position the candidate may have prior experience in any domain or industry (ranging from programmer to salesman) and thus the derived model exactly matches the model based on a candidate's total experience. In the case of the second job position, where only the relevant experience is taken into account, there is a slight difference in the consistency of the two approaches due to the small effect of the experience criterion to the overall score. In the last job position, where the weight of the experience criterion is increased, the difference in the correlation coefficient is clearly observed. More specifically, the values of the correlation coefficient are significantly improved (reaching up to 0.82 in the case of Support Vector Regression with PUK kernel) resulting in consistency values quite comparative to the other two job positions.

4   Conclusions

In this work we present a novel approach for evaluating job applicants in online recruitment systems, using machine learning algorithms to solve the candidate ranking problem and performing semantic matching techniques. The proposed scheme relics on objective criteria extracted from the applicants' LinkedIn profile and subjective criteria extracted from their social presence, to estimate applicants' relevance scores and infer their personality traits. Candidates that do not possess the required skills are filtered out of the selection process and for those remaining the relevant job experience is calculated using semantic matching techniques that allow significantly improved results. The implemented system was employed in a large-scale recruitment scenario, which included three different offered positions and 100 job applicants. The application of the approach in the real-world setting revealed that it is effective in calculating the applicants' suitability for a given job and ranking them accordingly.

References

1. Meo, P.D., Quattrone, G., Terracina, G., Ursino, D.: An xml-based multiagent system for supporting online recruitment services. IEEE Transactions on Systems, Man, and Cybernetics, Part A 37(4), 464–480 (2007)
2. Pande, S.: E-recruitment creates order out of chaos at SAT telecom: System cuts costs and improves efficiency. Human Resource Management International Digest 19(3), 21–23 (2011)
3. Kessler, R., Torres-Moreno, J.M., El-Bèze, M.: E-gen: automatic job offer processing system for human resources. In: Proc. Mexican International Conference on Advances in Artificial Intelligence, pp. 985–995 (2007)
4. Bizer, C., Heese, R., Mochol, M., Oldakowski, R., Tolksdorf, R., Eckstein, R.: The impact of semantic web technologies on job recruitment processes. In: Proc. Internationale Tagung Wirtschaftsinformatik (2005)
5. Radevski, V., Trichet, F.: Ontology-based systems dedicated to human resources management: An application in e-recruitment. In: OTM Workshops, vol. (2), pp. 1068–1077 (2006)
6. Dorn, J., Naz, T., Pichlmair, M.: Ontology development for human resource management. In: Proc. International Conference on Knowledge Management, pp. 109–120 (2007)
7. Mochol, M., Wache, H., Nixon, L.J.B.: Improving the Accuracy of Job Search with Semantic Techniques. In: Abramowicz, W. (ed.) BIS 2007. LNCS, vol. 4439, pp. 301–313. Springer, Heidelberg (2007)
8. Faliagka, E., Ramantas, K., Tsakalidis, A.K., Viennas, M., Kafeza, E., Tzimas, G.: An integrated e-recruitment system for cv ranking based on ahp. In: WEBIST, pp. 147–150 (2011)
9. Faliagka, E., Tsakalidis, A., Tzimas, G.: An integrated e-recruitment system for automated personality mining and applicant ranking. Internet Research (2012)
10. Amdouni, S., Ben Abdessalem Karaa, W.: Web-based recruiting. In: Proc. ACS/IEEE International Conference on Computer Systems and Applications, pp. 1–7 (2010)
11. Mochol, M., Paslaru, E., Simperl, B.: Practical guidelines for building semantic erecruitment applications. In: Proc. International Conference on Knowledge Management, Special Track: Advanced Semantic Technologies (2006)

12. Faliagka, E., Kozanidis, L., Stamou, S., Tsakalidis, A., Tzimas, G.: A Personality Mining System for Automated Applicant Ranking in Online Recruitment Systems. In: Auer, S., Díaz, O., Papadopoulos, G.A. (eds.) ICWE 2011. LNCS, vol. 6757, pp. 379–382. Springer, Heidelberg (2011)
13. Hall, M., Frank, E., Holmes, G., Pfahringer, B., Reutemann, P., Witten, I.H.: The weka data mining software: an update. SIGKDD Explor. Newsl. 11(1), 10–18 (2009)
14. Faliagka, E., Ramantas, K., Tsakalidis, A., Tzimas, G.: Application of machine learning algorithms to an online recruitment system. In: Proc. International Conference on Internet and Web Applications and Services (2012)
15. Liu, T.Y.: Learning to rank for information retrieval. Found. Trends Inf. Retr. 3(3), 225–331 (2009)
16. Gill, A., Nowson, S., Oberlander, J.: What are they blogging about? personality, topic and motivation in blogs (2009)
17. Mishne, G.: Experiments with mood classification in blog posts. In: Proc. Workshop on Stylistic Analysis of Text For Information Access (2005)
18. Pennebaker, J.W., King, L.A.: Linguistic styles: language use as an individual difference. Journal of Personality and Social Psychology 77(6), 1296–1312 (1999)
19. Schmid, H.: Improvements in part-of-speech tagging with an application to german. In: Lexikon und Text, pp. 47–50 (1995)
20. Mairesse, F., Walker, M.A., Mehl, M.R., Moore, R.K.: Using linguistic cues for the automatic recognition of personality in conversation and text. Journal of Artificial Intelligence Research 30, 457–500 (2007)

Fuzzy Energy-Based Active Contours Exploiting Local Information

Stelios Krinidis and Michail Krinidis

Information Management Department
Technological Institute of Kavala,
Ag. Loukas, 65404 Kavala, Greece
stelios.krinidis@mycosmos.gr, mkrinidi@gmail.com

Abstract. This paper presents a novel fast and robust model for active contours to detect objects in an image, based on techniques of curve evolution. The proposed model can detect objects whose boundaries are not necessarily defined by gradient, based on the minimization of a fuzzy energy. This fuzzy energy is used as the model motivation power evolving the active contour, which will stop on the desired object boundary. The fuzziness of the energy provides a balanced technique with a strong ability to reject "weak", as well as, "strong" local minima. Also, this approach differs from previous methods, since it does not solve the Euler-Lagrange equations of the underlying problem, but, instead, calculates the fuzzy energy alterations directly. So, it converges to the desired object boundary very fast. The theoretical properties and various experiments presented demonstrate that the proposed fuzzy energy-based active contour is better and more robust than classical snake methods based on the gradient or other kind of energies.

Keywords: Active contour, deformable curve, curve evolution, fuzzy logic, energy-based.

1 Introduction

Image segmentation is one of the first and most important tasks in image analysis and computer vision [1]. The design of robust and efficient segmentation algorithms is still a very challenging research topic, due to the variety and complexity of images.

Since the introduction of snakes [2], active contours have been applied to a variety of problems in image processing and computer vision such as segmentation and feature extraction, visual tracking, etc. The basic idea in active contour models or snakes is to evolve a curve, in order to detect objects in the image under consideration. For instance, starting with a curve around the object to be detected, the curve moves toward its interior normal and has to stop on the boundary of the object.

Geometric active contour models [3], where introduced shortly afterwards, based on curve evolution theory, which could also handle topology changes very

L. Iliadis et al. (Eds.): AIAI 2012, IFIP AICT 381, pp. 175–184, 2012.

naturally, when implemented using level set methods proposed by Osher and Sethian [4].

The geometric active contour model, which is most related to the original snake model, is probably the *geodesic active contour* model [5, 6], which has been based on the curvature of the image and on an *inflationary force* [7].

Cohen *et al.* [8] have proposed the minimal path technique, a semi-automatic method, which captures the global minimum of a contour energy between two fixed user-defined end points, exploiting an oriented graph characterized by its cost function. The object boundary detection becomes the optimal path search problem between two user-defined points in the graph. This approach leads the snake-like energy to a global minimum, avoiding any local minima.

Other implementations have also been proposed for capturing more global minimizers by restricting the search space. Dual snakes, proposed by Gunn and Nixon [9, 10], is such a method. This method uses two interlinked snakes instead of one. Similar methods were also proposed in [11–13]. These methods are restricted to detection of objects with simple topologies. Also, active contours have been combined with the optimization tool of graph-cuts [14], such as morphological dilation in order to restrict the search space for graph-cuts segmentation. However, this method cannot handle images with multiple objects simultaneously.

All these classical snakes and active contour models are known as "edge-based" models, since they rely on edge-functionals to stop the curve evolution. Also, these models can detect only objects with edges defined by gradient. The performance of the purely edge-based models is often inadequate. There has been much research into the design of complex region-based energy functionals [15–19] utilizing image information not only near the evolving contour, but image statistics inside and outside the contour as well in order to achieve better performance. Unfortunately, most of these region-based energy functionals assume highly constrained models for pixel intensities within each region.

This paper deals with the above mentioned problems. It presents a novel, fast and robust fuzzy energy-based active contour for image segmentation, which can handle objects whose boundaries are not necessarily defined by gradient. Generally, fuzzy methods provide more accurate and robust data clustering, thus, we combine it with active contour methodology, introducing here a model as a fuzzy energy-based minimization. The fuzziness of the energy provides a balanced technique with a strong ability to reject "weak", as well as, "strong" local minima. Furthermore, we formulate the model in terms of pseudo-level set functions, but instead of computing the associated Euler-Lagrange equations, we apply a direct method to solve the corresponding equations without numerical stability constraints.

The remainder of the paper is organized as follows. The description of the model and its fuzzy motivation energy is presented in Section 2. Experimental results are presented in Section 3 and conclusions are drawn in Section 4.

2   Fuzzy Energy-Based Model

Let us define the evolving curve C in the image domain Ω. The proposed approach is based on the minimization of a fuzzy energy-based segmentation. Firstly, let us assume that the image I is formed by two regions of approximately piecewise-constant intensities and the object to be detected is represented by a region and its boundary by C_0. Now, let us consider the following functionals:

$$F_1(C) + F_2(C)$$
$$= \int_\Omega [u(x,y)]^m \left[|I(x,y) - c_1|^2 + \int_\Omega w_{xy,x'y'} R_1(x',y') dx' dy' \right] dx dy$$
$$+ \int_\Omega [1 - u(x,y)]^m \left[|I(x,y) - c_2|^2 + \int_\Omega w_{xy,x'y'} R_2(x',y') dx' dy' \right] dx dy,$$

where c_1 and c_2 are constants depending on C, expressing the average prototypes of the image regions inside and outside respectively of C. The membership function $u(x,y) \in [0,1]$ is the degree of membership of $I(x,y)$ to the inside of C, and m is a weighting exponent on each fuzzy membership. The term $w_{xy,x'y'}$ incorporates the spatial dependence between the image points, that is, a term that quantifies the degree of spatial contiguity of the points $I(x,y)$ and $I(x',y')$. The spatial structure of a given data set is defined by using a matrix W:

$$w_{xy,x'y'} = \begin{cases} \frac{1}{d_{xy,x'y'}+1}, & \begin{array}{l} \text{if } I(x,y) \text{ and } I(x',y') \text{ are} \\ \text{neighbors and} \\ [x,y]^T \neq [x',y']^T, \end{array} \\ 0, & \text{otherwise,} \end{cases} \qquad (1)$$

where $d_{xy,x'y'}$ is the distance between points $I(x,y)$ and $I(x',y')$. Also, terms $R_1(x',y')$ and $R_2(x',y')$ are used for regularizing the data set points among the clusters, and given by:

$$R_1(x',y') = [1 - u(x',y')]^m |I(x',y') - c_1|^2,$$
$$R_2(x',y') = [u(x',y')]^m |I(x',y') - c_2|^2. \qquad (2)$$

In this simple case, it is obvious that the boundary of the object C_0, is the minimizer of the "fitting" term:

$$\inf_C \{F_1(C) + F_2(C)\} \approx 0 \approx F_1(C_0) + F_2(C_0). \qquad (3)$$

The proposed active contour is based on the minimization of the above fitting term, taking into account the length of the model C as a regularization term. Therefore, the energy functional $F(C, c_1, c_2, u)$ is introduced as:

$$F(C, c_1, c_2, u)$$
$$= \mu \cdot Length(C)$$

$$+\lambda_1 \int_\Omega [u(x,y)]^m \left[|I(x,y) - c_1|^2 + \int_\Omega w_{xy,x'y'} R_1(x',y') dx' dy' \right] dxdy \quad (4)$$

$$+\lambda_2 \int_\Omega [1 - u(x,y)]^m \left[|I(x,y) - c_2|^2 + \int_\Omega w_{xy,x'y'} R_2(x',y') dx' dy' \right] dxdy,$$

where $\mu \geq 0$, $\lambda_1, \lambda_2 > 0$ are fixed parameters. The curve C_0 that minimizes F:

$$F(C_0, c_1, c_2, u) = \inf_C F(C, c_1, c_2, u), \quad (5)$$

is the solution to the segmentation problem (object boundary). The first term, in the definition of F (4), accounts for smoothing of the curve C.

2.1   Pseudo Level-Set Formulation

Let us define a pseudo level set formulation, similar to the level set method[4], based on the membership values u, where $C \subset \Omega$ is represented by the pseudo zero level set of Lipschitz similar function $u : \Omega \to I\!R$, such that:

$$u(x,y) = \begin{cases} C & = \{(x,y) \in \Omega : u(x,y) = 0.5\}, \\ inside(C) & = \{(x,y) \in \Omega : u(x,y) > 0.5\}, \\ outside(C) & = \{(x,y) \in \Omega : u(x,y) < 0.5\}, \end{cases} \quad (6)$$

For more details, we refer the reader to Osher *et al.*[4].

We could express the regularization term of the energy F as $Length(C) = \int_\Omega |\nabla H(u(x,y) - 0.5)| dxdy$, exploiting the Heaviside function $H(s)$ [4]. Thus, the energy $F(C, c_1, c_2, u)$ (4) can be rewritten as:

$$F(C, c_1, c_2, u)$$

$$= \mu \int_\Omega |\nabla H(u(x,y) - 0.5)| dxdy$$

$$+\lambda_1 \int_\Omega [u(x,y)]^m \left[|I(x,y) - c_1|^2 + \int_\Omega w_{xy,x'y'} R_1(x',y') dx' dy' \right] dxdy \quad (7)$$

$$+\lambda_2 \int_\Omega [1 - u(x,y)]^m \left[|I(x,y) - c_2|^2 + \int_\Omega w_{xy,x'y'} R_2(x',y') dx' dy' \right] dxdy.$$

Keeping u fixed and minimizing the energy $F(C, c_1, c_2, u)$ (7) with respect to c_1 and and c_2, it is easy to express these constants functions of u by:

$$c_1 = \frac{\int_\Omega [u(x,y)]^m I(x,y) dxdy}{\int_\Omega [u(x,y)]^m dxdy},$$

$$c_2 = \frac{\int_\Omega [1 - u(x,y)]^m I(x,y) dxdy}{\int_\Omega [1 - u(x,y)]^m dxdy}. \quad (8)$$

Furthermore, keeping c_1 and c_2 fixed and minimizing the energy $F(C, c_1, c_2, u)$ (7) with respect to u, it is easy to express variable u in the following way:

$$u(x,y) = \frac{1}{1 + \left(\frac{\lambda_1 \left[|I(x,y) - c_1|^2 + \int_{\Omega} w_{xy,x'y'} R_1(x',y') dx' dy' \right]}{\lambda_2 \left[|I(x,y) - c_2|^2 + \int_{\Omega} w_{xy,x'y'} R_2(x',y') dx' dy' \right]} \right)^{\frac{1}{m-1}}}. \tag{9}$$

For simplicity, without losing the generality, the above minimization (9) has been considered without the length term ($\mu = 0$).

2.2   Numerical Approximation

In equation (7), the two fitting terms are easy to be computed directly. We can also, approximate the length term $\int_{\Omega} |\nabla H(u(x,y) - 0.5)| dx dy$ by:

$$\sum_{i,j} \sqrt{(H(u_{i+1,j} - 0.5) - H(u_{i,j} - 0.5))^2 + (H(u_{i,j+1} - 0.5) - H(u_{i,j} - 0.5))^2}, \tag{10}$$

where $u_{i,j}$ is the value of u at the (i,j) pixel. The summand can only take the values 0, 1, or $\sqrt{2}$, depending on whether the 3 distinct pair of points from the set $\{u_{i,j}, u_{i+1,j}, u_{i,j+1}\}$ belong to the same or different regions. Thus, the length term can be easily computed knowing only the $H(u - 0.5)$, and there is no need to know u.

The usual approach to solve a minimization problem as in (7), is to derive its Euler-Lagrange equation and then to use explicit time marching or implicit iteration. In the proposed method, the time step is not restricted as in the explicit time marching. The algorithm for the fuzzy energy model is:

1. Give an initial partition of the image, set $u > 0.5$ for one part and $u < 0.5$ for the other.
2. Compute c_1 and c_2 using (8).
3. Assume that the value of the current pixel is I_o and u_o its corresponding degree of membership. Calculate the new degree of membership u_n using (9) for the pixel I_o under consideration and then compute the difference between the new and the old energy ΔF defined as:

$$\Delta F =$$
$$G \left[\lambda_1(s_1(I_o - c_1) + c_1)^2 + \lambda_2(s_2(I_o - c_2) + c_2)^2 - c_1^2 - c_2^2 \right]$$
$$- 2\lambda_1 s_1(I_o - c_1) \sum_{\Omega} \left[u(i,j)^m \left(\sum_{\Omega} w_{ij,i'j'} [1 - u(i',j')]^m I(i',j') \right) \right]$$
$$- 2\lambda_2 s_2(I_o - c_2) \sum_{\Omega} \left[[1 - u(i,j)]^m \left(\sum_{\Omega} w_{ij,i'j'} u(i',j')^m I(i',j') \right) \right]$$
$$+ (u_n^m - u_o^m) \sum_{\Omega} w_{i_o j_o, i'j'} [1 - u(i',j')]^m \left[\lambda_1(I(i',j') - c_1 - s_1(I_o - c_1))^2 \right]$$
$$+ (u_n^m - u_o^m) \sum_{\Omega} w_{i_o j_o, i'j'} [1 - u(i',j')]^m \left[\lambda_2(I_o - c_2)^2(1 - s_2)^2 \right]$$

$$+ \left[(1 - u_n)^m - (1 - u_o)^m \right] \sum_\Omega w_{i_o j_o, i' j'} u(i', j')^m \left[\lambda_2 (I(i', j') - c_2 - s_2 (I_o - c_2))^2 \right]$$

$$+ \left[(1 - u_n)^m - (1 - u_o)^m \right] \sum_\Omega w_{i_o j_o, i' j'} u(i', j')^m \left[\lambda_1 (I_o - c_1)^2 (1 - s_1)^2 \right]$$

$$+ (u_n^m - u_o^m)(1 - s_1)(I_o - c_1)^2 + \left[(1 - u_n)^m - (1 - u_o)^m \right] (1 - s_2)(I_o - c_2)^2, \quad (11)$$

where terms G, s_1 and s_2 defined as:

$$
\begin{aligned}
G &= \sum_{i,j} u(i,j)^m \left[\sum_\Omega w_{ij, i'j'} \left[1 - u(i', j') \right]^m \right] \\
s_1 &= \frac{u_n^m - u_o^m}{\sum_{i,j} [u(i,j)]^m + u_n^m - u_o^m} \\
s_2 &= \frac{(1 - u_n)^m - (1 - u_o)^m}{\sum_{i,j} [1 - u(i,j)]^m + (1 - u_n)^m - (1 - u_o)^m}
\end{aligned}
\qquad (12)
$$

If $\Delta F < 0$, then change u_o with u_n value, else keep the old (u_o) one. If we consider the length term ($\mu \neq 0$), it is easy to be computed and incorporated into the above energy difference, since only four neighbor pixels will be affected, when we change the value of a pixel.

4. Calculate the new centers using equations:

$$
\begin{aligned}
\tilde{c}_1 &= c_1 + s_1 (I_o - c_1) \\
\tilde{c}_2 &= c_2 + s_2 (I_o - c_2).
\end{aligned}
\qquad (13)
$$

5. Repeat the step 3 using Jacobi iterations computing the total energy F of the image.
6. Repeat the steps 2 to 4 until the total energy F remains unchanged.

Although, the equation (11), that give us the difference between the old and the new energy value, is large, it is very easily calculated since its most parts are constants for the computation of each Jacobi iteration.

The proofs of the equations (11) and (13) are similarly extracted in a similar way as in [16, 20].

3   Experimental Results

In this Section, we show the performance of the proposed method on various synthetic and real images, with different types of contours and shapes. We show the active contour evolving in the original image Ω, and the associated piecewise-constant approximation of Ω (given by constants c_1 and c_2). In our numerical experiments, we generally choose the parameters to be $\lambda_1 = \lambda_2 = 1$. Only the length parameter μ, which has a scaling role, is not the same in all experiments. If we have to detect all or as many objects as possible and of any size, then μ should be small. If we have to detect only large objects and not to detect small objects (like points, due to noise), then μ has to be larger.

First, the segmentation results on a two-phase image (Figure 1) are presented. The length term is omitted ($\mu = 0$) since there is no noise. Five different initial conditions were used and all of them converged to the correct solution in a very limited number of iterations. In fact, it is hard to find an initial condition that it does not work. The interior contour was automatically detected without

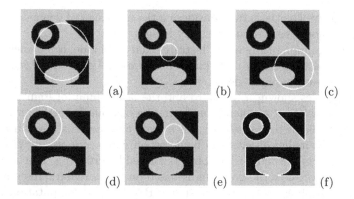

Fig. 1. Segmentation of a two-phase image. **(a)**, **(b)**, **(c)**, **(d)** and **(e)** are five different initial conditions, which have the same result after one sweep. **(f)** the result segmented image. The interior contour is automatically detected. In this experiment, the length parameter was omitted ($\mu = 0$).

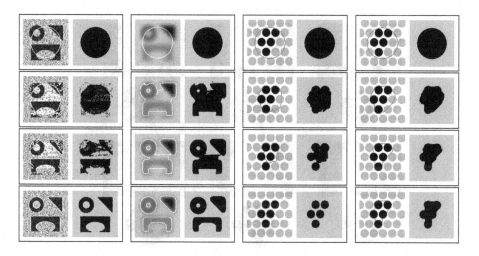

Fig. 2. Detection of different objects from a noisy image with various shapes and with interior contour (first column), blurred objects (second column), geometrically similar objects (third column) and grouping based on chromatic intensity (last column). In all examples, the model converges from the initial (first row) to the final position (last row) with two intermediate steps. The length parameter was set equal to $\mu = 3.66$, $\mu = 0.61$, $\mu = 0.12$ and $\mu = 12.21$ respectively.

considering a second initial model, something that shows the robustness of the algorithm.

The first column of Figure 2 shows how the model works on a noisy synthetic image, with various shapes and an interior contour. The contour is automatically detected, without considering a second initial curve. Jacobi iteration, as well as, the nature of the model allows the automatical change of the topology. Also, the length term as well as the terms considering the pixel spatial dependence render the proposed model very robust to noisy images.

The second column of Figure 2 illustrates that the model can detect different objects of different intensities and blurred boundaries. Again, the interior contour of the torus is automatically detected.

In the examples shown in the third and the last column of Figure 2, images with "contours without gradient" or "cognitive contours" (see Chan *et al.*[15]) are used. The role of the length term as a scale parameter is also illustrated: if μ is small, then smaller objects will be detected, while as μ getting larger, then only larger objects are detected, or objects formed by grouping. The last column of Figure 2 depicts how the grouping is based on the chromatic resemblance or identity, among objects of the same shape. Besides, the computation time for the proposed algorithm is much less (real time for images less than 256×256) than other algorithms that solves the Euler-Lagrange equations. Finally, Figure 3 demonstrates how the proposed algorithm could detect object boundaries on real images.

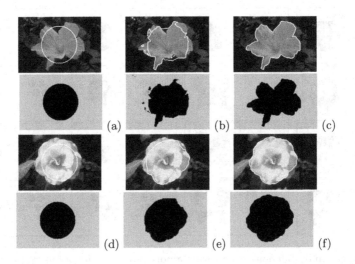

Fig. 3. Detection of object boundaries on real images. The model converges from the initial ((**a**) and (**d**)) to the final position ((**c**) and (**f**)) with an intermediate step ((**b**) and (**e**)). The length parameter was set equal to $\mu = 9.77$.

However, the proposed method shares a problem with the other classical methods that solves the Euler-Lagrange equations. There are objects, i.e., texture images, which cannot be detected using only the intensity average only. One way to overcome this difficulty, would be to use other information from the initial image Ω, like the curvature, the orientation of level sets, or any other discriminant. In this framework, we refer the reader to Lopez *et al.*[21].

4 Conclusion

In this paper, a novel fast and robust model for active contours to detect objects in an image was introduced. The model can detect objects whose boundaries are not necessarily defined by gradient, due to the fact that it is based on an energy minimization algorithm, and not on an edge-function as the most classical active contour models. This energy is based on fuzzy logic, which can be seen as a particular case of a minimal partition problem, and is used as the model motivation power evolving the active contour until to catch the desired object boundary. Furthermore, the stopping term of the model evolution does not depend on the gradient of the image, as most of the classical active contours, but instead is related to the image color and spatial segments. The fuzziness of the energy provides a balanced technique with a strong ability to reject "weak", as well as, "strong" local minima. Also, it is not needed to smooth the initial images, even if they are very noisy, since the model very well detect and preserve the locations of the boundaries. The interior contours of the objects can be automatically detected, starting only with the initial curve (model). The initial position of the model can be anywhere in the image, and it does not necessarily surrounds the objects to be detected. Finally, the small computation time of the evolution of the model renders the proposed method as a very promising tool even for real time applications. This lies in the fact that, the introduced method does not solve the Euler-Lagrange equation of the underlying problem, but, instead, calculates the fuzzy energy alterations directly without numerical stability constraints. So, it converges to the desired object boundary very fast.

References

1. Munoz, X., Freixenet, J., Cufi, X., Marti, J.: Strategies for image segmentation combining region and boundary information. Pattern Recognition Letters 24(1), 375–392 (2003)
2. Kass, M., Witkin, A., Terzopoulos, D.: Snakes: Active contour models. International Journal of Computer Vision 1(4), 321–331 (1988)
3. Malladi, R., Sethian, J.A., Vemuri, B.C.: Shape modeling with front propagation: A level set approach. IEEE Transactions on Pattern Analysis and Machine Intelligence 17(1), 158–175 (1995)
4. Osher, S., Sethian, J.A.: Fronts propagating with curvature-dependent speed: Algorithms based on hamilton-jacobi formulation. Journal of Computational Physics 79, 12–49 (1988)
5. Caselles, V., Kimmel, R., Sapiro, G.: Geodesic active contours. International Journal of Computer Vision 22(1), 61–79 (1997)
6. Yezzi, A., Kichenassamy, S., Kumar, A., Olver, P., Tannenbaum, A.: A geometric snake model for segmentation of medical imagery. IEEE Transactions on Medical Imaging 16(2), 199–209 (1997)
7. Tek, H., Kimia, B.: Image segmentation by reaction diffusion bubbles. In: Proceedings of International Conference on Computer Vision, pp. 156–162 (1995)
8. Cohen, L.: Multiple contour finding and perceptual grouping using minimal paths. Journal of Mathematical Imaging and Vision 14(3), 225–236 (2001)

9. Gunn, S., Nixon, M.: A robust snake implementation: A dual active contour. IEEE Transactions on Pattern Analysis and Machine Intelligence 19(1), 63–68 (1997)
10. Krinidis, S., Chatzis, V.: A physics-based dual deformable model. Journal of Information Hiding and Multimedia Signal Processing 3(1), 100–121 (2012)
11. Dawood, M., Jiang, X., Schäfers, K.P.: Reliable Dual-Band Based Contour Detection: A Double Dynamic Programming Approach. In: Campilho, A.C., Kamel, M.S. (eds.) ICIAR 2004. LNCS, vol. 3212, pp. 544–551. Springer, Heidelberg (2004)
12. Delgado-Gonzalo, R., Thevenaz, P., Seelamantula, C.S., Unser, M.: Snakes with an ellipse-reproducing property. IEEE Transactions on Image Processing 21(3), 1258–1271 (2012)
13. Erdem, C., Tekalp, A., Sankur, B.: Video object tracking with feedback of performance measures. IEEE Transactions on Circuits and Systems for Video Technology 13(4), 310–324 (2003)
14. Xu, N., Bansal, R., Ahuja, N.: Object segmentation using graph cuts based active contours. In: Proceedings of IEEE International Conference on Computer Vision and Pattern Recognition, vol. 2, pp. 46–53 (2003)
15. Chan, T., Vese, L.: Active contours without edges. IEEE Transactions on Image Processing 10(2), 266–277 (2001)
16. Krinidis, S., Chatzis, V.: Fuzzy energy-based active cotnours. IEEE Transactions on Image Processing 18(12), 2747–2755 (2009)
17. Mishra, A.K., Aloimonos, Y., Cheong, L.F., Kassim, A.: Active visual segmentation. IEEE Transactions on Pattern Analysis and Machine Intelligence 34(4), 639–653 (2012)
18. Samson, C., Blanc-Feraud, L., Aubert, G., Zerubia, J.: A level set model for image classification. International Journal of Computer Vision 40(3), 187–197 (2000)
19. Zheng, Q., Dong, E.Q., Cao, Z.L.: Graph cuts based active contour model with selective local or global segmentation. IET Electronis Letters 48(9), 490–491 (2012)
20. Song, B., Chan, T.: A fast algorithm for level set based optimization. Technical Report CAM 02-68, UCLA (2002)
21. Lopez, C., Morel, J.: Axiomatization of shape analysis and application to texture hyper discrimination. In: Proceedings of IEEE Conference on Computer Vision and Pattern Recognition, Berlin, Germany, pp. 646–647 (June 1993)

Fuzzy Friction Modeling
for Adaptive Control of Mechatronic Systems

Jacek Kabziński

Institute of Automatic Control, Technical University of Lodz
jacek.kabzinski@p.lodz.pl

Abstract. We discuss several fuzzy models to approximate friction and other disturbances in mechatronic systems, especially linear and rotarional electrical drives. Some methods of experimental identification of disturbance forces are presented. We consider several fuzzy models to compromise between model accuracy and complexity. Fuzzy model is used in an adaptive control loop. Several adaptive control algorithms are discussed and the influence of fuzzy model accuracy on the system performance is investigated.

Keywords: fuzzy modeling, adaptive control, motion control, friction compensation.

1 Introduction

It is well recognized that the presence of friction often destructs a performance of a precision motion control systems, especially servo drives realizing tracking tasks. The friction phenomenon is rather complicated and not yet completely understood, so existing friction models are also far from universality and accuracy. In this contribution we propose to connect fuzzy modeling with adaptive control. This approach allows to connect the simplicity of static friction models with the accuracy offered by adaptation to changing conditions, such as a lubricant temperature for example. As the experimental information about friction is usually corrupted and inaccurate we believe that using a flexible fuzzy model connected with adaptation of its parameters is an effective approach.

We consider the motion dynamics given by

$$\frac{dx}{dt} = v \qquad m\frac{dv}{dt} = F_e - F_{friction} - F_{ext}. \tag{1}$$

where m is a forcer mass, F_e is a thrust force, F_{ext} is external force, load (usually constant or slow-varying) and $F_{friction}$ represents all kinds of friction forces. The mover speed is v and position x. Although equation (1) is written according to linear motion convention, it may be also used for rotational movement description if we read m as a moment of inertia and consider torques instead of forces.

In this paper we shortly present the basic friction models and discuss the problem of experimental acquisition of the friction force data. We propose a TSK fuzzy

L. Iliadis et al. (Eds.): AIAI 2012, IFIP AICT 381, pp. 185–195, 2012.

model-based friction estimation structure that can be used for real-time nonlinear friction identification. We introduce a procedure to automatically decide the fuzzy model rules and starting parameters according to the desired modeling accuracy. Finally we apply friction fuzzy model in adaptive backstepping control assuring the position tracking stability without exact knowledge of all plant parameters, including the control gain coefficient. The presented contribution may be placed among many other concerning fuzzy adaptive control in presence of friction [1,2,3], but it develops a new and simpler (than for example in [1]) fuzzy model construction procedure and investigates new adaptive control approach.

2   Friction Models

Several models were proposed for friction forces. An excellent review is provided in [4]. As we claim that an approximated model should be connected with adaptive control approach, we mention only basic ideas here. Usually it is assumed that friction forces are speed dependent and are roughly described by the formula

$$F_{friction} = \left[f_c + (f_s - f_c)g(v) \right] sign(v) + Bv, \quad g(v) = e^{-\left(\frac{|v|}{v_s} \right)^{\delta}} . \quad (2)$$

where f_s is the level of static friction , f_c is the minimum level of Coulomb friction v_s is the lubricant parameter (so called Stribeck velocity), B - viscous friction parameter and δ is an even constant. The function $g(v)$ describing a characteristic of the Stribeck curve is only one of possibilities – several other are reported [4]. All parameters of this model are unknown and should be determined by empirical experiments, and still the model accuracy is doubtful. The simplified version of (2) takes into account only Coulomb and viscous friction:

$$F_{friction} = f_c sign(v) + Bv . \quad (3)$$

So called LuGre [1] dynamic friction model is supposed to capture most of the real friction behaviour, like Stribeck effect, hysteresis, spring-like characteristics, varying brake-away force. It is based on 'elastic bristles' model of contact surfaces. The average deflection z of the bristles is given by

$$\dot{z}(t) = v - \frac{|v|z}{g(v)} . \quad (4)$$

where $g(v)$ is a positive function. To describe Stribeck effect $g(v)$ is usually chosen as

$$g(v) = \frac{1}{\sigma} \left[f_c + (f_s - f_c)e^{-\left(\frac{|v|}{v_s} \right)^{\delta}} \right] . \quad (5)$$

Friction force is given by

$$F_{friction} = \sigma z + \tau \dot{z} + Bv .$$ (6)

where σ is the equivalent stiffness coefficient and τ is the equivalent damping coefficient of bristles. Several another models (more complicated, with bigger number of parameters and more difficult to identify) of friction forces are reported in literature [2,3]. As it follows from the above discussion friction and ripple forces are of very complicated nature, difficult to analyse and to model. In this paper we suggest modelling the sum of ripple and friction forces by a fuzzy inference system.

3   Acquisition of the Data for Friction Modeling

It is necessary to conduct some experiments to collect the data for the fuzzy model training.

One of possibilities is so called constant speed test. If we are able to produce a constant speed movement, it means that all the forces are balanced. If we are can measure or estimate the external force, calculate the thrust force (from measurement of motor currents for example), we are able to estimate the friction force.

Sporadically it is possible to apply a constant external force (from an another drive, or from a gravitational load), while the thrust force is zero. In this case we may try to tune the friction model parameters by curve fitting comparing measured position history with numerical solution of equation (1).

Both above methods are theoretically straightforward but difficult to implement in practice. Another possibility is to use a simple observer described by:

$$m_0 \frac{d}{dt} v_{est} = F_0 - F_{fric\,est} - K(v_{est} - v)$$
$$\frac{d}{dt} F_{fric\,est} = \Gamma(v_{est} - v)$$ (7)

where K and Γ are design parameters, $m_0 = m + \Delta m$ and $F_0 = F_e - F_{ext} + \Delta F$ are observer parameters assumed instead of real m and $F_e - F_{ext}$. If we denote the errors $e_v = v_{est} - v$, $e_F = F_{fric\,est} - F_{friction}$ and we measure $v + \Delta v$ instead of v and assuming $F_{friction} \neq const$ we get

$$\begin{bmatrix} m + \Delta m & 0 \\ 0 & 1 \end{bmatrix} \frac{d}{dt} \begin{bmatrix} e_v(t) \\ e_F(t) \end{bmatrix} = \begin{bmatrix} -K & -1 \\ \Gamma & 0 \end{bmatrix} \begin{bmatrix} e_v(t) \\ e_F(t) \end{bmatrix} + \begin{bmatrix} \Delta F - \Delta m \dfrac{dv}{dt} + K\Delta v \\ \dfrac{dF_{friction}}{dt} \end{bmatrix} .$$ (8)

As we see error dynamics is described by a linear system with disturbances. The eigenvalues s_1, s_2 of this system are connected with design parameters:

$$\frac{\Gamma}{m+\Delta m}=s_1 s_2, \quad \frac{K}{m+\Delta m}=-(s_1+s_2). \tag{9}$$

and so we may choose values of s_1, s_2 to obtain desired observer dynamics. We may tune observer parameters m_0 and F_0 to minimize $e_v = v_{est} - v$, as we know v_{est} and measure v. Equation (7) allows also to estimate the influence of $\dfrac{dF_{friction}}{dt}$ and Δv on the estimation error and to plan measurements properly. Special care must be taken to minimize Δv, as it is multiplied by K in (7) and its influence cannot be decreased by increasing K.

As we conclude from the above discussion the obtained triples (position - velocity – estimated friction), denoted by

$$\left(x_k = x_{1,k}, v_k = x_{2,k}\right) \rightarrow f_k \qquad k \in \{1,...,m\}. \tag{10}$$

will be corrupted by estimation method error and subject to estimation/measurement noise and outliers. We will develop special procedure to extract fuzzy rules to construct Takagi-Sugeno-Kang fuzzy model. Figure 1 presents about 300 triples of the data collected from an exemplary linear permanent magnet motor.

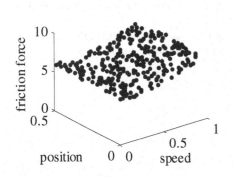

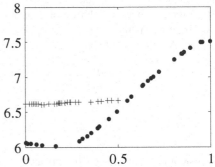

Fig. 1. Friction modelling data

Fig. 2. SIFM action curves for the data from fig. 1: + *position*, · *speed*

4   Fuzzy Model Construction

The proposed method of fuzzy friction modelling is based on One-dimensional Linear Local Prototypes (1dLLP) approach proposed in [5]. First we have to recognize if position is an irrelevant input or not. We consider two single-input fuzzy models (SIFM) described below:

- input - x_i ($i=1$ – position, $i=2$ – velocity), output - c_i,
- input linguistic categories: x_i IS $x_{i,k}$ $k \in \{1,...,m\}$,
- membership functions:

$$\varphi_{i,k}(x) = \frac{1}{1 + \left(\dfrac{x - x_{i,k}}{a}\right)^{2b}} , \tag{11}$$

- rules:

$$IF \quad x_i \quad IS \quad x_{i,k} \quad THEN \quad c_i = c_{i,k} = p_{i,k}x_i + q_{i,k}$$

$$p_{i,k} = \frac{f_k}{x_{i,k}}, \quad q_{i,k} = 0 \quad if \ x_{i,k} \neq 0 \tag{12}$$

$$p_{i,k} = 0, \quad q_{i,k} = f_k \quad if \ x_{i,k} = 0 \qquad k \in \{1,...,m\}.$$

The action curve given by the output c_i of this system for the input data - $x_{i,k}$ generalises information coded by $x_{i,k} \to f_k$ $i=1,2$ and the degree of this generalisation depends on membership function parameter a . Recommendations for the choice of a and b are given in [5]. The shape of each action curve is robust to outliers in the measured data and to the measurement noise. If the i-th input is inessential the curve generated by corresponding SIFM will be flat, if it is meaningful the curve will cover significant part of the range of $\{f_k, k \in \{1,...,m\}\}$. Fig. 2 depicts action curves for position and speed for the data presented in fig. 1. Its visible that in this case position was the irrelevant input for friction modelling.

Selection of membership functions for each input is based on piece-wise linear approximation of action curves derived above. Uniform or mean-square approach are both applicable. As the result of piece-wise linear approximation for the i-th significant input we obtain m_i linear local prototypes (LLP) defined on intervals

$$I_{i,j} = \left(x_{\min i,j}, x_{\max i,j}\right) \quad j \in \{1,2,\cdots,m_i\} \tag{13}$$

by linear polynomials

$$P_{i,j}(x_i) = p_{1i,j}x_i + p_{0i,j}, \quad x_i \in \left(x_{\min i,j}, x_{\max i,j}\right) \quad j \in \{1,2,\cdots,m_i\} \tag{14}$$

for each interval. The design parameter δ, which defines the approximation accuracy, governs the number of linear local prototypes. For each j we construct a bell-shaped functions $\mu_{i,j}(x)$ spanned over $I_{i,j}$ and centred at the middle point of $I_{i,j}$. The choice of the third parameter b_{ij} is arbitrary – usually $1.5 < b_{ij} < 2$ gives good results. The rules for the proposed TSK fuzzy model will be:

$$IF \quad x_i \quad IS \quad \mu_{i,j} \quad THEN \quad c_i = c_{i,j} = p_{1i,j} x_i + p_{0i,j}, \quad j \in \{1,2,...,m_i\}, \quad (15)$$

where starting values of parameters are taken from piecewise linear approximation results (14). The model will be trained by any suitable algorithm, we may choose a neural representation of the fuzzy system – ANFIS [6] and the appropriate training algorithm. Fig. 3 presents linear local prototypes obtained by linear mean-square approximation of the curve from fig.2 with $\delta = 0.005$ – in this case two "sticks" were enough. Smaller δ will impose bigger number of LLPs and so bigger number of rules.

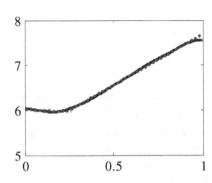
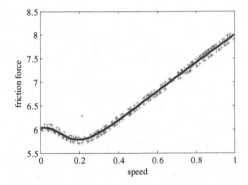

Fig. 3. SIFM action curve for speed (*solid*) and piecewise linear approximation (*dotted*) by two LLP

Fig. 4. Final fuzzy friction model (*solid*) and the data (++)

In this paper we concentrate on the case when the friction is velocity-dependent and it was possible to eliminate the position as the inessential input (although the procedure of final rule selection and model construction if both inputs are important is possible as presented in [5]). So starting from this point we model friction by m velocity-dependant rules

$$IF \quad v \quad IS \quad \mu_j \quad THEN \quad F_j = p_{1,j} v + p_{0,j} \quad , \quad j \in \{1,2,...,m\}, \quad (16)$$

with generalised bell-shaped functions $\mu_j = \mu_j(v; a_j, b_j, c_j)$ and the output of the fuzzy model calculated as:

$$F_{mdl}(v) = \frac{\sum_{j=1}^{m} \mu_j(v) F_j(v)}{\alpha(v)} = \theta^T \zeta(v), \quad \alpha(v) := \sum_{j=1}^{m} \mu_j(v), \quad (17)$$

where

$$\theta^T = [p_{1,1}, p_{0,1}, \cdots, p_{1,m}, p_{0,m}] \quad , \zeta(v) = \frac{1}{\alpha(v)} \begin{bmatrix} \mu_1(v)v \\ \mu_1(v) \\ \vdots \\ \mu_m(v)v \\ \mu_m(v) \end{bmatrix} . \tag{18}$$

5   Adaptive Control with a Fuzzy Friction Model

Several adaptive control techniques may be proposed for motion control of systems with friction, depending on details of the model description and the chosen method to prove the stability. In this contribution we describe adaptive backstepping position tracking. We also assume that the thrust force is proportional to the control variable (the motor current)

$$F_e(t) = \varphi \cdot i(t), \tag{19}$$

and coefficient φ is not known exactly. As the external disturbance F_{ext} may be compensated exactly the same way as the friction, we assume that $F_{ext} = 0$ in equation (1). Let us denote the desired smooth position trajectory by x_d, actual position by x and the tracking error by $e_1 = x_d - x$. The velocity will be 'virtual control' for position tracking. If we choose the desired velocity v_d according to

$$v_d = \dot{x}_d + k_1 \cdot e_1, \tag{20}$$

where $k_1 > 0_1$ is a design parameter, we will be able to describe the tracking error dynamics as

$$\dot{e}_1 = \dot{x}_d - \dot{x}_d - k_1 \cdot e_1 + e_2 = -k_1 \cdot e_1 + e_2, \tag{21}$$

$$e_2 = v_d - v = \dot{x}_d + k_1 \cdot e_1 - v \tag{22}$$

$$m_o \cdot \dot{e}_2 = m_o \cdot \dot{v}_d - m_o \cdot \dot{v} = m_o \cdot \dot{v}_d - i - \frac{1}{\varphi} F_{friction}, \quad m_o = \frac{m}{\varphi}, \tag{23}$$

$$\dot{v}_d = \ddot{x}_d + k_1 \cdot (-k_1 \cdot e_1 + e_2), \tag{24}$$

The control variable i has to compensate function

$$D = m_o \cdot \dot{v}_d - \frac{1}{\varphi} F_{friction},$$ (25)

and to assure fast tracking. We will use a model $\hat{D}$ for D, incorporating the fuzzy friction model (17,18). The general structure of $\hat{D}$ will be given by

$$\hat{D} = \hat{\mathbf{A}}^T \xi,$$ (26)

where $\hat{\mathbf{A}}^T$ is a vector of adaptive parameters and ξ is known. We have several possibilities to choose the number of adaptive parameters, for example:

1 adaptive parameter: $\hat{\mathbf{A}}^T = \hat{a}$, $\xi = \left[m_{oN} \cdot \dot{v}_d + \frac{1}{\varphi_N} F_{mdl} \right]$ (27)

where m_{oN}, φ_N are nominal values of m_o, φ ,

2 adaptive parameters: $\hat{\mathbf{A}}^T = \begin{bmatrix} \hat{m}_o, & \hat{k} \end{bmatrix}$ $\xi = \begin{bmatrix} \dot{v}_d \\ \frac{1}{\varphi_N} F_{mdl} \end{bmatrix}$,   (28)

$2m+1$ adaptive parameters: $\hat{\mathbf{A}}^T = \begin{bmatrix} \hat{m}_o, & \hat{\vartheta}^T \end{bmatrix}$ $\xi = \begin{bmatrix} \dot{v}_d \\ \frac{1}{\varphi_N} \zeta \end{bmatrix}$,   (29)

In (27) $\hat{a}$ is responsible for general model correction, in (28) $\hat{m}_o$ is supposed to adapt the changing inertia and $\hat{k}$ corrects $\frac{1}{\varphi_N} F_{mdl}$ to the actual value of $\frac{1}{\varphi} F_{friction}$, while in (29) $\hat{\vartheta}^T$ corresponds to $\frac{1}{\varphi} \theta^T = \frac{1}{\varphi} [p_{1,1}, p_{0,1}, \cdots, p_{1,m}, p_{0,m}]$, so corrects each fuzzy consequents' parameter separately.

Without loss of generality we may assume existence of "the best" adaptive parameters $\mathbf{A}^{*T}$ such that the model $D^* = \mathbf{A}^{*T} \xi$ gives bounded estimation error $\varepsilon = F - F^*$, $|\varepsilon| < \varepsilon_{max} < \infty$ and denote $\tilde{\mathbf{A}} = \mathbf{A}^* - \hat{\mathbf{A}}$.

If we choose the control law according to

$$i = \hat{D} + k_2 \cdot e_2 + e_1 \tag{30}$$

we get the tracking error dynamics

$$m_o \cdot \dot{e}_2 = \varepsilon + \tilde{\mathbf{A}}^T \xi - k_2 \cdot e_2 - e_1 . \tag{31}$$

To investigate the tracking stability we propose Lyapunov function

$$V = \frac{1}{2}\left(e_1^{\ 2} + m_o \cdot e_2^{\ 2} + \tilde{\mathbf{A}}^T \mathbf{\Gamma}^{-1}\tilde{\mathbf{A}} \right). \tag{32}$$

with positive definite symmetric matrix $\mathbf{\Gamma}$. Taking any of the adaptation laws

$$\dot{\hat{\mathbf{A}}} = e_2 \mathbf{\Gamma}\xi, \text{ or } \dot{\hat{\mathbf{A}}} = e_2 \mathbf{\Gamma}\xi - \delta\mathbf{\Gamma}\mathbf{A}, \text{ or } \dot{\hat{\mathbf{A}}} = e_2 \mathbf{\Gamma}\xi - \delta\sqrt{e_1^{\ 2} + e_2^{\ 2}}\ \mathbf{\Gamma}\mathbf{A} . \tag{33a,b,c}$$

with small positive δ, we are able to prove that the system derivative of (32) is negative outside a certain, bounded set, and so e_1, e_2 are uniformly ultimately bounded. For example with adaptation performed according to (33a) we get

$$\dot{V} = -k_1 \cdot e_1^{\ 2} + -k_2 \cdot e_2^{\ 2} + e_2 \cdot \varepsilon \leq -k_1 e_1^{\ 2} - \left(k_2 - \frac{1}{2} \right)e_2^{\ 2} + \frac{1}{2}\varepsilon^2 \tag{34}$$

and is negative outside

$$e_1^{\ 2} + e_2^{\ 2} > \frac{1}{k}\varepsilon_{max}^{\ 2}, \qquad k = \min\left(k_1, k_2 - \frac{1}{2} \right) \tag{35}$$

Several experiments with various adaptive controllers were conducted with the linear motor investigated in the previous sections, with fuzzy model presented in fig. 4 and adaptive laws (33a,b,c). All controllers perform correctly. More accurate fuzzy model results in smaller control signals. In fig. 5-7 we illustrate the performance of adaptive backstepping controller (28) with two adaptive parameters and adaptation law (33c).

the system was to track sinusoidal position trajectory $x_d(t) = 0.4\sin(0.8t + \frac{\pi}{4})$ with initial condition $x(0)=0$. Starting values of m_0 and k_i were subject to about 20% error. Adaptation was blocked while the current saturation (the saturation level was 1A) was active. As we notice the tracking accuracy is very high - it is limited by the encoder performance only. The adaptive gains are bounded and approach the desired values. The control input is bounded.

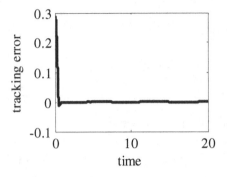

Fig. 5. The tracking error history

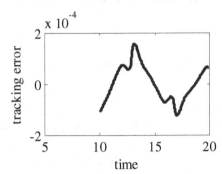

Fig. 6. The tracking error history – an enlarged part of fig.5

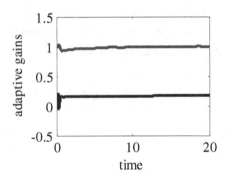

Fig. 7. Adaptive gains

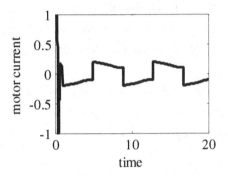

Fig. 8. Motor current

6   Conclusions

The proposed approach, connecting fuzzy modeling and adaptive control, may be used for any motion control problem with friction or other disturbances. The fuzzy model is robust against measurement noise and isolated outliers in the modeling data. On-line gain adaptation by the adaptive laws allows to use a static friction model and to obtain good control performance. The proposed procedure based on linear local prototypes allows to build a fuzzy model as simple as possible. The same approach may be used if friction is function of position and speed. Some other concepts of fuzzy models – for example TSK models with nonlinear consequences were also applied by the author with promising results.

References

1. Wang, Y., Wang, D., Chai, T.: Extraction and Adaptation of Fuzzy Rules for Friction Modeling and Control Compensation. IEEE Transactions on Fuzzy Systems 19, 682–693 (2011)
2. Wang, Y., Wang, D., Chai, T.: Modeling and control compensation of nonlinear friction using adaptive fuzzy systems. Mechanical Systems and Signal Processing 23, 2445–2457 (2009)
3. Lin, L., Lai, J.: Stable Adaptive Fuzzy Control with TSK Fuzzy Friction Estimation for Linear Drive Systems. Journal of Intelligent and Robotic Systems 38, 237–253 (2003)
4. Wojewoda, J., Stefanski, A., Wiercigroch, M., Kapitaniak, T.: Hysteretic effects of dry friction: modelling and experimental studies. Phil. Trans. R. Soc. A 366, 747–765 (2008)
5. Kabziński, J.: One-Dimensional Linear Local Prototypes for Effective Selection of Neuro-Fuzzy Sugeno Model Initial Structure. In: Papadopoulos, H., Andreou, A.S., Bramer, M. (eds.) AIAI 2010. IFIP AICT, vol. 339, pp. 62–69. Springer, Heidelberg (2010)
6. Jang, J.R.: ANFIS: Adaptive-network-based fuzzy inference system. IEEE Trans. Syst. Man Cybern. 23, 665–684 (1993)

Fuzzy Graph Language Recognizability

Antonios Kalampakas[1,*], Stefanos Spartalis[1], and Lazaros Iliadis[2]

[1] Department of Production Engineering and Management,
Laboratory of Computational Mathematics, School of Engineering,
Democritus University of Thrace, V. Sofias 12,
Prokat, Building A1, 67100Xanthi, Greece
akalampakas@gmail.com, sspart@pme.duth.gr

[2] Department of Forestry & Management of the Environment & Natural Resources,
Democritus University of Thrace, 193 Pandazidou St., 68200, Nea Orestiada, Greece

Abstract. Fuzzy graph language recognizability is introduced along the lines of the established theory of syntactic graph language recognizability by virtue of the algebraic structure of magmoids. The main closure properties of the corresponding class are investigated and several interesting examples of fuzzy graph languages are examined.

1 Introduction

Fuzzy models are becoming increasingly useful because of their ability to bridge the difference between the traditional numerical models used in engineering and sciences and the symbolic models used in formal systems and Artificial Intelligence. In this respect, fuzzy graphs are the appropriate mathematical object capable of modeling real time systems where the inherent level of information in them varies with different levels of precision. In this paper we present for the first time a syntactic recognizability theory for fuzzy hypergraph languages, analogously with the already established theory for crisp hypergraph languages [9, 13].

Directed hypergraphs consist of a set of vertices (nodes) and a set of hyperedges, just as ordinary directed graphs except that a hyperedge may have an arbitrary sequence of sources (incoming arrows) and an arbitrary sequence of targets (outgoing arrows), instead of only one source and one target as is the case for ordinary graph edges. Each hyperedge is labeled with a symbol from a doubly ranked alphabet Σ in such a way that the first (resp. second) rank of its label equals the number of its sources (resp. targets). Also, every hypergraph is equipped with a sequence of *begin* and *end* nodes. It is clear that ordinary directed graphs are obtained as a special case of directed hypergraphs i.e. in the case that each hyperedge has one source and one target and both the sequences of begin and end nodes are the empty word. *Directed fuzzy graphs* were

---

* The author has been co-financed by the European Social Fund and Greek national funds through the Operational Program "Education and Lifelong Learning" of the National Strategic Reference Framework - Research Funding Program: THALIS project "Algebraic Modeling of Topological and Computational Structures and Applications"

introduced by Rosenfeld in [19] by taking fuzzy subsets of the node and edge sets of a given directed crisp graph (see also [17]). In [15] Engelfriet and Verei-jken proved that every graph can be constructed from a finite set of elementary graphs by inductively using the operations of concatenation and sum. Since for every graph an infinite number of such expressions exist, at the same paper, the authors stated the open problem of finding a complete set of equations (rewriting rules) with the property that two expressions represent the same graph if and only if one can be transformed into the other by these equations. We solved this problem in [8] by appropriately adopting magmoids as the necessary algebraic structure for the representation of graphs and graph operations. This result led to the introduction of graphoids and in the construction, for the first time, of automata operating on arbitrary graphs (cf. [10],[16]). A magmoid is a doubly ranked set endowed with the operations of concatenation and sum which are associative, unitary, and compatible to each other ([1, 2]). They simulate the ordinary monoid structure and a natural regularity notion derives from this sim-ulation. More precisely, we say that a subset L of a magmoid M is recognizable whenever there exist a locally finite magmoid N and a morphism of magmoids $h : M \to N$, so that $L = h^{-1}(P)$ for some $P \subseteq N$ ([9]). We note that this is a global recognizability notion (sort independent), whereas the corresponding notion for many sorted algebras is local since it refers to a specified sort. As a consequence, although the set of all graphs is magmoid recognizable, this is not the case for the corresponding graph language in the framework of many sorted algebras. This recognizability mode is straightforwardly applied to graph languages and the class of recognizable graph languages is shown to be closed under boolean operations, inverse magmoid morphisms and sum operation.

The syntactic recognizability mechanism described above is analogous with the corresponding recognizability notion for string and tree languages. Fuzzy recognizability for strings and trees has already been introduced in [5–7] by virtue of syntactic monoids and syntactic algebras respectively. In the present paper we extend the magmoid recognizability mechanism to languages consist-ing of directed fuzzy graphs in the previously defined manner. In Section 2 we present the basic notations, definitions and facts that we will employ later on. In the same section we introduce fuzzy hypergraphs, their definition is derived from the one given by Rosenfeld for fuzzy graphs [19]. In the next section we in-troduce fuzzy graph language recognizability a characterization result obtained from their underlying magmoid structure is presented and their basic properties are investigated. It turns out that fuzzy graph language recognizability has nice closure properties, several interesting examples are examined and their relation with the corresponding case for crisp graphs is discussed.

2   The Algebraic Structure of Fuzzy Graphs

Given a finite set X we denote by X^* the set of all words over X, and for every word $w \in X^*$, $|w|$ denotes its length. A fuzzy set is a pair $X_\chi = (X, \chi)$ where $\chi : X \to [0, 1]$ is its *membership function*. A doubly ranked set, or doubly ranked alphabet, $(A_{m,n})_{m,n \in \mathbb{N}}$ is a set A together with a function $rank : A \to \mathbb{N} \times \mathbb{N}$, where $\mathbb{N}$ is the set of natural numbers. For $m, n \in \mathbb{N}$, $A_{m,n} = \{a \in A \mid rank(a) = (m, n)\}$. In what follows we will drop the subscript $m, n \in \mathbb{N}$ and denote a doubly ranked set simply by $A = (A_{m,n})$. A *fuzzy (m, n)-graph* $G = (V_\kappa, E_\lambda, s, t, l, begin, end)$ over the doubly ranked alphabet $\Sigma = (\Sigma_{m,n})$ consists of

- the fuzzy sets of nodes or vertices $V_\kappa = (V, \kappa)$ and edges $E_\lambda = (E, \lambda)$;
- the source and target functions $s : E \to V^*$ and $t : E \to V^*$;
- the labeling function $l : E \to \Sigma$ with $rank(l(e)) = (|s(e)|, |t(e)|)$ for all $e \in E$;
- the sequences of begin and end nodes $begin \in V^*$ and $end \in V^*$ with $|begin| = m$ and $|end| = n$.

with the additional requirement that for every $e \in E$,

$$\lambda(e) \leq min\{\kappa(v) \mid v \text{ appears in the words } s(e) \text{ or } t(e)\}. \tag{1}$$

For every $v \in V$ and every $e \in E$ the values $\kappa(v)$ and $\lambda(e)$ are called respectively the membership grade of v and e. The above is a generalization for hypergraphs of the definition of fuzzy graphs given by Rosenfeld in [19]. Notice that according to this definition vertices can be duplicated in the begin and end sequences of the graph and also at the sources and targets of an edge. For an edge e of a hypergraph G we simply write $rank(e)$ to denote $rank(l(e))$. The specific sets V and E chosen to define a concrete fuzzy graph G are actually irrelevant. We shall not distinguish between two isomorphic fuzzy graphs. Hence we have the following definition of an abstract graph. Two concrete fuzzy (m, n)-graphs $G = (V_\kappa, E_\lambda, s, t, l, begin, end)$ and $G' = (V'_{\kappa'}, E'_{\lambda'}, s', t', l', begin', end')$ over Σ are isomorphic if and only if there exist two bijections $h_V : V \to V'$ and $h_E : E \to E'$ commuting with $\kappa, \lambda, s, t, l, begin$ and end in the usual way. An *abstract fuzzy (m, n)-graph* is defined to be the equivalence class of a concrete fuzzy (m, n)-graph with respect to isomorphism. We denote by $FGR_{m,n}(\Sigma)$ the set of all abstract fuzzy (m, n)-graphs over Σ. Since we shall mainly be interested in abstract fuzzy graphs we simply call them graphs. Any graph $G \in FGR_{m,n}(\Sigma)$ with $E = \emptyset$ is called a *discrete (m, n)-graph*.

Given a fuzzy graph $G = (V_\kappa, E_\lambda, s, t, l, begin, end)$, the *fuzzy complement* G^{fc} of G is the fuzzy graph $(V_\kappa, E_{\lambda'}, s, t, l, begin, end)$ with

$$\lambda'(e) = min\{\kappa(v) \mid v \text{ appears in the words } s(e) \text{ or } t(e)\} - \lambda(e)$$

for all $e \in E$. This graph is well defined due to Eq. (1). We note that this is a generalization of the definition for the complement of a crisp graph. Given a fuzzy graph language $L \subseteq FGR(\Sigma)$ we set $L^{fc} = \{G^{fc} \mid G \in L\}$ i.e. L^{fc} consists of all the fuzzy complements of the elements of L.

A *magmoid* (cf. [1, 2]) is a doubly ranked set $M = (M_{m,n})$ equipped with two operations denoted by $\circ$ (circle) and $\square$ (box):

$$\circ : M_{m,n} \times M_{n,k} \to M_{m,k}, \qquad \square : M_{m,n} \times M_{m',n'} \to M_{m+m',n+n'}$$

for all $m, n, k, m', n' \geq 0$, which are associative in the obvious way and satisfy the distributivity law $(f \circ g) \, \square \, (f' \circ g') = (f \, \square \, f') \circ (g \, \square \, g')$ whenever all the above operations are defined. Moreover, it is equipped with a sequence of constants $e_n \in M_{n,n}$ $(n \geq 0)$, called units, such that

$$e_m \circ f = f = f \circ e_n, \quad e_0 \, \square \, f = f = f \, \square \, e_0$$

for all $f \in M_{m,n}$ and all $m, n \geq 0$, and the additional condition $e_m \, \square \, e_n = e_{m+n}$ holds true for all $m, n \geq 0$. Notice that, due to the last equation, the elements e_n $(n \geq 2)$ are uniquely determined by e_1. From now on e_1 will be simply denoted by e. Submagmoids, morphisms, congruences and quotients of magmoids are defined in the obvious way.

An elegant characterization of a congruence can be achieved by means of the notion of the context. In a magmoid M an *(m,n)-context* is a 4-tuple $\omega = (g_1, f_1, f_2, g_2)$, with $f_i \in M_{m_i, n_i}$ $(i = 1, 2)$, $g_1 \in M_{a, m_1+m+m_2}$, $g_2 \in M_{n_1+n+n_2, b}$, where $a, b \in N$. The set of all (m, n)-contexts is denoted $Cont_{m,n}(M)$. For any $f \in M_{m,n}$ and $\omega = (g_1, f_1, f_2, g_2)$ as above, we write $\omega[f] = g_1 \circ (f_1 \, \square \, f \, \square \, f_2) \circ g_2$; note that $\omega[f] \in M_{a,b}$. As it is shown in [9]

Proposition 1. *The equivalence $\sim = (\sim_{m,n})$ on the magmoid $M = (M_{m,n})$ is a congruence whenever, for all $m, n \geq 0$, $f, g \in M_{m,n}$ and all $\omega \in Cont_{m,n}(M)$*

$$f \sim_{m,n} g \text{ implies } \omega[f] \sim_{a,b} \omega[g].$$

The set of fuzzy graphs $FGR(\Sigma) = (FGR_{m,n}(\Sigma))$ can be organized into a magmoid by virtue of two operations: product and sum corresponding to $\circ$ and $\square$ respectively. If G is an (m, n)-graph and H is an (n, k)-graph represented respectively by $(V_\kappa, E_\lambda, s, t, l, begin, end)$ and $(V'_{\kappa'}, E'_{\lambda'}, s', t', l', begin', end')$, then their *product* $G \circ H$ is the (m, k)-graph $(V''_{\kappa''}, E''_{\lambda''}, s'', t'', l'', begin'', end'')$ obtained by taking the disjoint union of G and H and then identifying the ith end node v of G with the ith begin node v' of H, for all $i \in \{1, ..., n\}$; for the resulting node v'' we set $\kappa''(v'') = max\{\kappa(v), \kappa(v')\}$. Additionally, $begin'' = begin$ and $end'' = end'$. The *sum* $G \, \square \, H$ of arbitrary graphs G and H is their disjoint union with their sequences of begin nodes concatenated and similarly for their end nodes.

For instance let $\Sigma = \{a, b, c\}$, with $rank(a) = (2, 1)$, $rank(b) = (1, 1)$ and $rank(c) = (1, 2)$. In the following pictures, edges are represented by boxes, nodes by dots, and the sources and targets of an edge by directed lines that enter and leave the corresponding box, respectively. The order of the sources and targets of an edge is the vertical order of the directed lines as drawn in the pictures. We display two graphs $G \in FGR_{3,4}(\Sigma)$ and $H \in FGR_{4,2}(\Sigma)$, where the ith begin node is indicated by b_i, and the ith end node by e_i.

$$G \qquad\qquad\qquad\qquad H$$

Then their product $G \circ H$ and their sum $G \square H$ are respectively the $(3, 2)$ and the $(7, 6)$-graphs

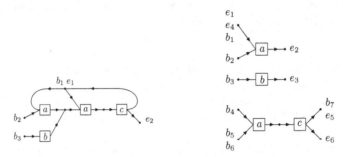

For every $n \in \mathbb{N}$ we denote by E_n the discrete (n, n)-graph with nodes $x_1, ..., x_n$, begin and end sequence $x_1 \cdots x_n$ and $\kappa(x_i) = 0$ for all i; we write E for E_1. It can be verified that $FGR(\Sigma) = (FGR_{m,n}(\Sigma))$ with the operations defined above is a magmoid, whose units are the graphs E_n, $n \geq 0$.

In what follows we will need the next concepts. Let $M = M_{m,n}$ be a magmoid. We say that a doubly ranked family $L = (L_{m,n})$ is a *subset* of M (notation $L \subseteq M$), whenever $L_{m,n} \subseteq M_{m,n}$ for all $m, n \in \mathbb{N}$. The boolean operations on subsets of M are defined in the obvious way. Given subsets L, L' of a magmoid M (with unit sequence e_n) we define their $\circ$ -*product* $L \circ L'$ by setting

$$(L \circ L')_{m,n} = \bigcup_{k \geqslant 0} L_{m,k} \circ L'_{k,n}, \qquad m, n \in \mathbb{N}$$

and their $\square$ -*product* $L \square L'$ by setting

$$(L \square L')_{m,n} = \bigcup_{\substack{\kappa+\kappa'=m \\ \lambda+\lambda'=n}} L_{\kappa,\lambda} \square L'_{\kappa',\lambda'}, \qquad m, n \in \mathbb{N}.$$

The subsets E and F of M given by $E_{m,n} = \{e_n\}$ if $m = n$ and $\emptyset$ else, while $F_{m,n} = \{e_0\}$ if $m = n = 0$ and $\emptyset$ else, are the units of the operations $\circ$ and $\square$ respectively. The reader will verify that the set of all subsets of M together with $\cup, \circ, \square$ is a double semiring.

3   Fuzzy Language Recognizability

Let $X = (X_{m,n})$ be a doubly ranked alphabet. We denote by $mag(X) = (mag_{m,n}(X))$ the free magmoid generated by X its elements are called *patterns* (see [9]). We say that two elements of a magmoid M are equivalent modulo the syntactic congruence of a subset $L \subseteq M$, whenever they have the same set of contexts with respect to L. More precisely, let L be a subset of the magmoid M and $f \in M_{m,n}$, we set $C_L(f) = \{\omega \mid \omega \in Cont_{m,n}(M), \omega[f] \in L\}$.

Proposition 2 (cf. [9]). *The equivalence* $\sim_L$ *on* M *defined by*

$$f \sim_{L,m,n} g, \quad whenever \quad C_L(f) = C_L(g)$$

is a congruence.

Given a magmoid M and a set $L \subseteq M$, $\sim_L$ is called the *syntactic congruence* of L and the quotient magmoid $M_L = M/ \sim_L$ is the *syntactic magmoid* of L. Thus, for all $m, n \geqslant 0$, the set $(M_L)_{m,n}$ can be identified with the set consisting of all distinct contexts of the elements of $M_{m,n}$, i.e., we may write $(M_L)_{m,n} = \{C_L(f) \mid f \in M_{m,n}\}$ whereas, the operations of M_L are given by the formulas:

$$C_L(f) \circ C_L(g) = C_L(f \circ g), \quad C_L(f) \,\square\, C_L(g) = C_L(f \,\square\, g).$$

The syntactic magmoid is characterized by the following universal property: for any magmoid epimorphism $H : M \to N$, such that $H^{-1}(H(L)) = L$, there exists a unique magmoid morphism $\bar{H} : N \to M_L$ such that $\bar{H} \circ H = H_L$, where $H_L : M \to M_L$ is the canonical projection onto the quotient. Thus M_L is unique up to isomorphism. For magmoids M, N, we write $M < N$ whenever M is a quotient of a submagmoid of N.

Proposition 3 (cf. [9]). *Let $F : M \to N$ be a magmoid morphism, $L \subseteq N$ and $L_1, L_2 \subseteq M$, it holds:*

i) $M_{L_1 \cap L_2} < M_{L_1} \times M_{L_2}$, $M_{L_1 \cup L_2} < M_{L_1} \times M_{L_2}$, $M_{L_1^c} = M_{L_1}$, *where L^c stands for the set-theoretic complement of L;*

ii) $M_{F^{-1}(L)} < N_L$, *if moreover F is surjective, then the syntactic magmoid $M_{F^{-1}(L)}$ is isomorphic with N_L.*

Recognizability of magmoid subsets is defined in [9] by suitably adapting monoid recognizability. We say that a congruence $\sim = (\sim_{m,n})$ on a magmoid $M = (M_{m,n})$ *saturates* $L \subseteq M$ whenever, for all $m, n \geq 0$, the subset $L_{m,n}$ is a union of $\sim_{m,n}$-classes. If, for all $m, n \geq 0$, the congruence $\sim_{m,n}$ has finite index (i.e., finite number of equivalence classes) we say that $\sim$ has locally finite index. Moreover, a magmoid $M = (M_{m,n})$ is said to be locally finite if, for all $m, n \geq 0$, the set $M_{m,n}$ is finite.

Definition 1. *A subset L of $FGR(\Sigma)$ is called recognizable if there exists a locally finite magmoid $N = (N_{m,n})$ (i.e., $N_{m,n}$ finite for all $m, n \in \mathbb{N}$) and a morphism $H : FGR(\Sigma) \to N$, so that $L = H^{-1}(P)$, for some $P \subseteq N$.*

We denote by $Rec(FGR(\Sigma))$ the class of all recognizable subsets of $FGR(\Sigma)$. The elements of $Rec(FGR(\Sigma))$ are called recognizable fuzzy graph languages. From this definition and from the construction of the syntactic magmoid, similarly with crisp graph language recognizability of [9], we obtain

Theorem 1. *Let $L \subseteq FGR(\Sigma)$, the following conditions are equivalent:*

1. *L is recognizable;*
2. *L is saturated by a congruence of a locally finite index;*
3. *$\sim_L$ has locally finite index;*
4. *the set $card\{C_L(G) \mid G \in FGR_{m,n}(\Sigma)\}$ is finite for all $m, n \in \mathbb{N}$;*
5. *the syntactic magmoid $FGR(\Sigma)_L$ is locally finite.*

Corollary 1. *The class $Rec(FGR(\Sigma))$ of all recognizable fuzzy graph languages is closed under finite union, intersection, complement and inverse morphisms of magmoids.*

Proof. Combine the above theorem with Proposition 3 of [9]. □

Proposition 4. *Let Σ be a doubly ranked alphabet and $a, b \in \Sigma$, $a \neq b$, the fuzzy graph language $L_k^{a,b} \subseteq FGR(\Sigma)$ consisting of all graphs that have an equal number of labels a and b on edges with memebeship grade greater or equal to k, $k \in [0, 1]$, is not recognizable*

Proof. For every $G \in FGR(\Sigma)$ and every $\omega = (G_1, F_1, F_2, G_2)$ we denote by $|G|_a$ the number of a's occurring as labels of edges with membership grade greater or equal to k in G and

$$|\omega|_a = |G_1|_a + |F_1|_a + |F_2|_a + |F_2|_a.$$

Let $G \in FGR(\Sigma)$, we observe that for every $\omega, \omega' \in C_{L_k^{a,b}}(G)$ it holds

$$|\omega|_a - |\omega|_b = |\omega'|_a - |\omega'|_b = |G|_b - |G|_a.$$

We can easily verify that the function

$$G \xmapsto{\phi_{m,n}} |G|_b - |G|_a, \qquad G \in GR_{m,n}(\Sigma)$$

is a bijection of the set $(FGR(\Sigma)_{L_k^{a,b}})_{m,n}$ on the set of integers $\mathbb{Z}$. Furthermore it holds

$$\phi(G \circ G') = \phi(G) + \phi(G'), \ \phi(G \,\square\, G') = \phi(G) + \phi(G') \ \text{ and } \ \phi(E_n) = 0_{n,n}$$

and thus, the syntactic magmoid of $L_k^{a,b}$ is isomorphic to the magmoid associated with the commutative monoid of the additive integers. Since this is locally infinite from Theorem 1 we derive that this language is not recognizable. □

Remark 1. Note that $L_0^{a,b}$ consists of all fuzzy graphs with an equal number of a's and b's. As we have shown in [9], the syntactic magmoid of the crisp graph language that consists of all graphs with an equal number of a's and b's in their labels, is also isomorphic with the same magmoid and in this respect the present result constitutes a generalization for fuzzy graph languages.

Proposition 5. *The fuzzy graph language $L_1 \subseteq FGR(\Sigma)$ consisting of all graphs that have exactly k edges ($k \geq 1$) with membership grade 1 is recognizable.*

Proof. For every $G \in FGR(\Sigma)$ let $|G|_1$ be the number of edges of G with membership grade 1 and for every $\omega = (G_1, F_1, F_2, G_2)$ we set

$$|\omega|_1 = |G_1|_1 + |F_1|_1 + |F_2|_1 + |F_2|_1.$$

It holds:

- $|G|_1 = 0$, whenever for every $w \in C_{L_1}(G)$, $|w|_1 = k$,
- $|G|_1 = 1$, whenever for every $w \in C_{L_1}(G)$, $|w|_1 = k - 1$,

$$\vdots$$

- $|G|_1 = k - 1$, whenever for every $w \in C_{L_1}(G)$, $|w|_1 = 1$,
- $|G|_1 = k$, whenever for every $w \in C_{L_1}(G)$, $|w|_1 = 0$,
- $|G|_1 \geq k + 1$, whenever $C_{L_1}(G) = \emptyset$.

The function $\phi_{m,n} : (FGR(\Sigma)_{L_1})_{m,n} \to \{0, 1, \ldots, k, \alpha\}$, sending the syntactic class of every graph $G \in FGR(\Sigma)_{m,n}$ to $0, 1, \ldots, k$ or α, whenever $|G|_1 = 0, 1, \ldots, k$ or $\geq k + 1$ respectively, is a bijection.

Now let $M(A) = (M(A)_{m,n})$ be the magmoid associated with the commutative monoid $A = \{0, 1, \ldots, k, \alpha\}$ whose operation is given by the following table.

$+$	0	1	$\ldots$	k	α
0	0	1	$\ldots$	k	α
1	1	2	$\ldots$	α	α
$\vdots$	$\vdots$	$\vdots$	$\ddots$	$\vdots$	$\vdots$
k	k	α	$\ldots$	α	α
α	α	α	$\ldots$	α	α

It holds:

$$\phi(G \circ G') = \phi(G) + \phi(G'), \ \phi(G \,\square\, G') = \phi(G) + \phi(G') \ \text{ and } \ \phi(E_n) = 0_{n,n}$$

and hence the syntactic magmoid of L_1 is isomorphic to $M(A)$. This is a locally finite magmoid and hence from Theorem 1 we deduce that $L_1 \in Rec(FGR(\Sigma))$.

Proposition 6. *Given a finite doubly ranked set Σ, the class $Rec(FGR(\Sigma))$ is closed under $\square$-operation.*

Proof. Similar with the corresponding proof for crisp graph languages (see [9]). $\blacksquare$

Given a fuzzy graph $G = (V_\kappa, E_\lambda, s, t, l, begin, end)$ we say that there exists a *path* from the node v_1 to the node v_k of G if there exist edges $e_1, \ldots, e_{k-1}$ and nodes $v_2, \ldots, v_{k-1}$ of G such that v_i appears in $s(e_i)$ and v_{i+1} appears in $t(e_i)$ for all $i = 1, \ldots, k - 1$.

Proposition 7. *The fuzzy graph language $L_{path} \subseteq FGR_{1,1}(\Sigma)$ that consists of all graphs that have at least one path from the begin node to the end node through edges with membership grade 1 is recognizable.*

Proof. We define the following equivalence on $FGR_{m,n}(\Sigma)$: $G_1 \sim_p G_2$ whenever the next two items are equivalent there exists a path from the ith begin node of G_1 to the jth end node of G_1 through edges with membership grade 1 if and only if there exists a path from the ith begin node of G_2 to the jth end node of G_2 through edges with membership grade 1. It holds:

$$G_1 \sim_p G_2 \text{ and } G_1' \sim_p G_2' \text{ implies } G_1 \,\square\, G_1' \sim_p G_2 \,\square\, G_2'$$

and similarly for $\circ$, hence $\sim_p$ is a congruence which trivially saturates L_{path} and thus by Theorem 1 we get that L_{path} is recognizable. $\blacksquare$

Fuzzy graph language recognizability can also be characterized through left derivatives in a result that is a generalization of the fundamental fact that a string language is recognizable, if and only if, it has finitely many left derivatives, if and only if, it has finitely many right derivatives (cf. [14]). Let $L \subseteq FGR(\Sigma)$ and $\omega \in Cont_{m,n}(FGR(\Sigma))$. The *left derivative* of L at ω is defined as

$$\omega^{-1}L = \{G \in FGR_{m,n}(\Sigma) \mid \omega[G] \in L\}.$$

Proposition 8. *The fuzzy graph language $L \subseteq FGR(\Sigma)$ is recognizable, if and only if, $card\{\omega^{-1}L \mid \omega \in Cont_{m,n}(FGR(\Sigma))\} < \infty$, for all $m, n \in \mathbb{N}$.*

Proof. As in the case of crisp graph languages (see Proposition 5 of [9])

By virtue of this proposition we prove the following result.

Proposition 9. *Let $L \subseteq FGR(\Sigma)$ be a fuzzy graph language consisting only of graphs that have nodes with membership grade 1, then L is recognizable if and only if L^{fc} is recognizable.*

Proof. Assume that $L \in Rec(FGR(\Sigma))$ and $m, n \in \mathbb{N}$, then by the previous proposition

$$card\{\omega^{-1}L \mid \omega \in Cont_{m,n}(FGR(\Sigma))\} < \infty.$$

Let $\omega_1^{-1}L, \ldots, \omega_k^{-1}L$ be representatives of the distinct left derivatives of L. We shall prove that $(\omega_1^{fc})^{-1}L, \ldots, (\omega_k^{fc})^{-1}L$ are all the distinct left derivatives of L^{fc}. For every $\omega = (G_1, F_1, F_2, G_2)$, we set

$$\omega^{fc} = (G_1^{fc}, F_1^{fc}, F_2^{fc}, G_2^{fc}).$$

Note that for any graph $G \in FGR(\Sigma)$ it holds $(G^{fc})^{fc} = G$. Now, let $\omega \in Cont_{m,n}(FGR(\Sigma))$, then for any graph $G \in FGR(\Sigma)$ it holds

$$\omega[G] \in L^{fc} \Leftrightarrow (\omega[G])^{fc} \in L \overset{*}{\Leftrightarrow} \omega^{fc}[G^{fc}] \in L.$$

Since we assumed that $\omega_1^{-1}L, \ldots, \omega_k^{-1}L$ are all the distinct left derivatives of L, from the last we deduce that there exists $1 \le i \le k$ such that

$$\omega_i[G^{fc}] \in L \Leftrightarrow (\omega_i[G^{fc}])^{fc} \in L^{fc} \overset{*}{\Leftrightarrow} \omega_i^{fc}[G] \in L^{fc}.$$

Hence the context ω is identified with one of $\omega_1^{fc}, \ldots, \omega_k^{fc}$, and thus L^{fc} has finite distinct left derivatives which by Proposition 8 concludes the proof. Notice that in the equivalences $\overset{*}{\Leftrightarrow}$ we used the equality $(\omega[G])^{fc} = \omega^{fc}[G^{fc}]$ which holds only in the case that the graph G has only nodes with membership grade 1.

4   Conclusion

We introduced a notion of fuzzy graph language recognizability based on the established concept of crisp graph language recognizability and similar with the corresponding theory for string and tree languages. In this respect, existing applications and methods in various areas including model checking and formal verification [3, 4], syntactic complexity [11, 12], and natural language processing [18] can be investigated in the framework of fuzzy graphs.

References

1. Arnold, A., Dauchet, M.: Théorie des magmoides. I. RAIRO Inform. Théor. 12, 235–257 (1978)
2. Arnold, A., Dauchet, M.: Théorie des magmoides. II. RAIRO Inform. Théor. 13, 135–154 (1979)
3. Blume, C.: Recognizable Graph Languages for the Verification of Dynamic Systems. In: Ehrig, H., Rensink, A., Rozenberg, G., Schürr, A. (eds.) ICGT 2010. LNCS, vol. 6372, pp. 384–387. Springer, Heidelberg (2010)
4. Blume, C., Bruggink, S., König, B.: Recognizable Graph Languages for Checking Invariants. In: Workshop on Graph Transformation and Visual Modeling Techniques, Electronic Communications of the EASST, vol. 29 (2010)
5. Bozapalidis, S., Bozapalidou, O.L.: On the recognizability of fuzzy languages I. Fuzzy Sets and Systems 157, 2394–2402 (2006)
6. Bozapalidis, S., Bozapalidou, O.L.: On the recognizability of fuzzy languages II. Fuzzy Sets and Systems 159, 107–113 (2008)
7. Bozapalidis, S., Bozapalidou, O.L.: Fuzzy tree language recognizability. Fuzzy Sets and Systems 161, 716–734 (2010)
8. Bozapalidis, S., Kalampakas, A.: An Axiomatization of Graphs. Acta. Inform. 41, 19–61 (2004)
9. Bozapalidis, S., Kalampakas, A.: Recognizability of graph and pattern languages. Acta. Inform. 42, 553–581 (2006)
10. Bozapalidis, S., Kalampakas, A.: Graph Automata. Theoret. Comput. Sci. 393, 147–165 (2008)
11. Brzozowski, J., Ye, Y.: Syntactic Complexity of Ideal and Closed Languages. In: Mauri, G., Leporati, A. (eds.) DLT 2011. LNCS, vol. 6795, pp. 117–128. Springer, Heidelberg (2011)
12. Brzozowski, J., Li, B., Ye, Y.: Syntactic complexity of prefix-, suffix-, bifix-, and factor-free regular languages. Theoret. Comput. Sci (2012), http://dx.doi.org/10.1016/j.tcs.2012.04.011
13. Courcelle, B.: The Monadic Second-Order Logic of Graphs. I. Recognizable Sets of Finite Graphs. Inform. and Comput. 85, 12–75 (1990)
14. Eilenberg, S.: Automata, Languages and Machines, vol. A. Academic Press (1974)
15. Engelfriet, J., Vereijken, J.J.: Context-free graph grammars and concatenation of graphs. Acta. Inform. 34, 773–803 (1997)
16. Kalampakas, A.: Graph Automata: The Algebraic Properties of Abelian Relational Graphoids. In: Kuich, W., Rahonis, G. (eds.) Algebraic Foundations in Computer Science. LNCS, vol. 7020, pp. 168–182. Springer, Heidelberg (2011)
17. Mordeson, J.N., Nair, P.S.: Fuzzy Graphs and Fuzzy Hypergraphs. Physica-Verlag, Heidelberg (2000)
18. Quernheim, D., Knight, K.: Towards Probabilistic Acceptors and Transducers for Feature Structures. In: Proc. 6th Workshop Syntax, Semantics and Structure in Statistical Translation. Association for Computational Linguistics (2012), http://www.ims.uni-stuttgart.de/~daniel/pub2/quekni12.pdf
19. Rosenfeld, A.: Fuzzy graphs. In: Zadeh, L.A., Fu, K.S., Tanaka, K., Shimura, M. (eds.) Fuzzy Sets and Their Applications to Cognitive and Decision Processes, pp. 77–95. Academic Press, New York (1975)

Adaptive Intuitionistic Fuzzy Inference Systems of Takagi-Sugeno Type for Regression Problems

Petr Hájek and Vladimír Olej

Institute of System Engineering and Informatics
Faculty of Economics and Administration, University of Pardubice, Studentská 84
532 10 Pardubice, Czech Republic
{petr.hajek,vladimir.olej}@upce.cz

Abstract. Recently, we have proposed a novel intuitionistic fuzzy inference system (IFIS) of Takagi-Sugeno type which is based on Atanassov's intuitionistic fuzzy sets (IF-sets). The IFIS represent a generalization of fuzzy inference systems (FISs). In this paper, we examine the possibilities of the adaptation of this class of systems. Gradient descent method and other special optimization methods are employed to adapt the parameters of the IFIS in regression problems. The empirical comparison of the systems is provided on several well-known benchmark and real-world datasets. The results show that by adding non-membership functions, the average errors may be significantly decreased compared to FISs.

Keywords: Intuitionistic fuzzy sets, intuitionistic fuzzy inference systems, adaptation, regression.

1 Introduction

Several generalizations of Zadeh's fuzzy set theory [1] have been developed to handle imprecision and uncertainty in a better way, e.g. IF-sets [2,3], L-fuzzy sets [4], interval-valued fuzzy sets (grey sets) [5], or interval-valued IF-sets [6], see [7] for a review. It was proven that interval-valued fuzzy set theory is equivalent to IF-set theory, which is equivalent to the vague set theory. In a similar manner, FISs have also been extended as interval-valued FISs [8], interval-valued type-2 FISs (IT2FISs) [9, 10] or IFIS [11, 12, 13].

The concept of IF-sets can be viewed as an alternative approach to define a fuzzy set in cases where available information is not sufficient for the definition of an imprecise concept by means of a conventional fuzzy set. This may hold true for many real world applications. For example, the uncertainty presented in air quality evaluation was modelled using hierarchical IFISs of Mamdani type in [13].

Recently, we presented a novel IFIS of Takagi-Sugeno type for time series prediction [12, 14]. It was shown that MIN t-norm (Gödel t-norm) provided the lowest error in terms of root mean squared error (RMSE) on testing data [14]. Furthermore, we developed IFIS with various levels of uncertainty represented by intuitionistic fuzzy index (IF-index). This class of systems presents a strong

L. Iliadis et al. (Eds.): AIAI 2012, IFIP AICT 381, pp. 206–216, 2012.
© IFIP International Federation for Information Processing 2012

possibility to express uncertainty (in the presence of imperfect facts and imprecise knowledge) and provides a good description of object attributes by means of membership functions $\mu_A(x)$ and non-membership functions $v_A(x)$.

In this paper, we compare several optimization algorithms used to adapt FISs and IFISs on selected artificial and real world datasets. We will also analyze the behaviour of the IFIS when the weights of the subsystems outputs (with membership functions $\mu_A(x)$ and non-membership functions $v_A(x)$) change. The rest of this paper is organized as follows. First, we briefly introduce FISs and IFIS of Takagi-Sugeno type. In this section, the difference between IFIS and related IT2FISs is mentioned. Next, we present the methods used for the adaptation of IFISs. In section 4, regression datasets are described. Finally, the results of experiments are provided and analyzed.

2   IFIS of Takagi-Sugeno Type

The FIS of Takagi-Sugeno type [15] is composed of the following steps. In the fuzzification process, the input variables are compared with the membership functions $\mu_A(x)$. Next, operators (AND, OR, NOT) are applied within the if-then rules. Thus, firing weight of each if-then rule is obtained. Further, the outputs of each if-then rule are generated. The output of each rule is a linear combination of input variables plus a constant term. The final output is the weighted average of each if-then rule's output.

Let $x \in X$, $X = \{x_1, x_2, \ldots, x_i, \ldots, x_n\}$ be input variables defined on the universes $X_1, X_2, \ldots, X_i, \ldots, X_n$ and let y be an output variable defined on the universe Y. Then FIS has n input variables and one output variable. Then the k-th if-then rule R^k, $k=1,2, \ldots, N$, in the FIS of Takagi-Sugeno type can be defined in following form

$$R_k: \text{if } x_1 \text{ is } A_{1,k} \text{ AND } x_2 \text{ is } A_{2,k} \text{ AND } \ldots \text{ AND } x_i \text{ is } A_{i,k} \text{ AND } \ldots \text{ AND } x_n \text{ is } A_{n,k} \quad (1)$$
then $y_k = f(x_1, x_2, \ldots, x_n)$, $i=1,2, \ldots, n$,

where $A_{1,k}, A_{2,k}, \ldots, A_{i,k}, \ldots, A_{n,k}$ represent fuzzy sets and $f(x_1, x_2, \ldots, x_n)$ can be a linear or polynomial function. In further considerations, we assume a linear function defined as $y_k = a_{1,k} x_1 + a_{2,k} x_2 + \ldots + a_{i,k} x_i + \ldots + a_{n,k} x_n + b$.

The FIS of Takagi-Sugeno type was designed in order to achieve higher computational effectiveness. This is possible as the defuzzification of outputs is not necessary. Its advantage lies also in involving the functional dependencies of output variable on input variables. The output level y_k of each the k-th if-then rule R_k is weighted by $w_k = \mu(x_1) \text{ AND } \mu(x_2) \text{ AND } \ldots \text{ AND } \mu(x_m)$. The final output y of the FIS Takagi-Sugeno type is the weighted average of all N if-then rule R_k outputs y_k, $k=1,2, \ldots, N$, computed as follows

$$y = \frac{\sum_{k=1}^{N} y_k w_k}{\sum_{k=1}^{N} w_k}. \quad (2)$$

Let a set X be a non-empty fixed set. An IF-set A in X is an object having the form $A=\{\langle x,\mu_A(x),\nu_A(x)\rangle|x\in X\}$, where the function $\mu_A:X\to[0,1]$ defines the degree of membership function $\mu_A(x)$ and the function $\nu_A:X\to[0,1]$ defines the degree of non-membership function $\nu_A(x)$, respectively, of the element $x\in X$ to the set A, which is a subset of X, and $A\subset X$, respectively; moreover for every $x\in X$, $0\le\mu_A(x)+\nu_A(x)\le1$, $\forall x\in X$ must hold [2,3]. The amount $\pi_A(x)=1-(\mu_A(x)+\nu_A(x))$ is called the hesitation part, which may cater to either membership value or non-membership value, or both. For each IF-set in X, we will call $\pi_A(x)$ as the IF-index of the element x in set A. It is obvious that $0\le\pi_A(x)\le1$ for each $x\in X$. The value denotes a measure of non-determinancy. The IF-indices $\pi_A(x)$ are such that the larger $\pi_A(x)$ the higher the hesitation margin of the decision maker. The k-th if-then rule R^μ_k in FIS$^\mu$ and R^ν_k in FIS$^\nu$ are defined as follows

$$R^\mu_k: \text{if } x_1 \text{ is } A^\mu_{1,k} \text{ AND } x_2 \text{ is } A^\mu_{2,k} \text{ AND } ... \text{ AND } x_i \text{ is } A^\mu_{i,k} \text{ AND } ... \text{ AND } x_n \text{ is}$$
$$A^\mu_{n,k} \text{ then } y^\mu_k= a^\mu_{1,k}x_1+a^\mu_{2,k}x_2+ ... +a^\mu_{i,k}x_i+ ... +a^\mu_{n,k}x_n+b^\mu, \tag{3}$$
$$R^\nu_k: \text{if } x_1 \text{ is } A^\nu_{1,k} \text{ AND } x_2 \text{ is } A^\nu_{2,k} \text{ AND } ... \text{ AND } x_i \text{ is } A^\nu_{i,k} \text{ AND } ... \text{ AND } x_n \text{ is}$$
$$A^\nu_{n,k} \text{ then } y^\nu_k= a^\nu_{1,k}x_1+a^\nu_{2,k}x_2+ ... +a^\nu_{i,k}x_i+ ... +a^\nu_{n,k}x_n+b^\nu.$$

The output y^μ of FIS$^\mu$ (the output y^ν of FIS$^\nu$) is defined in the same way as presented in equation (2) for firing weights w^μ_k and w^ν_k, respectively. The output y of the IFIS represents a combination of y^μ and y^ν

$$y=(1-\beta)y^\mu+\beta y^\nu, \tag{4}$$

where y is the output of the IFIS, y^μ is the output of the FIS$^\mu$ using the membership function $\mu_A(x)$, y^ν is the output of the FIS$^\nu$ using the non-membership function $\nu_A(x)$, and β is the weight of the output y^ν. In prior studies [13,14], it was assumed that $\beta=\pi$.

Like IFISs, IT2FISs also represent a generalization of the FISs. The membership degree of an interval type-2 fuzzy set is an interval (also known as the footprint of uncertainty) bounded from the above and below by two type-1 membership functions, $\bar{A}(x)$ (upper membership function) and $\underline{A}(x)$ (lower membership function). When related to IF-sets, $\underline{A}(x)=\mu_A(x)$ and $\bar{A}(x)=\mu_A(x)+\pi_A(x)$. This fact implies the differences between IFISs and IT2FISs. When applying AND operator (MIN t-norm) in if-then rules, the firing intervals $w_k=[\bar{w}_k,\underline{w}_k]$ are obtained. Notice that $\underline{w}_k=w^\mu_k=$ MIN$(\mu(x_1),\mu(x_2), ... ,\mu(x_m))$ but $\bar{w}_k=$MIN$(\mu(x_1)+\pi(x_1),\mu(x_2)+\pi(x_2), ... ,\mu(x_m)+\pi(x_m))$ while $w^\nu_k=$MIN$(\nu(x_1),\nu(x_2), ... ,\nu(x_m))$. Recently, IT2FISs have been adapted using Karnik-Mendel algorithm and its variants [16,17,18], back-propagation algorithm [19], fuzzy C-means algorithm [20], or genetic algorithms [21,22].

3   IFIS of Takagi-Sugeno Type Adaptation

There has been developed many approaches to adapt FISs in the literature, e.g. neural networks, genetic algorithms, Kalman filter, see e.g. [23,24]. In this paper, we use the following algorithms to adapt IFIS (and FISs): subtractive clustering algorithm (SCA) [25], Moore-Penrose pseudo-inverse (MPPI), Kalman filter [26], Kaczmarz algorithm [27,28], and gradient algorithm [23].

The identification of IFISs consists of the following steps. First, cluster centres are found to construct the number N of if-then rules and the antecedents of the if-then rules. Second, optimize the consequents of the rules, i.e. parameters $a^{\mu}_{1,k}, a^{\mu}_{2,k}, \ldots, a^{\mu}_{n,k}, b^{\mu}$ and $a^{v}_{1,k}, a^{v}_{2,k}, \ldots, a^{v}_{n,k}, b^{v}$. The number of cluster centres c is equal to the number of if-then rules N, $c=N$, and $k=1,2, \ldots, N$ is the index of the cluster centre. It is determined automatically using a SCA. The radius of influence of a cluster r_a is considered the most important parameter in establishing the number of cluster centres c. A large r_a results in fewer clusters, while a small r_a generates a large number of clusters and, thus, can lead to model over-fitting.

In the subtractive clustering algorithm, each data point is considered as a potential cluster centre. A measure of potential of data point $\mathbf{x}_k$ is defined as follows

$$P_k = \sum_{j=1}^{n} e^{-\alpha \|\mathbf{x}_k - \mathbf{x}_j\|}, \tag{5}$$

where $\alpha = 4/r_a^2$. Thus, a data point $\mathbf{x}_k$ with many neighbouring points $\mathbf{x}_j$ has a high potential value P_k. The data point with the highest potential represents the cluster centre of the first cluster. Then, an amount of potential from each data point is subtracted as a function of its distance from the first cluster centre

$$P_k \Leftarrow P_k - P_1^* e^{-\beta \|\mathbf{x}_k - \mathbf{x}_1^*\|^2}, \tag{6}$$

where $\mathbf{x}_1^*$ is the centre of the first cluster, P_1^* is the potential of $\mathbf{x}_1^*$, and $\beta = 4/r_b^2$. The positive constant r_b represents the radius defining the neighbourhood that will have measurable reductions in potential P_k.

Let each input vector $\mathbf{x}_k^*$ is decomposed into two component vectors $\mathbf{y}_k^*$ and y_k^*, where $\mathbf{y}_k^*$ contains first n elements of $\mathbf{x}_k^*$ (input data) and y_k^* contains the output component (for a multi-input and single-output system). Each cluster centre $\mathbf{x}_k^*$ is considered an if-then rule. The output y is represented by a weighted average of the output of each rule (2). Since we use a IFIS of Takagi-Sugeno type of the first order, i.e. $f(x_1, x_2, \ldots, x_m)$ is a linear function, we can compute the y_k^* as follows

$$y_k^* = \mathbf{G}_k \mathbf{y}_k + h_k, \tag{7}$$

where $\mathbf{G}_k$ is a constant vector and h_k is a constant. The estimation of the parameters of the given model can be understood as least squares estimate (LSE) in the form $\mathbf{AX=B}$, where $\mathbf{B}$ is a matrix of output values, $\mathbf{A}$ is a constant matrix, and $\mathbf{X}$ is a matrix of parameters to be estimated.

Let P is the set of linear parameters and $\mathbf{X}$ is an unknown vector whose elements are parameters in P. Then, we seek the optimal solution of $\mathbf{X}$ using a LSE $\mathbf{X}^*$. In this process, the squared error $\|\mathbf{AX-B}\|^2$ is minimized. The LSE $\mathbf{X}^*$ can be calculated using the pseudo-inverse of $\mathbf{X}$ as follows

$$X^*=(A^TA)^{-1}A^TB, \tag{8}$$

where A^T is a transpose of A, and $(A^TA)^{-1}A^T$ is the pseudo-inverse of A if A^TA is non-singular. A generalization of the inverse matrix A^{-1} is called MPPI A^+.

A sequential method of LSE (also known as recursive LSE) can be used to compute X^* [23]. This method is more efficient, especially for a low number of linear parameters. Let a_i^T is is the i-th input vector of matrix A and b_i^T is the i-th element of B. Then X can be calculated using the following formulas (online version)

$$X_{i+1}=X_i+S_{i+1}a_{i+1}(b^T_{i+1}-a^T_{i+1}X_i),\ S_{i+1}=1/\lambda[S_i-(S_ia_{i+1}a^T_{i+1}S_i)/(\lambda+a^T_{i+1}S_ia_{i+1})], \tag{9}$$

where S_i is the covariance matrix, λ is the forgetting factor, and $X^*=X_n$. The initial conditions are $X_0=0$ and $S_0=\gamma I$, where γ is a positive large number and I is the identity matrix. A small λ shows on the fast effects of old data decay. The LSE of X can be interpreted as the Kalman filter for the process [26]

$$X(k+1)=X(k),\ Y(k)=A(k)X(k)+e, \tag{10}$$

where e is noise, $X(k)=X_k$, $Y(k)=b_k$ and $A(k)=a_k$. Therefore, the sequential method of the LSE presented above is referred to as the Kalman filter algorithm. Kalman filters have been used to optimize the output function parameters FISs of Takagi-Sugeno type [29], to extract if-then rules from a given rule base of FISs [30], and to tune the input membership functions [31]. Another method used to compute the LSE X^* is the Kaczmarz algorithm. This algorithm is based on the following formula

$$X_{k+1} = X_k + \frac{b_k - a_kX_k^T}{a_k^Ta_k}a_k. \tag{11}$$

Another possibility to adapt an IFIS is represented by gradient algorithms. Let the i-th error is defined as $E_i=(y_i-o_i)^2$, where y_i is the actual output and o_i is the predicted output. Then, the total error E is given as $E=\Sigma E_i$. The derivative of the overall error measure E with respect to a generic parameter α is

$$\frac{\partial E}{\partial \alpha} = \sum_{i=1}^{n} \frac{\partial E_i}{\alpha}. \tag{12}$$

The update formula for the generic parameter α is defined as

$$\Delta\alpha = -\eta \frac{\partial E}{\partial \alpha}, \text{ where } \eta = \frac{h}{\sqrt{\sum_\alpha \left(\frac{\partial E}{\partial \alpha}\right)^2}}, \tag{13}$$

where η is the learning rate, h is the step size and the length of each gradient transition in the parameter space. Usually, we can change the value of h to vary the speed of convergence. Some practical difficulties associated with gradient descent are slow convergence and ineffectiveness at finding a good solution [31].

4   Experimental Results

The following regression datasets were selected for the modelling using IFISs: Friedman benchmark function [32], daily electricity energy [33], stock prices [34], and auto MPG dataset [35], for details see Table 1.

Table 1. Description of regression datasets

| | Dataset | | | |
	Friedman	Energy	Stock	AutoMPG
Origin	Artificial	Real world	Real world	Real world
Input variables	5	6	9	5
Real / integer / nominal	5/0/0	6/0/0	9/0/0	2/3/0
Instances	1200	365	950	392

Friedman dataset is a synthetic benchmark dataset where the instances are generated using the following method. Generate the values of $n=5$ input variables, $x_1, x_2, \ldots, x_5$ independently each of which uniformly distributed over [0.0, 1.0]. Obtain the value of the target variable y using the equation $y=10(\sin(\pi)x_1x_2)+20(x_3-0.5)^2+10x_4+5x_5+e$, where e is a Gaussian random noise $N(0,1)$.

The Energy problem involves predicting the daily average price of TkWhe electricity energy in Spain. The data set contains real values from 2003 about the daily consumption in Spain of energy from hydroelectric, nuclear electric, carbon, fuel, natural gas and other special sources of energy.

In the stock prices dataset, the data provided are daily stock prices from January 1988 through October 1991, for ten aerospace companies. The task is to approximate the price of the 10th company given the prices of the rest.

The AutoMPG dataset concerns city-cycle fuel consumption in miles per gallon (Mpg), to be predicted in terms of 1 multi-valued discrete and 5 continuous attributes (two multi-valued discrete attributes (Cylinders and Origin) from the original dataset are removed).

Datasets were divided into training and testing data in relation 1:1. This division was realized five times. The quality of regression was measured by RMSE on testing data. We conducted the experiments in Matlab Fuzzy Logic Toolbox using the adaptation algorithms available at [36].

The parameters of FISs and IFISs are set in the following way. The initial setting of FISs and IFISs was conducted using SCA. The number N of if-then rules (and the numbers of membership and non-membership functions at the same time) depends on the choice of parameter r_a in SCA primarily. In order to avoid over-fitting, we tested different values of $r_a=\{0.1, 0.2, \ldots, 0.9\}$. For the IFISs, the IF-index is set on $\pi=0.3$ (medium level of hesitancy recommended in [14]) and, therefore, the membership and non-membership functions are defined in the following way

$$\mu(x) = 0.7e^{-\frac{(x-b)^2}{2\sigma^2}}, \quad \nu(x) = (1-\pi(x))-\mu(x) = 0.7 - 0.7e^{-\frac{(x-b)^2}{2\sigma^2}}. \tag{14}$$

The settings of FISs and IFISs parameters are presented in Table 2. In our experiments, we used the following settings of adaptation methods:

- Gradient algorithm: maximum number of epochs was set to 500, step size h=0.01, step increasing rate to 1.1, and step decreasing rate to 0.9;
- Kaczmarz algorithm: maximum number of sweeps was set to 10;
- Kalman filter: data forgetting factor was set to λ=1.0 and its increasing factor to 1.0;
- MPPI: no parameters;

Table 2. Parameters of FISs and IFISs

Dataset	Friedman	Energy	Stock	AutoMPG
Type of $\mu_A(x)$ and $v_A(x)$	Gaussian	Gaussian	Gaussian	Gaussian
Radius r_a	0.6	0.7	0.5	0.7
N of if-then rules	26	7	12	4
t-norm	MIN	MIN	MIN	MIN

Experiments were realized for various weights β={0.0,0.1, ... ,1.0} of the subsystems with non-membership functions $v_A(x)$. Concretely, if β=0.0 the output of the IFIS is determined only by the subsystem with membership functions $\mu_A(x)$ while for β=1.0 only the output of the subsystem with non-membership functions $v_A(x)$ has impact on the IFIS output. The results of the experiments for β with minimum RMSE on testing data is presented in Table 3. The FISs and IFISs with adaptation achieved significantly lower RMSE (values in bold based on paired sample t-test at p=0.05) than those without adaptation (SCA). For all the datasets, using adapted IFISs resulted in the decrease in RMSEs.

Table 3. RMSE and its standard deviation on benchmark datasets

Method	Friedman	Energy	Stock	AutoMPG
FIS-SCA	1.538±0.020	0.494±0.030	2.378±0.584	3.405±0.254
IFIS-SCA	1.538±0.020	0.494±0.030	2.378±0.584	3.405±0.254
	(β=0.0)	(β=0.0)	(β=0.0)	(β=0.0)
FIS-gradient	**1.353±0.026**	7.443±1.579	1.423±0.227	3.702±0.211
IFIS-gradient	**1.332±0.032**	4.776±2.776	1.402±0.219	3.684±0.195
	(β=0.1)	(β=0.5)	(β=0.1)	(β=0.1)
FIS-Kaczmarz	1.759±0.099	0.619±0.246	1.888±0.267	3.991±0.355
IFIS-Kaczmarz	1.581±0.159	0.505±0.034	1.708±0.340	3.880±0.554
	(β=0.3)	(β=0.1)	(β=0.2)	(β=0.1)
FIS-Kalman	1.412±0.031	0.760±0.551	**0.944±0.046**	**2.881±0.113**
IFIS-Kalman	**1.353±0.039**	**0.437±0.031**	**0.914±0.044**	**2.825±0.100**
	(β=0.2)	(β=0.6)	(β=0.2)	(β=0.3)
FIS-MPPI	1.411±0.047	0.474±0.038	**0.943±0.045**	**2.866±0.113**
IFIS-MPPI	**1.352±0.049**	**0.435±0.027**	**0.913±0.044**	**2.821±0.101**
	(β=0.2)	(β=0.6)	(β=0.2)	(β=0.3)

For comparison, FISs and IFISs with corresponding adaptation methods are shown. In the case of Friedman dataset, the dependence of RMSE (testing data) on β is depicted in Fig. 1 (SCA) and Fig. 2 (gradient algorithm). When IFIS is identified using SCA only (i.e. it is not adapted subsequently), RMSE increases with rising β for all datasets (minimum RMSE for IFIS-SCA is achieved for β=0.0). In the case of gradient algorithm it is possible to reduce RMSE when combining the outputs of both subsystems with β=0.1. For the Friedman dataset (the most complex dataset in terms of N and the numbers of membership and non-membership functions), gradient algorithm provided the best result but for the Energy dataset this algorithm did not converge and for the Stock and AutoMPG datasets it was outperformed by the Kalman filter and MPPI, respectively. For the Energy dataset, the behaviour of IFIS differs from that one for the other datasets. The lowest RMSE is achieved for the Kalman filter and MPPI with β=0.6, i.e. both subsystems (with membership and non-membership functions) contribute with a similar magnitude.

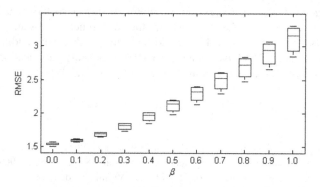

Fig. 1. Relation between β and RMSE for IFIS identified by SCA on Friedman dataset

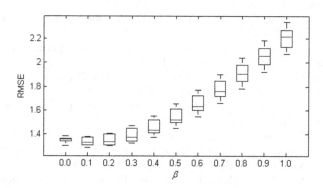

Fig. 2. Relation between β and RMSE for IFIS adapted by gradient algorithm on Friedman dataset

5   Conclusion

In this paper, we have introduced IFIS of Takagi-Sugeno type. We have proposed several methods to adapt the parameters of this class of systems. We conclude that Kalman filter and MPPI are suitable to adapt the linear functions in the consequents of the IFIS if-then rules. However, gradient algorithm may perform better than these methods for more complex problems. The antecedents parts of if-then rules, membership functions $\mu_A(x)$ and non-membership functions $\nu_A(x)$ were identified using SCA. For all the tested datasets, it was possible to significantly decrease RMSE on testing data when both an appropriate adaptation method was applied and the subsystems with membership and non-membership functions were combined properly.

In further research we plan to use a wider range of benchmark datasets to test the relation between data complexity and behaviour of IFISs (the value of IF-index especially). We also suggest comparing the two related concepts, IFISs and IT2FISs of Takagi-Sugeno type on regression problems (using also other performance measures such as R^2 or MAE) to show the differences between these two generalizations of FISs. Finally, IFISs should be tested on data with additional noise since generalizations of FISs have been proven to be especially beneficial in the cases where data was corrupted by measurement or estimated noise [37].

References

[1] Zadeh, L.A.: Fuzzy Sets. Inform. and Control 8, 338–353 (1965)
[2] Atanassov, K.T.: Intuitionistic Fuzzy Sets. Fuzzy Sets and Systems 20, 87–96 (1986)
[3] Atanassov, K.T.: Intuitionistic Fuzzy Sets. Physica-Verlag, Heidelberg (1999)
[4] Goguen, J.: L-fuzzy Sets. J. Math. Anal. Appl. 18, 145–174 (1967)
[5] Sambuc, R.: Fonctions Φ-floues. Application a L'aide au Diagnostic en Pathologie Thyroidienne. Ph.D. Thesis, Univ. Marseille, France (1975)
[6] Atanassov, K.T., Gargov, G.: Interval Valued Intuitionistic Fuzzy Sets. Fuzzy Sets and Systems 31, 343–349 (1989)
[7] Deschrijver, G., Kerre, E.E.: On the Relationship Between Some Extensions of Fuzzy Set Theory. Fuzzy Sets and Systems 133, 227–235 (2003)
[8] Gorzalczany, M.B.: A Method of Inference in Approximate Reasoning Based on Interval Valued Fuzzy Sets. Fuzzy Sets and Systems 21, 1–17 (1987)
[9] Karnik, N.N., Mendel, J.M.: Introduction to Type-2 Fuzzy Logic Systems. In: IEEE FUZZ Conf., Anchorage, AK (1998)
[10] Starczewski, J.T., Rutkowski, L.: Connectionist Structures of Type 2 Fuzzy Inference Systems. In: Wyrzykowski, R., Dongarra, J., Paprzycki, M., Waśniewski, J. (eds.) PPAM 2001. LNCS, vol. 2328, pp. 634–642. Springer, Heidelberg (2002)
[11] Montiel, O., Castillo, O., Melin, P., Sepúlveda, R.: Mediative Fuzzy Logic: A New Approach for Contradictory Knowledge Management. Soft Computing 20, 251–256 (2008)
[12] Olej, V., Hájek, P.: IF-Inference Systems Design for Prediction of Ozone Time Series: The Case of Pardubice Micro-region. In: Diamantaras, K., Duch, W., Iliadis, L.S. (eds.) ICANN 2010, Part I. LNCS, vol. 6352, pp. 1–11. Springer, Heidelberg (2010)

[13] Olej, V., Hájek, P.: Possibilities of Air Quality Modelling Based on IF-sets Theory. In: Mastorakis, N., Demiralp, N., Mladenov, M. (eds.) Computers and Simulation in Modern Science. Selected Papers from WSEAS Conferences, pp. 90–100. WSEAS Press (2010)

[14] Olej, V., Hájek, P.: Comparison of Fuzzy Operators for IF-Inference Systems of Takagi-Sugeno Type in Ozone Prediction. In: Iliadis, L., Maglogiannis, I., Papadopoulos, H. (eds.) EANN/AIAI 2011, Part II. IFIP AICT, vol. 364, pp. 92–97. Springer, Heidelberg (2011)

[15] Takagi, T., Sugeno, M.: Fuzzy Identification of Systems and its Applications to Modeling and Control. IEEE Trans. on Syst. Man and Cybern. 15(1), 116–132 (1985)

[16] Wu, D., Mendel, J.M.: Enhanced Karnik-Mendel Algorithms. IEEE Transactions on Fuzzy Systems 17(4), 923–934 (2009)

[17] Wu, D., Nie, M.: Comparison and Practical Implementation of Type-Reduction Algorithms for Type-2 Fuzzy Sets and Systems. In: IEEE Int. Conf. on Fuzzy Systems, Taipei, Taiwan (2011)

[18] Yeh, C.Y., Jeng, W.H., Lee, S.J.: An Enhanced Type-reduction Algorithm for Type-2 Fuzzy Sets. IEEE Trans. on Fuzzy Systems (2011) (in press)

[19] Lee, C.H., Lee, Y.H.: Nonlinear System Identification Using Takagi-Sugeno-Kang Type Interval-Valued Fuzzy Systems via Stable Learning Mechanism. IAENG International Journal of Computer Science 38(3), 1–11 (2011)

[20] Uncu, O., Turksen, I.B.: Discrete Interval Type-2 Fuzzy System Models Using Uncertainty in Learning Parameters. IEEE Trans. on Fuzzy Systems 15(1), 90–106 (2007)

[21] Lee, C.H., Hong, J.L., Lin, Y.C., Lai, W.Y.: Type-2 Fuzzy Neural Network Systems and Learning. International Journal of Computational Cognition 1(4), 79–90 (2003)

[22] Celikyilmaz, A., Turksen, I.B.: Genetic Type-2 Fuzzy Classifier Functions. In: 10th Annual Meeting of the North American Fuzzy Information Processing Society, NAFIPS, pp. 1–6 (2008)

[23] Jang, J.S.R.: ANFIS: Adaptive-Network-based Fuzzy Inference Systems. IEEE Transactions on Systems, Man, and Cybernetics 23(3), 665–685 (1993)

[24] Alcala, R., Casillas, J., Cordon, O., Herrera, F., Zwir, S.J.I.: Learning and Tuning Fuzzy Rule-based Systems for Linguistic Modeling and their Applications. In: Leondes, C.T. (ed.) Knowledge-Based Systems, vol. 3, pp. 889–933 (1999)

[25] Chiu, S.: Fuzzy Model Identification Based on Cluster Estimation. J. of Intell. and Fuzzy Syst. 2, 267–278 (1994)

[26] Jang, J.S.R.: Fuzzy Modeling Using Generalized Neural Networks Kalman Filter Algorithm. In: 8th National Conference on Artificial Intelligence (AAAI 1991), pp. 762–767 (1991)

[27] Kaczmarz, S.: Approximate Solution of Systems of Linear Equations. Int. J. Control 53, 1269–1271 (1993)

[28] Strohmer, T., Vershynin, R.: A Randomized Kaczmarz Algorithm with Exponential Convergence. Journal of Fourier Analysis and Applications 15(2), 262–278 (2007)

[29] Ramaswamy, P., Riese, M., Edwards, R., Lee, K.: Two Approaches for Automating the Tuning Process of Fuzzy Logic Controllers. In: IEEE Conf. on Decision and Control, San Antonio, TX, pp. 1753–1758 (1993)

[30] Wang, L., Yen, J.: Extracting Fuzzy Rules for System Modeling Using a Hybrid of Genetic Algorithms and Kalman Filter. Fuzzy Sets and Systems 101, 353–362 (1998)

[31] Simon, D.: Training Fuzzy Systems with the Extended Kalman Filter. Fuzzy Sets and Systems 132(2), 189–199 (2002)

[32] Friedman, J.: Multivariate Adaptive Regression Splines. Annals of Statistics 19(1), 1–141 (1991)

[33] Java Software Tool Named KEEL, Knowledge Extraction Based on Evolutionary Learning (2012), http://www.keel.es

[34] Altay, G.H., Bilkent, I.U.: University Function Approximation Repository (2000), http://funapp.cs.bilkent.edu.tr

[35] Quinlan, J.R.: C4.5 Programs for Machine Learning. Morgan Kaufmann (1993)

[36] Sidelnikov, P.K.A.: Sugeno-type FIS Output Tuning (2010), http://www.mathworks.ch/Matlabcentral/fileexchange/28458-sugeno-type-fis-output-tuning

[37] Liang, Q., Mendel, J.M.: Interval Type-2 Fuzzy Logic Systems: Theory and Design. IEEE Transactions on Fuzzy Systems 8(5), 535–550 (2000)

A Hybrid Method for Evaluating Biomass Suppliers – Use of Intuitionistic Fuzzy Sets and Multi-Periodic Optimization

Vassilis C. Gerogiannis, Vasiliki Kazantzi, and Leonidas Anthopoulos

Project Management Department,
Technological Education Institute of Larissa, 41110, Larissa, Hellas
{gerogian,kazantzi,lanthopo}@teilar.gr

Abstract. Evaluation of biomass suppliers is a time-dependent problem that requires assessment of different supply schemes in different periods. This paper presents a hybrid method for evaluating biomass suppliers that combines Intuitionistic Fuzzy Sets (IFS), linear programming (LP) and multi-periodic optimization (MPO). IFS allow evaluators to express their hesitation when they assess alternative suppliers. LP is used to estimate weights of evaluation criteria and calculate suppliers' ratings in a specific period. These ratings are utilized by a MPO model to determine what type and how much feedstock should be supplied by each supplier in each period.

Keywords: Biomass Supplier Evaluation, Intuitionistic Fuzzy Sets, Multi-Periodic Optimization.

1   Introduction

One way to deal with the seasonality of biomass supply requirements and achieve cost-effective supplies is via procurement from different suppliers in different time periods [1]. Alternative suppliers need to be evaluated regularly because of the time-dependency of most of the evaluation criteria. The problem is also characterized by high uncertainty [2] because of subjective judgments on biomass quality aspects, unexpected demand, variable raw material prices, variation in biomass availability and unstable procurement lead times.

The work presented in this paper emphasizes on evaluating biomass suppliers to maximize the total benefit for the supply chain and minimize the associated production and logistics costs. A multi-criteria method is suggested for supporting biomass supplier evaluators to solve two practical problems: (i) What is the optimal combination of biomass suppliers that can provide the required quantities of raw materials for bioenergy production to a conversion plant in certain periods of time? (ii) How much of the required biomass materials should be purchased from each supplier in each time period? The presented method addresses evaluators' judgments with Intuitionistic Fuzzy Sets (IFS) [3]. IFS consider, not only the uncertainty of evaluators to quantify (qualitative) evaluation criteria and determine suppliers' ratings on the identified criteria, but also evaluators' positive and negative judgments, which

L. Iliadis et al. (Eds.): AIAI 2012, IFIP AICT 381, pp. 217–223, 2012.

need not to be complementary. IFS include membership and non-membership of an element to a fuzzy set, as well as a third parameter that is called the hesitation degree. Expression of indeterminacy can be suitable to evaluate suppliers in a highly uncertain supply network, such as a biomass supply system. An IFS-based technique [4] is applied in combination with linear programming to derive weights for the evaluation criteria and suppliers' ratings. The technique results (supplier ratings) are further utilized by a multi-periodic optimization model that is proposed to determine what type and how much feedstock should be supplied, in each period, from each supplier.

2   Biomass Supplier Evaluation Criteria and Supplier Selection Methods

From reviewing the relevant literature, a set of criteria were identified for the evaluation of biomass suppliers [1] [2]: (i) *Reliability*. It includes, as sub-criteria, adherence to contract terms, ensuring on-time delivery and agreed biomass quality. Suppliers need to provide proper feedstock, at correct amounts, within right times, and at right conditions. (ii) *Responsiveness*. Unexpected demand variations may cause operational problems in bioenergy production. Biomass availability and seasonality are time-dependent characteristics that can lead to supply chain malfunctioning. (iii) *Flexibility*. In biomass systems, flexibility refers to the ability of a supplier to provide alternative biomass types that match the needs of the production process and facilitate uninterrupted biomass flow. (iv) *Cost*. Biomass market prices can be viewed as unstable parameters determined by changes in aggregate demand and supply conditions. (v) *Quality aspects*. Variability in quality of feedstock types is usually an undesirable aspect of provided sources. Quality aspects include biomass moisture, density and energy content, quality compliance, defect rates and quality certification possession. (vi) *Assets and infrastructure*. The economic performance of a supply chain is affected by the relative location of the biomass production site and existing transportation infrastructure. This category also includes facility and fleet size, warehouses number and capacity. (vii) *Environment and safety*. These criteria are related with a supplier's consideration to environmental and safety issues.

Most of these criteria are qualitative and unstable in nature and, therefore, a fuzzy-based multi-criteria decision making (MCDM) method can be proven beneficial for their characterization [5]. In the category of fuzzy-based MCDM, two classes of methods can be broadly identified. In the first class, there are methods which handle imprecise evaluation criteria and supplier ratings with fuzzy numbers, while in the second calls there methods making use of linguistic terms to evaluate criteria and suppliers. There are also hybrid methods combining fuzzy logic with techniques such as the Analytic Hierarchy Process - AHP [6] and the Technique for Order Performance by Similarity to Ideal Solution - TOPSIS [7]. Furthermore, some generalizations of fuzzy sets are valuable to deal with indeterminacy in a supplier selection setting. Such an extension is Intuitionistic Fuzzy Sets (IFS) [3] which allow evaluators to express hesitation degree or ambiguity, when they assess alternative suppliers and criteria weights. Representative examples of IFS applications in supplier selection problems can be found in [8] [9]. In the next section, we adopt and extend an IFS-based technique that was originally suggested by Li [4]. The main reason to utilize this technique in a biomass supplier evaluation setting is that it provides a comprehensive way to calculate optimal weights for the evaluation criteria, in each time period considered. The technique can be repeatedly applied for each period, to derive new criteria weights and new supplier ratings based on these weights. We present how this technique can further extended by using its results as inputs to a linear, multi-periodic optimization model. The final result is the optimal combination of suppliers with their respective optimal types and amounts of biomass supplies for each time period considered.

3   Description of the Method

Assume that there is a set of n biomass suppliers $S = \{ S_1, S_2, ..., S_n \}$. Each supplier has to be evaluated on m evaluation criteria $X = \{ X_1, X_2, ..., X_m \}$. There is also a group of k evaluators responsible to evaluate the suppliers, in each time period. In the first step of the method, the values of μ_{ij}, v_{ij} and π_{ij} are determined which specify, respectively, the degree of membership, non-membership and hesitation for a supplier S_j with respect to criterion X_i. These degrees represent the evaluation of the fuzzy concept "appropriateness of S_j offer with respect to X_i". One way to derive these values is by asking all k evaluators to express their judgment whether S_j offer is appropriate or not to fulfill criterion X_i. Suppose that from the k evaluators, k_1 consider S_j offer is strong, k_2 express that S_j offer is weak and k_3 give no answer, due to their indeterminacy ($k=k_1+k_2+k_3$). Then, μ_{ij}, v_{ij} and π_{ij} are calculated as follows:

$$\mu_{ij} = k_1 / k , v_{ij} = k_2 / k , \pi_{ij} = k_3 / k \tag{1}$$

Given the indeterminacy of evaluators, a hesitation in the degree of μ_{ij} exists, denoted by a lower μ^l_{ij} and an upper bound μ^u_{ij} which are expressed as follows:

$$\mu^l_{ij} = \mu_{ij} , \mu^u_{ij} = \mu_{ij} + \pi_{ij} = 1 - v_{ij} \tag{2}$$

The second step is to determine the weight of each criterion by asking each evaluator to assess the criteria with respect to their impact on the production. Each evaluator compares the criteria pair-wise (by following AHP) and k weights are derived, for each one of the m criteria. Each weight ω_i is a number that lies in the interval [ω^l_i, ω^u_i], where ω^l_i is the minimum and ω^u_i is the maximum weight for criterion X_i. To determine optimal weights, the following optimization model can be solved (eq. (3)) [4]:

$$max\left\{ \frac{\sum_{j=1i=1}^{n}\sum^{m}(\mu_{ij}^{u}-\mu_{ij}^{l})\omega_i}{n}\right\}, \text{ subject to: } \omega_i^l \le \omega_i \le \omega_i^u \ \forall i=1,...,m, \sum_{i=1}^{m}\omega_i =1 \qquad (3)$$

The computed weights can be utilized to calculate lower and upper bounds of the weighted rating for each supplier based on eqs. (4) & (5):

$$z_j^l = \sum_{i=1}^{m}\mu_{ij}^l \omega_i = \sum_{i=1}^{m}\mu_{ij}\omega_i \ \forall j=1,...,n \qquad (4)$$

$$z_j^u = \sum_{i=1}^{m}\mu_{ij}^u \omega_i = 1 - \sum_{i=1}^{m}v_{ij}\omega_i \ \forall j=1,...,n \qquad (5)$$

To obtain the final rating of supplier S_j, a comparison index ξ_j based on the TOPSIS method is calculated, as follows (eq. (6)):

$$\xi_j = \frac{D(A_j^0, B)}{D(A_j^0, B) + D(A_j^0, G)} \qquad (6)$$

In eq. (6), A_j^0, G and B represent the optimal rating for supplier S_j, the ideal alternative supplier and the negative ideal alternative supplier respectively. D stands for the Hamming Distance measure, as it was defined for IFS [10]. By using the Hamming Distance, the comparison index ξ_j can be computed as shown in eq. (9).

$$A_j^0 = \{ < S_j, z_j^l, 1-z_j^u > \} = \{ < S_j, \sum_{i=1}^{m}\mu_{ij}\omega_i, \sum_{i=1}^{m}v_{ij}\omega_i > \} \qquad (7)$$

$$G = \{ < g,1,0 > \} \text{ and } B = \{ < b,0,1 > \} \qquad (8)$$

$$\xi_j = \frac{z_j^u}{1 + z_j^u - z_j^l} \qquad (9)$$

The ranking derived for each supplier can be consequently utilized to answer the following questions: Which suppliers should be finally chosen based on the derived ratings and when? What type of and how much feedstock should be supplied by each supplier? How would total budget for the biomass purchasing function be allocated? What would be the maximum total purchasing value in each period? To answer these questions, we consider a decision-making horizon H, during which variations in market conditions are anticipated in terms of time-dependent changes in biomass quantities, types and prices of supply and demand. H is discretised into Nt periods (i.e., within each period purchasing conditions are assumed to be stable). By applying repeatedly the previous steps of the IFS-based technique, suppliers' ratings can be calculated per time period. An optimization model, using the suppliers' ratings as inputs (coefficients) of the objective function, can be specified, as follows.

Objective function: $$max(TPV_t) = \sum_{b \in B} \sum_{j_b \in J_b} w_{t,j} F_{t,j_b} \quad \forall t \in T \tag{10}$$

where: TPV_t=Total Purchasing Value in period t, j=supplier index, b=biomass type, j_b=index of supplier j providing biomass type b, $w_{t,j}$=rating of supplier j in period t, F_{t,j_b}=amount of biomass type b to be delivered from supplier j in period t, J_b=set of suppliers delivering feedstock type b, B=set of feedstock types and T = set of periods.

Capacity constraints: $$F_{t,j_b} \leq F_{t,j_b}^{max} \quad \forall t \in T, \forall j_b \in J_b, \forall b \in B \tag{11}$$

Demand Constraints: In each period t, the procured amounts of all biomass types b from all suppliers providing biomass b must sum up to the demand for feedstock, D_t:

$$\sum_{b \in B} \sum_{j_b \in J_b} F_{t,j_b} \leq D_t \quad \forall t \in T \tag{12}$$

Budget constraints: $$\sum_{b \in B} \sum_{j_b \in J_b} F_{t,j_b} c_{t,j_b} \leq C_t^{max} \quad \forall t \in T \tag{13}$$

C_t^{max} refers to the total purchasing budget in period t and c_{t,j_b} is the total cost for purchasing biomass type b from supplier j in period t.

Quality Constraints (optional): Quality constraints ensure that required production quality levels are maintained.

$$\sum_{j_b \in J_b} F_{t,j_b} q_{t,j_b} \leq D_t Q_{t,b} \quad \forall t \in T, \forall b \in B \tag{14}$$

$Q_{t,b}$ is the buyer's maximum acceptable defect rates for each biomass type b in period t and q_{t,j_b} is the defect rate of supplier j with respect to biomass type b at period t.

Non negativity constraints: $$F_{t,j_b} \geq 0 \quad \forall t \in T, \forall j_b \in J_b, \forall b \in B \tag{15}$$

To validate the method, we have considered a case problem of a bioenergy production plant in which three suppliers S_1, S_2 and S_3, offer specific biomass types; S_1 provides rapeseed (RP), S_2 provides both rapeseed (RP) and sunflower (SN) and S_3 delivers waste cooking oil (WCO). By using repeatedly the IFS-based technique (eqs. (1)-(9)), ratings of the three suppliers were calculated on a monthly basis. Capacity constraints for each supplier per month were also considered. If the total budget is 9000 €, total demand is 1000 tn/month and purchasing costs are 5.5€/tn for S_1, 8€/tn for S_2 and 6.5€/tn for S_3, the multi-periodic optimization problem was repeatedly solved (per month) by using eqs. (10)-(15). Suppliers' selection profiles and biomass amounts are depicted in Figure 1. Supplier S_2 and S_3 compete for being the best alternatives. This finding can be justified since S_3 represents a low-cost supply alternative (providing inexpensive WCO), whereas S_2 exhibits flexibility by offering adequate feedstock amounts of two biomass types ($b1$=RR and $b2$=SN).

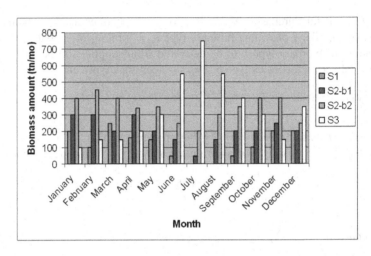

Fig. 1. Optimal supply profiles and purchased biomass schedule per month

4   Conclusions

The paper presented a multi-criteria decision making method for optimal supplier evaluation in a biomass supply network. High degree of time-dependency and uncertainty involved in biomass supply systems renders their management a dynamically evolved issue. The biomass supplier evaluation problem involves conflicting objectives that cannot be optimized simultaneously via a single multi-objective method. A hybrid method based on IFS and multi-periodic optimization was suggested. Our current research intention is to assess more the reliability of decision making by analyzing how sensitive the suppliers' ratings are with respect to the selected criteria and their respective weights.

References

1. Frombo, F., Minciardi, R., Robba, M., Rosso, F., Sacile, R.: Planning Woody Biomass Logistics for Energy Production: a Strategic Decision Model. Biomass and Bioenergy 33(3), 372–383 (2009)
2. Awudu, I., Zhang, J.: Uncertainties and Sustainability Concepts in Biofuel Supply Chain Management. Renewable and Sustainable Energy Reviews 16(2), 1359–1368 (2012)
3. Atanassov, K.T.: Intuitionistic Fuzzy Sets. Fuzzy Sets and Systems 20(1), 87–96 (1986)
4. Li, D.F.: Multiattribute Decision Making Models and Methods Using Intuitionistic Fuzzy Sets. Journal of Computer and System Sciences 70(1), 73–85 (2005)

5. Agarwal, P., Sahai, M., Mishra, V., Bag, M., Singhet, V.: A Review of Multi-Criteria Decision Making Techniques for Supplier Evaluation and Selection. International Journal of Industrial Engineering Computations 2(4), 801–810 (2011)
6. Chan, F., Kumar, N., Tiwari, M.K., Lau, H.C.W., Choy, K.L.: Global Supplier Selection: a Fuzzy-AHP Approach. International Journal of Production Research 46(14), 3825–3857 (2008)
7. Wang, J.W., Cheng, C.H., Huang, K.C.: Fuzzy Hierarchical TOPSIS for Supplier Selection. Applied Soft Computing 9(1), 377–386 (2009)
8. Boran, F.E., Genc, S., Kurt, M., Akay, D.: A Multi-Criteria Intuitionistic Fuzzy Group Decision Making for Supplier Selection with TOPSIS Method. Expert Systems with Applications 36(8), 11363–11368 (2009)
9. Ye, F.: An Extended TOPSIS Method with Interval-Valued Intuitionistic Fuzzy Numbers for Virtual Enterprise Partner Selection. Expert Systems with Applications 37(10), 7050–7055 (2010)
10. Szmidt, E., Kacprzyk, J.: Distances between Intuitionistic Fuzzy Sets. Fuzzy Sets and Systems 114(3), 505–518 (2000)

A Neural Network for Spatial and Temporal Modeling of *foF2* Data Based on Satellite Measurements

Haris Haralambous and Harris Papadopoulos

Computer Science and Engineering Department, Frederick University,
7 Y. Frederickou St., Palouriotisa, Nicosia 1036, Cyprus
{H.Haralambous,H.Papadopoulos}@frederick.ac.cy

Abstract. This paper presents the application of Neural Networks for the spatial and temporal modeling of (critical frequency) *foF2* data over Europe. *foF2* is the most important parameter in describing the electron density profile of the ionosphere since it represents the critical point of maximum electron density in the profile and therefore can be used to drive empirical models of electron density which incorporate *foF2* as an anchor point in the profile shape. The model is based on radio occultation (RO) measurements by LEO (Low Earth Orbit) satellites which provide excellent spatial coverage of *foF2* measurements.

Keywords: Ionosphere, radio occultation, F2 layer critical frequency.

1 Introduction

The ionosphere is defined as a region of the earth's upper atmosphere where sufficient ionisation can exist to affect radio waves in the frequency range 1 to 3 GHz. It ranges in height above the surface of the earth from approximately 50 km to 1000 km. The influence of this region on radio waves is accredited to the presence of free electrons.

The impact of the ionosphere on communication, navigation, positioning and surveillance systems is determined by variations in its electron density profile and subsequent electron content along the signal propagation path [1]. As a result satellite systems for communication and navigation, surveillance and control that are based on trans-ionospheric propagation may be affected by complex variations in the ionospheric structure in space and time leading to degradation of the accuracy, reliability and availability of their service.

The uppermost layer of the ionosphere is the F2 region which is the principal ionospheric region where electron density maximises and therefore introduces significant effects in transionospheric signals (navigation and communication) that penetrate the ionosphere. The maximum frequency that can be reflected at vertical incidence by this layer is termed the F2 layer critical frequency (*foF2*) and is directly related to the maximum electron density of the electron density profile (see Figure 1). The F2 layer critical frequency is therefore the most important parameter in characterising the ionospheric electron density profile. The maximum electron density

L. Iliadis et al. (Eds.): AIAI 2012, IFIP AICT 381, pp. 224–233, 2012.

of free electrons within the F2 layer and therefore *foF2* depend upon the strength of the solar ionising radiation which is a function of time of day, season, geographical location and solar activity [2,3,4].

This paper describes the development of a neural network model which describes the temporal and spatial variability of *foF2* data over a significant part of Europe. The model is based on approximately 53000 LEO satellite *foF2* values from RO measurements recorded from January 2007 to December 2010. This is the first attempt to develop a *foF2* model from RO satellite measurements. The significance of this model lies in its superior spatial interpolation properties due the fact that is based on *foF2* measurements of high spatial resolution as a result of the excellent geographical coverage provided by the COSMIC satellite mission. This spatial resolution is significantly higher than that provided by ionosonde measured *foF2* values which is the traditional method of probing the ionosphere and can also extend over the sea where ionosonde radars are impractical to operate for obvious reasons.

There have been other efforts for the introduction of the spatial aspect, both in the direction of long-term prediction and short-term forecasting models of *foF2* based on ground-based *foF2* measurements. Lamming and Cander (1999) [5] attempted to incorporate geographical latitude and longitude as model parameters to address the interpolation capability of a monthly median *foF2* Neural Network model between ionospheric stations. over Europe. Kumluca et al. (1999) [6] also explored this idea by including more than one ionospheric station to extend the work of Altinay et al. (1997) [7]. Oyeyemi et al. (2005) [8] used data from 40 worldwide ionospheric stations spanning the period 1964–1986 for training a global Neural Network for short-term forecasting of *foF2* including geographical latitude and other geographically related geophysical parameters as inputs. Oyeyemi et al. (2008) [9] extended their previous work by including ground based ionosonde data, from 84 global stations to propose a global long-term prediction Neural Network model of *foF2*.

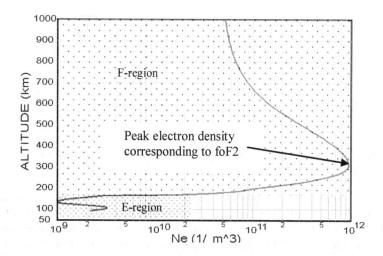

Fig. 1. Typical electron density altitude profile of the ionosphere

2   Measurement of F2 Layer Critical Frequency by Ground-Based and Satellite Techniques

Traditionally measurements of *foF2* were conducted by ionosondes which are special types of radar used for monitoring the electron density at various altitudes in the ionosphere up to the F2-layer peak electron density (corresponding to *foF2*). Their operation is based on a transmitter sweeping through the HF frequency range transmitting short pulses. These pulses are reflected at various layers of the ionosphere, and their echoes are received by the receiver giving rise to a corresponding plot of reflection altitude against frequency which is further analysed to infer the ionospheric plasma height-electron density profile (Figure 1). The maximum frequency at which an echo is received is called the critical frequency of the corresponding layer. Since the F2 layer is the most highly ionised ionosperic layer its critical frequency *foF2* is the highest frequency that can be reflected by the ionosphere.

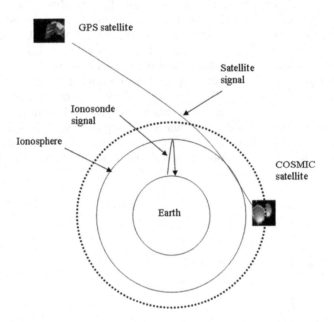

Fig. 2. Schematic illustrating a ground-based (ionosonde) and a space-based technique (satellite RO) for probing the ionosphere

A constellation of six satellites, called the Formosa Satellite 3-Constellation Observing System for Meteorology, Ionosphere, and Climate (COSMIC), was launched on April 15, 2006 to improve the global weather prediction and space weather monitoring [10,11]. The instrument that is of interest in this paper is the GPS receiver which is used to obtain atmospheric and ionospheric measurements through phase and Doppler shifts of radio signals. The Doppler shift of the GPS L-band (L1=1575.42 MHz, L2=1227.60 MHz) signals received by a LEO satellite is used to compute the amount of signal bending that occurs as the GPS satellite sets or rises through the earth's atmosphere as seen from LEO (Figure 1). The bending angles are related to the vertical gradients of atmospheric and ionospheric refractivity which is directly proportional to ionospheric electron density above 80 km altitude. Through the assumption of spherical symmetry, electron density profiles can be retrieved from either the bending angles or the total electron content data (computed from the L1 and L2 phase difference) obtained from the GPS radio occultations (RO) [12].We also need to emphasise that the RO occultation technique can be applied successfully in retrieving the ionospheric electron density profile only under the assumption of spherical symmetry in the ionosphere. This assumption is not always satisfied due to significant electron density gradients that give rise to horizontal electron fluxes. This violates the requirement for electron density profile inversion producing a very unrealistic profile. In order to overcome this limitation and concentrate on good quality electron density profiles a selection process was applied in order to exclude those measurements where the distortion of the profiles was excessive. In this paper the Neural Network *foF2* model is developed based on a training set of RO measurements across a part of Europe (shown by the shaded area in Figure 3) and evaluated on a test set of *foF2* values obtained by European ionosondes operating in Europe (in locations shown again in Figure 3).

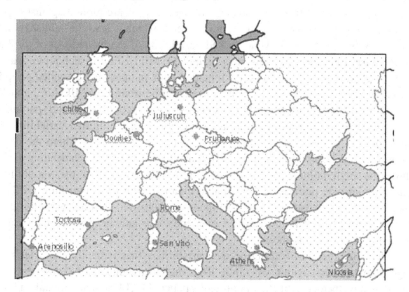

Fig. 3. A map of Europe illustrating the area within which RO measurements were considered in the model development and locations of ionosonde stations that provided the testing data sets for model evaluation

3   Temporal and Spatial Characteristics of the F2 Layer Critical Frequency and Model Parameters

The temporal variability of $foF2$ at a single location is well established and has been thoroughly described in previous papers [13,14] primarily based on ionosonde derived $foF2$ datasets. In short, ionospheric dynamics are governed principally by solar activity which in turn influences the electron density of the ionosphere. The electron density of the F2 layer exhibits variability on daily, seasonal and long-term time scales in response to the effect of solar radiation. It is also subject to abrupt variations due to enhancements of geomagnetic activity following extreme manifestations of solar activity disturbing the ionosphere from minutes to days on a local or global scale. There is also a spatial aspect of these variabilities which is depicted in Figures 4 and 6 for COSMIC $foF2$ values obtained at low ($35°$-$43°$), medium ($43°$-$51°$) and high ($51°$-$60°$) latitude European regions.

The most profound solar effect on $foF2$ is reflected on its daily variation as shown in Figure 4. As it is clearly depicted, there is a strong dependency on local time which follows a sharp increase of $foF2$ around sunrise and gradual decrease around sunset. This is attributed to the rapid increase in the production of electrons due to the photo-ionization process during the day and a more gradual decrease due to the recombination of ions and electrons during the night. The long–term effect of solar activity on $foF2$ follows an eleven-year cycle and is clearly shown in Figure 5 where values of $foF2$ obtained from ionosonde data over Cyprus are plotted against time as well as a modeled monthly mean sunspot number R which is a well established index of solar activity. We can observe a marked correlation of the mean level of $foF2$ and modeled sunspot number. Unfortunately using COSMIC $foF2$ for such a plot was not possible as the duration of the mission coincided with an unusually extended period of low solar activity that did not allow the correlation between COSMIC $foF2$ and R to clearly apperar. In addition to the effects of solar activity on both parameters mentioned above we can also identify a spatial effect on the diurnal variability. The spatial aspect of temporal variability is demonstrated in Figure 4 where the diurnal variation of $foF2$ is plotted for three different measurements corresponding to low, medium and high latitudes. It is evident from this figure that the variability is increased as latitude decreases. This spatial characteristic of diminishing $foF2$ with increasing latitude is also observed in Figure 6 where the seasonal variation of the median level of $foF2$ at noon recorded over the three latitude regimes over Europe (low, medium and high) is shown.

The plots in Figures 4-6 describe the variabilities that typically characterise the average temporal behaviour of $foF2$. The model parameters to describe these variabilities have been established in previous papers [13,14] and are annual and daily sinusoidal components as well as a modeled sunspot number which describes the level of solar activity. In addition to these parameters latitude and longitude are introduced as additional model parameters in this paper to express the spatial variability of $foF2$.

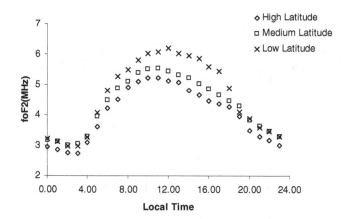

Fig. 4. Diurnal variability of foF2 at low, medium and high latitudes

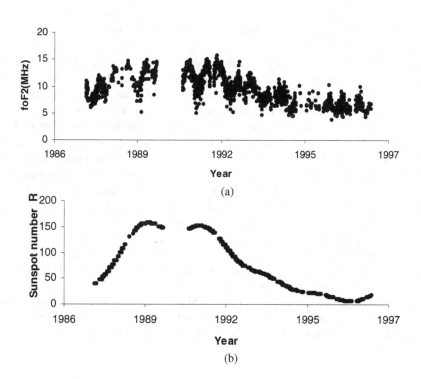

Fig. 5. Long-term *foF2* and solar activity variation with time over Cyprus

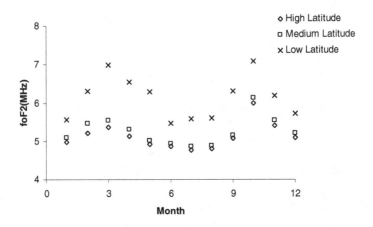

Fig. 6. Seasonal variation of *foF2* at 12:00 at low, medium and high latitudes

4   Experiments and Results

In all experiments the available *foF2* values from RO measurements between January 2007 and May 2010 were used for training the Neural Network and values from ionosond and RO measurements between June 2010 to December 2010 were used for testing it. The Neural Networks used had a fully connected two-layer structure, with 7 input, 10 hidden and 1 output neurons. Both their hidden and output neurons consisted of hyperbolic tangent sigmoid activation functions. The number of hidden neurons was determined by trying out the values 5, 10, 15 and 20. The training algorithm used was the Levenberg-Marquardt backpropagation algorithm with early stopping based on a validation set created from 20% of the training examples. In an effort to avoid local minima five NNs were trained with different random initialisations and the one that performed best on the validation set was selected for being applied to the test examples. The inputs and target outputs of the network were normalized setting their minimum value to -1 and their maximum value to 1. This made the impact of all inputs in the model equal and transformed the target outputs to the output range of the NN activation functions. The results reported here were obtained by mapping the outputs of the network for the test examples back to their original scale.

The example plots shown in Figure 7 show the ability of the model to approximate the diurnal behaviour of *foF2* at various locations over Europe. Apparently for lower latitude stations the approximation seems to be worse.

The obtained Root Mean Squared Error (RMSE) over all data for the model based on RO COSMIC measurements is shown for each site in table 1. Also RMSE values for a model based on ionosonde data and tested on COSMIC data is also given as a comparison model.

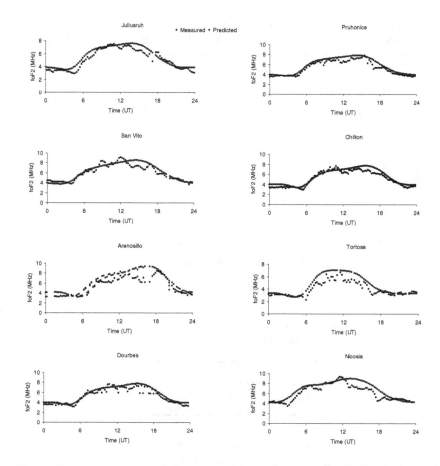

Fig. 7. Examples of measured (by ionosonde) and predicted (by COSMIC) *foF2* values at various locations over Europe

A worth noting result is the plot of RMSE versus latitude given in Figure 8 for the model based on COSMIC values which clearly demonstrates the increasing trend of RMSE with latitude. This can be possibly attributed to the fact that the ability of the algorithm to infer electron density profiles may deteriorate with latitude which was a possible suggestion in the literature in the recent past [15]. Another possible reason is the increased temporal and spatial variability which characterises lower latitude regions because of the complex morphology of the ionosphere in this area. This increased variability poses an increased challenge to any effort of modeling any ionospheric characteristic such as *foF2* even under geomagnetically quiet conditions which is reflected on the higher values of RMSE. Another important point to comment on is the fact that unlike Total Electron Content (TEC -which is another very important ionospheric characteristic) values that are measured from extended networks of ground-based GPS stations sampling adequately the European sector in space and time, measurements of *foF2* from ionosondes do not achieve the same spatial resolution and COSMIC measurements do not achieve the same temporal resolution. Therefore techniques that have been applied in the case of interpolating TEC data are not necessarily appropriate for *foF2* modeling.

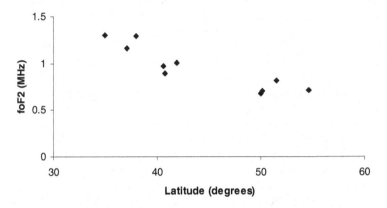

Fig. 8. RMSE versus latitude

Table 1. The Root Mean Squared Error (RMSE) of the Neural Network for the case of a COSMIC based and an ionosonde based model

Ionosonde location	RMSE (MHz)	
	COSMIC based	Ionosond based
Tortosa	0.788	**0.779**
SanVito	**0.524**	0.863
Rome	**0.752**	0.796
Nicosia	**0.714**	0.799
Dourbes	0.760	**0.647**
Athens	**0.740**	0.807
Arenosillo	**0.664**	1.027
Pruhonic	0.773	**0.645**
Chilton	**0.669**	0.788
Juliusruh	0.795	**0.639**

5   Conclusions and Future Work

In this paper we have presented the development of a neural network tool for the spatial and temporal modeling of (critical frequency) *foF2* data over Europe. The model has been developed based on a data set obtained during a period of approximately four years from COSMIC satellite mission. The tool has produced a good interpolation capability of the gaps in the *foF2* data series therefore providing a method to preserve the variability of *foF2*, a fact which is essential in the development of long-term time-series prediction models and procedures. As a next step the Neural Network approach to model *foF2* based on RO satellite measurements will be further extended to cover measurements obtained from other satellite missions (CHAMP,GRACE). The high solar activity period that is expected to reach is peak at around 2013 is expected to pose an additional challenge to such an effort. We also plan to explore the possibility of improving performance by employing more complex approaches such as an ensemble of neural networks.

References

1. Barclay, L.W.: Ionospheric Effects and Communication Systems performance. Keynote paper at the 10th Ionospheric Effects Symposium, Washington, DC (2002)
2. Goodman, J.: HF Communications, Science and Technology. Nostrand Reinhold (1992)
3. Maslin, N.: The HF Communications, a Systems Approach, San Francisco (1987)
4. McNamara, L.F.: Grid The Ionosphere: Communications, Surveillance, and Direction Finding. Krieger Publishing Company, Malabar (1991)
5. Lamming, X., Cander, L.R.: Monthly median foF2 modelling COST 251 area by neural networks. Phys. Chem. Earth (c) 24, 349–354 (1999)
6. Kumluca, A., Tulunay, E., Topalli, I., Tulunay, Y.: Temporal and spatial forecasting of ionospheric critical frequency using neural networks. Radio Sci. 34, 1497–1506 (2000)
7. Altinay, O., Tulunay, E., Tulunay, Y.: Forecasting of ionospheric critical frequency using neural networks. Geophys. Res. Lett. 24, 1467–1470 (1997)
8. Oyeyemi, E.O., Poole, A.W.V., McKinnell, L.A.: On the global model for foF2 using neural networks. Radio Sci. 40, RS6011 (2005), doi:10.1029/2004RS003223
9. Oyeyemi, E.O., McKinnell, L.A.: A new global F2 peak electron density model for the International Reference Ionosphere (IRI). Adv. Space Res. 42(4), 645–658 (2008), doi:10.1016/j.asr..10.031
10. Schreiner, W., Rocken, C., Sokolovsky, S., Syndergaard, S., Hunt, D.: Estimates of the precision of GPS radio occultations from the COSMIC/FORMOSAT-3 missio. Geophys. Res. Lett. 34, L04808 (2007), doi:10.1029/2006GL027557
11. Rocken, C., Kuo, Y.-H., Schreiner, W., Hunt, D., Sokolovsky, S., McCormick, C.: COSMIC system description. Terr. Atmos. Ocean Sci. 11, 21–52 (2000)
12. Hajj, G.A., Romans, L.J.: Ionospheric electron density profiles obtained with the Global Positioning system: Results from the GPS/MET experiment. Radio Sci. 33, 175–190 (1998)
13. Haralambous, H., Papadopoulos, H.: A Neural Network Model for the Critical Frequency of the F2 Ionospheric Layer over Cyprus. In: Palmer-Brown, D., Draganova, C., Pimenidis, E., Mouratidis, H. (eds.) EANN 2009. CCIS, vol. 43, pp. 371–377. Springer, Heidelberg (2009)
14. Haralambous, H., Ioannou, A., Papadopoulos, H.: A Neural Network Tool for the Interpolation of foF2 Data in the Presence of Sporadic E Layer. In: Iliadis, L., Jayne, C. (eds.) EANN/AIAI 2011, Part I. IFIP AICT, vol. 363, pp. 306–314. Springer, Heidelberg (2011)
15. Potula, B.S., Chu, Y.-H., Uma, G., Hsia, H.-P., Wu, K.-H.: A global comparative study on the ionospheric measurements between COSMIC radio occultation technique and IRI model. J. Geophys. Res. 116, A02310 (2011) doi:10.1029/2010JA015814

Detecting Glycosylations in Complex Samples

Thorsten Johl, Manfred Nimtz, Lothar Jänsch, and Frank Klawonn

Helmholtz Centre for Infection Research,
Inhoffenstraße 7, 38124 Braunschweig
Thorsten.Johl@helmholtz-hzi.de

Abstract. Glycoproteins are the highly diverse key element in the process of cell – cell recognition and host – pathogen interaction. It is this diversity that makes it a challenge to identify the glyco-peptides together with their modification from trypsin-digested complex samples in mass spectrometry studies. The biological approach is to isolate the peptides and separate them from their glycosylation to analyse both separately. Here we present an in-silico approach that analyses the combined spectra by using highly accurate data and turns previously established knowledge into algorithms to refine the identification process. It complements the established method, needs no separation, and works on the most readily available clinical sample of them all: Urine.

Keywords: proteomics, glycosylation, mass spectrometry.

1   Introduction

Glycosylations are complex post-translational modifications of proteins and the resulting glycoproteins are often found as part of the cell membrane where the glycosylations extend from the plasma membrane into the intracellular space and form the glycocalyx. Due to a high variability in molecular structure, they play a vital role in cell – cell interactions and immune responses [1] comparable to a key that fits only a specific lock, and several pathogens exploit this system to gain entry into specific cell types. Other roles include helping in protein folding, as transport molecule, lubricant or providing frost resistance. Current estimates state that about half of the eukaryotic proteome is glycosylated, making it the most abundant of modifications and one of the most versatile. Glycosylations most commonly attach either to the asparagine (N) amino acid via nitrogen, or to serine (S) or threonine (T) via oxygen and are hence called N- or O-linked glycosylations. These two types are made up of the same basic mono-saccharides, which can be broken down into four groups of identical masses (Fig. 1 A). These basic components are arranged into link type-dependent core structures of highly variable overall complexity (Fig. 1 B). This complexity hampers mass spectrometric (MS) analyses of glycosylated samples, because MS relies on a known total peptide mass for identification before it compares the most intense signals of a fragmentation pattern (MS²) of this peptide to a number of theoretical signals produced by known peptides of similar mass from a reference database [2]. The best

L. Iliadis et al. (Eds.): AIAI 2012, IFIP AICT 381, pp. 234–243, 2012.
© IFIP International Federation for Information Processing 2012

fitting spectrum is then assumed to be the correct interpretation. But these theoretical spectra do not contain any modification unless each is specified, and to predict all possible glycosylation patterns would soon reach incalculable complexity.

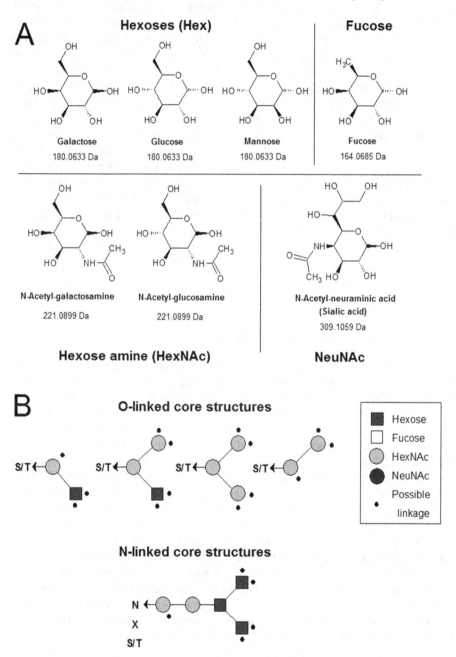

Fig. 1. Overview of saccharide makeup und structure

The general accepted solution to this problem is to separate the oligo-saccharide side chain from the peptide by using endoglycosidases specific to N-linked, and strong bases for both side chains [3] [4]. The resulting components are then analyzed separately, with a single mono-saccharide remaining attached to the peptide, which identifies the modified amino acid in MS experiments. This method works best on samples of low complexity where it is clear which oligo-saccharide originated from which peptide, but it is unsuited for large-scale proteomic studies. Yet glycosylations are ideally suited to be analyzed by MS workflows. First of all, they are readily ionized and thus result in good signal to noise ratios in the mass spectra. Secondly, the mono-saccharides produce distinct signals in the low mass region of such a spectrum. These identify spectra of glycosylated peptides with absolute certainty. And thirdly, the signals stemming from amino acid and saccharide residues in conjunction have a characteristic atomic mass that sets them apart from pure amino acid residues of similar mass. The reason for this is the greater amount of oxygen contained in saccharides which weighs in at just under sixteen Dalton, while every other atom prevalent in amino acids has a decimal value just above a full Dalton (C,H,N,S) [5]. It will be these criteria that are exploited by the program introduced herein in order to identify the peptides and their glycosylations from the same spectrum. This paper will demonstrate the feasibility of this new approach that does not rely on a reference database for glycosylated peptides like UniProt[14] or Glycosuite[15].

2   Methods

2.1   Sample Preparation and Acquisition

3x 2.5ml urine were desalted in turn with a PD-10 column, united, and then concentrated to 400µl with a speedvac and prepared according to the "glycoprotein isolation kit, WGA" from Thermo Scientific with minor adjustments. Incubation time was increased to 30min, the elution time to 15min. Subsequent digestion was performed over night with 20µg trypsin. Two MS measurements were performed in turn, one injecting 1µl onto the loading column of a Dionex Ultimate 3000 HPLC, the other 3µl. The samples were separated over 60 min on a reverse phase LC at 350µl/min flow rate with 80% Acetonitrile as mobile phase. Ions were introduced into the Thermo Fisher MS Orbitrap Velos device via an electrospray ion source. The FTMS Orbitrap was used as the detector for all scans. Survey scans were performed at a resolution of 60000 in the range of 700 to 2000 and the 5 most intense signals were chosen with an isolation width of 4 Da for MS² fragmentation in the HCD collision cell at normalized collision energy of 40 for 40ms. The minimum signal threshold was set at 10000 and the default dynamic exclusion list was enabled, removing signals for 90 sec from the selection process after having been measured twice. Concluding MS² scans were performed at a resolution of 7500 and a range of 100 to 2000. All preprocessing steps like baseline estimation and noise removal were performed by Xcalibur 2.7 at the time of acquisition with the device inherent parameters.

2.2   Program Setup

The program was created using Java programming language from Oracle and Eclipse as programming environment. Readw.exe [6] from the trans-proteomic pipeline [7] was used in conjunction with files from ThermoFinnigan for centroidization and to convert Thermo Fisher RAW files into mzxml. Mzxml in turn was converted to mgf using JRAP [8]. Alternative mzML parsing is provided by jmzML [9]. Settings are stored as xml file using the Apache commons parser [10]. The program requires the JRE and has been tested on Windows 32 and 64bit as stand-alone application. Note that the readw.exe only works on 32bit windows operating systems, and thus 64bit machines can only process mzxml, mzml and mgf files.

2.3   Spectra Preparation and Identification of Glycosylated Spectra

All signals of a deconvoluted spectrum were checked for isotopic C13 satellites allowing for 20ppm mass tolerance, and the appropriate charge state was deduced from the distance between the isotopes. All following steps were then performed on the mono-charged mass of the most intense isotopic signal. All spectra were searched in turn for any two of the following marker oxonium-ion masses: dehydro-N-acetyl hexose (204.079 Da), a dehydro di-saccharide of N-acetyl hexose and a hexose (366.132 Da) or di-dehydro sialic acid (274.085 Da). The search was performed with a mass tolerance of 20ppm and a minimum marker signal intensity of 10% of the most intense signal in the spectrum. Every signal that was found during identification and, if present, an additional mono-dehydro sialic acid (291.095 Da) signal were marked as signals of saccharide origin. All spectra that met the identification criteria were then considered glycosylated and passed on to the following processing steps.

2.4   Signal Makeup

Any mass from top to bottom, beginning with the total mass of the modified peptide, was then considered as the starting point of a tree of saccharide masses. In order to qualify as such a seed, all mono-saccharide masses from Fig. 1 A where in turn checked to find a smaller signal in the estimated mass range of starting signal minus saccharide mass. The steps are repeated from these seeds onward until no further signals are found that can be explained by the loss of a saccharide. There are additional criteria that influence the signal selection process:

Firstly, the peaks are only considered if the mass delta between the actual and expected decimal fractions of a peptide (Mi) [5] increases from higher peak (Pi) mass to lower peak mass (Pi+1), denoting the loss of an oxygen-rich monosaccharide from the remaining molecule. The logic allows for a small mass-dependent inaccuracy (Δ) of 15ppm (1).

$$f(x) = \begin{cases} true, & M_i - P_i < M_{i+1} - P_{i+1} - \Delta \\ false, & otherwise \end{cases} \tag{1}$$

Secondly, the program allows for a shift in the most abundant isotope form, searching for a second peak at +/- 1.00335 Da distance with 15ppm tolerance if no primary peak can be found. Such shifts are marked, and block a further shift into the same direction, which is highly unlikely given the standard distribution of C13 isotopes. It may, however, shift back to the original isotope pattern once to accommodate for inaccuracies in the measured intensities.

Thirdly, signals without an isotope pattern that denote the charge state are considered to be of any charge state up to the charge state of the mother mass. Such signals are called jokers, and may be assigned only one charge during the enlargement of the saccharide trees. Signals that pass all criteria and thus carry a glycosylation are marked and the saccharide tree is stored for further analyses.

2.5   Typecasting Spectra

The kind of glycosylation found in any given set of spectra can be determined depending on the completeness of the saccharide tree. Four different core structures as introduced in Fig. 1 B and three further N-linked sub types are known to be expressed in human tissue. At least four saccharides from the start of the tree are necessary to discern between N and O-linked glycosylations. The more complex N-linked core structure is checked first. This is done by a pattern recognition that performs a depth-first traversal of the saccharide tree and checks first if there are one or two leading HexNAc saccharides, and then if there are three consecutive hexoses following that. The traversal for any sub-tree is aborted if the child does not match expectations, and the next child is considered until the pattern is found, or no children are left. There is some allowance for a single extra saccharide attached to the leading HexNAc, and either the last hexose or the first and its connected HexNAc may be substituted by a Hex-HexNAc di-saccharide.

If the algorithm fails to find the key elements for an N-linked core structure, it checks the less complex O-linked core structures in the following order: 2 HexNAc and a hexose, 3 HexNAc, 2 HexNAc, HexNAc and a hexose. The leading HexNAc may be missing in all cases.

2.6   Determining True Peptide Mass

All signals that have lost a saccharide are removed from the spectrum. The remaining signals are then considered from top to bottom. If they have a relative intensity of 5% of the most intense signal ever present in the spectrum, the algorithm tries to fit any combination of saccharide masses into the mass delta of this signal and the total mother ion mass. This is achieved by beginning with just one saccharide, trying each in turn before adding a second saccharide and checking all possible mass combinations of these. The method stops adding more saccharide elements once it has found one combination that fits, or multiple combinations of the smallest mass have become larger than the mass delta it was trying to fit. Signals that can be explained by the loss of a combination of saccharides are stored. All spectra are then grouped by their mother mass and charge state, and the stored signal intensities are added up. The four most intense combined signals are then chosen in turn and proposed as the true peptide mass and new mother mass of the spectrum retaining the charge the spectrum originally carried, resulting in up to four new spectra for each that was identified as containing saccharides. The spectra are then stored separately and can be submitted to the Mascot server for identification.

2.7   Mascot Search

The Mascot search (V 2.3.02, Matrix Science) was submitted via Mascot Daemon against a UniProt database (Release 2012_02) restricted to human proteins with a peptide mass tolerance of 20ppm, a fragment mass tolerance of 0.4 Da and no missed cleavages. Mother ion charge was limited to up to 4 charges. Allowed modifications were Carbamidomethyl (fixed) and Oxidation (M) (variable).

3   Results

8615 and 9577 spectra, respectively, were processed from the 1µl and 3µl samples. About a third of these were identified as glycosylated and resulted in more than 11000 proposed spectra per sample (Table 1). Roughly 34% of all spectra were found to carry a glycosylation. Only 5% of these could be identified as carrying an N-linked glycosylation tree. The majority is estimated to carry the less complex O-linked glycosylation. Less than 20% resulted in an unknown type of glycosylation. The final number of spectra proposed to Mascot by the program is nearly equal to four times the amount of glycosylated spectra, meaning that almost every spectrum had enough signals that qualified it as one of four new possible true peptide masses. Only 13 spectra in total could not be assigned a new mass and thus resulted in no new spectra (not shown).

Table 1. Overview of glycosylated Spectra

Sample	Original Spectra	Glycosylated	N-linked	O-linked	Unknown	Proposed Spectra
1µl	8615	3086	139	2439	508	11598
3µl	9577	3255	207	2475	573	12328

The results of the identification by Mascot are summarized in Table 2. The identity score cut offs for the 1µl and 3µl samples were 19 and 20, respectively, and the significance (homology) threshold was a p-value of 0.05. The false discovery rate (FDR) of Mascot was 5.13 (7.5) above identity (above homology) for the 1µl sample, and 1.83 (5.17) in the 3µl sample. The table shows all proteins that were found in both samples by Mascot with their respective score. Shown next are the number of matching spectra that were associated with this protein in total and the number of unique sequences. The affirmed glycosylations are the number of unique sequences already known and found in this analysis. The number in parenthesis for all three groups is the number of hits above homology. The numbers of predicted and known glycosylations are taken from the UniProt entry and show the total number of listed glycosylations followed by the number of confirmed ones in parenthesis. Asterisks mark entries where more than one glycosylation occurs on the same tryptic peptide and the location is not unambiguous. Values were taken from the protein overview tab of Mascot and UniProt Release 2012_02.

Table 2. Comparison of identified and known glycosylations

Uniprot Name	Score		Matches		Sequences		Affirmed Glycosylations		Uniprot
	1µl	3µl	1µl	3µl	1µl	3µl	1µl	3µl	
HEG1	140	326	22(7)	27(17)	5(2)	5(3)	1(0)	1(0)	9(3)
CSPG2	179	267	17(7)	21(13)	4(2)	6(3)	0	0	23(1)
CSF1	275	137	14(9)	8(7)	2(1)	1(1)	1*(0)	0	4(2)
X3CL1	45	263	14(3)	26(16)	3(1)	4(3)	2(1)	3(2)	4(3)
EGF	140	125	9(6)	9(5)	3(1)	2(1)	0	0	10*(1)
FBLN2	188	40	8(6)	3(3)	2(1)	1(1)	0	0	3*(1)
YIPF3	102	89	6(5)	4(3)	1(1)	1(1)	1(1)	1(1)	1(1)
NCAN	79	108	9(4)	5(4)	2(1)	2(1)	0	0	4(0)
NID1	46	112	5(2)	14(7)	3(2)	3(2)	0	0	1(0)
IGF2	59	70	4(3)	11(7)	1(1)	1(1)	2*(2)	2*(2)	3(3)
PGCB	107	21	3(3)	1(1)	1(1)	1(1)	0	0	4(0)
FA5	79	41	4(4)	2(2)	1(1)	1(1)	0	0	26(5)
APOF	45	57	3(2)	2(2)	1(1)	1(1)	1(1)	1(1)	2(1)
CF072	26	48	2(1)	2(2)	1(1)	1(1)	1(1)	1(1)	4(0)
VASN	18	52	1(1)	2(2)	1(1)	1(1)	0	0	5(3)
IC1	46	22	5(2)	4(1)	1(1)	1(1)	2*(2)	2*(2)	15(0)
GOLM1	30	30	4(2)	4(3)	2(1)	1(1)	0	0	3(1)
P3IP1	25	29	4(2)	4(2)	1(1)	1(1)	1(1)	1(1)	2(1)
OSTP	27	23	5(3)	6(3)	2(1)	3(3)	0	0	7(7)

The 3µl sample proved to be the more reliable and was consecutively used for further analyses of the saccharide trees and glycosylation sites. 7 of the identified 19 proteins were detected with at least one peptide with a known glycosylation site above homology. All of these sites, except the one from CF072, were related to O-glycosylations. All of them were identified by the program as O-glycosylated at least once per sequence.

Three spectra are shown in Fig. 2. They exemplify different O-glycosylations that were correctly identified by the program, and demonstrate the potential to reliably identify the glycosylation site from glycosylated spectra. The notable exception is CF072_HUMAN, which is predicted to have an N-linked glycosylation at the identified site 191, but was identified as O-type. The identified spectra in question only contain unlinked Hex-HexNAc and Hex, and thus do not permit the direct identification of an N-linked glycosylation, although the peptide could be identified with an adjusted mono-charged mother mass of 1461.77 Da (down from 1206.06, z=2).

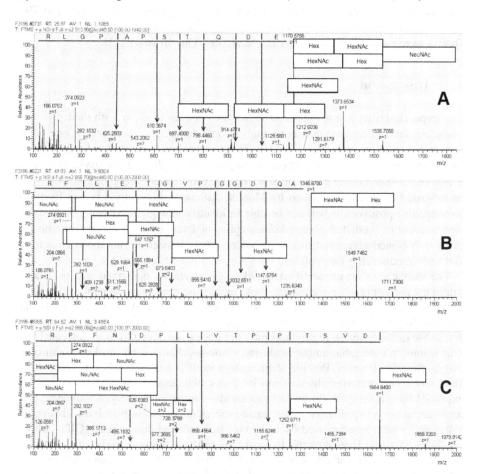

Fig. 2. Saccharide tree and peptide y-ion series

(A) Is taken from P3IP1_HUMAN. A consecutive saccharide tree links the total mass of the mother ion over three steps with the true peptide mass of 1170.58 Da. Three further saccharide trees were found that stem from different dissociation events. One of them is linked directly to S39, the reported glycosylation site. Spectrum (B) shows a glyco-peptide that was associated with site T183 of Protein X3CL1_HUMAN. The lower half of the spectrum shows several saccharide ions from the b-series, in one case extending the mass 274 used for identification by NeuNAc and HexNAc. The true peptide mass 1346.67 Da is again linked to the total mother mass. (C) Shows two saccharide trees in the lower mass spectrum. They originate from the same fragment, which dissociated from the peptide and were found complementary forward and backward. Note the double charged di-saccharide that was found and extends the mono charged saccharide that is attached to 1252.67 Da. Neither the total mother mass nor the actual site is directly attached to a saccharide. It was nonetheless identified as IGF2_HUMAN, which could be modified either at T96 or at T99. The HexNac-Hex tree connected to proline (P) suggests that it is site T96. Spectra were taken from Thermo Fisher Xcalibur – QualBrowser [13].

4   Discussion

This paper describes a new approach to analyze peptides together with their glycosylation. As the current standard is to analyze both separately, there is just one other project that uses a similar approach [12]. That approach uses the ProteinScape 3.0 and GlycoQuest programs in conjunction. It compares the unadjusted glycosylated spectra with a list of theoretical spectra of dissociated glyco-peptides, very similar to the database-based approach performed by Mascot. Although this method is less likely to provide false positives, it will not be able to identify new glycosylations. Indeed the high number of predicted glycosylations apparent from Table 2, especially with regards to N-linked glycosylations, demonstrates there is need for an approach similar to de-novo sequencing for peptides.

The example of an unidentified N-linked glycosylation described above demonstrates the problems when identifying large saccharide trees in combination with large peptides in mass spectrometric studies. The variety of possible dissociation events and multiple fragmentations within the tree means that no specific fragment reaches a detectable quantity in the highly accurate, but not as sensitive Orbitrap detector [11]. This results in detectable amino acids and mono-saccharides, but no larger consecutive saccharide elements. Peptide identification itself is based largely on correct total mass and some preferred dissociation locations like proline. The situation may be improved by allowing for di- and tri-saccharides during peak analysis, however, this could decrease the specificity by increasing the chance of false positives. This problem may, in turn, be addressed by using collision-induced dissociation (CID), which generates only two fragments per ion and thus can generate a more gradual distribution of signals. The results may be even better when coupled to the more sensitive but less accurate ion trap detector. On the other hand, this would produce even less pure amino acid ions, resulting in less ion intensity. As is, many of the 3000+ original spectra did not lead to peptide identification, even when allowing four possible interpretations of the remaining non-glyco signals. The reason for this may be found in the more readily ionized saccharides. The remaining amino acid part is less likely to be ionized, provides weaker signals when fragmented, and eventually vanishes into the background noise. A possible solution to this is to perform a third dissociation and detection cycle (MS³) of the most likely true peptide mass. The evident workflow for this would be as follows:

After the first data acquisition cycle, the peptides would have to be analyzed by the program and the mother mass, estimated true peptide mass and retention time recorded. A second data acquisition cycle would follow the first, using the recorded data as an inclusion list for MS³ experiments. The results would then be sent to Mascot for identification. It is, however, presently unclear whether such a general approach is possible and would result in spectra of the required intensity.

As demonstrated above, the basic principle works. The program uses a shotgun approach to reliably identify O-linked glycosylations and their site – but further development is necessary to extend this to N-linked glycosylations, which is already underway at the time of this writing.

The program is available from the author upon request.

References

1. Marino, K., Bones, J., Kattla, J., Rudd, P.: A systematic approach to protein glycosylation analysis: a path through the maze. Nat. Chem. Biol. 6, 713–723 (2010)
2. Perkins, D., Pappin, D., Creasy, D., Cottrell, J.: Probability-based protein identification by searching sequence databases using mass spectrometry data. Electrophoresis 20(18), 3551–3567 (1999)
3. Patel, T., Bruce, J., Merry, A., Bigge, C., Wormald, M., Jaques, A., Parekh, R.: Use of hydrazine to release in intact and unreduced form both N- and O-linked oligosaccharides from glycoproteins. Biochemistry 32, 679–693 (1993)
4. Pan, S., Chen, R., Aebersold, R., Brentnall, T.: Mass Spectrometry Based Glycoproteomics - From a Proteomics Perspective. Mol. Cell Proteomics 10(1) (2011)
5. Dodds, E., An, H.J., Hagerman, P.J., Lebrilla, C.B.: Enhanced peptide mass fingerprinting through high mass accuracy: Exclusion of non-peptide signals based on residual mass. J. Proteome Res. 5, 1195–1203 (2006)
6. ReadW, http://tools.proteomecenter.org/wiki/index.php?title=Software:ReAdW
7. Pedrioli, P.: Trans-proteomic pipeline: a pipeline for proteomic analysis. Methods Mol. Biol. 604, 213–238 (2010)
8. JRAP Extended, http://javaprotlib.sourceforge.net/packages/io/jrap/
9. JMZML, http://code.google.com/p/jmzml/
10. The Apache XML Project, http://xerces.apache.org/xml-commons/
11. Scigelova, M., Makarov, A.: Orbitrap Mass Analyzer - Overview and Applications in Proteomics. Practical Proteomics 2, 16–21 (2006)
12. Neue, K., Kiehne, A., Meyer, M., Macht, M., Schweiger-Hufnagel, U., Resemann, A.: Straightforward N-glycopeptide analysis combining fast ion trap data acquisition with new ProteinScape functionalities, http://www.bdal.com/library/literature-room/detail-view/article/straightforward-n-glycopeptide-analysis-combining-fast-ion-trap-data-acquisition-with-new-proteinsca.html
13. Xcalibur Thermo Scientific, http://www.thermoscientific.com/ecomm/servlet/productsdetail_11152_L11240_80588_11961721_-1 (April 20, 2012)
14. UniProt, http://www.uniprot.org/
15. Cooper, C.A., Joshi, H.J., Harrison, M.J., Wilkins, M.R., Packer, N.H.: GlycoSuiteDB: a curated relational database of glycoprotein glycan structures and their biological sources. 2003 update. Nucleic Acids Res. 31(1), 511–513 (2003)

Experiments with Face Recognition
Using a Novel Approach Based on CVQ Technique

Arman Mehrbakhsh[1] and Alireza Khalilian[2]

[1] Sama Technical and Vocational Training College, Islamic Azad University
Andisheh Branch, Andisheh, Tehran, Iran
[2] School of Computer Engineering, Iran University of Science and Technology, Tehran, Iran
ar_mehr@damavandiau.ac.ir, khalilian@comp.iust.ac.ir

Abstract. Face recognition techniques attempt to identify faces according to the patterns of mouth, lip, eyes and so on. However, the effectiveness of existing approaches degrades in presence of uncontrolled conditions such as variations of background light and image sizes. To deal with this problem, we propose a novel approach based on Classified Vector Quantization (CVQ) technique. The new approach divides images into some blocks and each block is classified into several patterns. Then, the Vector Quantization (VQ) technique is applied on the vectors of each pattern. In order to evaluate our approach, we have conducted a family of experiments on some standard image databases, MIT, YALE, and AR. The results demonstrate that the new approach is steadily capable of identifying faces in different situations.

Keywords: Face Recognition, Classified Vector Quantization (CVQ), Vector Quantization (VQ).

1 Introduction

Currently, due to the increment of violence and crime rate, the usage of systems to establish security and safety is a critical requirement to the current human's life [1]. To deal with this requirement, face recognition of convicts is considered as the major purpose in development of such systems [6]. In addition, since the events of September 11, the process of development and implementation of automated biometric systems have been noticeably increased [6, 7]. Moreover, there are many practical situations in which we need to the techniques of pattern recognition, and in particular face recognition. The wide range usage of these techniques in commercial and law enforcement applications, such as biometric authentication [7], video surveillance [39] and information security [1] has made it a popular and significant area of research. Nowadays, we are capable to identify people through automatic recognition techniques based on physiological and characteristic behaviours, such as finger print [28, 29], iris [2, 3, 4], ear [5], vein face [27], and so on. Although many techniques have been proposed for pattern recognition [30], especially face recognition [31, 32, 33], and a number of studies [34, 35, 36] have been performed in the literature to evaluate and compare the existing techniques, there is still a long road ahead to achieve optimal approaches.

L. Iliadis et al. (Eds.): AIAI 2012, IFIP AICT 381, pp. 244–253, 2012.

There are various natural characteristics in human's body such as finger print, ear, iris, and similar other parts whose structure are unique for each human [7]. The uniqueness of the structure for the mentioned body features can be used to recognize and to distinguish people from each other [37]. Ear recognition [5, 38] is a new class of biometrics that has certain advantages over most of the established biometrics; it has a rich and stable structure that is preserved from birth into the old ages [5]. Iris recognition [2, 4, 9, 8] is one of the most accurate biometric systems when a high level of security is required. Therefore, designing such systems has attracted the attention of a large number of researches [2, 4, 8, 9]. Unlike of other biometric methods, recognition based on human's face does not have any reliable and dependable mechanism that controls any undesirable conditions in the image of the face. These conditions include (1) changing the position of the face, (2) changing the size of the image, (3) the lack of sufficient light, etc. In ear recognition method, we need a half face image of each person. Hence, it is not an appropriate technique for facial expression recognition [5, 38]. In iris recognition method, by closing the eyes, the recognition process fails and we still need a reliable control mechanism. In addition, if the recognition relies only on the geometric structure of face [25, 26], it will also need a reliable control mechanism because face in various angles loses its potential to be recognized properly [10].

In this paper, we propose a novel approach for face recognition based on Classified Vector Quantization (CVQ) [17, 18, 23, 24] technique. At first step, this approach divides the face image into some blocks, converts each block to a vector, and classifies the vectors into some predefined patterns. These patterns are the major and common curves on the faces of different people. At second step, it uses Vector Quantization (VQ) [11, 13, 14, 15, 16, 19] on the classified vectors of each pattern. The selection of the predefined patters significantly affects the performance of this approach. To evaluate the functionality of the proposed approach, we have conducted experiments on some standard image databases used in the experimental studies in the literature. The overall results showed that the proposed approach is capable to recognize images with an appropriate recognition rate.

The remainder of this paper is organized as follows. In Section 2, we introduce the face recognition problem. Vector quantization techniques and classified vector quantization technique are described in Sections 3 and 4 respectively. In Section 5 the proposed approach is presented. The experimental setup and the obtained results are discussed in Sections 6. Finally, some concluding remarks are outlined in Section 7.

2   Face Recognition

Face recognition has been used in different fields of security and protection, namely automated crowd surveillance [39], access control [6], identification of convicts [40], face reconstruction [43, 44, 45] and so on. Thus, it has been one of the challenging issues in the last decade and it is a necessity in our current life.

In computer vision, there are two important methods whose purpose is face recognition. The first group is holistic appearance base methods, including PCA [46, 47], and LDA [48, 49]. In these methods, a facial image is considered as an instance in N dimensional feature space, where N is the number of pixels in the image. The second group is local facial based methods such as elastic bunch graph matching. In these methods, a set of orthogonal basis vectors, that maximize the variance of facial image data, are obtained by Eigen decomposition of the scatter matrix of facial images. Combination of the two above methods, that are named hybrid methods, can be used for face recognition purpose.

Face recognition algorithms have focused on some unique properties of face [7] such as iris, ear, skin, etc. Among the different components of face, brow is the most important for face recognition purpose. According to the existing studies [16, 17, 18, 19], all of the face features or components do not have equal effect in the effectiveness of face recognition process. Most of the experimental results and researches [37] showed that in the field of face recognition, nose and mouth are more important than eyes. Although brows have equal effect in face recognition as compared with other face components, less attention has been considered toward the effects of brows against the other face features [11]. Thus, an algorithm that utilizes all of these features together would perform better than existing ones. The proposed method in this paper takes different features of the face into account to improve the efficacy of face recognition.

3   Vector Quantization

The Vector Quantization (VQ) technique [22] was first developed for image processing to compress the images [13, 14]. Codebook is a 2D array which is initialized randomly by the vectors of some of the training images. The number of columns in the codebook depends upon the size of the blocks of image and is usually considered as 9 or 16. In the decompressing phase, each code word by index i is replaced by the code word with index i in the codebook. One of the benefits of this technique is its low computational complexity. However, it is considered as a lossy technique. The CVQ [17, 18, 23, 24] and TSVQ [41, 42] are two major extensions of this technique.

When applying the VQ technique in face recognition, each block of the original image is identified with an index (number of the nearest code word in codebook). The response time is an important factor in real time applications [15] and VQ-based recognition provides reasonable response time in these situations.

In Vector Quantization (VQ) technique, the original image is partitioned into several blocks with the size of n × m. Then, then each block is arranged in the form of a vector using the row major or the column major method. In the next step, the codebook is updated using the obtained vectors of the image. Each row of this matrix is called a code word. Next, the difference between each vector of the original image and each code word in the codebook is computed by using of Euclidean distance in Equation (1):

$$D(B_i, C_j) = \sqrt{\sum_{k=1}^{n} (B_{i,k} - C_{j,k})^2}$$

(1)

In this equation, B_i is a vector from the original image, C_j is a code word from the codebook, $B_{i,k}$ is the kth element of the ith vector, and $C_{j,k}$ is the kth element of the jth code word. By this computation, the nearest code word to each vector with respect to the Euclidean distance is found. Finally, code words are updated by the centroid of all training vectors, which were mapped during coding [11]. Finding the optimized codebook is the major goal in VQ technique. It has been shown [12] that the design of VQ is optimal. This design is illustrated in Figure 1.

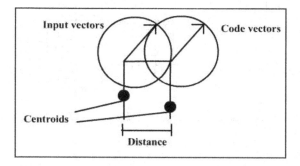

Fig. 1. Optimality of VQ design

4   Classified Vector Quantization

Edge is a very significant feature perceptually in an image. In our proposed method, we have used edge information for face recognition. Each coding technique that preserves the edge information is preferable because it has been proved [50] that the human's eye can recognize objects through their edges and this is adapted in image processing methods. The Classified Vector Quantization (CVQ) technique has been proved [17] to be an efficient technique for lossy image compression at low bit rates. CVQ technique can be used to reduce the computational complexity of VQ technique [11]. In this technique, each input vector is classified into a class. Then VQ is applied on vectors of each class [24]. Using this technique, there would be two indexes for each input vector: One for specifying the number of class and another for specifying the index of the nearest code word in codebook. This idea is shown in Figure 2. With CVQ technique, when applied to the face recognition, input vectors can be partitioned into some predefined patterns and then quantization is used for all vectors of each pattern. Since CVQ technique classifies the vectors in several classes, it is more precise than VQ.

5   The Proposed Approach

VQ technique can be very reliable in face recognition based on human's skin because of the correlation of human's skin. But the main problem is when the number of training vectors of images is increased, VQ cannot distinguish among vectors because many of the vectors happen to be similar and the differences can be hardly observed. As a result, the efficiency of this technique degrades and it is the only limitation in the usage of this technique. Now, each block of the test image is indexed with the number of a pattern. When all of the image blocks were indexed, VQ technique is applied over all vectors that were classified for each pattern.

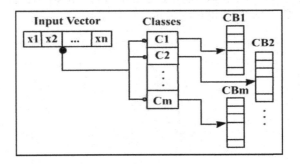

Fig. 2. A schematic of CVQ design

In order to deal with the issues of the VQ technique, CVQ can be applied. In our proposed approach, each block of the original image, the human's face, is compared with several predefined patterns. These patterns are defined according to the curves that are most seen in faces of each human. They can be extracted from a random number of faces using a pattern recognition approach like neural network. Figure 3 depicts some samples of patterns.

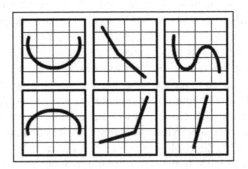

Fig. 3. Some samples of patterns

Several studies [11, 20, 21] have shown that different features and components of the face can have unequal and various effects to the face recognition. Hence, taking all features of the face together into account has much potential for a face recognition technique to lead to better results. As a result, edge detection in face and also skin correlation can be considered together to enhance the face recognition process and improve the results. The Classified VQ technique enables us to consider the mentioned two features of the face simultaneously. This capability exposes the importance of CVQ technique in face recognition.

According to the above descriptions, we can present our method in the following steps:

1. The image is cropped to fit in the desired size.
2. The image is converted to the grayscale.
3. We need to detect the edges of the faces related to each original image. For example, Sobel filter can be used for this purpose.
4. According to the previous step, we can classify each block of the original image to one of the predefined patterns. Suppose that all blocks have been classified into classes C_1 to C_n. Each class corresponds to a certain pattern. So, there would be a number for each block between 1 to n, so that it can describe the class number of the block.
5. After that, we can do vector quantization (VQ) technique on all *original* blocks which have the same class number. Note that there is a code book for each of the n classes of patterns.

6   Experimental Studies

6.1   VQ Results

In order to implement the VQ technique to be applied in face recognition, we have used a low pass filter (2D- Moving average with mask 9×9) on each of face image at the first step. The images were selected from the standard image databases, AR [51], MIT [52], and ORL [53]. It can remove the noises of image and is able to detect the components of the face. Then, we applied a code book with 128 code vectors. Next, we tested the code books of different sizes and we found that code book with size 128 is the optimal.

Finally, for each training image, we created a histogram based on the number of similar vectors of each code word. Figure 4 shows a sample histogram for a face image. As a result, we will have a histogram for each of the training images. In order to find the output image, the histogram of input image and existing images in the database have been compared and the closest histogram was found.

We implemented this method using Delphi language, on standard image databases such as ORL, AR, Yale using a PC system with CPU 2.19 GHz and 1.87 GB of Ram. We observed the recognition rate between 95% and 97% (depending on the size of used mask). In this experiment, Error Recognition Rate (ERR) has been measured to 2.6%.

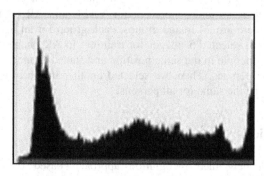

Fig. 4. A sample of face histogram obtained with the proposed approach

6.2   CVQ Results

Observing the results obtained from the implementation of CVQ technique demonstrates that it has much potential for face recognition at high accuracy. As mentioned before, VQ is a robust technique for face recognition, especially in real time systems. However, when the number of input vectors of images is increased, it is no longer beneficial. Nevertheless, CVQ classifies the vectors into some patterns and then VQ can be used for vectors of each class. Moreover, by using this technique, we can utilize the two unique characteristics in human face, geometric structure of the face, and skin correlations. Hence, recognition rate is improved in comparison with other existing face recognition methods. The proposed method was implemented and tested using MIT-CBCL, AR, Yale and ORL image databases. At the first step, the faces in images are cropped. Then a filter such as Sobel filter is applied on each cropped face image in order to detect edges in faces. This enables us to classify each block of the size $n \times n$ in each of predefined patterns. Finally, VQ technique is used for all the blocks of the same pattern. We have implemented this method using the Java language, on a system with CPU 2.4 GHz and 512 MB of Ram. The recognition rate has been observed to be between 93% and 100% (depending on the type of database). Table 1 shows the recognition rate in some major face recognition methods and that of our proposed approach with respect to MIT image database.

Table 1. Recognition rate of some face recognition method along with our proposed approach

Method	Percent of Recognition Rate
Global PZM	82.91
AWPPZMA	92.31
Eigen face(k=20)	55.56
Eigen face(k=60)	69.23
Eigen face(k=117)	76.07
Modular PCA(k=64)	76.06
Our method	93

The required time for pre-processing has been measured as 12 msec and for recognition it has been measured at most 25 msec. Therefore, the total time would be 37 msec. The reason why we could achieve small processing time is that the proposed algorithm avoids from complex calculations. This will consequently improve the overall performance of the recognition with higher accuracy. We achieved the recognition rate of 45.83 and 37 percent for YALE and AR database images. For YALE database, there are 24 image groups, each group for an individual person. In each group, we have selected 5 images for training. In AR image database, there are images of different people in the same position and state. We have selected one image for each of the 65 persons. Then, we selected another image of each person to test such that it would be the same for all persons.

7   Conclusions

In this paper, a new face recognition approach based on Classified Vector Quantization (CVQ) has been proposed. We extended the concept of this technique in order to classify components of face images into some patterns and then we used Vector Quantization in order to create a code book for each pattern. In Comparison with other methods, our algorithm can be implemented without using hard mathematical computations.

References

1. Zhao, W., Chellappa, R., Phillips, P.J., Rosenfeld, A.: Face recognition: A literature survey. ACM Comput. Surv. 35(4), 399–458 (2003)
2. Estudillo-Romero, A., Escalante-Ramirez, B.: The Hermite Transform: An Alternative Image Representation Model for Iris Recognition. In: Ruiz-Shulcloper, J., Kropatsch, W.G. (eds.) CIARP 2008. LNCS, vol. 5197, pp. 86–93. Springer, Heidelberg (2008)
3. Bevilacqua, V., Cariello, L., Carro, G., Daleno, D., Mastronardi, G.: A face recognition system based on Pseudo 2D HMM applied to neural network coefficients. Soft Comput. 12(7), 615–621 (2008)
4. Roy, K., Bhattacharya, P.: Optimal Features Subset Selection Using Genetic Algorithms for Iris Recognition. In: Campilho, A., Kamel, M.S. (eds.) ICIAR 2008. LNCS, vol. 5112, pp. 894–904. Springer, Heidelberg (2008)
5. Attarchi, S., Nosrati, M.S., Faez, K.: A New Ear Recognition Approach for Personal Identification. In: Huang, D.-S., Wunsch II, D.C., Levine, D.S., Jo, K.-H. (eds.) ICIC 2008. LNCS, vol. 5226, pp. 356–363. Springer, Heidelberg (2008)
6. Butavicius, M., Mount, C., MacLeod, V., Vast, R., Graves, I., Sunde, J.: An Experiment on Human Face Recognition Performance for Access Control. In: Lovrek, I., Howlett, R.J., Jain, L.C. (eds.) KES 2008, Part I. LNCS (LNAI), vol. 5177, pp. 141–148. Springer, Heidelberg (2008)
7. Blackburn, T., Butavicius, M., Graves, I., Hemming, D., Ivancevic, V., Johnson, R., Kaine, A., McLindin, B., Meaney, K., Smith, B., Sunde, J.: Biometrics Technology Review, DSTO Systems Science Laboratory, Australia (2002)
8. Lim, S., Lee, K., Byeon, O., Kim, T.: Efficient Iris Recognition through Improvement of Feature Vector and Classifier. ETRI Journal 23(2), 61–70 (2001)
9. Ma, L., Tan, T., Wang, Y., Zhang, D.: Efficient iris recognition by characterizing key local variations. IEEE Trans. Image Proc. 13(6), 739–750 (2004)
10. Lu, Y.Z.: A Novel Face Recognition Algorithm for Distinguishing Faces with Various Angles. International Journal of Automation and Computing 5(2), 193–197 (2008)
11. Gersho, A., Gray, R.M.: Vector Quantization and Signal Compression. Kluwer Academic Publishers, Norwell (1991)
12. Jushi, M.A.: Digital image processing An algorithmic approach. Prentice-Hall Publishing (2006)
13. Hsieh, Y.P., Chang, C.C., Liu, L.J.: A two-codebook combination and three-phase block matching based image-hiding scheme with high embedding capacity. Pattern Recogn. 41(10), 3104–3113 (2008)
14. Kekre, H.B., Sarode, T.K.: Fast codevector search algorithm for 3-D Vector Quantized codebook. International Journal of Computer and Information Science and Engineering, 235–239 (2008)
15. Chang, C., Wu, W.: Fast plannar-oriented ripple search algorithm for Hyperspace VQ codebook. IEEE Trans. Image Proc. 16(6), 1538–1547 (2007)
16. Ramamurthi, B., Gersho, A.: Classified Vector Quantization of images. IEEE Trans. Communucations 34(11), 1105–1115 (1986)

17. Quweither, M.K., Farison, J.B.: Classified Vector Quantization using Principal components. Electronics Letters 34(6), 538–540 (1998)
18. Kim, J.W., Lee, S.U.: A transform domain Classified vector Quantizer for image coding. IEEE Trans. Circuits and Systems for Video Technology 2(1), 3–14 (1992)
19. Kim, S., Lee, S.U.: Image Vector Quantizer based on a classification in the DCT domain. IEEE Trans. Communications 39(4), 549–556 (1991)
20. Young, A.W., Hay, D.C., McWeeny, K.H., Flude, B.M., Ellis, A.W.: Matching familiar and unfamiliar faces on internal and external features. Perception 14(6), 737–746 (1985)
21. Davies, G., Ellis, H., Shepherd, J.: Cue saliency in faces as assessed by the "Photofit" technique. Perception 6(3), 263–269 (1977)
22. Nasrabadi, N., King, R.: Image coding using vector quantization: A review. IEEE Trans. Communications 36(8), 957–971 (1988)
23. Tseng, H., Chang, C.: A Very Low Bit Rate Image Compressor Using Transformed Classified Vector Quantization. Informatica, 335–342 (2005)
24. Kim, J.W., Lee, S.U.: A transform domain classified vector quantizer for image coding. IEEE Trans. Circuits and Systems for Video Technology 2(1), 3–14 (1992)
25. Kanade, T.: Picture processing by computer complex and recognition of human faces. Ph.D. dissertation (1973)
26. Brunelli, R., Poggio, T.: Face recognition through geometrical features. In: European Conference on Computer Vision (1992)
27. Heenaye-Mamode Khan, M., Subramanian, R.K., Ali Mamode Khan, N.: Representation of Hand Dorsal Vein Features Using a Low Dimensional Representation Integrating Cholesky Decomposition. In: 2nd International Congress on Image and Signal Processing, CISP 2009 (2009)
28. Hong, L., Jian, A., Pankanti, S., Bolle, R.: Fingerprint enhancement. In: Proceedings 3rd IEEE Workshop on Applications of Computer Vision, WACV 1996 (1996)
29. Ito, K., Morita, A., Aoki, T., Higuchi, T., Nakajima, H., Kobayashi, K.: A fingerprint recognition algorithm using phase-based image matching for low-quality fingerprints. In: IEEE International Conference on Image Processing, ICIP 2005 (2005)
30. Jain, A.K., Duin, R.P.W., Mao, J.: Statistical Pattern Recognition: A Review. IEEE Transactions on Pattern ANALYSIS and Machine Intelligence 22(1) (January 2000)
31. Guo, G.-D., Zhang, H.-J.: Boosting for fast face recognition. In: Proceedings of the IEEE ICCV Workshop (2001)
32. Watanabe, E., Kodate, K.: Ultra-fast facial recognition system based on VanderLugt Correlator. In: CLEO/Pacific Rim 2003 - The 5th Pacific Rim Conference on Lasers and Electro-Optics (2003)
33. Momeni, H., Sadeghi, M.T., Abutalebi, H.R.: Fast face recognition using a combination of image pyramid and hierarchical clustering algorithms. In: International Conference on Wireless Communications & Signal Processing, WCSP 2009 (2009)
34. Sani, M.M., Ishak, K.A., Samad, S.A.: Evaluation of face recognition system using Support Vector Machine. In: 2009 IEEE Student Conference on Research and Development, SCOReD (2009)
35. Phillips, P.: The FERET Evaluation Methodology for Face-Recognition Algorithms. IEEE Transaction 22 (2000)
36. Utsumi, Y., Iwai, Y., Yachida, M.: Performance Evaluation of Face Recognition in the Wavelet Domain. In: 2006 IEEE/RSJ International Conference on Intelligent Robots and Systems (2006)

37. Dewi Agushinta, R., Suhendra, A., Madenda, S., Suryadi, H.S.: Face Component Extraction Using Segmentation Method on Face Recognition System. Journal of Emerging Trends in Computing and Information Sciences (2011)
38. Ali, M., Javed, M.Y., Basit, A.: Ear Recognition Using Wavelets. In: Proceedings of Image and Visioin Computing (2007)
39. Zhong, Z., Ding, N., Wu, X., Xu, Y.: Crowd surveillance using Markov Random Fields. In: IEEE International Conference on Automation and Logistics, ICAL 2008 (2008)
40. Daygman, J.G.: High Confidence Visual Recognition of persons by a test of Statistical Independence. IEEE Transaction (1993)
41. Makwana, M.V., Nandurbarkar, A.B., Joshi, S.M., Coll, L.E.: Image Compression Using Tree Structured Vector Quantization with Compact Codebook. In: International Conference on Computational Intelligence and Multimedia Applications (2007)
42. Chen, J.-Y., Bouman, C.A., Allebach, J.P.: Fast image database search using tree-structured VQ. In: Proceedings of the International Conference on Image Processing (1997)
43. Park, U., Jain, A.K.: 3D Face Reconstruction from Stereo Video. In: The 3rd Canadian Conference on Computer and Robot Vision (2006)
44. Zhao, M., Chua, T.-S., Sim, T.: Morphable Face Reconstruction with Multiple Images. In: Proceedings of the 7th International Conference on Automatic Face and Gesture Recognition (2006)
45. Wang, S.-F., Lai, S.-H.: Reconstructing 3D Face Model with Associated Expression Deformation from a Single Face Image via Constructing a Low-Dimensional Expression Deformation Manifold. IEEE Transactions on Pattern Analysis and Machine Intelligence (2011)
46. Weingessel, A., Hornik, K.: Local PCA algorithms. IEEE Transactions on Neural Networks (2000)
47. Diamantaras, K.I., Kung, S.Y.: Principal Component Neural Networks: Theory and Applications. John Wiley & Sons, Inc. (1996)
48. Lu, J., Plataniotis, K.N., Venetsanopoulos, A.N.: Face Recognition Using LDA Based Algorithms. IEEE Transactions on Neural Networks (2002)
49. Martinez, A.M., Kak, A.C.: PCA versus LDA. IEEE Transactions on Pattern Analysis and Machine Intelligence (2001)
50. Lindeberg, T.: Edge Detection and Ridge Detection with Automatic Scale Selection. In: 1996 IEEE Computer Society Conference on Computer Vision and Pattern Recognition (1996)
51. Martinez, A.M., Benavente, R.: The AR Face Database. CVC Technical Report #24 (1998)
52. Pissarenko, D.: Face databases (2003)
53. ORL face database,
    http://www.uk.research.att.com/facedatabase.html

Novel Matching Methods
for Automatic Face Recognition Using SIFT

Ladislav Lenc and Pavel Král

Department of Computer Science and Engineering,
University of West Bohemia, Plzeň, Czech Republic
{llenc,pkral}@kiv.zcu.cz

Abstract. The object of interest of this paper is Automatic Face Recognition (AFR). The usual methods need a labeled corpus and the number of training examples plays a crucial role for the recognition accuracy. Unfortunately, the corpus creation is very expensive and time consuming task. Therefore, the motivation of this work is to propose and implement new AFR approaches that could solve this issue and perform well also with few training examples. Our approaches extend the successful method based on the Scale Invariant Feature Transform (SIFT) proposed by Aly. We propose and evaluate two methods: the Lenc-Kral matching and the SIFT based Kepenekci approach [7]. Our approaches are evaluated on two face data-sets: the ORL database and the Czech News Agency (ČTK) corpus. We experimentally show that the proposed approaches significantly outperform the baseline Aly method on both corpora.

Keywords: Automatic Face Recognition, Czech News Agency, Scale Invariant Feature Transform.

1 Introduction

Automatic Face Recognition (AFR) consists of automatic identification of a person from a digital image or from a video frame by a computer. This field became intensively studied in the last two decades. Concerning other biometrics methods, AFR seems to be one of the most important ones.

The spectrum of applications utilizing AFR is really broad: access control to restricted areas, surveillance of persons, various programs for sharing and labeling of photographs, social networks and many others.

The most of the current AFR approaches perform well when high quality images available (well aligned, unified pose, etc). Unfortunately, the performance of such methods degrades significantly, when this assumption is violated. This issue is often handled by using more training examples/person (also our case). For creation of the correct face models a training corpus with enough training examples is necessary. Unfortunately, the corpus creation is very expensive and time consuming task. Our motivation is thus to propose and implement new AFR approaches that have high recognition accuracy also with few training examples

L. Iliadis et al. (Eds.): AIAI 2012, IFIP AICT 381, pp. 254–263, 2012.

(close to one). The main contribution of this work consists in proposing new matching methods and their integration to the SIFT algorithm.

The outcomes of this work shall be used by the Czech News Agency (ČTK) as follows. ČTK disposes a large database of photographs. A certain number of photos is manually annotated (i.e. the photo identity is known). However, another photos are unlabeled; the identities are thus unknown. The main task of our application consists in the automatic labeling of the unlabeled photos. Note that only few labeled images of every person are available.

The rest of the paper is organized as follows. The following section presents a short review of automatic face recognition approaches. Section 3 describes the SIFT algorithm and the proposed matching approaches. Section 4 evaluates the approaches on two corpora. In the last section, we discuss the results and we propose some future research directions.

2   Related Work

Thanks to the intensive research in the past years, many successful AFR methods were developed. The first attempts were based upon simple measures between important facial features [3]. The main drawback of such methods is the need of manual face labeling. Later, several approaches reducing the facial vector dimensionality were developed. One of such methods is the successful Eigenfaces approach [13,14] which is based on the Principal Component Analysis (PCA). Another method belonging to this group are Fisherfaces [2]. This approach uses Linear Discriminant Analysis (LDA) and Independent Component Analysis [12].

In the last ten years, a lot of attention was given to the feature based methods. The core of such methods is creating of a feature representation of the face. The facial image is inspected and the points of interest are detected. Then the features are created in the detected points. Some of those methods utilize Gabor wavelets to extract the features (e.g. Elastic Bunch Graph Matching (EBGM) [15] and the Kepenekci method [4]).

Recently, the Scale Invariant Feature Transform (SIFT) [9] proposed by David Lowe has been also used to create the facial features leading to high recognition accuracy. It has the ability to detect and describe local features in images. The features are invariant to image scaling, translation and rotation. The algorithm is also partly invariant to changes in illumination. The SIFT algorithm was originally developed for object recognition. The features of the reference and test images are compared using the Euclidean distance of their feature vectors.

2.1   SIFT for Face Recognition

One of the first applications of this algorithm for the AFR is proposed in [1] by Aly. It takes the original SIFT algorithm and creates the set of descriptors as described in Section 3. Each image is represented by the set of descriptors corresponding to the features.

First, the feature vectors are extracted from all gallery images. The test face is then matched against the faces stored in the gallery. The face, that has the largest number of matching features is identified as the closest one. The feature is considered to be matched if the difference between similarities of two most similar gallery features is higher than a specified threshold. In this work, the ORL and Yale databases are used for testing. It is reported that the recognition rate is 96.3% and 91.7% respectively. The results are compared with Eigenfaces [13,14] and Fisherfaces [2].

In [5], another approach using SIFT is presented. This method is called Fixed-key-point-SIFT (FSIFT). Contrary to the previous method, the SIFT keys are fixed in predefined locations determined in the training step as follows.

In the training step, the key-point candidates are localized in the same manner as in the original SIFT. A clustering algorithm is then applied to this key-point candidate set. The number of clusters is set to 100. The centroids of the clusters are used as the fixed key-point locations. The number of features thus remains constant. The distance between faces can be computed as the sum of Euclidean distances between the corresponding features. The reported recognition rate for the Extended Yale Database is comparable to the previously described approaches.

The proposed approaches, which use the features resulting from the SIFT algorithm, are described in more detail in the sequel.

3   Method Description

The first step is the determination of extrema in the image filtered by the Difference of Gaussian (DoG) filter. The input image is gradually down-sampled and the filtering is performed in several scales. Figure 1 demonstrates the process of creation of the DoG filters at the different scales [10].

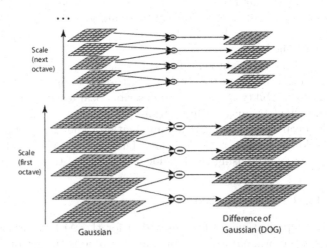

Fig. 1. Difference of Gaussian filters at the different scales [10]

In the next step, the detected key-points are further examined to choose the "best" candidates. Only points with high enough contrast are used and also points near edges are discarded. Then, orientation is assigned to each key-point. The resulting set of points is then used for creation of feature vectors (descriptors). Each descriptor contains a vector of the length 128 and also the coordinates of the point.

Figure 2 shows how two images of the same object (a face) with varying scale and orientation are matched.

Fig. 2. Matched key-points in two different views of the same object (face)

3.1   SIFT Features Extraction

The SIFT algorithm has basically four steps: extrema detection, removal of key-points with low contrast, orientation assignment and descriptor calculation [5].

To determine the key-point locations, an image pyramid with re-sampling between each level is created. It ensures the scale invariance. Each pixel is compared with its neighbours. Neighbours in its level as well as in the two neighbouring (lower and higher) levels are examined. If the pixel is maximum or minimum of all the neighbouring pixels, it is considered to be a potential key-point.

For the resulting set of key-points their stability is determined. Locations with low contrast and unstable locations along edges are discarded.

Further, the orientation of each key-point is computed. The computation is based upon gradient orientations in the neighbourhood of the pixel. The values are weighted by the magnitudes of the gradient.

The final step is the creation of the descriptors. The computation involves the 16×16 neighbourhood of the pixel. Gradient magnitudes and orientations are computed in each point of the neighbourhood. Their values are weighted by a Gaussian. For each sub-region of size 4×4 (16 regions), the orientation histograms are created. Finally, a vector containing 128 (16×8) values is created.

Figure 3 shows the SIFT features detected in the example images from the ČTK face corpus.

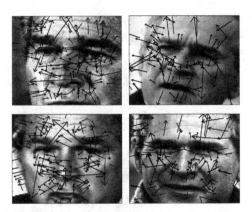

Fig. 3. Examples of detected SIFT features with orientation

The SIFT algorithm is described in detail in [9,10,5]. An implementation example can be found in [11].

3.2 Aly Matching

The first approach computes the number of the gallery image feature vectors that are matched against the test face feature vectors. For each test feature vector the similarities to all of the gallery feature vectors are computed. The cosine similarity of two feature vectors f_1 and f_2 is computed as follows:

$$S(f_1, f_2) = \frac{f_1 \cdot f_2}{\|f_1\|\|f_2\|} \tag{1}$$

The two most similar gallery feature vectors are determined. If the difference between these two similarities is higher than a prespecified threshold the feature vector is considered to be matched. For each gallery face, the number of matched feature vectors is computed. The recognized face is the one with highest number of matched feature vectors.

3.3 Lenc-Kral Matching

The first proposed approach computes a sum of similarities between pairs of image feature vectors. For each feature vector of the test face the most similar feature vector of the gallery face is identified. The sum of the highest similarities is computed and is used as a measure of similarity between two faces.

Speaking in more mathematical terms, let T be a test image represented by m feature vectors $t_1, t_2, .., t_m$. Let G be a gallery of images composed of N images $G_1, G_2, .., G_N$. Let every gallery image G_i be represented by n_i feature vectors $g_1, g_2, .., g_{n_i}$. Similarity of two feature vectors $S(t, g)$ is computed by the cosine similarity (see Equation 1). For each feature vector t_i of the recognized face T we determine the most similar vector $g^j_{max_i}$ of one gallery image G_j:

$$g^j_{max_i} = \arg\max_{G_j}(S(t, g)) \tag{2}$$

The sum of those similarities is computed as follows:

$$D(T, G_j) = \sum_{i=1..m} g_{max_i} \tag{3}$$

where m is the number of test image feature vectors. The recognized face is then determined by the following equation:

$$\hat{G}_i = \arg\max_{G_i}(D(T, G_j)) \tag{4}$$

3.4   Kepenekci Matching

This approach has been initially used by Kepenekci in [4] with Gabor wavelets. Author shows that this approach exhibits high recognition accuracy. Therefore, we decided to adapt this approach and integrate it with the SIFT.

Kepenekci combines two methods of matching and uses a weighted sum of the two values as a result. The cosine similarity is employed for vector comparison.

Let us call T a test image and G a gallery image. For each feature vector t of the face T we determine a set of relevant vectors g of the face G. Vector g is relevant iff:

$$\sqrt{(x_t - x_g)^2 + (y_t - y_g)^2} < distanceThreshold \tag{5}$$

where x and y are coordinates of the feature vector points.

If no relevant vector to vector t is identified, vector t is excluded from the comparison procedure. The overall similarity of two faces OS is computed as an average of similarities between each pair of corresponding vectors as:

$$OS_{T,G} = mean\ \{S(t, g), t \in T, g \in G\} \tag{6}$$

Then, the face with the most similar vector to each of the test face vectors is determined. The C_i value informs how many times the gallery face G_i was the closest one to some of the vectors of test face T. The similarity is computed as C_i/N_i where N_i is the total number of feature vectors in G_i. Weighted sum of these two similarities is used for similarity measure:

$$FS_{T,G} = \alpha OS_{T,G} + \beta\frac{C_G}{N_G} \tag{7}$$

The face is recognized as follows:

$$\hat{FS}_{T,G} = \arg\max_{G}(FS_{T,G}) \tag{8}$$

4 Experimental Setup

4.1 Corpora

ORL Database. The ORL database was created at the AT & T Laboratories[1]. The pictures of 40 individuals were taken between April 1992 and April 1994. For each person 10 pictures are available. Every picture contains just one face. They may vary due to three following factors: 1) time of acquisition; 2) head size and pose; 3) lighting conditions. The images have black homogeneous background. The size of pictures is 92×112 pixels. A more detailed description of this database can be found in [8].

Czech News Agency (ČTK) Database. This corpus is composed of the images of individuals in uncontrolled environment that were randomly selected from the large ČTK database. All images were taken during a long time period (20 years or more). The detection of faces was made automatically utilizing the OpenCV library. They were automatically resized to the size 92 × 92 pixel and transformed to grayscale. The resulting corpus contains images of 63 individuals, 8 images for each person. Note that orientation, lighting conditions and background of images differ significantly. Performing accurate face recognition using this dataset is thus very difficult.

Figure 4 shows one example from this corpus. This corpus is available for free for the research purpose upon request to the authors.

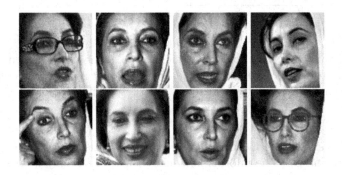

Fig. 4. Examples of one face from the ČTK face corpus

4.2 Experiments

All experiments were performed on two datasets: the ORL dataset and the ČTK corpus mentioned before. Previously, we tested the ČTK database using the Eigenfaces approach. It exhibited very low accuracy. Therefore, we use the SIFT based methods for experiments. We used the successful Aly method (see Section 3.2) as a baseline. In all cases the cross-validation was used to ensure more reliable results.

---

[1] http://www.cl.cam.ac.uk/research/dtg/attarchive/facedatabase.html

We made a series of experiments for each dataset. The size of the training set is gradually increased from 1 image/person to $N-1$ images/person (N is the total number of images/person). We used 9 different set-ups for the ORL dataset and 7 set-ups for the ČTK dataset. To allow a straightforward comparison of these methods, we evaluated each set-up with three previously described matching schemes.

Table 1. Recognition rate of the different matching schemes for the ORL dataset according to the different training set size

Matching scheme	Aly	Lenc-Kral	Kepenekci
Training Set	Recognition rate (%)		
1 of 10	61.25	78.75	80.56
2 of 10	78.72	88.24	90.15
3 of 10	85.36	92.46	94.24
4 of 10	88.83	95.67	97.25
5 of 10	92.42	96.75	97.92
6 of 10	95.27	97.86	97.86
7 of 10	96.88	98.65	98.65
8 of 10	98.36	98.86	99.17
9 of 10	99.00	99.00	99.25

Table 1 shows the recognition rates of the different test set-ups for the ORL dataset. This table shows that the scores of the proposed Lenc-Kral approach are significantly higher than the original Aly method especially where not enough training examples available. The second proposed approach (SIFT based Kepenekci method) have slightly better recognition accuracy than both other approaches.

Table 2. Recognition rate of the different matching schemes for the ČTK corpus according to the different training set size

Matching scheme	Aly	Lenc-Kral	Kepenekci
Training Set	Recognition rate (%)		
1 of 8	9.78	12.95	19.73
2 of 8	14.18	19.11	27.78
3 of 8	16.90	24.29	31.75
4 of 8	20.40	28.89	37.10
5 of 8	22.93	31.92	41.18
6 of 8	24.12	34.27	43.85
7 of 8	25.79	36.71	46.63

Table 2 shows the recognition accuracy of the experiments on the ČTK corpus. The recognition accuracy is significantly lower than such in the case of the ORL database probably due to the different orientation of the images (see Figure 4).

This table also shows that both proposed methods significantly outperform the baseline Aly approach for all training examples in all cases.

5   Conclusions and Perspectives

In this paper, we presented two new AFR methods: namely Lenc-Kral matching and SIFT based Kepenekci approach. Both methods are based on the SIFT features. The experiments show that both proposed methods outperform the baseline Aly approach. The recognition accuracy on the ORL corpus is significantly higher particularly when the training set is small. In the case that only one training example per person is used, the Lenc-Kral and the Kepenekci matching increase the recognition rate respectively by 17% and by 19%, over the baseline. The results on the ČTK dataset show the difficulties of the face recognition in the real conditions. However, the recognition accuracy is in all cases significantly higher than in the Aly method.

The first perspective consists in combining this method with another successful method in order to further improve the recognition accuracy. Particularly, the adapted Kepenekci approach [7] based on the Gabor wavelets could be a suitable choice due to its high recognition accuracy. Another perspective consists in the use of confidence measures in the post-processing step [6]. The confidence measure technique will be used to detect and remove incorrectly recognized examples from the result set.

Acknowledgements. This work has been partly supported by ČTK and by the UWB grant SGS-2010-028 Advanced Computer and Information Systems. We also would like to thank ČTK for providing the photographic data.

References

1. Aly, M.: Face recognition using sift features (2006)
2. Belhumeur, P.N., Hespanha, J.A.P., Kriegman, D.J.: Eigenfaces vs. fisherfaces: Recognition using class specific linear projection. In: IEEE Transactions on Pattern Analysis and Machine Intelligence (1997)
3. Bledsoe, W.W.: Man-machine facial recognition. Tech. rep., Panoramic Research Inc., Palo Alto, CA (1966)
4. Kepenekci, B.: Face Recognition Using Gabor Wavelet Transform. Ph.D. thesis, The Middle East Technical University (2001)
5. Krizaj, J., Struc, V., Pavesic, N.: Adaptation of sift features for robust face recognition (2010)
6. Lenc, L., Král, P.: Confidence measure for automatic face recognition. In: International Conference on Knowledge Discovery and Information Retrieval, Paris, France, October 26-29 (2011)
7. Lenc, L., Král, P.: Gabor wavelets for automatic face recognition. In: 38th International Conference on Current Trends in Theory and Practice of Computer Science, Špindlerův Mlýn, Czech Republic (January 2012)
8. Li, S., Jain, A.: Handbook of face recognition. Springer (2005)

9. Lowe, D.G.: Object recognition from local scale-invariant features. In: International Conference on Computer Vision (1999)
10. Lowe, D.G.: Distinctive image features from scale-invariant keypoints. International Journal of Computer Vision 2 (2004)
11. Lowe, D.: Software for sift (2004), http://people.cs.ubc.ca/~lowe/keypoints/
12. Shakhnarovich, G., Moghaddam, B.: Face recognition in subspaces. In: Handbook of Face Recognition, pp. 141–168. Springer (2004)
13. Sirovich, L., Kirby, M.: Low-dimensional procedure for the characterization of human faces. Journal of the Optical Society of America 4 (1987)
14. Turk, M.A., Pentland, A.P.: Face recognition using eigenfaces. In: IEEE Computer Society Conference on in Computer Vision and Pattern Recognition (1991)
15. Wiskott, L., Fellous, J.M., Krüger, N., von der Malsburg, C.: Face recognition by elastic bunch graph matching. Intelligent Biometric Techniques in Fingerprint and Face Recognition (1999)

Exploring the Design Space of a Declarative Framework for Automated Negotiation: Initial Considerations*

Alex Muscar and Costin Bădică

University of Craiova, Blvd. Decebal, nr. 107, RO-200440, Craiova, Romania
{amuscar,cbadica}@software.ucv.ro

Abstract. In this paper we present our results on the exploration of the design space of a declarative framework for automated negotiation by: (1) identifying a minimal yet viable generic negotiation protocol and a declarative way of representing rules and constraints specific to negotiation mechanisms and strategies; (2) identifying the need of a set of basic concepts for describing negotiations that form a core negotiation ontology that the agents can use to reason about the negotiations; (3) proposing Belief-Desire-Intention (BDI) agents as an implementation model of our framework. We introduce both a conceptual framework for declarative specification of automated negotiations as well as a prototype implementation using the Jason agent programming language.

1 Introduction

An essential feature of agents is their ability to communicate by exchanging meaningful information for improving their own goals and society goals. Sometimes agents behave selfishly and their goals are only partly overlapping or can even be in conflict. In such cases agents must interact to reach an agreement between all or some of the involved agents. These interactive processes are broadly called negotiations. Negotiations carried out between computer systems are known as automated negotiations.

There are two components of negotiation: mechanism and strategy. The *negotiation mechanism* states the interaction rules that must be obeyed by the participants in order to meet some objectives. The mechanism, sometimes known as protocol, is public. The *negotiation strategy* describes the behavior of a negotiation participant and it is directed towards reaching his or her private goals, usually to maximize his or her gain [1], [2].

Many models of automated negotiations were proposed by researchers. The models can be parameterized according to different criteria, like: number and roles of negotiation participants, structure and properties of the negotiation subject, presence or not of a mediator, a.o [3]. Some well-known automated negotiation models are: bargaining, auctions [4], multi-criteria negotiation [5], multi-commodity negotiations.

Based on our literature survey, we observed an important deficiency of existing research approaches to automated negotiation: *they do not properly address the reusability*

---

* This work was supported by the strategic grant POSDRU / CPP107 / DMI1.5/S/78421, Project ID 78421 (2010), co-financed by the European Social Fund – Investing in People, within the Sectoral Operational Programme Human Resources Development 2007 – 2013, as well as by the *K-SWAN: An Interoperable Knowledge-based Framework for Negotiating Semantic Web Agents* Bilateral Research Project Greece – Romania, co-financed by UEFISCDI.

L. Iliadis et al. (Eds.): AIAI 2012, IFIP AICT 381, pp. 264–273, 2012.

and extensibility of the research results which are so much required for their usefulness in open environments such as the Internet and the Web. Specific research questions immediately derived from this general problem are: (1) how can we define a reusable representation of automated negotiation protocols and strategies?; (2) how can agents define and interpret their behaviors based on the formal representation of a negotiation mechanism?; (3) how can we capture and encode agents' knowledge about negotiation mechanisms such that they can adjust dynamically their strategies depending on needs?

In this paper we propose the following initial answers: (1) inspired by [6], we identify a minimal (our prototype protocol allows only six actions, see Sec. 3.3) yet viable generic negotiation protocol (the protocol actions can be used to model various negotiation types), as well as a declarative approach of representing rules and constraints specific to negotiation types such that by combining them we can derive the basis of a generic, reusable negotiation framework. This would allow agents to specify and understand customizable negotiation protocols and strategies; (2) we identify a set of core concepts related to negotiations that should be part of a core negotiation ontology that can be used by agents to reason about negotiations; (3) we propose the Belief-Desire-Intention (BDI) model [7] as the basis for the design of our solution due to its suitable level of abstraction that can appropriately incorporate declarative specification of negotiation agents' behavior, thus offering a richer framework than existing proposals [8].

While we acknowledge that our goal is very ambitious, we think that the research effort is worthwhile, as our endeavor can clearly contribute to the reusability of negotiation protocols and strategies, an issue that in our opinion was insufficiently explored by the research community.

The main results consist of a conceptual framework for development of automated negotiations, as well as an initial prototype developed using Jason agent programming language [9]. While the implementation still lacks some of the features of our proposed conceptual framework, we believe it is a good starting point for the further exploration of the design space of declarative specifications of automated negotiations. Our choice for Jason as implementation language is motivated by: (i) Jason is probably the best known example of the BDI camp in the agent programming community; (ii) Jason supports a declarative programming style, closer to logic programming. We believe that the mix of declarative, goal-oriented, knowledge representation and meta-programming features of Jason make it a good candidate implementation language for our prototype.

The paper is structured as follows. In Sec. 2 we set the stage by looking at some of the related research in this area. In Sec. 3 we outline our system architecture and go into more details regarding our proposed approach and at the same time illustrate our presentation with code samples from a prototype implementation. If Sec. 3 looks at the system as a whole, Sec. 4 is dedicated to exploring the details of single agents in our framework. We conclude in Sec. 5 and we also outline future research directions.

2 Background and Related Work

Albeit the plethora of formal representations originating from researches in Semantic Web and Software Engineering ([10], [11]), the current results do not go beyond simple XML-based representations of specific negotiation mechanisms.

Authors of [6] proposed a generic software framework for automated negotiations. Although very interesting for our research, this proposal only addressed the problem from the perspective of the authority that controls the negotiation, i.e. the *Auction Host*. This approach is limited because, unlike a human negotiator, an artificial negotiator would have to be *a priori* designed to understand certain negotiation mechanisms. This is a considerable limitation of artificial agents as compared to humans, which leads to the impossibility of an artificial agent to act on a market whose negotiation mechanism is not known and understood before the agent design. Therefore, the perspective of the artificial agent acting as a participant in a negotiation should be also taken into account.

Authors of [8] proposed the AB3D software framework for auction development. The AB3D system provides an auction specification framework, as well as a runtime system for the agents that enact the auction specifications. While this approach is interesting as it acknowledges the importance of generic, parameterizable auction specifications, we believe that the AB3D scripting language lacks with respect to flexibility when specifying participant strategies. The question of how a participant agent should dynamically understand the specification of a negotiation protocol (auction in particular) in order to define his strategy is not addressed by that work.

A recent and interesting example of BDI being used for automated negotiations is presented in [12]. The author proposes a multi-strategy automated negotiation framework based on the BDI model. While the presented approach is somewhat similar to our own in that it uses a specialization of the BDI model for the agents in the system, it differs in that the author mostly focuses on individual agent's decision-making mechanism. While we acknowledge that this is an important aspect of an automated negotiation framework, we believe that the holistic approach we propose can lead to a truly generic solution. Nevertheless, the paper remains interesting, because it goes into much more detail w.r.t individual decision-making than the approach we are proposing and it might prove a valuable source of inspiration in our further efforts.

3   Negotiation Model and System Architecture

An automated negotiation can be observed and analyzed from two perspectives: the perspective of the authority that controls the negotiation and the perspective of the negotiation participants, their preferences and their private strategies. Each perspective addresses one of the two components of an automated negotiation: negotiation mechanism and respectively negotiation strategy.

The initial source of inspiration for our negotiation model is [6]. Its authors focused on auctions and performed a deeper analysis of the parametrization of the negotiation protocol by conceptually decomposing it into several rule sets for: admission of participants to negotiation, proposal validity, protocol enforcement, updating negotiation status and informing participants, agreement formation, and controlling the lifecycle of the negotiation process. Although very interesting, their proposal is biased towards the perspective of the auction authority, while the implications of this classification onto the participating agents is omitted from their analysis. In order to understand this point, it is useful to observe that a declarative specification can be utilized in two ways: (i) as an enforcer, to constrain the agents as well as (ii) a generator of agents' permissible actions depending on the specific negotiation context. For example, let us consider the rules for proposal validity that constrain the submitted proposals to be consistent with a given template. According to [6], the auction authority will perform a consistency check of each proposal submitted by an auction participant. However, a participant agent can use this template to generate, whenever this is allowed by the protocol rules, a new proposal that is compliant with the proposal validity rules.

Additionally, considering the rules for protocol enforcement, they might state that at some point a participant can either bid or leave the auction. The auction authority will use this information to check the messages received from a participant, recognizing and allowing only the actions of bidding or withdrawing (while an action for updating a previously submitted bid would be rejected as not allowed), thus acting as enforcing mechanism for the auction semantics. The participant, on the other hand, will consider its options and, based on its custom strategy, it will pick up one of the two possible actions. From its perspective the negotiation mechanism will act as action generator. Incidentally, the custom strategy will act by additionally filtering the generated available options. We will get back to this point in Sec. 4.

3.1 System Architecture

We are going to use the architectural model introduced in [13] as a foundation for our system architecture. The framework proposed by the authors features several types of agents (note that the initial framework targeted auctions, so we have used the term 'auction' interchangeably with 'negotiation'):

Auction Service (AS). Usually it implements a specific type of auction. Its main purpose is to manage auction-related activities (e.g. creation, termination) and to coordinate the auction participants. AS registers with an Auction Service Directory (ASD) that can be queried for a certain AS. An AS contains an Auction Directory (AD), which keeps track of all the ongoing auctions which are represented by Auction Instances (AI);

Auction Host (AH). It is an arbitrator agent responsible with coordinating the agents participating in a single AI. In order to accomplish this task is uses a specific mechanism for each auction type (e.g. English, Dutch).

Auction Participant (AP). This agent participates by bidding in an AI. Out of all the participants one is distinguished as the Auction Initiator Participant (AIP).

We envision that a negotiation agent will use the framework to either initiate a new negotiation or to register for bidding in an existing negotiation. For achieving these functions, the agent can be pre-designed to understand the negotiation mechanism (we already claimed that this is a considerable limitation) or it can download the declarative description of the negotiation mechanism from the *Auction Host* and apply it to define its private strategy. This second option is more interesting for the purpose of this paper.

For a more detailed presentation of this architecture please consult reference [13].

3.2   Conceptual Model

We propose to conceptualize the declarative specification of a negotiation as composed of three essential ingredients: *Generic Negotiation Protocol* (GNP), *Declarative Negotiation Mechanism* (DNM) and *Custom Negotiation Strategy* (CNS).

GNP. Defines and governs the interaction between the participant agents and the host agent that are part of the system. It is the same for all participants independently of their specific role in the negotiation (i.e. buyer, seller). However, we will have a GNP part for the AH role, as well as a GNP part for the AP role; see Sec. 3.3 for more details);

DNM. Is specific to a given negotiation type (e.g. English Auction, Continuous Double Auction, Iterative Bargaining). The DNM is usually defined using a declarative, Prolog-like language. The DNM serves to customize the GNP for representing the conditions and events that enable the permissible actions of negotiation agents (see Sec. 3.4 for more details); and

CNS. Is specific to a given AP and must be consistent with the DNM. It is used by the AP to select and configure a specific negotiation action that could be most useful in a given negotiation context (see Sec. 3.5 for more details).

Based on the three features described above we can define the following metaphorical equations that more succinctly describe the agents present in our framework:

$$AH = GNP_{role-host} + DNM_{host}$$
$$AP = GNP_{role-participant} + DNM_{participant} + CNS$$

These ingredients can be consistently bound into descriptions of AH and respectively AP behaviors by referring a common core vocabulary of terms for defining and parameterizing the space of negotiation types. This vocabulary is called *Core Negotiation Ontology* – CNO. The CNO can contain generic parameters, and actions. The former can be used to customize the auction, while the latter can be used by APs to choose corresponding actions with a certain semantics for the auction. Another element of the CNO could be the types of rules used to guide action generation/selection during the negotiation process. The development of a comprehensive CNO is a complex task that is part of our future research and it is outside the scope of this paper.

We are going to use the English Auction as a running example for the following subsections, to give more insight into the three components introduced here. Therefore, for the purpose of this paper we are going to use a sample set of terms, mainly referring to negotiation actions for English Auctions, that will be introduced in Sec. 3.4.

3.3   Generic Negotiation Protocol

The GNP describes the permissible negotiation conversations involving a set of agents comprising the following two generic roles: (i) role of AH; there is a single agent playing this role in a negotiation; and (ii) role of AP; there are one or more agents playing this role in a negotiation. Thus the GNP is agnostic of any details that are specific to a particular negotiation type. Consequently, it must only expose the basic negotiation actions that are common to as many negotiation types, leaving unspecified the specific

```
state(uninitialised).                      // When the negotiation finishes
                                           +!close
// When registering an agent                   // and we have a winner
+register[source(A)] : can_register(A)         : check_winner
    // check if the agent can register         <- -+state(closed);
    // (e.g. the negotiation hasn't started)      ?initiator(I);
    <- +registered(A);                           ?quote(A, Offer);
        // get the current quote                 // notify the initiator
        ?quote(_, Quote);                        .send(I, tell,
        // and send it to the freshly added          winner(A, Offer));
        // participant                          // and the participant
        .send(A, tell, quote(Quote)).          .send(A, tell, winner).

// When receiving a bid check if bids       // Otherwise notify the initiator
// are accepted and if the bid is an        // that there is no winner
// improvement                              +!close
+bid(Offer)[source(A)] : can_bid(A) &           <- -+state(closed);
    check_progress(Offer)                        ?initiator(I);
    // Update the current quote                  .send(I, tell, no_winner).
    <- -+quote(A, Offer);
        // and notify all the other participants
        .findall(X, registered(X), Registered);
        .send(Registered, tell, quote(Offer)).
```

Listing 1: The *Auction Host* agent which implements the generic negotiation protocol

details of a certain negotiation type. Those details can be captured and declaratively represented with the help of the DNM.

GNP is inspired by our previous work [13]. As part of the GNP we identified a set of actions as sufficient for introducing the proof-of-concept implementation:

register used by an AP to register itself with a specific auction host
bid used to place a bid
tell/ask depending on the semantics of the push/pull semantics desired the protocol can expose information to the participants via tell or ask actions
fold used by the participant to get out of an auction
close used by the auction host to end the auction
winner used by the auction host to notify the winning participants

The Jason implementation of the AH role of the GNP is presented in listing 1. We refer the reader to [9] for details on the syntax and semantics of the language.

The code in listing 1 closely follows the definition of the GNP outlined above. Very briefly, the plan for +**register** registers a participant and sends it the current *quote* (i.e. the value of the outstanding bid) – this presumes push semantics; the +bid plan updates the quote and pushes messages with the new value to *all* the participants; the +close plan ends the auction and notifies any winner if there is one. All the plans keep track of the current state of the agent.

Note that the plans make use of rules that are included in the definition of the DNM (this will be covered in the next subsection). While an AS conceptually implements a type of auction, an AH is spawned for each new auction started by the AS. During its initialization the AH loads the DNM specific to the negotiation type supported by the AS and initialized with the parameters received from the AIP.

```
// A negotiation can be initialized       // An offer needs to be an inprovement of
// only if it hasn't been previously      // the current quote
// initialized                            check_progress(Offer)[state(bidding)] :-
can_init[state(uninitialised)] :-             increment(Increment) &
    state(uninitialised).                     quote(_, Quote) &
                                              Offer >= Quote + Increment.
// A participant can register only
// after the negotiation was initialized   // The winner needs to offer more
// and if it hasn't already registered     // than the reserved price
can_register(A)[state(bidding)] :-         check_winner[state(bidding)] :-
    state(bidding) &                           quote(A, Offer) & A \== "" &
    not(registered(A)).                        min_price(MinPrice) &
                                               Offer >= MinPrice.
// Only registered agents can bid
can_bid(A)[state(bidding)] :-
    state(bidding) &
    registered(A).
```

Listing 2: *Declarative Negotiation Mechanism* for English Auction

3.4   Declarative Negotiation Mechanism

We chose a rule-based representation for our DNM, which builds on our previous approach presented in [14]. Moreover, this representation was mapped naturally to Jason.

Listing 2 shows the rules that we used to customize the *Auction Host* for our running example. Note that these rules illustrate the DNM from the AH perspective.

Besides their formal parameters, the rules are also implicitly parameterized by the current state of the negotiation process – here we are using the annotations mechanism from Jason to achieve this, e.g. the [state(bidding)] annotation on the can_bid rule makes it applicable only when the agent is in the bidding state.

One important aspect, that we haven't detailed yet, is the sample CNO for this particular example. This is implicitly represented both in the rules presented in listing 2 and in the *Auction Host* presented in listing 1. min_price, increment, quote and state are part of the CNO. They are either parameters that must be filled in by the AIP – the first two – or state variables that are maintained by the *Auction Host* – the former two. In our prototype implementation the CNO is expressed as s set of beliefs. Note, though that alternative approaches are possible, like explicitly specifying the ontology with the aid of a specialized language like OWL. With such an approach the OWL specification of the CNO must be compiled into agent beliefs before the start of a new negotiation.

3.5   Custom Negotiation Strategy

The CNS must be implemented by each individual AP agent. In particular, this component allows each AP to adopt different risk attitudes toward the auction participation (e.g. eager, aggressive, neutral). As mentioned before, this is a key aspect of our approach, since it allows to model much richer negotiation scenarios. So, in principle, CNS should be not only understandable and adjustable by the AP itself, but also externally configurable by a human user which can be present behind a certain AP.

```
product(macbook).                    +!bid(Quote) : amount(Amount) & Amount > 100
amount(1200).                            <- ?auction_host(AuctionHost);
current_offer(0).                            // Eagerly bid as long as the
                                             // agents has money left
!start.                                      -+current_offer(Quote + 100);
                                             -+amount(Amount - 100);
+!start : product(Product)                   .send(AuctionHost, tell,
    <- .wait(100);                                bid(Quote + 100)).
       // Look for an auction featuring
       // the product                    +auctions_for(Product, [auction(_, AH)|_])
       .send(auction_service, tell,          <- +auction_host(AH);
            find_auction(Product)).              .send(AH, tell, register).


+quote(Quote) : current_offer(Offer)
    <- !bid(Quote).

+winner <- .print("I won!").
```

Listing 3: *Auction Participant* that acts as a buyer

Conceptually, the CNS is built on top of the GNP, so the strategy designer is not exposed to the lower level encoding of the negotiation process. Instead he or she can focus on the more relevant aspects of the definition of an appropriate strategy – like improving his or her outcome from the negotiation participation.

Listing 3 shows an eager buyer for our running example of an English auction. Conceptually, the plans on the left side of the listing would be the only ones that the auction designer has to write. The plans on the right would be part of the framework, and as such, it would not be exposed to the implementer. For our prototype, this could be achieved by customizing the Jason agent to internally handle the GNP related events (e.g. +quote messages) and triggering the relevant GNS plans (in our case +!bid) which would act like "hooks" into the framework for the auction designer.

4   Agent Architecture

The GNP, DNM and CNS fit naturally to the generic BDI model, and in particular to the Jason programming model. We chose BDI as our underlying architecture because it offers certain advantages:

- The GNP, DNM and CNS map naturally to components of the BDI agents' architecture: beliefs, goals, events, and plans, that are also supported by Jason.
- The BDI model is an established paradigm for the development of autonomous rational agents. So we expect that it can be useful for more complicated negotiation scenarios, for example involving agents that simultaneously participate in multiple auctions possibly sharing partly or entirely the negotiation subject.

Other recent approaches of developing automated negotiations using BDI agents were reported by [15,12].

The reasoning cycle of a BDI agent [7, p. 7] can be utilized to materialize the *enactment* and *enforcement* of the GNP for both AH and AP agents. The option generation and selection phase corresponds to finding all possible actions at each step of the negotiation. Selecting an intention to be executed based on the available options corresponds to executing a step of the negotiation. Both steps use the DNM to filter out actions that are not achievable. Additionally, APs use the CNS to further constrain the actions that are going to be performed during each step of the negotiation process. Finally, getting the new external events maps to receiving messages that form the GNP.

Note that although our conceptual model is mature enough, the proposed implementation using Jason is just an initial prototype. Its further extension and evaluation are required as part of our future work. Nevertheless, the decision to use Jason as an implementation language has proven a good choice for the following reasons.

First, Jason's Prolog-like rule language fit well with the DNM specification. It can be used to write declarative rules that serve to enforce negotiation protocol semantics on top of the GNP. This aspect of Jason's rule language corresponds to checking generic parameters defined in the CNO. By using Jason's built-in inference engine, the rule language could be used to guide the generation/selection of actions defined in the CNO. This approach is appealing because it would allow a complete (it supports both generic parameters and negotiation actions) and uniform (the same language is used for both generic parameters and actions) representation of the DNM. Due to these reasons this is one of the next steps we are going to take in further developing our prototype.

Second, we can leverage Jason's features (e.g. first class beliefs and goals, BDI architecture) to build our framework. Since our abstract framework could be seen as a specialization of BDI, this is a natural approach. Nevertheless, more investigation is required to check the appropriateness of the matching between Jason's internal architecture and our proposed conceptual framework.

Third, Jason was engineered as an extensible system, offering many customization points. Together with its metaprogramming capabilities this reduced the development time of our prototype considerably. We will further investigate using these customization mechanisms to address the problem mentioned earlier.

Overall, the development of the prototype on top of the BDI model has proven the viability of our proposal. It remains to be seen if the final representation language will be Jason or another – possibly custom – language. Ideally, programs should read like an executable specification of the implemented negotiation.

5   Conclusions and Future Work

In this paper we have presented our initial ideas of an original approach for the development of a generic, declarative negotiation framework. The first important aspect is the proposal of a novel dichotomy between the auction host and the auction participant when considering the declarative negotiation mechanism. The second innovative aspect is the description of the mapping of our conceptual framework to the BDI agent architecture. Our proposal is supported by a sample proof-of-concept implementation using Jason agent programming language. We claim that our initial results make the development of a generic framework for automated negotiations a feasible endeavor.

Our future directions are related to expanding the negotiation framework in order to allow it to express a richer set of negotiation types. Important aspects are related to: (i) sharing specifications of negotiation mechanisms in open environments, like the Internet and the Web; (ii) developing a proper *Core Negotiation Ontology*; and (iii) evaluating the framework by developing specifications of different negotiation types and negotiation agents.

References

1. Kraus, S.: Automated Negotiation and Decision Making in Multiagent Environments. In: Luck, M., Mark, V., Åtepnkov, O., Trappl, R. (eds.) ACAI 2001. LNCS (LNAI), vol. 2086, pp. 150–172. Springer, Heidelberg (2001)
2. Jennings, N., Faratin, P., Lomuscio, A., Parsons, S., Wooldridge, M., Sierra, C.: Automated negotiation: Prospects, methods and challenges. Group Decision and Negotiation 10, 199–215 (2001)
3. Buttner, R.: A classification structure for automated negotiations. In: Proceedings of the 2006 IEEE/WIC/ACM International Conference on Intelligent Agent Technology – Workshops, pp. 523–530. IEEE Computer Society, Los Alamitos (2006)
4. Wurman, P.R., Wellman, M.P., Walsh, W.E.: A parametrization of the auction design space. Games and Economic Behavior 35, 304–338 (2001)
5. Scafeş, M., Bădică, C.: Computing equilibria for constraint-based negotiation games with interdependent issues. In: Proceedings of Federated Conference on Computer Science and Information Systems - FedCSIS 2011, pp. 597–603 (2011)
6. Bartolini, C., Preist, C., Jennings, N.R.: A Software Framework for Automated Negotiation. In: Choren, R., Garcia, A., Lucena, C., Romanovsky, A. (eds.) SELMAS 2004. LNCS, vol. 3390, pp. 213–235. Springer, Heidelberg (2005)
7. Rao, A.S., Georgeff, M.P.: Bdi agents: From theory to practice. In: Lesser, V.R., Gasser, L. (eds.) Proceedings of the First International Conference on Multiagent Systems, ICMAS 1995, pp. 312–319. The MIT Press (1995)
8. Lochner, K.M., Wellman, M.P.: Rule-based specification of auction mechanisms. In: Proceedings of the Third International Joint Conference on Autonomous Agents and Multiagent Systems, AAMAS 2004, vol. 2, pp. 818–825. IEEE Computer Society, Washington, DC (2004)
9. Bordini, R.H., Hbner, J.F., Vieira, R.: Jason and the golden fleece of agent-oriented programming. In: Bordini, R.H., Dastani, M., Dix, J., Fallah-Seghrouchni, A.E. (eds.) Multi-Agent Programming. Multiagent Systems, Artificial Societies, and Simulated Organizations, vol. 15, pp. 3–37. Springer (2005)
10. Tamma, V., Phelps, S., Dickinson, I., Wooldridge, M.: Ontologies for supporting negotiation in e-commerce. Engineering Applications of Artificial Intelligence 18, 223–236 (2005)
11. Dong, H., Hussain, F., Chang, E.: State of the art in negotiation ontologies for enhancing business intelligence. In: 4th International Conference on Next Generation Web Services Practices, NWESP 2008, pp. 107–112 (2008)
12. Cao, M.: Multi-strategy selection supported automated negotiation system based on bdi agent. In: Proc. 45th Hawaii International International Conference on Systems Science (HICSS-45 2012), pp. 638–647. IEEE Computer Society, Los Alamitos (2012)
13. Dobriceanu, A., Biscu, L., Bădică, A., Bădică, C.: The design and implementation of an agent-based auction service. IJAOSE 3(2/3), 116–134 (2009)
14. Bădică, C., Giurca, A., Wagner, G.: Using Rules and R2ML for Modeling Negotiation Mechanisms in E-Commerce Agent Systems. In: Draheim, D., Weber, G. (eds.) TEAA 2006. LNCS, vol. 4473, pp. 84–99. Springer, Heidelberg (2007)
15. Le Dinh, B.C., Seow, K.T.: Unifying distributed constraint algorithms in a bdi negotiation framework. In: Proceedings of the 6th International Joint Conference on Autonomous Agents and Multiagent Systems, AAMAS 2007, pp.117:1–117:8. ACM, New York (2007)

Hybrid and Reinforcement Multi Agent Technology for Real Time Air Pollution Monitoring

Andonis Papaleonidas and Lazaros Iliadis

Democritus University of Thrace,
Department of Forestry & Management of the Environment & Natural Resources,
193 Pandazidou St., 68200 N Orestiada, Greece
papaleon@sch.gr, liliadis@fmenr.duth.gr

Abstract. This paper describes the design and implementation of a modular hybrid intelligent model and system, for monitoring and forecasting of air pollution in major urban centers. It is based on Multiagent technologies, Artificial Neural Networks (ANN), Fuzzy Rule Based sub-systems and it uses a Reinforcement learning approach. A multi level architecture with a high number of agent types was employed. Multiagent's System modular and distributed nature, allows it's interconnection with existing systems and it reduces its functional cost, allowing its extension by incorporating decision functions and real time imposing actions capabilities.

1   Introduction

The problem of air quality especially in urban sites is a major and composite task. Nearly all of the air pollutants have been blamed by several epidemic and clinical studies for their direct or indirect involvement in the cause of several serious diseases (Brunekreef and Holgate, 2002) (Nafstad P. et al., 2003)(Lisabeth LD et al., 2008) after a long term exposure. The risk limits and the right of the citizens to have access in vital information related to air quality and to pollutants' concentration levels, have led to the establishment (by the European Commission) of specific safety limits, whereas public awareness services were also created (European Commission, 1990) (European Commission, 1992) (European Commission, 1996) (European Commission, 2000). Even platforms like app store, supply the public with applications that provide real time air quality data in mobile phones in various parts of major cities (diMobile, 2011) (Aratos, 2012). Due to the importance of the problem, various Soft Computing applications have been developed recently for air quality modeling or monitoring, whereas some of them also provide real time proposals when the pollutants' concentration is too high (Triantafilou A.G et. all, 2011), (Iliadis and Papaleonidas, 2009), (Wahab and Alawi, 2002), (Paschalidou et al., 2007), (Iliadis et al., 2007). These efforts use mainly ANN or Multi Agent Systems (MAS). However they only try to face the problem under a monolithic point of view of forecasting or monitoring.

The MAS described in this paper not only records data and presents the actual real time situation (in terms of air pollution) but it also tries to estimate the evolution of the problem in a short term scale, by employing a hybrid approach that satisfies the needs of a modular wide architecture perceptive design (Estrin et al., 2002).

L. Iliadis et al. (Eds.): AIAI 2012, IFIP AICT 381, pp. 274–284, 2012.

A main advantage of the proposed system is that it does not work as a stand alone application in a specific place. It rather uses a large number of independent agents of various types that are distributed around several points of an urban center. Each agent is assigned a distinct task depending on the pollutants that characterize the specific area. The agents interact and exchange messages in order to accomplish the task of air quality monitoring. The whole process is supported by the computational power of ANN that have the ability to estimate missing values and the decisions are taken by a fuzzy rule based model. The Reinforcement learning has been applied to enhance the learning ability of the system. It is an integrated holistic approach.

2   General System's Architecture

The system is based on the Jade multiagent platform (Bellifemine F. et al., 2007). Its main advantage is that it can operate under any environment and operation system regardless the processors and the network type. This is due to the fact that it uses Java for the construction of the agents and MySQL in the database mechanism which are both open source environments (under GPL license) and they both allow programming of modular and distributed credible applications of low cost. The system has been divided in five sub-systems with high cohesion level which are assigned distinct functions and as it is shown in the figure1. This architecture enhances the independence and the personal perspective of the comprising agents.

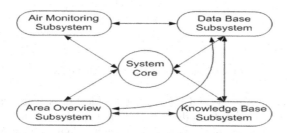

Fig. 1. Overall system's architecture

2.1   General Description of the Subsystems

The Air Monitoring Subsystem (AMS) aims in capturing the instant perception of the environment and in transferring it to the system's core. It comprises of a set of typical air pollution sensors which record the actual measurements and they transfer them to the Sensor Intelligent Agents (SIA) that store them in the Data Base (DB) subsystem acting in a proxy mode. SIA are also responsible for the correct function of the typical sensors, the removal of the improper values, the arrangement of the new measurements' time interval and the update of the administrator for a potential malfunction. The structure of the AMS is seen in figure 2.

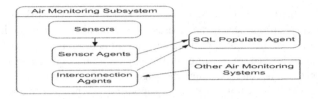

Fig. 2. Air Monitoring Subsystem architecture

The next level comprises of the Interconnection Agents (IA) which allow the import of data from other information systems activated in the area under study. Given the high cost of buying and supporting several air pollution sensors, a mechanism was applied that allows the performance with non dedicated sensors. This is done by connecting the system with sensors of other systems aiming in data transfer. Also the IA allow the interconnection between systems located in different areas.

The Data Base Subsystem performs the management, storage and retrieval of data. Its structure can be seen in the following figure 3.

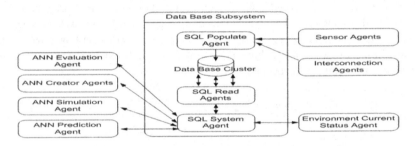

Fig. 3. Data Base Subsystem architecture

The SQL Populate Agents (SQLPA) are the simplest ones. Their basic function is to receive the measurements from the SIA and the IIA, to check their validity and to store them after considering the Data Base schema. The SQL System Agent (SQLSA) connects the Data Base Subsystem with the rest, by receiving requests from the agents that require data. It decides to forward the request to the retrieval agents based on the number of queries and the load of the database. No other agent can access the actual data if the SQLSA do not approve. For the same reason, the logic of data base clustering with multiple data bases communicating with a series of SQL Read agents (SQLRA) was selected. In this way the performance of the system was enhanced. The necessity for this approach will be understood further in the next chapter.

3   The Role of Reinforcement ANN

ANN employing the Reinforcement approach were developed and incorporated into the multi agent system in order to offer reliable estimation of values when the sensors were malfunctioning and the obtained data were unacceptable. Before referring to the ANN the whole process of data retrieval and storage must be clarified.

3.1   Data Manipulation

The SQLRA undertake the task of retrieving data from the data base in order to answer in the request that was diverted by the SQL system agent. Technically, every SQLRA is a compiler which converts the requests from the descriptive language that was developed for the system to SQL query statements, in order to produce the desired data set. The requests are implemented according to the FIPA (Foundation for Intelligent Physical Agents) protocol (FIPA, 2002) and the answers are sent with a message of Inform type (Bellifemine F. et al., 2007). The text in the request message, informs the agent on the data that have to be retrieved and it comprises of a string that describes the data in the following form:

1001d*dd*01;01h*Kor_SO2*01;
24h*mar_pm10*04;01f*pen_tair*03;11p*oin_o3*02;
Command1. Sample of a retrieval command sent to an SQL Read Agent

The above sample command creates a text file, where each line contains 14 elements and corresponds to a record of the data base. This means that if the DB had only correct hourly measurements for the whole 2005, then the output file would have 365*24=8760 lines with totally 8760*14=122640 values.

3.2   Agents Handling ANN Storage

The Knowledge Base subsystem manages the development of the ANN which will be potentially used by the system, in order to forecast the evolution of a phenomenon. The structure of the KBS is shown in the following figure 4.

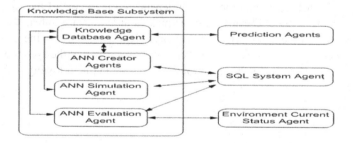

Fig. 4. Knowledge Base Subsystem architecture

The Knowledge Base Agent stores all ANN used by the system, together with their characteristics and the *overall score* of each and decides which one of them will be used every time that a forecast is requested. The *characteristics* are obtained by rules determining in which cases each network can be used (e.g. other ANN is reliable for the winter period, other for the summer). The choice of the proper ANN is done by the application of filters on the data. These filters consider the criteria like: a) The number of cases in which the network had the best estimation b) The average error of each c) The overall score of each one which is estimated during the network validation process by the Evaluation Agent (EVA). The network choice mechanism checks based on the characteristics of the available ANNs, which ones can be used in each specific case and from the emerging candidates it nominates the ANN with the highest score. The validation and assessment of the neural networks is done with the cooperation of the ANN Simulation Agent (ANNSA) and the EVA.

The score for every network is obtained based on a Reinforcement approach in combination with fuzzy logic. According to the Reinforcement technique every ANN receives merit or it is punished by raising or reducing its score, every time that the ANNSA is executed. The amount of penalty or benefit is determined by the employment of a fuzzy algebra model.

It is really important that based on the system's philosophy the ANN which are available in the knowledge base are not executed only when there is a request for data but they are re-adjusted every time that the actual perception of the environment is refreshed. The ANNSA runs all available networks after each new measurement obtained, regardless if this was done after a request or not.

In this point the Knowledge Base is informed on which network had the best performance per estimated value, so that the table with the *characteristics* is updated after the EVA estimates the new scores.

4   Fuzzy Modeling

For each parameter under study four boundary values A, B, C, D were stored in the database. These values were used by two semi trapezoidal fuzzy membership functions (Iliadis, 2007) to define the linguistics (fuzzy sets) *high error, small error*, between the estimated by the ANN and the actual values of each feature (Iliadis et al., 2008). The numbers A and B correspond to error quite close to zero, whereas C and D are related to higher errors and their range depends on the range of the values of the actual feature under study. Even if someone picks a little higher numbers for A,B,C,D the slight change in the fuzzy sets does have any significant effect in the membership values. The design of the fuzzy sets has been done is a way that the system is flexible and adaptable. These membership functions can be seen in the following figure 5.

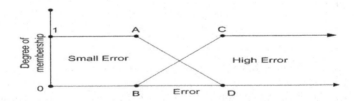

Fig. 5. Semi Trapezoidal definition of small and high error

The degree of membership (DOM) of each ANN to every error fuzzy set is the score based on which it will be rewarded or punished. The DOM to the *small error* linguistic is added to the overall score of each ANN, whereas its DOM to the *high error* fuzzy set is subtracted. By using this flexible approach we avoid to change the overall score of an ANN in a rare case of an extremely bad or extremely good performance and thus the final score reflects the whole performance of the network.

5   Agents for the Development and Assessment of the ANN

The ANN Creator Agents (ANNCA) offer the capability of an automatic creation and initial assessment of an ANN. The system accepts as input by the user, the ever recorded width of the values of each considered parameter and potential restrictions which are also incorporated in the data base, the number of hidden neurons and the number of hidden sub layers, potential transfer functions and the minimum score under which the networks are not considered credible.

The ANNCA follow a trial and error approach by offering a range of values to the above parameters. In this way they automatically construct multiple ANN which cover the set of all potential combinations. This approach leads to the creation of a vast number of ANN in a combinatorial explosion mode and thus computational power is required. As it will be clarified in the results section where the system was executed for a specific case, a huge number of 2200 networks were automatically created. A threshold value of R^2 was used as the minimum criterion of acceptance for a network to be recorded in the knowledge base and it will not be subjected to continuous assessment as it was mentioned before. The Air Overview sub system is responsible for the visualization of the data and of the forecasts and it comprises of two agent types, namely the Area Monitor Agents and the Prediction Agents Its structure is shown in the following figure 6.

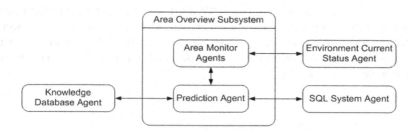

Fig. 6. Structure of the Air Overview subsystem

The Area Monitor Agents (AMA) constitute an awareness area of the system. By the term area the system understands any group of data and forecasts that it can manage and it can be an air pollution measurement station, a hospital service that monitors the evolution of an air pollutant concentration or a wider area represented by dozens measurements stations, or a simple user that needs to monitor specific data in certain locations. It other words it can combine any number of features to construct a logical area of study. The visualization of the situation in a specific area can be done by using fuzzy sets (Linguistics) of the type *"Low"*, *"High"* *"Critical"*. This is shown in figure 7.

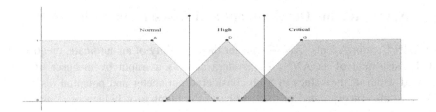

Fig. 7. Fuzzy determination of the situation in a specific area

The Prediction Agent (PREA) is responsible to provide forecasting when an area asks for one. When it receives a request for a forecast it communicates with the Knowledge Database Agent (KDA) and asks the nominated ANN to perform estimation for a specific parameter. Then, the PREA asks the SQL System Agent to provide the data required by the network in order to perform forecast. In the rare case of an ANN that cannot function due to the fact that the SQL system Agent did not find the necessary data, the Prediction agent will contact the KDA again and it will require the next available network. This process repeats in an iterative manner until a proper solution to be found.

6　System Core Agents

The last subsystem is the System Core (SYSCO). The SYSCO handles the general attendance of the system. It provides access rights to all agents that try to connect to the platform and it offers an overall image for the situation of the system and for the area under study. It also informs the administrator for potential malfunctions and connection failures of an agent.

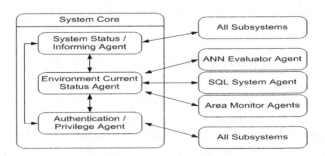

Fig. 8. System Core architecture

The System status Agent monitors if the system performs functional tests and informs the administrator accordingly. It communicates with all of the system's agents to check their functional status. It also keeps the list with the minimum available agents required for the proper operation of each subsystem.

The Environment Current Status Agent is responsible for the temporary storage of the current status of all factors monitored by the system. Its usefulness is in the fact that it can update all agents on the current status of the environment without using the database.

Finally the Authentication/Privilege Agent defines the level of information for which each agent will have access. It does not allow the access in the database of Sensor agents which are not declared in the system and in the database schema and it defines the credentials of the Area Agents when they are not public.

7   Results and Discussion

In order to check the proper function and the performance of the system, a preliminary testing was performed by using actual historic data. The target was the short term forecast and also the current, plus 1,2,3 and 6 hours ahead estimation of Ozone (O_3) in the measurements station "*Athinas*" located in the center of Athens. Actual hourly data records were used coming from 11 surrounding stations (9 of them measuring air pollutants and 2 meteorological data) located in the center of Athens. The data were measured on a 24 hours basis for the whole annual period and only when the measuring stations functioned properly. When a station was malfunctioning the obtained value was defined to be as high as -9999.00 and it was never considered in any analysis. The data come from the website of the Greek ministry of environment and Climate change (Minenv, 2012) and they are historical data from the establishment of each station until year 2010. The stations used and the date of their first operation plus the parameters employed can be seen in the following table 1.

Table 1. Description of measurement stations

Name	CODE	Type	Established	Data
Ag Paraskevis	AGP	Air Pollution	2000	O_3,NO, NO_2
Amarousion	MAR	Air Pollution	1987	O_3,NO,NO_2, CO
Peristeriou	PER	Air Pollution	1990	O_3,NO,NO_2, CO, SO_2
Athinas	ATH	Air Pollution	1984	O_3,NO,NO_2, CO, SO_2
Pathsion	PAT	Air Pollution	1990	O_3,NO,NO_2, CO, SO_2
Aristotelous	ARI	Air Pollution	1994	NO, NO_2
Geoponikis	GEO	Air Pollution	1984	O_3,NO,NO_2, CO, SO_2
Peiraias	PIR	Air Pollution	1984	O_3,NO,NO_2, CO, SO_2
N Smyrnis	SMY	Air Pollution	1984	O_3,NO,NO_2, CO, SO_2
Pendelis	PEN	Meteorological	1999	Temp, Wspeed, SunTime, Wdirection, Radiation, RH
Thiseiou	THI	Meteorological	1985	Temp, Wspeed, SunTime, Wdirection, Illumin, RH

The Ozone for the "*Athinas*" station was used as output whereas the data of the rest 10 stations except of their Ozone values were used as input. The assessment process of the system included the following steps:

A) Creation of the Database schema according to the data management system's needs. The following table 2 presents the structure of the database and the number of records.

Table 2. Description of the Database

Table	Records	Type	Size
array01h	236,754	InnoDB	82.6 MB
day_names	7	InnoDB	16.0 KB
factors01h	51	InnoDB	32.0 KB
measurements01h	237,168	InnoDB	15.0 MB
values01h	12,196,690	InnoDB	1.2 GB

InnoDB engine was selected not only because it is the default storage engine for MySQL, as of MySQL 5.5, providing a transaction safe environment with rollback and crash-recovery capabilities in order to protect data but as at the same time provides high speed data access (MySQL, 2012).

B) Creation of an Interconnection Agent which aimed in reading the text files in order to send the data in the proper form to the Populate agent that would store them in the database.

C) Creation of an ANN Creator Agent which automatically implemented the required networks. It was asked from the ANNCA to construct all potential networks' combinations, using data at least from one meteorological station and two air pollution ones, until all combinations between the stations described in table 1 were done. The above combinations produced 741 ANN. For the best control of the system it was asked by each network to train and test itself using a variable number of hidden neurons, providing the system with three values (0.5, 0.75, 1) which would be multiplied with the number of the proposed hidden neurons in order to produce their actual number. The ANN Creator Agent produced automatically the data files used for training and testing. These files contained values related to each factor from the first year of the station's establishment until 2009. Each of the 741 ANN asked the SQL System Agent the creation of a measurements' file using the right format. To show the computational complexity we must add that totally there were 10,132,450,485 SQL queries performed and the whole process used a cluster with 8 dataset servers (based on average pc configuration). For all of the process it took less than 7 days. This was actually the reason for choosing the solution of clustering in the data base. After the implementation of the ANN the training and their registration in the Knowledge Database Agent followed.

D) The last step in the system's testing was the use of the 2010 data for the total simulation of its function based on the knowledge base created in the previous step.

A Simulation Agent was created which ran the system with hourly values from the 01/01/2010 until 31st of December 2010. Also the ANNSA and the ANN Evaluation Agent were used. Totally the system was executed 8,760 times (365*24) and managed to obtain results for the 8,738 of them. For the 22 cases there was a lack of data due to simultaneous interrupt of function of both meteorological stations. In 167 cases it was not possible to use the ANN Evaluation Agent due to the fact that for these hours the "Athinas" station was not functioning and thus we could not estimate the ANN error. As it was mentioned before in this chapter the ANN had 5 output values whereas the error during simulation and testing of the system was estimated only for the current value. When the system will be fully executed the error will be estimated separately for each output.

The following table 3 presents the results for the simulation of the year 2010.

Table 3. Results for the simulation of the year 2010

Output	R	R^2	Mean Square Error (MSE) (Root MSE)
Current estimation	0,9146	0,8365	139,712 (11.820)
1 hour forecast	0,9057	0,8203	154,735 (12.439)
2 hours forecast	0,8909	0,7937	173,413 (13.169)
3 hours forecast	0,8754	0,7663	193,302 (13.903)
6 hours forecast	0,8441	0,7125	230,007 (15,166)

The system in this phase covers a wide range of requirements and potentials, whereas in its direct extensions it will have Agents that will materialize mechanisms of retraining the ANN when they appear to have a reduced performance and a continuous reduction in their overall score. In this way the ANN will be automatically adjustable without the interference of the user in systemic changes (e.g. close a specific area to automobiles) which change the weight of the used parameters in an ANN. Additionally a complementary decision support subsystem could be implemented that will use rules and it will process the available data to impose actions that will improve the situation in the environment. This of course (in an iterative manner) will readjust the system under the new improved environment.

References

Triantafyllou, A.G., Skordas, S., Diamantopoulos, C., Topalis, E.: The New Dynamic Air Quality Information System and its Application in West Macedonia, Hellas. In: 4th Environmental Conference of Macedonia, March 18-20 (2011)

Abdul-Wahab, S.A., Al-Alawi, S.: Assessment and prediction of tropospheric ozone concentration levels using artificial neural networks. Environmental Modelling & Software 17(3), 219–228 (2002)

Aratos, Ozone Index Saves Earth By Aratos Technologies S.A. (2012), http://itunes.apple.com/us/app/ozone-index-saves-earth/id468073482?mt=8 (valid at April 15, 2012)

Brunekreef, B., Holgate, S.: Air pollution and health. The Lancet 360(9341) (2002)

Council Directive 90/313/EEC of 7 June 1990 on the freedom of access to information on the environment. Official Journal L 158, 0056–0058 (1990)

Council Directive 92/72/EEC of 21 September 1992 on air pollution by ozone. Official Journal L 297, 0001–0007 (1992)

Council Directive 96/62/EC of 27 September 1996 on ambient air quality assessment and management. Official Journal L 296, 0055–0063 (1996)

Council Directive COM, 613 final, directive of the European parliament and of the council relating to ozone in ambient air (2000)

diMobile, Ozone Position By diMobile (2011), http://itunes.apple.com/ke/app/ozone-position/id439508448?mt=8 (valid at April 15, 2012)

Estrin, D., Culler, D., Pister, K., Sukhatme, G.: Instrumenting the Physical World with Pervasive Networks. IEEE Pervasive Computing 1(1), 56–69 (2002)

Bellifemine, F., Caire, G., Greenwood, D.: Developing Multi-Agent Systems with JADE. John Wiley & Sons, New York (2007)

FIPA, FIPA ACL Message Structure Specification (2002), http://www.fipa.org/specs/fipa00061/index.html (valid at April 15, 2012)

Iliadis, L., Spartalis, S., Tachos, S.: Application of fuzzy T-norms towards Artificial Neural Networks' evaluation: A case from wood industry. Journal Information Sciences (Informatics and Computer Science Intelligent Systems Applications) 178(20), 3828–3839 (2008)

Iliadis, L., Spartalis, S., Paschalidou, A., Kassomenos: Artificial Neural Network Modeling of the surface Ozone concentration. International Journal of Computational and Applied Mathematics 2(2), 125–138 (2007)

Iliadis, L.: Intelligent Information systems and Applications in risk estimation. Stamoulis A. Publishing, Thessaloniki (2007)

Iliadis, L., Papaleonidas, A.: Intelligent Agents Networks Employing Hybrid Reasoning: Application in Air Quality Monitoring and Improvement. In: Palmer-Brown, D., Draganova, C., Pimenidis, E., Mouratidis, H. (eds.) EANN 2009. CCIS, vol. 43, pp. 1–16. Springer, Heidelberg (2009)

Lisabeth, L.D., Escobar, J.D., Dvonch, J.T., Sánchez, B.N., Majersik, J.J., Brown, D.L., Smith, M.A., Morgenstern, L.B.: Ambient air pollution and risk for ischemic stroke and transient ischemic attack. Annals of Neurology (2008), doi:10.1002/ana.21403

Ministry of Environment, Energy & Climate Change. Air Pollution Measurements (2012), http://www.ypeka.gr/Default.aspx?tabid=495&language=el-GR (valid at April 15, 2012)

MySQL, Storage Engines, ch 14 (2012), http://dev.mysql.com/doc/refman/5.0/en/storage-engines.html (valid at April 15, 2012)

Nafstad, P., Håheim, L.L., Oftedal, B., Gram, F., Holme, I., Hjermann, I., Leren, P.: Lung cancer and air pollution: a 27 year follow up of 16 209 Norwegian men. Thorax 58, 1071–1076 (2003)

Paschalidou, A., Iliadis, L., Kassomenos, P., Bezirtzoglou, C.: Neural Modeling of the Tropospheric Ozone concentrations in an Urban Site. In: Proceedings of the 10th International Conference Engineering Applications of Neural Networks, pp. 436–445 (2007)

Rule-Based Behavior Prediction
of Opponent Agents Using Robocup 3D Soccer
Simulation League Logfiles

Asma Sanam Larik and Sajjad Haider

Artificial Intelligence Lab., Faculty of Computer Science,
Institute of Business Administration, Garden Road, Karachi-74400, Pakistan
asma.sanam@khi.iba.edu.pk, sahaider@iba.edu.pk

Abstract. Opponent modeling in games deals with analyzing opponents' behavior and devising a winning strategy. In this paper we present an approach to model low level behavior of individual agents using Robocup Soccer Simulation 3D environment. In 2D League, the primitive actions of agents such as Kick, Turn and Dash are known and high level behaviors are derived using these low level behaviors. In 3D League, however, the problem is complex as actions are to be inferred by observing the game. Our approach, thus, serves as a middle tier in which we learn agent behavior by means of manual data tagging by an expert and then use the rules generated by the PART algorithm to predict opponent behavior. A parser has been written for extracting data from 3D logfiles, thus making our approach generalized. Experimental results on around 6000 records of 3D league matches show very promising results.

Keywords: Robocup Soccer, PART algorithm, opponent modeling, machine learning.

1 Introduction

Behavior mining[1] of agents can lead us to some insight into how agents interact in a dynamic environment. RoboCup Soccer [2] is a research initiative that uses the game of soccer to advance research within artificial intelligence, cognitive robotics, multi-agent systems and other related fields. The RoboCup Soccer competition comprises of several robot leagues namely humanoid league, standard platform league, small-size league, middle-size league and simulation league. The simulation league consists of both 2D and 3D agents and it aims to simulate a robot soccer match. Matches are played in a client/ server environment. The platform provides a test bed to analyze behavior of agents and teams.

During a soccer game, we want to predict our opponent's strategy and then adjust our own accordingly. If we are able to analyze the logs of past games played by teams and distinguish their game play based on their behaviors then it can aid us in devising our own team strategy. Much work has been done within the Simulation 2D league

L. Iliadis et al. (Eds.): AIAI 2012, IFIP AICT 381, pp. 285–295, 2012.
© IFIP International Federation for Information Processing 2012

due to the simplicity of available actions such as Kick, Turn and Dash. In Simulation 3D league, however, such primitive actions are not available, thus it becomes difficult to perceive high level behavior. This paper presents a novel approach that aims to predict the behavior of agents in Simulation 3D league. Our approach is the first attempt to connect low level primitive actions to high level behaviors; thus enabling us to discover strategic activity from raw data. Our goal is to develop a mechanism that discovers key behavior of an agent and translates multi-agent action sequences and observations into a rule based representation. The proposed approach is divided into three phases: data preparation, rule generation and prediction. The first phase comprises of extracting raw data from 3D simulation league log files. We have created a tool that performs post-hoc offline analysis of the past matches and extracts agent and ball locations. We also derive distances of the ball/agent and their velocity. The extracted data is manually tagged for behavior identification by an expert by pressing appropriate key as he/she observers the game. In this paper we have restricted our focus to learning the skills of an attacker. The skill set includes two actions: Approach Ball and Dribble. The extracted and (manually) tagged data is then utilized in the second phase for learning rules. The rules learn particular behaviors depending upon the features used in the training dataset. Once rules have been learnt, these rules are applied to predict the behavior of opponent player in the third phase.

The rest of the paper is organized as follows. Section 2 provides a brief overview of RoboCup Soccer Simulation League. Section 3 provides a literature survey of the past efforts in opponent modeling. The proposed approach is explained in detail in Section 4 while Section 5 outlines the design of experiment and results. Finally, Section 6 concludes the paper and provides future research directions.

2   Robocup Soccer Simulation Leagues

RoboCup Soccer Simulation League provides a multi-agent system in which teams of autonomous agents play soccer in a simulated environment. All agents can move and act independently as long as they comply with the soccer rules. Agents can also have limited communication among each other. The league is further classified into 2D and 3D leagues. In Simulation 2D league, shown in Figure 1a, two teams of eleven autonomous wheeled robots play soccer in a two-dimensional virtual soccer stadium. The agents can perform low level actions such as turn, kick or dash to influence the environment. Simulation 3D league, on the other hand, adds the complexity of locomotion and localization and hence a lot of research has been focused on handling these issues. In Simulation 3D, shown in Figure 1b, two teams of 9 humanoid robots play soccer in a simulated environment. Unlike 2D, here a robot can only perceive objects that are in their own field of vision.

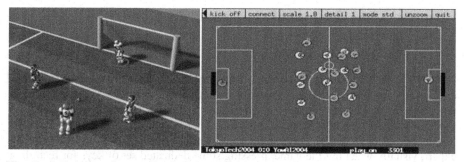

Fig. 1. a) Robocup Soccer Simulation   **Fig. 1.** b) Robocup Soccer Simulation Field 2D
Field 3D

3   Related Work

For our work we have studied papers contributed in three domains namely data extraction from log files, opponent modeling and papers on Robocup simulated coaching competitions. Although the area of data extraction in 2D RoboCup simulations has been extensively investigated, there is no work reported for 3D simulation environments. Within the 2D data extraction domain, T. Nakashima and H.Ishibuchi[3] have tried to mimic dribble trajectory of a player by taking snap shot of log files in which the player is actually performing the dribble action. The dribble intervals extracted create a set of training patterns for the neural network. T.Nakashima et al [4], have used offline learning for pass prediction behavior and predicted the new position of opponent player by training a neural network. Both the approaches are related to ours however we are using rule based learning for behavior prediction while they create neural network for this task. Within the opponent modeling domain, Ball and Wyeth[6] used Robocup small sized league to predict opponent behavior. Agapito et al.[7] presented OMBO, an opponent modeling approach based on observations. They first build a classifier to label opponent actions. In the next phase, they placed a dummy player in the field for the purpose of recording opponent actions and finally they predicted actions of opponent based on the training dataset. For the coaching competition, many simulated coaches have been presented; Fathzadeh et al.[8], Agapito et al.[9], Peter Stone at al[10][11] being some of them. The focus of these coaches is to learn normal base patterns of team plays and then predict the strategy with which the team is playing. The coach that recognizes more patterns wins. For this purpose they are using different data structures and pattern identification mechanisms. The coach advice is given in a particular language namely Clang[12].

4   Proposed Approach

This section presents our approach to discover and model low level behavior of opponent soccer agents. The approach is divided into the following three distinct phases namely:

- Extraction of information from 3D log files
- Rules learning
- Agent behavior prediction

In the first phase a parser is written that maintains a queue data structure that serves as a repository for storing noiseless contextual information. A log file is passed through the parser that extracts player's position and ball's position and populates it in separate lists. In the next step of this phase, behavior features, missing from the logfile, are incorporated by an expert. A simple application is written that helps an expert, visually watching the game, pressing some dedicated set of keys for recording the agent's behavior. This is essential for rules learning since the low level behavior (such as kicking, dribbling, etc.) is not recorded in the log files and is evident only to the individual watching the game. In the second phase we use the extracted (and tagged) data for learning a set of rules. Finally, in the third phase we parse an arbitrary log file (not in the training data set) and start predicting the agent behavior and verify visually whether the rules learnt are correct or not. For rule learning we have used the PART[13] algorithm. All phases are described in detail in the subsections below.

4.1   Extraction of Data from Logfiles

In Simulation 3D league, the server continuously records game state in a logfile during a match.   The data from these logfiles can be extracted and analyzed to construct a model of opponent.   The first phase focuses on data extraction from logfiles.

4.1.1   Data Format in Logfiles

Messages from server to agent and vice versa use S-Expression. The basic idea of S-Expression is that they are simple and are best known for their use in the Lisp family of programming languages. An advantage of using S-exp over other data formats is that it provides an easy to parse and compact syntax that to some extent is also readable by human for debugging purpose. Figure 2 shows an excerpt from a 3D logfile.

```
((FieldLength 21)(FieldWidth 14)(FieldHeight 40)(GoalWidth 2.1)(GoalDepth 0.6)(GoalHeight 0.8)(BorderSize 0)
((time 0))(RDS 0 1)((nd(nd))(nd(nd))(nd(nd StaticMesh (setVisible 1)))(nd(nd StaticMesh (setVisible 1)))(nd(
((FieldLength 21)(FieldWidth 14)(FieldHeight 40)(GoalWidth 2.1)(GoalDepth 0.6)(GoalHeight 0.8)(BorderSize 0)
((time 0))(RDS 0 1)((nd(nd))(nd(nd))(nd(nd StaticMesh))(nd(nd StaticMesh))(nd(nd(nd StaticMesh))(nd)(nd)(nd)
((time 0))(RDS 0 1)((nd(nd))(nd(nd))(nd(nd StaticMesh))(nd(nd StaticMesh))(nd(nd(nd StaticMesh))(nd)(nd)(nd)
```

Fig. 2. Sample Logfile Header

The messages recorded in the log are environment information messages, game state messages, Ruby scene graph header and scene graph contents. The scene graph is a structure that arranges logical and spatial representation of a graphical scene. In Simspark [14], the scene graph is a tree with a root node defined to be at the origin. Each node has one or more children. The nodes are further classified as base node, transformation nodes, geometry nodes, static mesh nodes, light nodes, etc. A header expression is sent initially that contains information regarding environment, game state and scene graph contents. These variables are stored in the form of nodes with opening and closing brackets distinguishing among them. The initial header contains information messages including field length, field width, field height, goal width, goal depth, goal height, border size, free kick distance, wait before kickoff time, radium of the agent, ball radius, ball mass, , play mode (such as goal kick, play on, side kick, etc.), time, score, , SLT (single linear transform), light nodes, TRF (transformation matrices), etc. A new header is sent whenever the scene changes. For instance, loading of a player, removal of a player from field, etc., result in the transmission of a new header. However, if only the player or ball changes its position (and orientation) then no header is resent. Instead, only minor modifications in the nodes are represented by another s-expression that contains the time stamp and the change information that has been modified as evident in Figure 2. RDS attribute indicates that the scene has changed partially and only few nodes have changed. As a result, we traverse the entire S-Expression to find the node that has changed and update our values accordingly.

4.1.2 Parser for Data Extraction
In order to analyze the log data we first need a parser that can traverse each and every S-Expression and identify nodes for ball and players. The main challenges in writing a parser are: Extraction of node information from the header, Computing position of players and ball from transformation nodes, Extraction of updated nodes from the timestamp information

4.1.2.1 Extraction of Node Information from Header. To extract required information, we traverse the entire S-Expression, break the header information into tokens information and store tokens in a list T. During this tokenization process, if we encounter any node "**nd**" we store its index in the list **I**. After the completion of tokenization process we populate the list of nodes **N** and players **P** respectively using the indexes stored in the list **I**.

4.1.2.2 Computing Positions of Players and Ball. The major challenge in extracting position information lies in the fact that in 3D this data is stored in the form of transformation matrix specified by token "**TRF**". A transformation matrix **T** is a 4*4 matrix defined as:

$$T = \begin{bmatrix} nx & ox & ax & px \\ ny & oy & ay & py \\ nz & oz & az & pz \\ 0 & 0 & 0 & 1 \end{bmatrix}$$

Where: $\overleftarrow{n}, \overleftarrow{o}$ and $\overleftarrow{a}$ vectors represent orientation information,
$\overleftarrow{p}$ represents the position information in x, y and z space.

In a Scene Graph there are a number of frames so a particular node can have many transformation matrices. To obtain correct position information we need to multiply all the specified 4*4 transformation matrices as demonstrated using the following example. Let **"nd"** be a ball node containing the image information **"soccerball.obj"** and let **"TRF"** be its transformations and **SLT** be the single linear transformation. We get the the following information from the logfile:

(nd TRF (SLT 1 0 0 0 0 1 0 0 0 0 1 0 0 0 0 1))(nd TRF (SLT 1 0 0 0 0 1 0 0 0 0 1 0 0 0 0.0402764 1)(nd StaticMesh (setVisible 1) (load models/soccerball.obj)

In the above expression, we have two matrices T_1 and T_2:

$$T_1 = \begin{bmatrix} 1 & 0 & 0 & 0 \\ 0 & 1 & 0 & 0 \\ 0 & 0 & 1 & 0 \\ 0 & 0 & 0 & 1 \end{bmatrix} \qquad T_2 = \begin{bmatrix} 1 & 0 & 0 & 0 \\ 0 & 1 & 0 & 0 \\ 0 & 0 & 1 & 0.0402764 \\ 0 & 0 & 0 & 1 \end{bmatrix}$$

When we multiply the two matrices we get the desired matrix and from that we can extract the position vector $\vec{p}$ with $(p_x, p_y, p_z) = (0, 0, 0.0402764)$.

4.1.2.3 Extraction of Updated Nodes from the Timestamp Information Sent. When we receive a timestamp and RDS node, this means that there has been a partial change in the scene and position/orientation of one or more nodes. In the S-Expression, the node structure is preserved and only the changed nodes contain the new transformation matrix. Thus, during the update process all node indexes remain the same and the node which contains some new **"SLT"** information is updated. A formal extraction process of the current and the previous subsection is described in detail in Table1.

Table 1. Algorithm

<u>**Data Extraction Algorithm**</u>
　Read a new line from .logfile
　While (not end of file)
　　Begin
　　　Parser algorithm
　　　Read Another line
　　End
　Write to csv file

<u>**Parser Algortihm:**</u>

Let $T = \{t_1, t_2, \ldots, t_n\}$ be a list of tokens

Let $P = \{p_1, p_2, \ldots p_n\}$ be a list of players in the game
Let $I = \{i_1, i_2, \ldots i_n\}$ be a list of node indexes
Let $N = \{ n_1, n_2, \ldots n_n \}$ be list of nodes
　where each node n_i contains the following attributes:
　　$n_i = \{$ **pos** = node position, // 3*1 vector denoting (x,y,z) coordinates
　　　name = node name,
　　　id = node id,
　　　mat = denoting 4*4 transformation matrix,
　　　type = type of node e.g SLT, SMN, StaticMesh, Light etc
　　　child = list of child nodes
　　　$\}$

Begin
1. **Initialization Process:** Set $T = P = N =$ Null denoting Empty Lists
2. **Tokenization of file**
    Foreach (element e in file separated by space)
        Begin
            T. Add(e) // Add to tokens list
            If (e is a node) then
                I. Add(e) // Add to node indexes list
        End
3. If (RSG= true) // The scene has changed totally
    Begin

        Foreach (element e in T)
        Begin
            If (e is a node) then
                Create n_i with attributes initialized
            If (e is a SLT node) then
                Compute transformation matrix of node
                N.Add(n_i)
            If (e is BallNode ball node) then
                Preserve its index in I
            If (e is a PlayerNode) then
                Preserve its index in I
        End
        Add these nodes in I into player list P
        Parse first player to identify team and player Id
        Foreach (player p_i in P)
            Compute their positions in the field
        Compute the position of ball in the field
    End
4. if (RDS= true) // the scene has partially changed
    Begin
        Foreach (node n_i in N)
            Update postion of n_i
        Foreach (index i in I)
            Recompute node indexes
        Foreach (player p_i in P)
            Recompute player positions in the field
        Recompute the position of ball in the field
    End

4.1.3 Preprocessing of Extracted Data and Derivation of Attributes

Given a logfile, our parser is able to extract positions of all the eighteen players of both the teams and the ball position. This extracted data is in the form of a csv file. Although the parser extracts data for all the players but for the scope of this paper we have limited our focus to the generation of data related to only one player, that is, opponent's attacker. We have used Roboviz [15] application for logfile generation with a single player, that is, the attacker taking the ball towards the goal. The parser extracts the following features:

- Timestamp
- Position of the opponent attacker
- Position of the ball

In addition to the extracted data, we derive the following attributes from the raw data:

- Distance travelled by ball (DTB)
- Distance travelled by player (DTP)
- Distance between player and ball at time t (DBPBat t)
- Distance between player and ball at time t-1 (DBPBat t-1)

4.1.4 Behavior Identification

As mentioned previously, logfiles do not provide us any details about the primitive actions executed by a player. In 3D we only learn about specific motion of hinge joints but the information is not enough to categorize this data into behaviors such as kick, approach ball, dribble, clear ball, etc. This information needs to be inferred by observing an agent's behavior. The information is also needed to label each record extracted in the previous step and to learn a classifier. To obtain this data we have written a piece of code that helps an expert, observing a game, to tell the behavior that he/she is observing. By manually pressing a key, an specific behavior performed by the robot agent a particular timestamp can be recorded. For the scope of this work, we have limited our attention to two behaviors: Dribble and Approach Ball.

- *Dribble*: is an event in which the ball and player both are moving and they are at a considerably shorter distance from each other.
- *Approach Ball*: In this event, the ball is stationary and at a farther distance, and the player is moving towards the ball.

Thus, we tag/label data with the help of expert. This behavior information is then merged with the previous features. The combined data can then used to generate rules in the next step.

4.2   Rule Generation

The csv file generated at the end of previous phase would be utilized in creation of rule base for agent behavior identification. The stronger the rule base, the greater would be the accuracy with which we would be able to predict agent behavior. Numerous algorithms that efficiently search large databases for rules have previously been developed. We use standard software named WEKA[16] and utilize PART[13] algorithm for rule generation. The reason for selection of this particular algorithm is that it generates rules in the form of decision tree thus making them quite understandable.

4.3   Agent Behavior Prediction

Once the rules have been learned, we can use them to predict the behavior of a player. To test the accuracy of the proposed approach, we record the logfile of a new game using RoboViz[15]. We use the parser to extract raw data. The behavior information would be tagged manually by visual verification. This combined data would serve as a test data.

5   Design of Experiment and Results

In our experiments, we placed a goalie of our team (Karachi Koalas) in the field whose sole purpose is to help us in recording the data. We ran attacker from another team whose job is to take the ball towards our goal. It must be mentioned that our goalie does not interfere in this process as it is primarily standing there as an observer. An expert watches the game and records key presses referring to events such as **"D"** for Dribble and **"A"** for Approach Ball. If the expert feels that some erroneous activity is being performed, for example, the agent has fallen, the agent is unable to locate the ball, etc. then the expert presses "E" to eliminate these erroneous episodes during data preprocessing phase. In a similar manner, we ran ten different games with five teams and recorded the attacker behavior of each and every team. In the next step, the logfiles were processed by the parser. The parser extracted the timestamp, opponent attacker's position and ball's position. The behavior information is merged with this data. The behavior became the class variable that we are going to predict. To obtain better rules, a pre-processing of data was performed as follows:

- Erroneous episodes such as getting up from back, falling, back walk recorded in the log and behavior files were removed
- All the pre-kickoff records were removed.
- To reduce the number of records, instances in which the play was at a stop or a player was idle were also eliminated.
- It was also observed that an event takes multiple cycles to execute thus the transactions were divided by 10. This made sure that very similar records do not disturb the correct identification of event.

We used feature selection algorithms provided by Weka[16] and after several experiments, the following attributes were retained: ballX, ballY, playerX, playerY, DBPBat t (Distance between player and ball at time t), DTP (Distance travelled by Player) and DTB(Distance travelled by Ball) respectively. Table 2 shows some tuples from the csv file.

Table 2. Demonstration of some training instances

Time	ballX	ballY	playerX	player	DBPBatt	DTB	DTP	Behavior
6.26	5.369	1.505	0.687	0.300	4.834	0.002	0.009	ApproachBall
6.46	5.425	1.488	0.843	0.327	4.727	0.020	0.033	ApproachBall
11.66	5.577	1.520	4.009	0.890	1.690	0.001	0.003	Dribble
11.86	5.580	1.492	4.139	0.922	1.550	0.006	0.033	Dribble

Next, we used the PART[13] algorithm for rules generation from approx 6000 tuples pertaining to ten games played by five different teams. Its configuration was set to a confidence factor of 0.25 and support of 30 rules after successive experimentations by varying these values. Some of the learned rules are shown below:

Rule 1:
  If (DBPBat t > 1.615657 AND   ballX > 5.354948 AND
      ballX <= 5.757669) then ApproachBall
Rule 2:
  If (DTP > 0.000391 AND   ballY <= 0.148716 AND
      ballY > -0.859947 AND ballX > 3.953067 AND
      playerY > -1.286594 AND playerY <= -0.0246) then DribbleTowardsGoal

The rules proved to be 84% accurate when we used training data. The results obtained on training data are described in Table 3. Next we used training instances for a single game of approx 1000 records and obtained an accuracy of 80.4%. The results on test data are shown in Table 4.

Table 3. Results on training dataset

Behavior (class)	Instances (N)	Correctly classified (C)	Incorrectly classified (I)	Precision (P)	Recall (R)	F-measure (F)
Approach Ball	2456	1936	520	0.78	0.792	0.78
Dribble	3966	3459	507	0.87	0.869	0.869

Table 4. Results on test dataset

Behavior (class)	Instances (N)	Correctly classified (C)	Incorrectly classified (I)	Precision (P)	Recall (R)	F-measure (F)
Approach Ball	640	515	73	0.80	0.87	0.78
Dribble	423	298	125	0.70	0.80	0.746

6   Conclusion and Future Work

The paper presented an approach for rule based behavior classification of opponent agents in Robocup Soccer Simulation 3D environment. The behaviors classified are pertinent to the skills of the attacker namely Approach Ball and Dribble the ball towards goal. The proposed approach used a parser for positional data extraction and expert guidance for behavior identification and rule generation. The rules thus generated aided in predicting the behavior of an agent. The approach is first of its kind as it creates an opponent model based on 3D soccer simulation logfiles. The approach, being tested on approximately 6000 records, seems very promising and has generated good results. In the future we wish to extend this approach to classify if a team is playing in a defensive or an offensive manner. Similarly, it can also be used to distinguish between strong and weak teams. Furthermore, we can also learn the skills of goal keeper, defender and supporter using the same technique. In addition, currently the generated rules have crisp boundaries that somehow restrict the proposed approach. We aim to fuzzify the rules so that they become generalized and better readable by humans.

References

[1] Symeonidis, A., Mitkas, P.: A Methodology for Predicting Agent Behavior by the Use of Data Mining Techniques. In: Gorodetsky, V., Liu, J., Skormin, V.A. (eds.) AIS-ADM 2005. LNCS (LNAI), vol. 3505, pp. 161–174. Springer, Heidelberg (2005)

[2] Robocup official website, http://www.robocup.org

[3] Nakashima, T., Ishibuchi, H.: Mimicking Dribble Trajectories by Neural Networks for RoboCup Soccer Simulation. In: IEEE 22nd International Symposium on Intelligent Control, ISIC 2007, pp. 658–663. IEEE (2007)

[4] Nakashima, T., Uenishi, T., Narimoto, Y.: Off-line learning of soccer formations from game logs. In: World Automation Congress (WAC), pp. 1–6. IEEE (2010)

[5] Faria, B.M., Reis, L.P., Lau, N., Castillo, G.: Machine Learning algorithms applied to the classification of robotic soccer formations and opponent teams. In: 2010 IEEE Conference on Cybernetics and Intelligent Systems (CIS), pp. 344–349. IEEE (2010)

[6] Ball, D., Wyeth, G.: Classifying an opponents behavior in robot soccer. In: Proceedings of the Australasian Conference on Robotics and Automation, Australia (2003)

[7] Ledezma, A., Aler, R., Sanchis, A., Borrajo, D.: OMBO: An opponent modeling approach. AI Communications 22(1), 21–35 (2009)

[8] Fathzadeh, R., Mokhtari, V., Kangavari, M.R.: Opponent Provocation and Behavior Classification: A Machine Learning Approach. In: Visser, U., Ribeiro, F., Ohashi, T., Dellaert, F. (eds.) RoboCup 2007. LNCS (LNAI), vol. 5001, pp. 540–547. Springer, Heidelberg (2008)

[9] Iglesias, J.A., Ledezma, A., Sanchis, A.: CAOS Coach 2006 Simulation Team: An Opponent Modelling Approach. Computing and Informatics Journal 28(1), 57–80 (2009)

[10] Kuhlmann, G., Knox, W.B., Stone, P.: Know thine enemy: A champion RoboCup coach agent. In: Proceedings of the Twenty-First National Conference on Artificial Intelligence, pp. 1463–1468 (2006)

[11] Kuhlmann, G., Stone, P., Lallinger, J.: The UT Austin Villa 2003 Champion Simulator Coach: A Machine Learning Approach. In: Nardi, D., Riedmiller, M., Sammut, C., Santos-Victor, J. (eds.) RoboCup 2004. LNCS (LNAI), vol. 3276, pp. 636–644. Springer, Heidelberg (2005)

[12] Robocup Simulation Coach Competition, http://www.cs.utexas.edu/~ml/wasp/robocup-clang.html

[13] Eibe, F., Witten, I.H.: Generating Accurate Rule Sets without Global Optimization. In: Proceedings of the 15th International Conference on Machine Learning, San Francisco, USA (1998)

[14] Simspark, http://simspark.sourceforge.net/wiki/index.php/Main_Page

[15] RoboViz official webiste, https://sites.google.com/site/umroboviz/usage/startup

[16] Holmes, G., Donkin, A., Witten, I.H.: WEKA: A Machine Learning Workbench. In: Proceedings of Second Australia and New Zealand Conference on Intelligent Information Systems, Brisbane, Australia (1994)

An Ontology-Based Model for Student Representation in Intelligent Tutoring Systems for Distance Learning

Ioannis Panagiotopoulos, Aikaterini Kalou,
Christos Pierrakeas, and Achilles Kameas

Educational Content, Methodology and Technology Laboratory (e-CoMeT Lab.)
Hellenic Open University, Patras, Greece
{gpanagiotopoulos,kalou,pierrakeas,kameas}@ecomet.eap.gr

Abstract. An Intelligent Tutoring System (ITS) offers personalized education to each student in accordance with his/her learning preferences and his/her background. One of the most fundamental components of an ITS is the student model, that contains all the information about a student such as demographic information, learning style and academic performance. This information enables the system to be fully adapted to the student. Our research work intends to propose a student model and enhance it with semantics by developing (or via) an ontology in order to be exploitable effectively within an ITS, for example as a domain-independent vocabulary for the communication between intelligent agents. The ontology schema consists of two main taxonomies: (a) student's academic information and (b) student's personal information. The characteristics of the student that have been included in the student model ontology were derived from an empirical study on a sample of students.

Keywords: Ontology, intelligent tutoring systems, stereotypes, personalized learning, student model.

1 Introduction

Intelligent Tutoring Systems (ITSs) are complex systems that can be adapted easily to each student's cognitive features, characteristics and learning progress [1]. These systems use a large amount of educational knowledge and many of them also employ pedagogical methodologies.

Traditional ITSs consist of the following four modules: (a) the domain module, which contains all the knowledge (educational content), (b) the student model, (c) the pedagogical module, which contains all the information relevant to the various pedagogical decisions and (d) the user interface which enables communication between the user and the system [2]. Especially, in multi-agent architectures the communication between these modules is achieved through the communication of intelligent agents assigned to each module. So, for example a learner model agent is responsible for answering queries from other agents about learner's information, which information is included in the student model.

L. Iliadis et al. (Eds.): AIAI 2012, IFIP AICT 381, pp. 296–305, 2012.

The characteristics and progress of the students are captured in the student model. This is achieved by using AI techniques to represent pedagogical decisions, domain knowledge, and personal information about the student [3]. Since there are many candidate characteristics of a user that can be included in the student model, the selection of the appropriate characteristics is a very challenging and significant procedure. Consequently, we have to obtain a tradeoff of the completeness of the model so that the systems can be adapted successfully and their performance is not affected. Some of the basic student characteristics maintained in the model are: (a) demographic information, (b) knowledge of the teaching domain, (c) background and interests, (d) learning styles and interaction preferences and (e) learning goals and specificities that can affect the learning procedure. From the above characteristics (a) is a basic feature when describing a user, while (b) and (c) are essential parameters in every educational process. Characteristics (d) and (e) are considered as crucial for a user-centered intelligent tutoring system in order to deliver the appropriate educational material, according to the individual's needs.

In this paper, we propose to use an ontology to represent all the above-mentioned information in a student model. In addition, the model includes the learner characteristics, as they were identified after an empirical study conducted on a sample of students of the Hellenic Open University.

The rest of the paper is structured in the following way. In section 2, we discuss the student modeling approaches based on ontologies. In Section 3, we elaborate on the proposed student model, in terms of the learning styles, the modeling approaches and the basic characteristics of the student. In Section 4, we outline the ontology that represents the proposed student model. Finally, in Section 5, we discuss future work and summarize our conclusions.

2   Related Work

Ontologies have been widely used for student modeling mainly for two reasons: (a) ontologies support the formal representation of abstract concepts and properties in a way that they can be reused by many tasks or extended if needed and (b) they enable the extraction of new knowledge by applying inference mechanisms (e.g. reasoner) on the information presented in the ontology.

Therefore, a plethora of ontology-based approaches for student modeling have been proposed in the field of ITSs. Paneva [4] proposes an ontology-based student model for eLearning systems that adopts technologies and standards from the Semantic Web. Chen et al. [5] describe a domain-independent student model for a multi-agent intelligent educational system (IES). In [6], they propose a student model ontology for an e-Learning system. The ontology is based on the representation of prior knowledge of the student and his/her learning style. Jeremic et al. [7] describe a student model for the Design Pattern intelligent tutoring system. Similarly, many other approaches have been presented in literature (i.e. [8], [9]). Additionally, in [10] the authors propose a student modeling mechanism for Intelligent Virtual Environments for Training (IVETs). They divide the student information in three major categories: (a) student profile (personal data), (b) state of student's progress and (c) trace of student's activity.

Furthermore, many attempts have been made in order to model the learner data in a more formal way and have been resulted in a number of standards, such as PAPI (Public and Private Information) [11], IMS LIP (Learner Information Package) [12], eduPerson [13], Dolog LP [14,15], FOAF (Friend of a Friend) [16] etc. Even if these models share a set of common learner characteristics, they vary on their main purpose and the way in which a system may use their embedded information. It is a usual practice to produce a learner profile for a learner system combining different learner standards and profiting from their unique benefits.

It is a common belief that PAPI and LIP are the most significant and important among the known standards due to their extended use and the benefits that they provide when used jointly. In [17], the presentation of the main characteristics of the aforementioned standards and the comparison of them denote the importance and the completeness of PAPI and LIP.

Nevertheless, both standards have some shortcomings. For example, the IMS LIP standard is based on the notion of a classic CV, while the PAPI standard considers student's performance as the most important information. However, in the context of our work, we took them into consideration and incorporated some of their basic notions to our proposed student model so that it conforms to these international standards.

Regarding the aforementioned approaches, our idea adopts the basic principles of the student model described in [10] suitably adapted to the needs and characteristics of an adult learner (e.g. time for study, previous experience, educational level and learning goals) in a distance learning educational framework. For example, in [4] the author does not include student's learning style in their model, while in [8] they take into consideration only student's performance and his/her interaction with the system. Furthermore, in [6] the authors do not include student's preferences, learning goals and motivation state in their model.

3   Description of the Student Model

In this section, we shall give a brief description of the basic components of the proposed approach and some of the basic characteristics of the model.

3.1   Learning Style

One of the most important components of the student model in an ITS is the personal learning style of the learner. The term *"learning style"* is used to describe the individual differences in the learning process. It is based on the assumption that each person has a unique and distinctive way to learn, i.e. to collect, process and organize information [18].

Among the models and theories presented in the literature, we have adopted the Felder-Silverman theory for student modeling. Most existing learning-style based theories classify students into few coarse grained groups, whereas Felder and Silverman describe the learning styles of a student in more detail, distinguishing between preferences on four dimensions [19].

According to the Felder-Silverman model, the learning types are categorized in the following four dimensions: (a) *active/reflective*, (b) *sensing/intuitive*, (c) *visual/verbal*, and (d) *sequential/global*.

3.2   Modeling Approach

The most common representation of a student model is the *overlay model*. The overlay model represents a learner's knowledge as a subset of the domain knowledge (expert's knowledge). Therefore, the system provides the learner with educational material until learner's knowledge coincides with the expert's knowledge [20]. Another approach which is widely used is the *buggy model*. Systems that use such models record and represent the most common/frequent mistakes made by learners based on statistics. Finally, one widely adopted approach for student modeling is the use of *stereotypes* [21]. New learners are classified into distinct categories and the system adjusts its performance based on the category assigned to the learner.

In the context of our work, we adopt a combination of the stereotype and overlay techniques. A fully stereotype-based model was excluded as a choice because (1) the initialization of the system derived from students descriptions or questionnaires may not be accurate for every knowledge domain and (2) the system would adapt to the learner's needs very slowly. So we developed a model where some attributes of the student profile (e.g. previous knowledge, experience in a specific knowledge domain) are initialized based on a stereotype. In addition, dynamic attributes related to the learning process are represented with an overlay model. After the initialization phase, the profile is dynamically modified, as the overlay model is updated with the information gathered by the interaction between learner and system.

3.3   Basic Characteristics of the Students

The users' classification in categories, called *stereotypes* constitutes a technique that has been widely used in user modeling systems. Stereotypes can be specified according to the following criteria: age, gender, educational level, working experience etc.

An empirical study was conducted by the Educational Content, Methodology and Technology Laboratory[1] among students of the Hellenic Open University[2] (HOU) in order to extract the basic characteristics and formulate the corresponding stereotypes of the student. The HOU was founded in 1992 and provides open distance learning at both undergraduate and postgraduate level.

---

[1] http://eeyem.eap.gr
[2] http://www.eap.gr/index_en.php

Table 1. Students' characteristics and their corresponding stereotypes

Characteristic	Stereotype
Learning style	active/reflective – sensing/intuitive – visual/verbal – sequential/global
Use of technology	adaptable - adaptive
Computer literacy	novice – beginner – advanced
Previous experience	novice – beginner – advanced
Time for study	no time – little – much
Reasons for education	career development –career change – self improvement
Academic literacy	poor – good – excellent
Socialization style	lonely – collaborative

The students who participated in the study were chosen based on their different characteristics such as different gender, age, educational background and current course. The study included (a) personal interviews with the students and (b) observation of the face to face meetings, by a social scientist. In particular, from the 13 students who were interviewed, 5 were male and 8 female, 8 of them pursued undergraduate studies and 5 postgraduate, 10 of them are working in the public sector, 2 are unemployed and 1 is working as a freelancer. Table 1 summarizes the characteristics and the stereotypes that came up from the empirical study.

Besides the modeling approach that defines the specialization of the model, a few more model characteristics have been taken into account: (a) it is a *dynamic model* that can change over time as the system collects information about the individual, (b) it is a *long-term model* that keeps generalized information regarding the user-system interactions and (c) it is a combination of *"active"* and *"passive"* user model, i.e. in the beginning the user provides directly information about him/her and then the system indirectly collects more information.

4   Student Model Ontology

In this section, we thoroughly describe the *Student Model* ontology that has been developed in order to capture the main concepts presented in Section 3. The focus of our attempt is not restricted on modeling the static profile of the user, but encompasses both permanent and dynamic characteristics. Moreover, the developed ontology complies partly with well-known standards for student modeling, i.e. IEEE PAPI Learner [11] and IMS Learner Information Package (LIP) [12].

In order to build the ontology, we followed a widely-adopted methodology, proposed in [22]. As far as its formal representation is concerned, we adopted the Web Ontology Language (OWL), which is a W3C standard. More specifically, our ontology falls into the OWL DL sublanguage, which provides the maximum expressiveness, while maintains computational completeness (all the conclusions are measurable and all calculations are terminated in finite time). The development process of the ontology was accomplished with the aid of Protégé[3] tool.

---

[3] http://protege.stanford.edu/

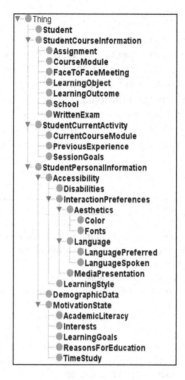

Fig. 1. The Student Model ontology as displayed in Protégé

In the proposed ontology *Student Model*, we define a set of four upper level classes, namely *Student, StudentCourseInformation, StudentCurrentActivity* and *StudentPersonalInformation*. The class hierarchy of the ontology, as displayed in Protégé, is depicted in Figure 1. The class *Student* represents any student. The *StudentCourseInformation* class comprises information relevant to the student's performance during the overall educational process and has a number of subclasses that are listed below, together with a brief explanation:

1. *Assignment* - the written assignments that the student has to submit during a course module
2. *CourseModule* - the course modules of the course program
3. *FaceToFaceMeeting* - face to face meetings during a course module
4. *LearningObject* - the learning objects that the student has been taught
5. *LearningOutcome* - the learning outcomes succeeded by the student as indicated by the learning objects
6. *School* - the school for which the student is registered
7. *WrittenExams* – the written exams that the student has to participate during a course module

In order to capture any detail in terms of student's activity for the current academic year, we define the class *StudentCurrentActivity*. Student's activity for the current academic year can be specified by the following three axes: (i) current chosen course modules (class *CurrentCourseModule*), (ii) the experience on a specific course module that the student has previously gained (class *PreviousExperience*) and (iii) the student's goals on a specific course module (class *SessionGoals*).

The more compact class in the proposed ontology, *StudentPersonalInformation*, is defined so as to represent mostly static and permanent student information, describing not only simple data, like demographic data, but more complex characteristics that concern student's interaction with the e-learning system. Table 2 lists the subclasses that exist under the upper level class *StudentPersonalInformation*. The table gives also a brief description of the entities that are represented by these classes.

Table 2. Description of the class *StudentPersonalInformation*

Class Name	Class Description
Accessibility	The overall set of features that characterizes the student's behavior during his interaction with the e-learning system
Disabilities	The set of student's disabilities that could affects the educational process
DemographicData	Student's demographic data
InteractionPreferences	Student's preferences regarding interaction with the e-learning system
MediaPresentation	Student's preferences regarding the presentation of learning objects
Language	Student's preferences regarding the language of the learning objects
LanguageSpoken	Student's native languages
LanguagePreferred	Language that the student prefers for the presentation of learning objects
Aesthetics	Aesthetic factors such as the use of highly interactive sensory and visual communication
Color	Student's preferences regarding the coloring scheme of learning system's environment
Fonts	Student's preferences regarding the fonts used by the learning system's environment
LearningStyle	Student's learning style - This class will be further divided to the sub classes according to the Felder-Silverman theory
MotivationState	Student's motivation during the educational process
LearningGoals	Overall goals set by the student
ReasonsForEducation	The reasons why the student desires to engage in the educational process
AcademicLiteracy	Student's previous formal educational experiences
Interests	Student's interests
TimeStudy	The average time per day that the student can use for studying

On the other hand, relationships between instances (members of classes) are modeled as object properties. In this context we define a set of object properties (mostly of the *hasA* kind). This kind of object property is used for expressing the association of the aforementioned characteristics with students. It links an instance of the class *Student* to instances of classes that reflects student characteristics such as *StudentCourseInformation*, *StudentCurrentActivity*, *PreviousExperience*, *Demographic*

Data, StudentPersonalInformation, Learningstyle, InteractionPreferences, Disabilities, Interests, TimeStudy, ReasonsForEducation and *AcademicLiteracy*. Furthermore, datatype properties, that link individuals to data values, have been set in order to define more effectively the classes.

In addition, as foresaid, we have adopted a scheme inspired by the Felder-Silverman Learning Style Model [23], in order to infer the student's learning style. The eight proposed learning styles are captured as individuals of the class LearningStyle, in the *Student Model* ontology (see Section 3.1 for a learning styles description).

The proposed ontology has been enriched with a set of rules in order to enable inference mechanisms (i.e. reasoner) to automatically classify the students into different stereotypic profiles (Table 1). As foresaid, these rules and the stereotypic profiles have resulted from an empirical research on a sample of adult learners. All the rules are expressed in Semantic Web Rule Language[4] (SWRL) and a subset are given in Table 3. For example, the first rule in the table indicates that *"if the student is female, over 50 years old and doesn't have a bachelor degree, then has little familiarity with computers"*.

Table 3. SWRL Rules

#		**Rule Body**	**Rule Head**
1	IF	x **is-a** *DemographicData*	
	AND	y **is-a** *Student*	
	AND	y **hasDemographics** x	
	AND	x **age** z	x **computer_literacy** "beginner"
	AND	x **educational_level** "secondary"	
	AND	x **gender** "female"	
	AND	z **greaterThan** "50"^^integer	
2	IF	y **is-a** *DemographicData*	
	AND	x **is-a** *Student*	
	AND	z **is-a** *TimeStudy*	
	AND	x **hasDemographics** y	
	AND	y **age** w	x **hasTime** z
	AND	y **educational_level** "secondary"	
	AND	y **gender** "female"	
	AND	z **time_for_study** "much"	
	AND	w **greaterThan** "50"^^integer	
3	IF	x **is-a** *DemographicData*	
	AND	y **is-a** *Student*	
	AND	z **is-a** *ReasonsForEducation*	
	AND	y **hasDemographics** x	
	AND	x **age** w	
	AND	x **educational_level** "secondary"	y **hasReasons** z
	AND	x **gender** "female"	
	AND	z **reasons_for_education** "general knowledge"	
	AND	w **greaterThan** "50"^^integer	

---

[4] http://www.w3.org/Submission/SWRL/

5   Conclusions and Future Work

We proposed in this paper an ontology-based approach to model student profiles especially for distance learning students. The student profile ontology we developed can be used as an integral ITS module, while it can be easily accessed from a web-based application. The proposed approach collects the characteristics of an adult student which are considered important for an ITS in order to be fully adapted to the needs of the learner. This model is a combination of international standards in user modeling and the results of an empirical study on a group of HOU students. One of the main advantages of the proposed model is the integration of semantic rules. These rules combined with inference mechanisms classify learners into stereotypic profiles which are already incorporated in the ontology and thus produce additional knowledge. The most challenging part of our research has been the selection of the characteristics to be included in the ontology. We further plan to add more rules in the student model ontology in order to cover as many stereotypic profiles as possible. Moreover, as a future work is the implementation of a web application which will communicate directly with the *Student Model* ontology and allow users to create their own profile.

Acknowledgment. This research described in this paper was partly funded by the National Strategic Reference Framework programme 2007-2013, project MIS 296121"Hellenic Open University".

References

1. Beck, J., Stern, M., Haugsjaa, E.: Applications of AI in Education. ACM, Crossroads (1996)
2. Shimic, G., Gasevic, D., Devedzic, V.: Classroom for the Semantic Web. In: Intelligent Learning Infrastructure for Knowledge Intensive Organizations: A Semantic Web Perspective, pp. 251—283 (2005)
3. Polson, M.C., Richardson, J.J. (eds.): Foundations of intelligent tutoring systems. Routledge (1988)
4. Paneva, D.: Use of Ontology-based Student model in Semantic-oriented Access to the Knowledge in Digital Libraries. In: Proceedings of the Open Workshop Semantic Web and Knowledge Technologies Applications, Bulgaria, pp. 31–41 (2006)
5. Chen, W., Mizoguchi, R.: Commincation Content Ontology for Learner Model Agent in Multi-agent Architecture. In: Workshop on Ontologies for Intelligent Educational Systems (AI-ED 1999), France (1999)
6. Pramitasari, L., Hidayanto, N.A., Aminah, S., Krisnadhi, A.A., Ramadhanie, A.M.: Development of Student Model Ontology for Personalization in an E-Learning System based on Semantic Web. In: International Conference on Advanced Computer Science and Information Systems (ICACSIS 2009), Indonesia, December 7-8, pp. 434–439 (2009)
7. Jeremic, J., Devedzic, V.: Design Pattern ITS: Student Model Implementation. In: Proceedings of the IEEE International Conference on Advanced Learning Technologies (ICALT 2004), pp. 864–865 (2004)
8. Henze, N., Dolog, P., Nejdl, W.: Reasoning and Ontologies for Personalized E-Learning in the Semantic Web. Educational Technology and Society 7, 82–97 (2004)

9. Muñoz, L.S., Palazzo, J., Oliveira, M.: Applying Semantic Web Technologies to Achieve Personalization and Reuse. In: Proceedings of the SWEL Workshop at Adaptive Hypermedia, pp. 348–353 (2004)
10. Clemente, J., Ramirez, J., de Antonio, A.: A proposal for student modeling based on ontologies and diagnosis rules. Expert Systems with Applications Journal, 8066–8078 (2011)
11. LTSC Learner Model Working Group of the IEEE (2000) IEEE p1484.2/d7, 2000-11-28 Draft Standard for Learning Technology - Public and Private Information (PAPI) for Learners (PAPI Learner), Technical report (2000)
12. Smythe, C., Tansey, F., Robson, R.: IMS Learner Information Package Information Model Specification. Technical report (2001)
13. EduPerson Specification. Document: Internet2-mace-dir-eduPerson-200312. Copyright © 2003 by Internet2 and/or the respective authors. Internet2 Middleware Architecture Committee for Education, Directory Working Group (December 2003)
14. Dolog, P., Nejdl, W.: Challenges and Benefits of the Semantic Web for User Modelling (2003)
15. Dolog, P.: Identifying Relevant Fragments of Learner Profile on the Semantic Web. In: Proceedings of the SWEL 2004 Workshop at ISWC 2004 Conference, Hiroshima, Japan (2004)
16. Brickley, D., Miller, L.: FOAF Vocabulary Specification -Namespace Document - ('Pages about Things' Edition) (July 27, 2005)
17. Ounnas, A., Davis, H.C., Millard, D.E.: Towards semantic group formation. In: The 7th IEEE International Conference on Advanced Learning Technologies (ICALT 2007), Niigata, Japan (2007)
18. Jones, D.C., Mungai, D.: Technology-enabled teaching for maximum learning. International Journal of Learning 10, 3491–3501 (2003)
19. Felder, R.M., Silverman, L.: Learning and Teaching Styles in Engineering Education. Engineering Education Journal 78, 674–681 (1988)
20. Conlan, O., O'Keeffe, I., Tallon, S.: Combining Adaptive Hypermedia Techniques and Ontology Reasoning to Produce Dynamic Personalized News Services. In: Wade, V.P., Ashman, H., Smyth, B. (eds.) AH 2006. LNCS, vol. 4018, pp. 81–90. Springer, Heidelberg (2006)
21. Kay, J.: Stereotypes, Student Models and Scrutability. In: Gauthier, G., VanLehn, K., Frasson, C. (eds.) ITS 2000. LNCS, vol. 1839, pp. 19–30. Springer, Heidelberg (2000)
22. Noy, N., McGuiness, D.: Ontology Development 101: A Guide to Creating Your First Ontology, Stanford Knowledge Systems Laboratory Technical Report KSL-01-05 and Stanford Medical Informatics Technical Report SMi-2001-0880 (2001)
23. Graf, S., Viola, S.R., Kinshuk, Leo, T.: Representative Characteristics of Felder-Silverman Learning Styles: An Empirical Model. In: Proceedings of the International Conference on Cognition and Exploratory Learning in Digital Age (CELDA 2006), Spain, pp. 235–242 (2006)

Assistant Tools for Teaching FOL to CF Conversion

Foteini Grivokostopoulou, Isidoros Perikos, and Ioannis Hatzilygeroudis

School of Engineering
Department of Computer Engineering & Informatics
University of Patras
26500 Patras, Hellas, Greece
{grivokwst,perikos,ihatz}@ceid.upatras.gr

Abstract. The FOL to CF system is an interactive web-based system for learning to convert first order logic (FOL) formulas into Clause Form (CF). FOL to CF conversion is a fundamental part of using FOL for making inferences. In this paper, we present two tutor assistant tools integrated with that system. The first, called tutoring manager, helps the tutor to manage the teaching material and monitor the progress of students. It helps tutors to investigate students' answers and errors made by providing useful statistics. Also, it gives a graphical view of them for an easier understanding of difficulties that students face. The second tool, the difficulty estimating expert system, aims at helping tutors in determining the difficulty level of a formula's conversion process. This is based on the complexity of the FOL formula. Experimental results show that the difficulty estimating system is quite successful.

Keywords: Tutor assistant tool, Student progress statistics, Difficulty level estimation, First Order Logic, Clause Form.

1 Introduction

Logic is one of the fundamental topics taught in computer science and/or engineering departments. In most such departments, logic is taught as a means for constructing formal proofs in a natural deduction style. However, teaching logic as a knowledge representation and reasoning (KR&R) vehicle is also basic in all introductory artificial intelligence (AI) courses. A basic KR language is First-Order Logic (FOL), the main representative of logic-based representation languages [3], which is part of almost any introductory AI course. To make automated inferences, Clause Form (CF), a special form of FOL, is used. Students usually find difficulties in converting complex FOL formulas into CF [7].We have constructed tools for helping tutors in teaching logic as a KR&R language and more specifically tools for learning the conversion of NL sentences into FOL ones [12][11].

There are several systems [1][2][8][13] for teaching propositional logic (PL), but most of them teach how to construct formal proofs using natural deduction. Logic-ITA [14] deals with propositional logic in the same sense as above, but it is addressed to both students and teachers and uses intelligent techniques to automatically adapt to their needs. None of the above deals with PL as a KR&R language and cannot determine the difficulty level of a formula's conversion.

L. Iliadis et al. (Eds.): AIAI 2012, IFIP AICT 381, pp. 306–315, 2012.

In a previous work [6], we dealt with teaching the FOL to CF conversion using a simple tool implemented in Java. However, that system does not offer any tools for helping the tutors.

In this paper, we introduce two assistant tools that aim at assisting tutors in their tasks. The first one helps tutors in managing and updating the teaching material of the system. In addition, it can help them monitor student's progress and trace learning gaps and difficulties the students may face. The second tool is an expert system that aims at helping the tutor in determining the difficulty level of a formula's conversion process. The determination of the difficulty of the exercises material is a vital aspect for an e-learning system. However, in most systems the difficulty levels of the exercises are determined by the tutor, when inserting them into the system. The determination of an exercise difficulty level is a time-consuming task for the tutor and to some degree non-consistent. Therefore, such a system can be of great help.

The paper is organized as follows. In Section 2, related work is presented. Section 3 deals with the FOL to CF conversion, by presenting the process through an example. Section 4 presents the revised architecture of the FOL to CF system. Section 5 and Section 6 focus on the assistant tools. Section 7 presents and discusses experimental results. Finally, Section 8 concludes and provides directions for future research.

2   Related Work

There are several tools for teaching logic and logic-based reasoning, but most teach natural deduction. A number of them are characterized as logic educational software. ProofWeb [8] is a system for teaching logic as natural deduction, based on the higher order proof assistant Coq [4]. Moreover, some systems provide strategies for proving propositions and use those strategies to provide hints or worked out examples (e.g. Fitch [1], AProS [13] and Pandora [2]).

Logic-ITA [14] is an intelligent teaching assistant system for Logic. It deals with propositional logic and provides students with an environment to practice formal proofs. It expands Logic Tutor (an earlier version) with new tools which are designed to assist the tutor. The teacher can use those tools to manage the teaching configuration settings and the teaching material. Also, they help him/her monitor the students' results and their progress. P-Logic Tutor [10] is a kind of intelligent tutoring system aiming at teaching students fundamental aspects of PL and theorem proving. P-Logic tutor also provides an environment in which the tutor can track student learning activities. However, none of them explicitly deals with the FOL to CF conversion and provides no information about the difficulty of the FOL formula's conversion into CF.

In [11], a work that deals with the difficulty of the conversion of NL into FOL is presented. It is an expert system that automatically determines the difficulty level of a sentence's conversion process. It takes as input the corresponding FOL formula of a NL sentence and gives as output an estimation of the difficulty of its conversion process. The estimation of the difficulty level is based on a set of parameters, like the number and the type of the quantifier(s), the number of the implications and the different connectives of the FOL expression.

However, according to our knowledge, there isn't any effort to determine the difficulty level of a FOL formula's conversion process into CF.

3   FOL to CF Conversion

FOL is the most widely used logic-based knowledge representation formalism. FOL is a KR language used for representing knowledge in a knowledge base, in the form of logical formulas. To make automated inferences using the resolution principle (the strongest inference rule), their Clause Form (CF) is used. The FOL to CF conversion is a well-defined process that can be automated within a computer. However, most of existing systems perform it in one step manner so that, one cannot see all the steps of the conversion. Therefore, they are not suitable for use in teaching the conversion process. The conversion process in our system includes six steps, as presented below through an example: the conversion of FOL sentence

$$\text{"}(\forall x)\ dog(x) \Rightarrow (\exists y)(cat(y) \wedge chases(x,y))\text{"}.$$

<u>Step1</u>: Eliminate connectives $\Leftrightarrow$ and $\Rightarrow$, using the equivalences: $P \Leftrightarrow Q \equiv (P \Rightarrow Q) \wedge (Q \Rightarrow P)$ and $P \Rightarrow Q \equiv \neg P \vee Q$
Result: $(\forall x)\ \neg dog(x) \vee (\exists y)(cat(y) \wedge chases(x,y))$

<u>Step2</u>: Reduce negation scope using the following equivalences:
$\neg(P \vee Q) \equiv \neg P \wedge \neg Q,\ \neg(P \wedge Q) \equiv \neg P \vee \neg Q,\ \neg\forall xP \equiv \exists x \neg P,\ \neg\exists xP \equiv \forall x \neg P,\ \neg \neg P \equiv P$
Not applicable here

<u>Step3</u>: Rename variables (so that each quantifier has its own unique variable name) and transform into Prenex Normal Form (move all quantifiers to the left of the formula)
Result: $(\forall x)\ (\exists y)\ \neg dog(x) \vee (cat(y) \wedge chases(x,y))$

<u>Step4</u>: Remove existential (Skolemization) and universal quantifiers. The process is as follows:
  (a) if there are universal quantifiers whose scope includes the scope of an existential quantifier replace each occurrence of its variable with a (Skolem) function whose arguments are the variables of those universal quantifiers,
  (b) If there are no such universal quantifiers, replace each occurrence of its variable with a (Skolem) constant.
  (c) After skolemization, remove all universal quantifiers.
Result: $\neg dog(x) \vee (cat(sk\text{-}f(x)) \wedge chases(x, sk\text{-}f(x)))$

<u>Step5</u>: Transform the produced formula into CNF (Conjunctive Normal Form), using the equivalence:
$(P \wedge Q) \vee R \equiv (P \vee R) \wedge (Q \vee R)$
$(\neg dog(x) \vee cat(sk\text{-}f(x))) \wedge (\neg dog(x) \vee chases(x, sk\text{-}f(x)))$

<u>Step6</u>: Extract clauses and rename variables.
  As many clauses as the number of conjunction connectives in the produced formula plus one are extracted.
Result: $\{\neg dog(x), cat(sk\text{-}f(x))\}$
  $\{\neg dog(x), chases(x, sk\text{-}f(x))\}$

4   System Architecture

The architecture of the system is depicted in Fig. 1. It consists of four units: the Tutor Interface (TI), the Tutoring Manager (TM), the Difficulty Estimating System (DES) and the System Database (SD). Through TI, the tutor interacts with tutoring manager (TM) unit. TM is responsible for the teaching process and helps the tutor in examining students' progress.TM unit offers to the tutor the capability of exercise management. So, the tutor can create new exercises-sentences and also edit or even delete existing ones. Also, the tutor can monitor the knowledge level of students and the common errors of student's answers. Moreover, the system can present useful statistics about the student learning process.

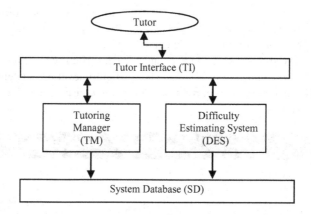

Fig. 1. Architecture of the Tutoring Assistant System

The Difficulty Estimating System (DES) is an expert system that aims to help the tutor to determine the difficulty level of FOL to CF conversion. DES automatically determines the difficulty level of a formula without tutor intervention. Tutor can accept or modify the result of DES.

Finally, in System Database (SD) the FOL sentences and their corresponding estimated difficulty levels are stored.

5   Tutoring Manager

The tutoring manager (TM) offers the capability of the tutor to manage the exercise-FOL formula. An exercise consists of the following parts:

- Its name (Natural Language)
- Its body (FOL formula)
- Its difficulty level (1-5)
- Its correct answer (CF formula (s))

The tutor can create a new exercise-FOL formula and also edit or even delete existing ones. To create a new question-answer pair, the tutor has to insert a new FOL formula. The difficulty level is automatically determined. The tutor can modify it. Also, the correct answers are automatically determined and inserted into SD. Also, the tutor can see all the exercises and all information concerning them. So, the tutor can change the exercises' information at any time according to his/her teaching needs.

Another facility that is available to the tutor is monitoring a student's learning process. As mentioned above, while the student is using the FOLtoCF system, all the necessary information is stored in SD. The tutor can monitor the errors of the student made during the FOL to CF process, the exercises that the student has tried, the time spent and the knowledge level obtained. The system can present some statistics about the students, which are the following:

- The percentage of correct answers to exercises per difficulty level per student.
- The number of errors made per sentence.
- The type of errors made per sentence per difficulty level per student.
- The percentage of errors made per difficulty level per student.
- The percentage of correct answers to exercises per difficulty level for all students.

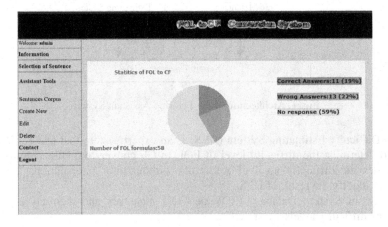

Fig. 2. Statistics of Students

In Fig. 2, some statistics about students is illustrated. As mentioned above, the tutor can evaluate not only the students, but also the exercises-formulas and indirectly the teaching process. So, the tutor can check the quality (e.g difficulty level) of each exercise that he/she has designed for the evaluation of students.

6   Difficulty Estimating System

We have developed the Difficulty estimating system (DES), which is an expert system. It aims at helping the tutor in determining the difficulty level of a FOL to CF conversion. DES takes as input a set of values corresponding to difficulty estimation parameters and automatically determines the conversion difficulty of the corresponding FOL formula. Values are extracted from that formula by an analysis process. The structure of DES is illustrated in Fig. 3. DES is a rule-based expert system implemented in Jess, which an expert system shell [5]. It consists of the Difficulty Parameters Fact Base (DFB), the Difficulty Estimation Rule Base (DRB) and the Jess inference engine (JESS IE). DFB contains the values of the estimation parameters and the DRB contains the rules for estimating difficulty levels. DFB and DRB constitute the Knowledge Base (KB) of the system. FA is a tool that analyses a FOL formula and automatically extracts the values of the estimation parameters for a FOL formula, which are stored as facts in DFB and used as input to DES.

The process of estimating the difficulty level of the FOL to CF process of a FOL formula is as follows:

1. The FOL formula is analyzed via the FOL Analyzer tool (FA) to extract values for the difficulty estimation parameters
2. The extracted values are transformed into Jess facts and stored in DFB
3. Jess IE is triggered and deduces the difficulty level of the corresponding FOL to CF conversion process

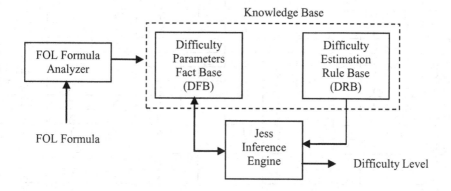

Fig. 3. The structure of Difficulty Estimating System

The development of DES is based on expert-tutor. The expert offers knowledge based on experience. Most tutors empirically estimate the difficulty of a FOL sentence's conversion. In co-operation with the expert-tutor, we tried to specify which factors/parameters have an impact on the difficulty level of a sentence's FOL to CF conversion. Finally, we came up with the following difficulty estimation parameters:

• The number, the type, the order and the position of quantifier (s)
• The number of the implication symbols
• The number of the different connectives

- The type of negation(s)
  - Negation in front of a group of atoms
  - Negation in front of a quantifier.
- Whether the LHS of an implication is a group of atoms

where by "connectives" we mean $\{\wedge, \vee, \neg\}$.

Table 1. Rules for determining difficulty levels

Rules	Num of ∀, ∃	∀	∃	Order ∀∃	PNF	Num of ⇒	Num of different ∧, ∨, ¬	¬ in front of group	¬ in front of ∀, ∃	Group at LHS of ⇒	Difficulty Level
1.	0	N	N	*	*	0	≤1	*	*	*	VE
2.	1	N	Y	*	*	0	0	*	*	*	VE
3.	1	N	Y	*	*	0	≥1	*	*	*	VE
4.	≤2	Y	Y	*	Y	0	≥1	N	N	*	E
5.	≤2	Y	-	*	*	0	≥1	N	N	*	E
6.	≤2	-	Y	*	*	0	≥1	N	N	*	E
7.	1	Y	-	*	*	1	≤1	N	N	N	E
8.	0	N	N	*	-	1	≥0	*	*	Y	E
9.	1	Y	N	*	*	1	≥1	Y	Y	*	E
10.	2	Y	Y	Y	*	0	≥1	Y	Y	*	E
11.	2	N	Y	*	Y	1	≥1	N	N	*	M
12.	≥2	Y	N	*	Y	1	≥1	*	*	*	M
13.	≥2	Y	Y	N	N	1	≥1	Y	N	N	M
14.	≥2	Y	Y	Y	Y	1	≥1	N	Y	N	M
15.	2	Y	Y	Y	Y	1	≥1	N	N	Y	M
16.	2	Y	Y	Y	N	1	≥1	N	N	N	M
17.	<3	Y	-	*	Y	1	≥1	N	N	Y	M
18.	<3	Y	Y	Y	*	1	≥1	Y	N	N	D
19.	<3	Y	Y	Y	*	1	≥1	Y	N	N	D
20.	3	Y	Y	Y	N	≥1	≥1	N	Y	Y	A
21.	3	Y	Y	N	Y	≥1	≥1	Y	Y	Y	A

Afterwards, we consulted the expert-tutor to acquire the necessary rules for the difficulty estimation based on the above parameters. The results of knowledge acquisition are depicted in Table 1. There are 21 rules for determining the difficulty level. The difficulty level is classified in one of the following five classes: (a) very easy (VE), (b) easy (E), (c) medium (M), (d) difficult (D) and (e) advanced (A). Also, in Table 1, PNF stands for "Prenex Normal Form" (all quantifiers are already at the left hand side of the formula), LHS for "Left Hand Side", "Y" for "Yes", "N" for "No" and "*" stands for "don't care".

In Table 2, some example FOL formulas/sentences and corresponding difficulty levels of their conversion processes as produced by DES are presented.

Table 2. Examples of difficulty levels estimated by DES

First Order Logic (FOL)	Clause Form	Difficulty Level	Rule applied
¬happy(john) ∨¬happy(peter)	{¬happy(john),¬happy(peter)}	Very Easy	1
(∃x) apple(x) ∧red(x)	{apple(sc_x)} {red(sc_x)}	Very Easy	3
(∀x) gardener(x) ⇒likes(x,sun)	{¬gardener(x),likes(x,sun)}	Easy	7
(∀x)(∀y)(∀z) (ancestor (x, y) ∧ancestor (y,z) ⇒ancestor(x,z))	{¬ancestor (x, y) , ¬ancestor (y,z),ancestor(x,z)}	Medium	12
(∀x)(student (x) ⇒ (∃y)(student(y) ∧ loves(x,y)))	{¬student(x), student(sk-f(x))} {¬student(x), loves (x, sk-f(x))}	Medium	16

7 Experimental Results

We used DES for a number of 94 FOL formulas, found in textbooks and web resources. DES results were compared to the results of the expert-tutor on the basis of the extracted factual information. To evaluate DES, we used four metrics, commonly used for this purpose: *accuracy, precision, sensitivity* and *specificity*. The metrics for two output classes are defines as following:

$$Acc=\frac{TP+TN}{TP+FN+TN}, \quad Prec=\frac{TP}{TP+FP}, \quad Sen=\frac{TP}{TP+FN}, Spec=\frac{TN}{TN+FP}$$

Where *TP, TN, FP* and *FN* denote the number of the true positives, true negatives, false positive and false negatives, respectively. By "positive", we mean that a case belongs to the class of the corresponding difficulty level and by negative that it doesn't.

In case of multi-class classification, as ours, the above metrics are calculated as follows:

$$acc=\frac{\sum_{i=1}^{m} acc_i}{m}, \quad prec=\frac{\sum_{i=1}^{m} prec_i}{m}, \quad Sen=\frac{\sum_{i=1}^{m} sen_i}{m}, \quad Spec=\frac{\sum_{i=1}^{m} spec_i}{m}$$

The evaluation results are presented in Table 3 and show an acceptable performance of DES.

Table 3. Evaluation metrics for DES

Metrics	Difficulty Class					Average
	Very Easy	Easy	medium	Difficult	Advanced	
Accuracy	0.9	0.89	0.89	0.89	0.96	0.906
Precision	0.55	0.92	0.83	0.5	0.5	0.66
Sensitivity	1	0.75	0.91	0.2	0.5	0.672
Specificity	0.92	0.99	0.96	0.89	0.97	0.946

The results show a very good performance of DES. From the corpus of 94 FOL formulas that were used the system correctly identified the difficulty of 77 FOL formulas.

8   Conclusion and Future Work

The FOL to CF system is a web based interactive system for helping students to convert First order logic (FOL) formulas into Clause Form (CF). In this paper, two teaching assistant tools are presented. The first one is the Teaching Manager tool, which is developed to help the tutor to manage and update the teaching material of the system. Moreover, can help him/her monitor student's progress and trace learning gaps and difficulties that the students may face. Also the tool offers sets of graphs and statistics and based on them the tutor can accordingly reconfigure the contents of the system by adding specific sentences.

The second is the Difficulty estimation expert system, which is used to automatically determine a FOL formulas difficulty concerning its conversion into CF. DES is a rule-based expert system that takes as input a FOL formula, analyses it in terms of connectives, quantifiers, negations types and order specification and based on theses parameters determines the FOL formulas conversion difficulty-complexity into CF.

A more interesting direction is to use a method similar to that in [9] could be investigated. This means to use a student-based approach instead of a sentence-based approach or a combination of them.

Acknowledgement. This work was supported by the Research Committee of the University of Patras, Greece, Program "Karatheodoris", project No C901.

References

1. Barwise, J., Etchemendy, J.: Language, Proof and Logic, Center for the Study of Language and Information (2002)
2. Boda, K., Ma, J., Sinnadurai, G., Summers, A.: Pandora: A reasoning Toolbox using natural Deduction Style. Logic Journal of the IGPL 15(4), 293–304 (2007)
3. Brachman, R.J., Levesque, H.J.: Knowledge Representation and Reasoning. Elsevier, Amsterdam (2004)
4. Coq Development Team. The Coq Proof Assistant User's Guide. Version 8.3 (2010)

5. Friedman-Hill, E.: Jess in Action: Rule-Based Systems in Java. Manning Publications Company (2003)
6. Hatzilygeroudis, I., Giannoulis, C., Koutsojannis, C.: A Web Based Education System for Predicate Logic. In: Proceedings of the IEEE International Conference on Advanced Learning Technologies (ICALT 2004), pp. 106–110 (2004)
7. Hatzilygeroudis, I.: Teaching NL to FOL and FOL to CL Conversions. In: Proceedings of the 20th International FLAIRS Conference, Key West, FL, pp. 309–314. AAAI Press, Menlo Park (2007)
8. Hendriks, M., Kaliszyk, C., Van Raamsdonk, F., Wiedijk, F.: Teaching logic using a state-of-the-art proof assistant. Acta Didactica Napocensia 3(2), 35–48 (2010)
9. Koutsojannis, C., Beligiannis, G., Hatzilygeroudis, I., Papavlasopoulos, C., Prentzas, J.: Using a hybrid AI approach for exercise difficulty level adaptation. International Journal of Continuing Engineering Education and Life-Long Learning 17(4-5), 256–272 (2007)
10. Lukins, S., Levicki, A., Burg, J.: A tutorial program for propositional logic with human/computer interactive learning. In: SIGCSE 2002, pp. 381–385. ACM, New York (2002)
11. Perikos, I., Grivokostopoulou, F., Hatzilygeroudis, I., Kovas, K.: Difficulty Estimator for Translating Natural Language into First Order Logic. In: Proceedings of the Third International Conference on Intelligent Decision Techologies (KES-IDT 2011), vol. 10(pt. I), pp. 135–144 (2011)
12. Perikos, I., Grivokostopoulou, F., Hatzilygeroudis, I.: Teaching assistant tools for NL to FOL Conversion. In: Proceedings of the IADIS International Conference e-Learning 2011, Rome, Italy, July 20-23, pp. 337–345 (2011)
13. Sieg, W.: The AProS project: Strategic thinking & Computational logic. Logic Journal of the IGPL 15(4), 359–368 (2007)
14. Yacef, K.: The Logic-ITA in the classroom: a medium scale experiment. International Journal on Artificial Intelligence in Education, 41–60 (2005)

Effective Diagnostic Feedback
for Online Multiple-Choice Questions

Ruisheng Guo, Dominic Palmer-Brown, Sin Wee Lee, and Fang Fang Cai

Faculty of Computing, Londonmet University, 166-220 Holloway Road, London N7 8DB
{r.guo1,ff.cai,d.palmer-brown}@londonmet.ac.uk,
s.w.lee@uel.ac.uk

Abstract. When students attempt MCQs (Multiple-Choice Questions) they generate invaluable information which can form the basis for understanding their learning behaviours. In this research, the information is collected and automatically analysed to provide customized, diagnostic feedback to support students' learning. This is achieved within a web-based system, incorporating the SDNN (Snap-drift neural network) based analysis of students' responses to MCQs. This paper presents the results of a large trial of the method and the system which demonstrates the effectiveness of the feedback in guiding students towards a better understanding of particular concepts.

Keywords: learning behavior, diagnostic feedback, neural networks, on-line multiple-choice questions.

1 Introduction

In recent years, e-learning has become commonplace in higher education. The involvement of intelligent e-learning systems has the potential to make higher education accessible with increasing convenience, efficiency and quality of study. According to the National Student Survey reports (2007- 2010) [1], in England, only about half of students believe that: 1, feedback on their work has been prompt; 2, feedback on their work has helped to clarify things they did not understand; 3, they have received detailed comments on their work. These reports reveal that the feedback and its related fields are one of the weakest areas in higher education in England. This research investigates the relative effectiveness of different types of feedback, and how to optimize feedback to facilitate deep learning. It compares and contrasts several methods in order to investigate the effectiveness of using intelligent feedback towards modeling the stages of students' knowledge. The investigation will lead to an understanding of the potential of the on-line diagnostic feedback across different subject areas.

The Virtual Learning Environment presented in this paper provides a generic method for intelligent analysis and grouping of student responses that applicable to any area of study. This tool offers important benefits: immediate feedback, significant time-saving evaluating assignments, and consistency in the learning process. The time taken to create the feedback is well spent not only because this feedback can be reused, but also it is made available through the system to large numbers of students.

L. Iliadis et al. (Eds.): AIAI 2012, IFIP AICT 381, pp. 316–326, 2012.
© IFIP International Federation for Information Processing 2012

2   Background and Review of Previous Work

Rane and Sasikumar [2] pointed out that, intelligent tutoring systems attempt to simulate a teacher, who can guide the student's study based on the student's level of knowledge by giving intelligent instructional feedback. According to Blessing, Gilbert, Ourada and Ritter [3], the intense interaction and feedback achieved by intelligent tutoring systems can significantly improve student learning gains. To make the feedback effective and meaningful, a range of quality attributes need to be achieved. Hatziapostolou and Paraskakis [4] summarized the work by Race (2006), Irons (2008) and Juwah et al (2004) and suggested that in order to improve learning gains, formative feedback should address as many as possible of the following attributes, including constructive [5], motivational [4], personal [6], manageable [4], timely [7] and directly related to assessment criteria and learning outcomes [8].

Many researches investigating the effect of different types of feedback in web-based assessments showed the positive results of using MCQs in online test for formative assessment (e.g. [9] [10] [11]). Higgins and Tatham [9] studied the use of MCQs in formative assessment in a web-based environment using WebCT for a level 1 unit on undergraduate law degree. They summed that they could forecast all the possible errors for a question and write a general feedback for this question. However, using this type of feedback, it could be difficult to predict all the possible errors and produce the general feedback for a combination of questions, and it would be impossible for a large test banks (e.g. 3 questions with 5 answers would require 125 answer combinations; 5 questions with 5 answers require 3125 combinations, etc.). Payne et al [11] assessed the effectiveness of three different forms of feedback (corrective, corrective explanatory, and video feedback) used in e-learning to support students' learning. This type of feedback shows exactly which questions are answered correctly or not, with further corrective explanation and video feedback. Our approach to feedback is different from the above. The intelligent diagnostic feedback we present is concept-oriented instead of question-oriented. The learners are encouraged to review the concepts they misunderstood through the feedback in order to retake the test again and study further. It is important that each category of answers is associated with carefully designed feedback based on the level of understanding and prevalent misconceptions of that category-group of students so that every individual student can reflect on his or her learning level and certain mistakes using this diagnostic feedback. In addition, when students retake the test they receive new feedback according to his or her knowledge state, which in turn leads to more self-learning. Moreover, concept-based feedback can also prevent the student from guessing the right answers; if the students do not read the diagnostic feedback carefully, they may not even know which questions were answered incorrectly. According to our current research, there are no other reported studies on MCQs and formative web-based assessment which have used any similar form of using intelligent agent to analyse the students' response in order to provide diagnostic feedback.

3   Multiple-Choice Questions Online Feedback Systems (M-OFS)

To analyse the students' answers, and integrate over a number of questions to gain insights into the students' learning needs, a snap-drift neural network (SDNN) approach is proposed. SDNN provides an efficient means of discovering a relatively small and therefore manageable number of groups of similar answers. In the following sections, an e-learning system based on SDNN is described.

3.1   Snap-Drift Neural Networks (SDNNs)

One of the strengths of the SDNN is the ability to adapt rapidly in a non-stationary environment where new patterns are introduced over time. The learning process utilises a novel algorithm that performs a combination of fast, convergent, minimalist learning (snap) and more cautious learning (drift) to capture both precise sub-features in the data and more general holistic features. Snap and drift learning phases are combined within a learning system that toggles its learning style between the two modes. On presentation of input data patterns at the input layer F1, the distributed SDNN (dSDNN) will learn to group them according to their features using snap-drift (Lee et al., [12]). The neurons whose weight prototypes result in them receiving the highest activations are adapted. Weights are normalised weights so that in effect only the angle of the weight vector is adapted, meaning that a recognised feature is based on a particular ratio of values, rather than absolute values. The output winning neurons from dSDNN act as input data to the selection SDNN (sSDNN) module for the purpose of feature grouping and this layer is also subject to snap-drift learning.

The learning process is unlike error minimisation and maximum likelihood methods in MLPs and other kinds of networks. These perform optimization for classification or equivalents by for example pushing features in the direction that minimizes error, without any requirement for the feature to be statistically significant within the input data. In contrast, SDNN toggles its learning mode to find a rich set of features in the data and uses them to group the data into categories. Each weight vector is bounded by snap and drift: snapping gives the angle of the minimum values (on all dimensions) and drifting gives the average angle of the patterns grouped under the neuron. Snapping essentially provides an anchor vector pointing at the 'bottom left hand corner' of the pattern group for which the neuron wins. This represents a feature common to all the patterns in the group and gives a high probability of rapid (in terms of epochs) convergence (both snap and drift are convergent, but snap is faster). Drifting, which uses Learning Vector Quantization, tilts the vector towards the centroid angle of the group and ensures that an average, generalised feature is included in the final vector. The angular range of the pattern-group membership depends on the proximity of neighbouring groups (natural competition), but can also be controlled by adjusting a threshold on the weighted sum of inputs to the neurons. The output winning neurons from dSDNN act as input data to sSDNN module for the purpose of feature grouping and this layer is also subject to snap-drift learning.

3.2   Training Neural Network

The E-Learning Snap-Drift Neural Network (ESDNN) is trained with the students' responses to questions on a particular topic in a course. The responses are obtained from the previous cohorts of students. Before training, each of the responses from the students is encoded into binary form in preparation for presentation as input patterns for

ESDNN. Table 1 shows examples of a possible format of questions for five possible answers and some encoded responses. This version of ESDNN is a simplified unsupervised version of the snap-drift algorithm (Lee et al., [12]) as shown in Fig. 1.

Table 1. Example of input patterns for ESDNN

Codification	A:00001	B:00010	C:00100	D:01000	E:10000	N/A:00000
Response	Recorded Response					
[C,D,B,A]	[0,0,1,0,0,0,1,0,0,0,0,0,1,0,0,0,0,0,1]					
[E,B]	[1,0,0,0,0,0,0,0,1,0]					

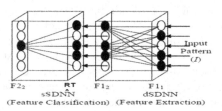

Fig. 1. E-learning SDNN architecture

During training, on presentation of an input pattern at the input layer, the dSDNN will learn to group the input patterns according to their general features. In this case, 5 F12 nodes, whose weight prototypes best match the current input pattern, with the highest net input are used as the input data to the sSDNN module for feature classification. In the sSDNN module, a quality assurance threshold is introduced. If the net input of an sSDNN node is above the threshold, the output node is accepted as the winner; otherwise a new uncommitted output node will be selected as the new winner and initialised with the current input pattern. For example, for one group, every response might have in common the answer C to question 2, the answer D to question 3, the answer A to question 5, the answer A to question 6, the answer B to question 8, and the answer A to question 10. The other answers to the other questions will vary within the group, but the group is formed by the neural network based on the commonality between the answers to some of the questions (four of them in that case). From one group to another, the precise number of common responses varies in theory between 1 and X, where X is the number of questions. In this experiment, where there are 10 questions in 1st English trial (Section 5), the groups had between 5 and 8 (Trial 1) common answers. More details of the steps that occur in ESDNN and the ESDNN learning algorithm are given in (Lee et al., [12]). The training relies upon having representative training data. The number of responses required to train the system so that it can generate the states of knowledge varies from one domain to another. When new responses create new groups, more training data is required. Once new responses stop creating new groups, it is because those new responses are similar to previous responses, and sufficient responses to train the system reliably are already available. The number of groups formed depends on the variation in student responses.

3.3   How the System Guides Learning

The feedback is designed by academics so that it does not identify which questions were incorrectly answered. The academics are presented with the groups in the form of tem-plates of student responses. For example, "A/D B mix" represents a group characterized by all the students answering A or D to question 1, B to question 2, and mixed answers to question 3. Hence, the educator can easily see the common mistakes in the groups of the student answers highlighted by the tool. The feedback texts are associated with each of the pattern groupings and are composed to address misconceptions that may have caused the incorrect answers common to that pattern group. The student responses, recorded in the database, can be used for monitoring the progress of the students and for identifying misunderstood concepts that can be addressed in subsequent face-to-face sessions. The collected data can be also used to analyze how the feedback influences the learning of individual students by following a particular student's progress over time and observing how that student's answers change after reading the feedback. Student responses can also be used to retrain the neural network and see whether refined groupings are created, which can be used by the educator to improve the feedback. Once designed, MCQs and feedbacks can be reused for subsequent cohorts of students.

4   Approach

In order to evaluate the performance and effectiveness of this novel e-learning system, target-oriented testing of the system needs to be carried out in different fields. Furthermore, we also aim to enhance this system to overcome its deficiencies during practical applications. Thus, this study is composed of three main parts. Firstly, we evaluated the M-OFS system by collecting and analysing a large number of testing data reflecting the students' learning gains by using this system as well as the survey and interview data reflecting the students' satisfaction and attitudes towards this system. Secondly, the investigation leads to an understanding of the potential of the on-line diagnostic feedback approach across different subject areas. Thirdly, this research should also produce guidelines for the design principles of on-line MCQs in the context of diagnostic feedback learning environments. The details of this experiment which are conducted to assess the use of M-OFS during academic year 2010-2011 are reported below. Four hypotheses are formulated: (H1) students are satisfied with using M-OFS; (H2) students improved their understanding by reading given feedback; (H3) in a separate MCQs paper test, students get higher mark in the first test than the second trial by learning from the M-OFS; (H4) in the final examination, the average score of the experimental group is higher than the average score of the control group.This research used 5 instruments: 1, four previous MCQ test (1 conducted in 2008 and 3 in 2009) results are collected to train the ESDNN; 2, two separate MCQs paper tests are applied before and after the system trial; 3, compare and contrast the scores between first and last attempt during the system trial; 4, compare and contrast the final examination grades between experimental group and control group; 5, survey and interview to assess learner's satisfaction and motivation.

4.1   Data Collection

Data of Six trials were collected in total. It includes three English trials, two Math trials and one Plagiarism trial. In this paper, it will present the details of data collection of 1st English trial as below:

The data for training is collected from three previous year's MCQs tests (2008-2010). For these three tests, 94 students' answers were used to training. The trials data were collected during academic year 2010-2011. The data of two separate MCQ paper tests and final examination results were gathered. 83 students entered the survey and 16 students were randomly selected for interview. The states of knowledge of students were achieved by using ESDNN.

4.2   English Experiments

To investigate and evaluate how the M-OFS guide and support students to learn, three English experiments were under taken by level 2 and level 3 students at JinQiao University and Kunming Technology University in China during the academic year 2010-2011. The 1st experiment is introduced below.

In the first experiment, data was collected from 148 students taking English language courses whom were randomly separated into two groups. The experimental group of 83 students used M-OFS, and the control group of 65 students received the same training but without using M-OFS. The system trial includes 10 MCQs with 4 potential answers, related to English grammar. The duration of this trial is flexible. When students were using M-OFS, they were encouraged to answer the MCQs (submit their answers) as many times as they wish until they got all the correct answers or gave up (students were not given answers or how many answers were correct in their feedback, except that they answered all correct answers). Two MCQ paper tests with different questions from system trials were applied to 116 students, and 83 students participated in both paper test and system trial. 83 students completed survey after second paper test. System trial, paper test and survey were completed in practice lessons in computer room at JinQiao University.

5   Empirical Study

This section discusses the results from the first experiment in order to evaluate the effectiveness of using M-OFS to support students' deep learning.

The survey and interview were conducted after the system trial. 83 (100%) students conducted the survey. 16 (19%) students were randomly chosen for interview. For the survey, 71.1% students are satisfied with using system. 84.4% students think the feedback is what they need. Using M-OFS to learn were positively evaluated by students, illustrate that the hypotheses H1 is supported. 90.4% students would like to use the system again. 92.8% students would like to recommend the system to a friend or classmate in the future. 81.9% students have never used similar system before. For interviews, most students (94%) feel this system is useful and helps them to improve their knowledge, it indicates the hypotheses H1 is supported as well; moreover, 69% students want the exact answers in the feedback in the end. Students also want a picture of their learning process which can point out their weakness and a suggestion of how to improve their English. Some students feel that if they tried many time but cannot find the correct answer, they will lose patience in the end.

5.1   Experiment and Result

148 students are involved in the first experiment. 116 students completed the separate MCQs paper test before and after using the system. 83 students participated in system trial, and separate MCQs paper tests.

For system trials, a total of 1118 answers/attempts were submitted and a total 2143 minutes were spent by 83 participants. All of the students submitted their answers at least once. The maximum number of attempts was 106 times and the minimum was 1. The average attempts for each student is 13.5 times. The average time spent by each student is 25.8 minutes and the average time of each attempt is 1.92 minutes. 2 students (2.4%) spent more than 60 minutes. 35 (42.2%) students spent more than the average time. No students achieved the all correct answers at the beginning. 55 (66.3%) students increased their scores by an average of 12.77%, whilst 1 student increased his score by 70%. In this trial, with 10 questions and 4 possible answers, there are more than 1 million possible combinations of answers, thus the students are unlikely to make improvement by guessing answers; hence, the results show the feedback had a positive impact which partially supports hypotheses H2.

For separate MCQs paper tests, the average score before system trial is 51.6%, and the average score after system trial is 59.15%. One student (Student no. 200916031222) increased his score by 40%. 74% students increased their scores. In this test, the students were not given any answers or feedback between first (before system trial) and second (after system trial) test; furthermore, the first trial were applied 3 hours before the system trial and the second test were conducted 30 minutes after system trial; hence, the students are only learnt by using M-OFS but not any other ways; thus the results above are confident, therefore partially supporting hypotheses H3. In addition, this result also can partially support hypotheses H2.

For final examination, both the experimental group and the control group enter the same 4 days final examination. The experimental group got 79.52% and control group got 71.28% in English grammar module. This result confirms the hypotheses H4; furthermore, it also supports hypotheses H2.

5.2   Some Group Behavioural Characteristics

Previous work has made an initial investigation of the behavioural characteristics of students during their learning interaction with a diagnostic feedback system [13]. In order to explore the characteristics of students, and relate these to student responses and performance in the tests, five behavioural variables were analysed: the number of attempts (submissions), the average score changed between attempts, the average score at the end of trial, the amount of time spent to make each attempt, and the learning duration. Fig. 2 illustrates a learning behaviour of this group of students by analysing the relationship between average scores increased and learning duration. Each blue point represents average scores increased of all students used the same learning time, and its coordinate of x-axis represents student's learning duration, and its coordinate of y-axis represents average scores increased. It can be achieved from this figure that average scores increased when students spent more time on studying from the system.

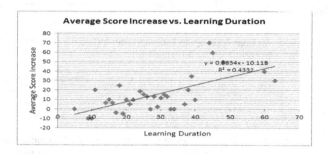

Fig. 2. Average Score Increased vs. Learning Duration

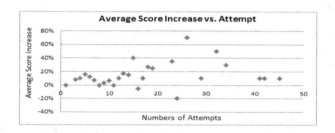

Fig. 3. Average Score Increase vs. Attempt

Fig. 3 illustrates a learning behaviour of this group of students by analysing the relationship between average scores increased and number of attempts. Each blue point represents average scores increased of all students did the same number of attempts, and its coordinate of x-axis represents number of students' attempts, and its coordinate of y-axis represents average scores increased. It can be achieved from this figure that average scores increased when students did more attempts before peak, and average scores no longer increased when students did 26 attempts, and average scores decreased by doing more attempts after peak.

Table 2. Short *learning duration*: time spent on learning < 25.6 minutes; *Long learning duration*: time spent on learning > 25.6 minutes

Behavioral Group	Average score changed	Average score at the end	Number of students
Short learning duration	9.17%	48.33%	48
Long learning duration	17.14%	61.43%	35

Table 3. Many *attempts*: number of attempt >13.4, *Few attempts*: number of attempt <13.4

Behavioral Group	Average score changed	Average score at the end	Number of students
Many attempts	20.77%	63.07%	26
Few attempts	8.77%	49.65%	57

Table 4. Slow *attempt*: average time spent on each attempt >1.92 minutes; *Rapid attempt*: average time spent on each attempt <1.92 minutes

Behavioral Group	Average score changed	Average score at the end	Number of students
Slow attempt	12.96%	53.52%	54
Rapid attempt	11.72%	54.48%	29

Table 5. Rapid *few attempt*: average time spent on each attempt <1.92 minutes and number of attempt <13.4; *Few slow attempts*: number of attempt <13.4 and average time spent on each attempt >1.92 minutes; *Many rapid attempts*: number of attempt >13.4 and average time spent on each attempt <1.92 minutes; *Slow many attempts*: average time spent on each attempt >1.92 and number of attempt >13.4

Behavioral Group	Average score changed	Average score at the end	Number of students
Rapid few attempts	2.22%	48.89%	9
Few slow attempts	10%	50%	48
Many rapid attempts	16%	57%	20
Slow many attempts	36.67%	81.67%	6

Table 6. Slow, *few attempt and short learning duration*: average time spent on each attempt >1.92 minutes, number of attempt <13.4, and time spent on learning <25.6 minutes; *Rapid, many attempts and short learning duration*: average time spent on each attempt <1.92 minutes, number of attempt >13.4, and time spent on learning <25.6 minutes; *Rapid, many attempt and long learning duration*: average time spent on each attempt <1.92 minutes, number of attempt >13.4, and time spent on learning >25.6 minutes; *Slow, few attempt and long learning duration:* average time spent on each attempt >1.92 minutes, number of attempt <13.4, and time spent on learning >25.6 minutes

Behavioral Group	Average score changed	Average score at the end	Number of students
Slow, few attempt and short learning duration	9.35%	48.71%	31
Rapid, many attempts and short learning duration	15.7%	54.29%	7
Rapid, many attempt and long learning duration	16.15%	58.46%	13
Slow, few attempt and long learning duration	11.18%	52.35%	17

Many attempts and long learning time are consistently associated with good score increases, and hence represent successful learning strategies amongst the students.

6   Summary

Six trials in three totally different subject areas have been carried out in three universities in two countries which are the UK and Chinese: 3 English trials, 2 Mathematics trials and 1 Java Programming trial. The English trials are very successful with 500 students participated. A large volume of data has been captured during the trials. The results of the first English system trials show that the average score is increased by 12.8% at the end. The results of the separate MCQ paper test present the average mark of the group is increased by 7.6%. In addition, the final examination, the average mark of experimental group is 8.2% higher than the control group. Furthermore, student surveys show that 71.1% students are satisfied with our e-learning system and 84.4% students feel that the intelligent diagnostic feedback is what they need.

7   Conclusion and Future Work

In this paper, a novel method for using snap-drift in a diagnostic tool to provide intelligent diagnostic feedback is presented. There are several innovative features of the work: this is the first time that the neural network diagnostic feedback approach in MCQ has been systemically applied to large cohorts of students and evaluated across a range of different subject areas; and the use of a neural network to discover groups of similar answers that represent different knowledge states of the students. The feedback targets the level of knowledge of individuals, and their misconceptions, guiding them toward a greater understanding of particular concepts. The results of the experiment demonstrate that an improvement in the learning process can be achieved.

In future work, it is intended to compare the effects of the feedback to the effects of other types of feedback already studied in the literature. Another promising avenue for further investigation is the extension of the tool to support knowledge state transition diagram construction and statistical data collection, which could help instructors to analyze the difficulty of the MCQs and to track students though the developmental stages of their learning.

References

1. Hefce national student survey, HEFCE, London, U.K. (2007-2010),
   http://www.hefce.ac.uk/learning/nss/
2. Rane, A., Sasikumar, M.: A constructive learning framework for language tutoring. In: Iskander, M. (ed.) Innovations in e-Learning, Instruction Technology, Assessment, and Engineering Education. Springer, The Netherlands (2007)
3. Blessing, S., Gilbert, S., Ourada, S., Ritter, S.: Lowering the bar for creating model-tracing intelligent tutoring systems. In: Luckin, R., et al. (eds.) Artificial Intelligence in Education. IOS Press (2007)
4. Hatziapostolou, T., Paraskakis, I.: Enhancing the Impact of Formative Feedback on Student Learning through an Online Feedback System. Electronic Journal of e-Learning 8(2), 111–122 (2010), http://www.ejel.org
5. Nelson, M.M., Schunn, C.D.: The Nature of Feedback: How Different Types of Peer Feedback Affect Writing Performance. Instructional Science 37(4), 375–401 (2009)
6. Garber, P.R.: Giving and Receiving Performance Feedback. HRD Press, Canada (2004)
7. Race, P.: The Lecturer's Toolkit – A Practical Guide to Assessment, Learning and Teaching, 3rd edn. Routledge, London (2006)

8. Springgay, S., Clarke, A.: Mid-Course Feedback on Faculty Teaching: A Pilot Project. In: Darling, L.F., Erickson, G.L., et al. (eds.) Collective Improvisation in a Teacher Education Community, ch. 13, pp. 171–185. Springer, The Netherlands (2007)

9. Higgins, E., Tatham, L.: Exploring the potential of multiple choice questions in assessment. Learn. Teach. Action 2(1) (2003)

10. Kuechler, W.L., Simkin, M.G.: How well do multiple choice tests evaluate student understanding in computer programming classes? J. Inf. Syst. Educ. 14(4), 389–399 (2003)

11. Payne, A., Brinkman, Wilson, F.: Towards effective feedback in e-learning packages: The design of a package to support literature searching, referencing and avoiding plagiarism. In: Proceedings of HCI 2007 Workshop: Design and Use and Experience of e-Learning Systems, pp. 71–75 (2007)

12. Lee, S.W., Palmer-Brown, D., Draganova, C.: Diagnostic Feedback by Snap-drift Question Response Grouping. In: Proceedings of the 9th WSEAS International Conference on Neural Networks (NN 2008), pp. 208–214 (2008)

13. Alemán, J.L.F., Palmer-Brown, D., Jayne, C.: Effects of Response-Driven Feedback in Computer Science Learning. IEEE Trans. Education 54, 501–508 (2011)

A Fast Hybrid k-NN Classifier Based on Homogeneous Clusters

Stefanos Ougiaroglou* and Georgios Evangelidis

Department of Applied Informatics, University of Macedonia
156 Egnatia St., 54006 Thessaloniki, Greece
{stoug,gevan}@uom.gr

Abstract. This paper proposes a hybrid method for fast and accurate Nearest Neighbor Classification. The method consists of a non-parametric cluster-based algorithm that produces a two-level speed-up data structure and a hybrid algorithm that accesses this structure to perform the classification. The proposed method was evaluated using eight real-life datasets and compared to four known speed-up methods. Experimental results show that the proposed method is fast and accurate, and, in addition, has low pre-processing computational cost.

Keywords: nearest neighbors, classification, clustering.

1 Introduction

The k-Nearest Neighbor (k-NN) classifier [4] makes predictions by searching in the available Training Set (TS) for the k nearest items (neighbors) to a new item. The latter is assigned to the most common class among the retrieved k nearest neighbors. This method, in its simplest form, must compute all distances between the new item and all items in TS. Thus, the computational cost of searching depends on the size of TS and it may be prohibitive for large datasets and time-constrained applications.

The reduction of the cost of k-NN classifier remains an important open research issue that has attracted the interest of many researchers. Many methods have been proposed to speed-up k-NN searching. A possible categorization of these methods is: (i) Multi-attribute Indexes, (ii) Data Reduction Techniques (DRTs), and, (iii) Cluster-Based Methods (CBMs). DRTs, contrary to the other two categories, have the extra benefit of the reduction of storage requirements. The effectiveness of indexes [16] highly depends on the data dimensionality. In dimensions higher than ten, the curse of dimensionality may render their performance even worse than that of sequential search.

DRTs [18,5,20,9,6,2,13] reduce the computational cost of classification by building a small representative set of the initial training data, called the Condensing Set (CS). The idea behind DRTs is to apply the k-NN classifier over

---

* S. Ougiaroglou is supported by a scholarship from the Greek Scholarships Foundations (I.K.Y.).

L. Iliadis et al. (Eds.): AIAI 2012, IFIP AICT 381, pp. 327–336, 2012.

this small set attempting to achieve accuracy as high as when using the original TS. Most DRTs produce their CS by keeping or generating for each class, many representative items (or prototypes) for the close-class-borders data areas and removing the items of the "internal" data areas.

DRTs can be divided into two main categories: (i) selection, and, (ii) abstraction algorithms. Both have the same motivation but differ on the way that they build the CS. Selection algorithms select some "real" TS items as prototypes. A typical example of this category is the well-known CNN-rule [7]. In contrast, abstraction algorithms generate prototypes by summarizing similar items. Examples of this category is the Chen and Jozwik method [3] and its variations (RSP algorithms [17]). Selection and abstraction algorithms are reviewed categorized and compared to each other in [5] and [18].

Contrary to DRTs, CBMs [8,21,10,19] do not reduce the size of TS. They preprocess the training items and group them into clusters. For each new item, they dynamically form an appropriate training subset of the initial TS (or reference set) that is then used to classify the new item.

In our previous work [15], we demonstrated that DRTs and CBMs can be combined in a hybrid classification method to achieve the desirable performance. In particular, we proposed a pre-processing algorithm to construct a data structure and a fast algorithm to classify new items by accessing this structure. The main disadvantage of our method was that both algorithms were parametric and required a trial-and-error procedure to properly adjust their parameters.

Our motivation for this paper was the development of a non-parametric and fast nearest neighbor classification method for large and high dimensional data that combines two speed-up strategies, namely, DRTs and CBMs. We have extensively evaluated the proposed method and compared it to well-known DRTs and CBMs using eight real-life datasets through a cross-validation schema.

The rest of the paper is organized as follows. Section 2 considers in detail the proposed classification method. Section 3 presents the experimental results of its evaluation against other k-NN classification methods. Finally, Section 4 concludes the paper and gives some future directions.

2   The Proposed Method

The proposed classification method includes two major stages: (i) pre-processing, that is applied on the TS items in order to construct the Speed-up Data Structure (SUDS), and, (ii) classification, that uses the SUDS and applies the proposed hybrid classifier. In this section, we present the pre-processing algorithm as well as the hybrid classifier.

The pre-processing algorithm builds SUDS by finding homogeneous clusters in TS. A cluster is homogeneous if it contains items of a specific class only. The SUDS Construction Algorithm (SUDSCA) repetitively executes the well-known k-Means clustering algorithm [12] until all of the identified clusters become homogeneous. SUDS is a two-level data structure. Its first level is a list of centroids (or representatives) of the identified homogeneous clusters. Each one represents a

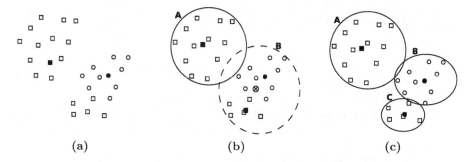

(a)   (b)   (c)

Fig. 1. SUDS construction by finding homogeneous clusters

data area of a specific class and indexes the "real" cluster items which are in the second level of SUDS. Figure 1 shows how SUDS is constructed and Algorithm 1 summarizes the steps of the corresponding algorithm.

Initially, SUDSCA finds the mean items (class-centroids) of each class in TS by averaging its items (Figure 1(a)). Then, it executes the k-Means clustering algorithm using these class centroids as initial means. Thus, for a dataset with M classes, SUDSCA initially identifies M clusters (Figure 1(b)). SUDSCA continues by analyzing the M clusters. If a cluster is homogeneous, it is added to SUDS. On the other hand, for each non-homogeneous cluster X, k-Means is executed on its items and identifies as many clusters as the number of distinct classes in X following the aforementioned procedure(Figure 1(c)). The repetitive execution of k-means terminates when all constructed clusters are homogeneous. Practically, SUDSCA constructs large clusters for internal class data areas, and small clusters for close-class-border data areas.

SUDSCA can be easily implemented using a simple queue data structure that stores the unprocessed clusters. Initially, the whole TS constitutes an unprocessed cluster and it becomes the head of the queue (line 1 in Algorithm 1). In each iteration, SUDSCA checks if cluster C in the head of the queue is homogeneous or not (line 4). If it is, the cluster is added to SUDS (lines 5-7). Otherwise, the algorithm computes a mean item for each class (*ClassCentroids*) present in C (lines 9-13). SUDSCA continues by calling the k-Means clustering algorithm for the items of C (line 14). This procedure returns a list of clusters (*NewClusters*) that are added to the queue structure (line 15) as unprocessed clusters. This procedure is repeated until the queue becomes empty (line 17).

Contrary to the pre-processing algorithm proposed in [15], SUDSCA is non-parametric. It determines the length of SUDS automatically (i.e., the number of clusters) based on the dataset used. SUDSCA extends on the idea of a previous work of ours that introduced a fast DRT called Reduction through Homogeneous Clusters (RHC) [14]. Here, our propose is not the development of a DRT, but the development of a hybrid, non-parametric method that combines DRTs and CBMs.

The second part of the proposed method is a classifier that uses SUDS. It is called Hybrid Classification Algorithm based on Homogeneous Clusters

Algorithm 1. SUDS Construction Algorithm

Input: TS **Output:** $SUDS$

1: Enqueue($Queue$, TS)
2: **repeat**
3:     $C \leftarrow$ Dequeue($Queue$)
4:     **if** C is Homogeneous **then**
5:         Compute the mean vector (class-centroid) M of C
6:         Put M into the first level of $SUDS$
7:         Put the items of C into the second level of $SUDS$ and associate them to M
8:     **else**
9:         $ClassCentroids \leftarrow \varnothing$
10:         **for** each Class L in C **do**
11:             $Centroid_L \leftarrow$ Compute the mean vector of items that belong to L
12:             $ClassCentroids \leftarrow ClassCentroids \cup Centroid_L$
13:         **end for**
14:         $NewClusters \leftarrow$ k-Means(C, $ClassCentroids$)
15:         **for** each cluster X in $NewClusters$ **do**
16:             Enqueue($Queue$, X)
17:         **end for**
18:     **end if**
19: **until** $Queue$ is empty
20: **return** $SUDS$

(HCAHC) and is described in Algorithm 2. When a new item x arrives and must be classified (line 1 in Algorithm 2), HCAHC initially scans the first level of the SUDS and retrieves the Rk nearest representatives to x (lines 2-4). We call this scan a first level search. If all Rk retrieved representatives vote a specific class, x is classified to this class (lines 5-6). Otherwise, HCAHC goes to the second level of SUDS and x is classified by searching the k "real" items within the data subset dynamically formed by the union of the clusters of the Rk representatives (lines 8-10). We call this search a second level search.

Obviously, a second level search involves higher computational cost than a first level search. However, even in this case, HCAHC searches only a small subset of the initial TS data. For instance, suppose that SUDSCA has built a SUDS with 200 nodes and we have set Rk=8. HCAHC performs the first level search and retrieves the eight nearest representatives. Suppose that not all eight of them belong to the same class. As a result, HCAHC searches for the k nearest neighbors in the union of the eight clusters that correspond to the eight representatives and performs the classification. Even in this case, HCAHC avoids searching in the rest 192 clusters.

A new item can be classified via either a first or a second level search. Practically, the first level search is an abstraction DRT, while the second level search is a CBM. That is why HCAHC is a hybrid method. Furthermore, when HC-AHC performs a second level search, it accesses an almost noise-free subset of the initial TS. Since each cluster contains items of a specific class only, the subset (union of the Rk clusters) will not contain noisy items of other irrelevant

---

Algorithm 2. HCAHC Algorithm

Input: $SUDS$, Rk, k

1: **for** each new item x **do**
2:     Scan 1st level of $SUDS$ and retrieve the Rk Nearest Representatives (NR) to x
3:     Find the majority class MC_1 of the Rk NR (ties are resolved by 1-NR)
4:     $MCC \leftarrow$ COUNT(representatives of the majority class)
5:     **if** $MCC = Rk$ **then**
6:         Classify x to MC_1
7:     **else**
8:         Scan within the set formed by the union of clusters of the Rk representatives
            and retrieve the k Nearest Neighbors (NNs) to x {Second level search}
9:         Find the majority class MC_2 of the k NNs (ties are resolved by 1-NN)
10:        Classify x to MC_2
11:    **end if**
12: **end for**

---

classes, i.e., classes which are not represented by the Rk representatives. Thus, classification performance is not affected as much by noisy data. Of course, the length of SUDS depends on the level of noise. The more noisy items in TS, the higher the final number of homogeneous clusters (or length of SUDS).

Since we aim to a non-parametric method, we must find a way to automatically determine Rk. In the experiments of the following section, we have tested the effect of the value of Rk on the performance of our method. In addition, we use the empirical rule: $Rk = \lfloor \sqrt{|SUDS|} \rfloor$, where $|SUDS|$ is the number of nodes (clusters) in SUDS.

3 Performance Evaluation

3.1 Experimental Setup

The proposed classification method was tested using eight real life datasets distributed by KEEL Repository[1][1] (see Table 1). For comparison purposes, we evaluated: (i) CNN-rule [7], the first and one of the most popular selection DRTs, (ii) RSP3 [17], the well-known abstraction DRT, (iii) the CBM proposed by Hwang and Cho (Hwang's method) [8], and, (iv) our abstraction DRT called RHC [14]. All methods were implemented in C and evaluated using 5-fold cross validation. For each dataset, we used the five already constructed pairs of Training/Testing sets hosted by KEEL repository. These sets are appropriate for 5-fold cross validation. Furthermore, we used the Euclidean distance as the distance metric.

CNN-rule, RSP3 and RHC are non-parametric methods, that is, they do not use user-defined parameters in order to reduce the data. On the other hand, Hwang's method is parametric. In addition to parameter k (number of nearest neighbors to search), which is used by all methods during the classification step,

---

[1] http://sci2s.ugr.es/keel/datasets.php

Table 1. Dataset description - Conventional k-NN Classifier performance

Dataset	Size	Attr.	Classes	Conv-k-NN	
				Acc. (%)	Cost (M)
Letter Recognition (LR)	20000	16	26	96.01	64.00
Magic Gamma Telescope (MGT)	19020	10	2	99.37	19.34
Pen-Digits (PD)	10992	16	10	91.22	6.63
Landsat Satellite (LS)	6435	36	6	81.32	57.88
Shuttle (SH)	58000	9	7	99.82	538.24
Texture (TXR)	5500	40	11	99.02	4.84
Phoneme (PH)	5404	5	2	90.10	4.67
Ring (RNG)	7400	20	2	74.69	8.76

it uses three extra parameters. For two of those parameters we used the values suggested by Hwang and Cho in their experiments. The third parameter, C, is used during the pre-processing phase and is related to the number of clusters that are constructed by the k-Means clustering algorithm. We built eight Hwang's classifiers using different C values. More specifically, each classifier $i=1,\ldots,8$, used $C = \lfloor\sqrt{\frac{n}{2^i}}\rfloor$, where n is the number of TS items. The first classifier, i.e. $i=1$ is based on the rule of thumb $C = \lfloor\sqrt{\frac{n}{2}}\rfloor$ [11].

Although SUDSCA is non-parametric, HCAHC is a parametric classifier. In addition to k, it uses parameter Rk. We built 29 HCAHC classifiers, for $Rk=2,3,\ldots,30$, and we also considered the automatic determination of Rk (see Section 2). We refer to that classifier as HCAHC-sqrt. For both HCAHC and Hwang's method, we report only the most accurate classifiers for each cost.

During the classification step, all methods involve parameter k. The DRTs perform k-NN classification using their CS, while Hwang's method does this over a small reference set that is dynamically formed for each new item. Finally, HCAHC searches for k nearest neighbors when it performs a second level search. We used the best k values for each method and dataset, i.e., the value that achieved the highest classification accuracy. In effect, we ran the cross validation many times for different k values and kept the best one.

3.2   Pre-processing Comparisons

Table 2 presents the pre-processing computational costs in terms of millions (M) distance computations (how many distances were computed during the pre-processing). As we expected, SUDSCA and RHC were executed very fast in comparison to the other approaches. This happened because: (i) the construction of SUDS and of the RHC condensing set are based on the repetitive execution of the fast k-Means clustering algorithm, and, (ii) in both cases, k-Means uses the mean items of the classes as initial centroids, and thus, clusters are consolidated very quickly. It is worth mentioning that SUDSCA and RHC pre-processing could become even faster had we used a different k-Means stopping criterion

Table 2. Preprocessing Cost (in million of distance computations)

Dataset	CNN	RSP3	Hwang's method				RHC/
			i=1	i=3	i=5	i=7	SUDSCA
LR	163.03	326.52	88.88	63.66	26.35	10.89	41.85
MGT	277.88	511.67	142.99	80.62	21.10	12.83	4.09
PD	11.76	94.80	28.80	11.27	5.97	1.70	2.88
LS	18.59	37.70	16.74	12.44	4.54	0.81	1.69
SH	45.40	17597.68	744.82	399.23	105.13	34.78	16.83
TXR	5.57	27.63	14.86	7.43	3.89	0.83	3.63
PH	13.47	20.32	9.87	3.70	1.33	0.74	0.65
RNG	29.63	43.42	18.48	12.35	5.50	2.83	2.00
Avg.	70.58	2332.47	133.18	73.84	21.73	8.18	9.20

than the full clusters consolidation (no item move from one cluster to another during a complete algorithm pass).

Concerning the other methods, RSP3 was the most time consuming approach. This is because RSP3 repetitively executes a costly procedure for finding the most distant items in data groups. Hwang's method for $i \geq 5$ is executed very fast. However, in real applications, the user must perform a trial-end-error procedure for determining the parameters. This may render pre-processing a hard and extremely time consuming procedure. Although CNN-rule is quite faster than RSP3, its pre-processing cost remains at high levels.

3.3 Classification Performance Comparisons

We performed the classification step by using seven classification methods on the eight datasets without any previous knowledge about the data (like distribution of classes, shape and size of the class data areas). The methods used were: (i) Conventional k-NN (conv-k-NN), (ii) HCAHC, (iii) HCAHC-sqrt, (iv) RHC, (v) CNN, (vi) RSP3, and, (vii) Hwang's method.

The performance measurements of conv-k-NN are shown in Table 1 while the measurements of the speed-up methods are depicted in Figure 2. In particular, figure 2 includes one diagram for each dataset. The eight diagrams present the cost measurements (in terms of millions or thousands distance computations) on the x-axis and the corresponding accuracy on the y-axis. The cost measurements indicate how many distances were computed in order to classify all testing items. Since, we used a cross-validation schema, cost measurements are average values.

Almost in all cases, HCAHC and HCAHC-sqrt achieved noteworthy performances. In some cases, they even reached the accuracy level of conv-k-NN. All diagrams show that rule $Rk = \lfloor \sqrt{|SUDS|} \rfloor$ is a good choice for the determination of Rk. With the exception of the SH dataset, HCAHC achieved better classification performance than all DRTs. On the other hand, although HCAHC and HCAHC-sqrt achieved higher accuracy than Hwang's method in all datasets, for the MGT, SH, and PH datasets, Hwang's method may be preferable because

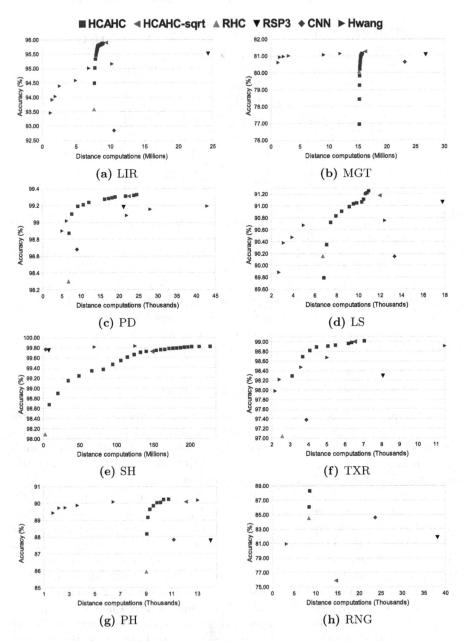

Fig. 2. Classification performance (Accuracy vs Computational cost)

it achieved accuracies close to those of HCAHC and HCAHC-sqrt at a lower computational cost.

Contrary to all other datasets, RNG was the only dataset where low Rk parameter values performed better than higher values (in Figure 2(h), HCAHC classifiers were built with $Rk=2$ and $Rk=3$. HCAHC-sqrt was built using a higher Rk value. Consequently, the corresponding performance measurements were quite bad.

Concerning the SH dataset, CNN and RSP3 built very small Condensing Sets. Thus, the k-NN classifiers that executed over these CSs, were not only accurate but very fast as well. HCAHC and HCAHC-sqrt were able to achieve the accuracy levels of CNN and RSP3 at a higher computational cost. Finally, we should consider the fact that SH is an imbalanced dataset. It has two very rare classes. Despite that, HCAHC and HCAHC-sqrt were able to classify testing items of the rare classes with high accuracy.

4 Conclusions

Speeding-up distance based classifiers is a very important issue in data mining. In this paper, we presented and evaluated a hybrid classification method. The motivation of our work was the development of a non-parametric method that has low pre-processing cost and is able to classify new items fast and with high accuracy. We presented a non-parametric fast pre-processing algorithm that builds a two-level data structure, and a hybrid classifier that makes predictions by accessing either the first or the second level of this structure. The proposed classifier is parametric since it uses parameter Rk (number of cluster representatives to use in a first level search). However, we demonstrated how Rk can be automatically determined and render the proposed method non-parametric. Experimental results based on eight real-life datasets showed that our method achieved the aforementioned goals.

Our future work includes the incremental execution of the clustering pre-processing procedure. Although the pre-processing algorithm has low computational cost, rebuilding the structure from scratch may be inadequate for dynamic environments.

References

1. Alcalá-Fdez, J., Sánchez, L., García, S., del Jesús, M.J., Ventura, S., Guiu, J.M.G., Otero, J., Romero, C., Bacardit, J., Rivas, V.M., Fernández, J.C., Herrera, F.: Keel: a software tool to assess evolutionary algorithms for data mining problems. Soft Comput. 13(3), 307–318 (2009)
2. Brighton, H., Mellish, C.: Advances in instance selection for instance-based learning algorithms. Data Min. Knowl. Discov. 6(2), 153–172 (2002)
3. Chen, C.H., Jóźwik, A.: A sample set condensation algorithm for the class sensitive artificial neural network. Pattern Recogn. Lett. 17, 819–823 (1996)
4. Dasarathy, B.V.: Nearest neighbor (NN) norms: NN pattern classification techniques. IEEE Computer Society Press (1991)

5. Garcia, S., Derrac, J., Cano, J., Herrera, F.: Prototype selection for nearest neighbor classification: Taxonomy and empirical study. IEEE Transactions on Pattern Analysis and Machine Intelligence 34(3), 417–435 (2012)
6. Grochowski, M., Jankowski, N.: Comparison of Instance Selection Algorithms II. Results and Comments. In: Rutkowski, L., Siekmann, J.H., Tadeusiewicz, R., Zadeh, L.A. (eds.) ICAISC 2004. LNCS (LNAI), vol. 3070, pp. 580–585. Springer, Heidelberg (2004)
7. Hart, P.E.: The condensed nearest neighbor rule. IEEE Transactions on Information Theory 14(3), 515–516 (1968)
8. Hwang, S., Cho, S.: Clustering-Based Reference Set Reduction for k-Nearest Neighbor. In: Liu, D., Fei, S., Hou, Z., Zhang, H., Sun, C. (eds.) ISNN 2007, Part II. LNCS, vol. 4492, pp. 880–888. Springer, Heidelberg (2007)
9. Jankowski, N., Grochowski, M.: Comparison of Instances Seletion Algorithms I. Algorithms Survey. In: Rutkowski, L., Siekmann, J.H., Tadeusiewicz, R., Zadeh, L.A. (eds.) ICAISC 2004. LNCS (LNAI), vol. 3070, pp. 598–603. Springer, Heidelberg (2004)
10. Karamitopoulos, L., Evangelidis, G.: Cluster-based similarity search in time series. In: Proceedings of the Fourth Balkan Conference in Informatics, BCI 2009, pp. 113–118. IEEE Computer Society, Washington, DC (2009)
11. Mardia, K., Kent, J., Bibby, J.: Multivariate Analysis. Academic Press (1979)
12. McQueen, J.: Some methods for classification and analysis of multivariate observations. In: Proc. of 5th Berkeley Symp. on Math. Statistics and Probability, pp. 281–298. University of California Press, Berkeley (1967)
13. Olvera-López, J.A., Carrasco-Ochoa, J.A., Martínez-Trinidad, J.F., Kittler, J.: A review of instance selection methods. Artif. Intell. Rev. 34(2), 133–143 (2010)
14. Ougiaroglou, S., Evangelidis, G.: Efficient data-set size reduction by finding homogeneous clusters. In: Procendings of the Fifth Balkan Conference in Informatics, BCI 2012. ACM (to appear, 2012)
15. Ougiaroglou, S., Evangelidis, G., Dervos, D.A.: An Adaptive Hybrid and Cluster-Based Model for Speeding Up the k-NN Classifier. In: Corchado, E., Snášel, V., Abraham, A., Woźniak, M., Graña, M., Cho, S.-B. (eds.) HAIS 2012, Part II. LNCS, vol. 7209, pp. 163–175. Springer, Heidelberg (2012)
16. Samet, H.: Foundations of multidimensional and metric data structures. The Morgan Kaufmann series in computer graphics. Elsevier/Morgan Kaufmann (2006)
17. Sánchez, J.S.: High training set size reduction by space partitioning and prototype abstraction. Pattern Recognition 37(7), 1561–1564 (2004)
18. Triguero, I., Derrac, J., García, S., Herrera, F.: A taxonomy and experimental study on prototype generation for nearest neighbor classification. IEEE Transactions on Systems, Man, and Cybernetics, Part C 42(1), 86–100 (2012)
19. Wang, X.: A fast exact k-nearest neighbors algorithm for high dimensional search using k-means clustering and triangle inequality. In: The 2011 International Joint Conference on Neural Networks (IJCNN), pp. 1293–1299 (August 2011)
20. Wilson, D.R., Martinez, T.R.: Reduction techniques for instance-based learning algorithms. Machine Learning 38(3), 257–286 (2000)
21. Zhang, B., Srihari, S.N.: Fast k-nearest neighbor classification using cluster-based trees. IEEE Trans. Pattern Anal. Mach. Intell. 26(4), 525–528 (2004)

A Spatially-Constrained Normalized Gamma Process for Data Clustering

Sotirios P. Chatzis[1], Dimitrios Korkinof[2], and Yiannis Demiris[2]

[1] Cyprus University of Technology, Limassol, Cyprus
[2] Imperial College London, UK

Abstract. In this work, we propose a novel nonparametric Bayesian method for clustering of data with spatial interdependencies. Specifically, we devise a novel normalized Gamma process, regulated by a simplified (pointwise) Markov random field (Gibbsian) distribution with a countably infinite number of states. As a result of its construction, the proposed model allows for introducing spatial dependencies in the clustering mechanics of the normalized Gamma process, thus yielding a novel nonparametric Bayesian method for spatial data clustering. We derive an efficient truncated variational Bayesian algorithm for model inference. We examine the efficacy of our approach by considering an image segmentation application using a real-world dataset. We show that our approach outperforms related methods from the field of Bayesian nonparametrics, including the infinite hidden Markov random field model, and the Dirichlet process prior.

1 Introduction

Nonparametric Bayesian modeling techniques, especially Dirichlet process mixture (DPM) models, have become very popular in statistics over the last few years, for performing nonparametric density estimation [1,2,3]. This theory is based on the observation that an infinite number of component distributions in an ordinary finite mixture model (clustering model) tends on the limit to a Dirichlet process (DP) prior [2,4]. Eventually, the nonparametric Bayesian inference scheme induced by a DPM model yields a posterior distribution on the proper number of model component densities (inferred clusters) [5], rather than selecting a fixed number of mixture components.

Markov random fields (MRFs) [6] are a classical methodology for modeling spatially-interdependent data. In essence, MRFs impose a Gibbsian distribution over the allocation of the modeled data into states (clusters), which enforces the belief that spatially adjacent data are more likely to cluster together. As the Gibbsian prior imposed by MRFs entails complex calculations that make it intractable in real-world problems dealing with large datasets, efficient approximations of the full MRF distribution are usually employed. For example, a pointwise simplification of the MRF prior based on the *mean-field principle* from statistical mechanics [7] was employed in [8]. Recently, MRFs have also

L. Iliadis et al. (Eds.): AIAI 2012, IFIP AICT 381, pp. 337–346, 2012.
© IFIP International Federation for Information Processing 2012

been used in the context of Bayesian nonparametrics yielding the infinite hidden Markov random field (iHMRF) model [9,10].

Inspired by these advances, in this paper we come up with a different approach towards clustering data with spatial interdependencies. We propose a spatially-adaptive random measure, coined the Markov random field normalized Gamma process (MRF-NGP). Our model is based on the introduction of a normalized Gamma process (NGP) controlled by an additionally postulated pointwise Markov random field imposed over the data allocation into model states, obtained by application of the mean-field principle [9]. We evaluate the efficacy of our approach considering an unsupervised image segmentation application using benchmark data.

2   Theoretical Background

2.1   The Dirichlet Process

Dirichlet process (DP) models were first introduced in [11]. A DP is characterized by a base distribution G_0 and a positive scalar α, usually referred to as the innovation parameter, and is denoted as $\mathrm{DP}(\alpha, G_0)$. Essentially, a DP is a distribution placed over a distribution. Let us suppose we randomly draw a sample distribution G from a DP, and, subsequently, we independently draw M random variables $\{\Theta_m^*\}_{m=1}^M$ from G:

$$G|\alpha, G_0 \sim \mathrm{DP}(\alpha, G_0) \tag{1}$$

$$\Theta_m^*|G \sim G, \quad m = 1, \dots M \tag{2}$$

Integrating out G, the joint distribution of the variables $\{\Theta_m^*\}_{m=1}^M$ can be shown to exhibit a clustering effect. Specifically, given the first $M-1$ samples of G, $\{\Theta_m^*\}_{m=1}^{M-1}$, it can be shown that a new sample Θ_M^* is either (a) drawn from the base distribution G_0 with probability $\frac{\alpha}{\alpha+M-1}$, or (b) is selected from the existing draws, according to a multinomial allocation, with probabilities proportional to the number of the previous draws with the same allocation [12].

2.2   Markov Random Fields

We consider an alphabet $Q = \{1, \dots, K\}$. Let S be a finite index set, $S = \{1, \dots, N\}$; we shall refer to this set, S, as the set of sites or locations. Let us consider for every site $j \in S$ a finite space $\mathcal{Z}_j$ of states z_j taking on Q. The product space $\mathcal{Z} = \prod_{j=1}^N \mathcal{Z}_j$ will be denoted as the space of the configurations of the state values of the considered sites set, $\boldsymbol{z} = (z_j)_{j \in S}$. A strictly positive probability distribution, $p(\boldsymbol{z})$, $\boldsymbol{z} \in \mathcal{Z}$, on the product space $\mathcal{Z}$ is called a random field [13].

Let ∂ denote a neighborhood system on S, i.e. a collection $\partial = \{\partial_j : j \in S\}$ of sets, such as $j \notin \partial_j$ and $l \in \partial_j$ if and only if $j \in \partial_l \; \forall l, j \in S$. Then, the previously considered random field, $p(z)$, is a Markov random field with respect to the introduced neighborhood system ∂ if [14] $p(z_j|z_{S-\{j\}}) = p(z_j|z_{\partial_j}) \; \forall j \in S$. The distribution $p(z)$ of a Markov random field can be shown to be of a Gibbsian form [15]:

$$p(z) \triangleq \frac{1}{W(\gamma)} \exp\left(-\sum_{c \in C} V_c(z|\gamma)\right) \tag{3}$$

where γ is the inverse temperature of the model, $W(\gamma)$ is the (normalizing) partition function of the model, $V_c(z|\gamma)$ are the clique potentials of the model, and C is the set of the cliques included in the model neighborhood system.

A significant problem of MRF models concerns computational tractability, as the normalizing term $W(\gamma)$ is hard to compute in applications dealing with large datasets. One way to resolve these issues is the mean-field approximation [7,16]. It is based on the idea of neglecting the fluctuations of the sites interacting with a considered site, so that the resulting system behaves as one composed of independent variables for which computation becomes tractable. That is, given an estimate $\hat{z}$ of the unknown site labels vector z, obtained by means of a stochastic restoration criterion, such as the iterative conditional modes (ICM) or the marginal posterior modes (MPM) algorithm (see, e.g., [14,16]), we make the hypothesis [17]

$$p(z) = \prod_{j=1}^{N} p(z_j|\hat{z}_{\partial_j}; \gamma) \tag{4}$$

where

$$p(z_j = i|\hat{z}_{\partial_j}; \gamma) = \frac{\exp(-\sum_{c \ni j} V_c(\tilde{z}_{ij}|\gamma))}{\sum_{h=1}^{K} \exp(-\sum_{c \ni j} V_c(\tilde{z}_{hj}|\gamma))} \tag{5}$$

$\tilde{z}_{ij} \triangleq (z_j = i, \hat{z}_{\partial_j})$, $\hat{z}_{\partial_j}$ is the estimate of the jth site neighborhood, and the indexes c refer to the cliques that contain the jth site.

3 Proposed Approach

3.1 Model Formulation

Let us consider a set of observations $Y = \{y_n\}_{n=1}^{N}$, $y_n \in \mathcal{Y}$, measured over a set of sites $S = \{1, ..., S\}$ on which a neighborhood system ∂ is defined. Let us denote as $X = \{x_n\}_{n=1}^{N}$, $x_n \in S$, the sites where the observed data points $\{y_n\}_{n=1}^{N}$ were measured. Let us introduce the latent variables $\{z_n\}_{n=1}^{N}$ denoting the model state (cluster) where an observed data point y_n measured at the location x_n is assigned by our model. Motivated by the merits and the theory of the DP discussed in the previous section, to derive the sought model, we make the key-assumption, based on the mean-field-based approximation of the MRF distribution, that for any given site x_n, we have available an estimate $\hat{z}_{\partial_n}$ of the value $z_{\partial_n} \triangleq (z_m)_{m \in \partial_n}$ of the latent cluster assignment variables of the observations measured at sites in the neighborhood of site x_n. Apparently, this assumption entails obtaining an initial estimate of the latent variables $\{z_n\}_{n=1}^{N}$ for the modeled data, as discussed in Section 2.2. Further, we consider the following predictor (location)-dependent random measure

$$G(x) = \sum_{i=1}^{\infty} \varpi_i(x)\delta_{\Theta_i} \tag{6}$$

where

$$\varpi_i(x) = \frac{\Lambda_i(x)}{\sum_{j=1}^{\infty} \Lambda_j(x)} \tag{7}$$

the random variables Λ_i follow a Gamma distribution as

$$\Lambda_i | x_n \sim \mathcal{G}(\alpha k_i(x_n; \hat{z}_{\partial_n}), 1) \tag{8}$$

α is the innovation parameter of the process, $k_i(x_n; \hat{z}_{\partial_n})$ is the probability of the nth site being assigned to the ith cluster as computed by the employed pointwise MRF distribution

$$k_i(x_n; \hat{z}_{\partial_n}) \triangleq p\left(z_n = i \middle| \hat{z}_{\partial_n}; \gamma\right) = \frac{\exp(-\sum_{c \ni x_n} V_c(\tilde{z}_{ni}|\gamma))}{\sum_{h=1}^{\infty} \exp(-\sum_{c \ni x_n} V_c(\tilde{z}_{nh}|\gamma))} \tag{9}$$

$\tilde{z}_{ni} \triangleq (z_n = i, \hat{z}_{\partial_n})$, $\hat{z}_{\partial_n}$ is the current estimate of the nth site neighborhood, $V_c(\cdot)$ are the employed clique potential functions, and the indexes c refer to the cliques that include the nth site, x_n. The utility of the pointwise MRF distribution $k_i(x_n; \hat{z}_{\partial_n})$ in our model consists in reducing the probability (discounting) of clusters that seem rather unlikely from the viewpoint of the postulated neighborhood system. We dub this random probability measure $G(x)$ the MRF-NGP process. A proof that the normalizing constant in the denominator of (7) is finite almost surely is provided in the Appendix.

3.2 Variational Bayesian Inference

Let us a consider a set of observations $Y = \{y_n\}_{n=1}^{N}$ with corresponding locations $X = \{x_n\}_{n=1}^{N}$. We postulate for our observed data a likelihood function of the form

$$p(y_n | z_n = i) = p(y_n | \theta_i) \tag{10}$$

while for the latent assignment variables z_n we consider

$$p(z_n = i | x_n) = \varpi_i(x_n) \tag{11}$$

where the $\varpi_i(x)$ are given by (7), with the prior over the $\Lambda_i(x)$ given by (8). Regarding the likelihood parameters θ_i, we impose a suitable conjugate exponential prior over them; for instance, in case of a Gaussian likelihood function

$$p(y_n | \theta_i) = \mathcal{N}(y_n | \mu_i, R_i) \tag{12}$$

we impose a Normal-Wishart prior over the likelihood parameters $\theta_i = \{\mu_i, R_i\}$, i.e.

$$p(\mu_i, R_i) = \mathcal{NW}(\mu_i, R_i | \lambda_i, m_i, \omega_i, \Omega_i) \tag{13}$$

Regarding the MRF temperature parameter γ, and the innovation parameter α, we choose to optimize them as model hyperparameters, as part of the variational inference procedure discussed next.

Our variational Bayesian inference formalism consists in derivation of a family of variational posterior distributions $q(.)$ which approximate the true posterior distribution over the infinite sets $\{z_n\}_{n=1}^{N}$, $\{\Lambda_i(x_n)\}_{i,n=1}^{\infty,N}$, and $\{\theta_i\}_{i=1}^{\infty}$. Apparently, under this infinite dimensional setting, Bayesian inference is not tractable. For this reason, we fix a value K and we let the variational posterior over the $\Lambda_k(x)$ have the property $q(\Lambda_{k>K}(x) = 0) = 1$, $\forall x \in \mathcal{S}$. In other words, we set $\varpi_k(x)$ equal to zero for $k > K$, $\forall x \in \mathcal{S}$. Note that, under this setting, the treated model involves a full MRF-NGP prior; truncation is not imposed on the MRF-NGP prior itself, but only on the variational distribution to allow for a tractable inference procedure [18].

Letting W be the parameters of our model, and Ξ the entailed hyperparameters, including γ, α, and the hyperparameters of the prior over the likelihood parameters, variational Bayesian inference consists in derivation of an approximate posterior $q(W)$ by maximization (in an iterative fashion) of the variational free energy

$$\mathcal{L}(q) = \int dW q(W) \log \frac{p(X,Y,W|\Xi)}{q(W)} \tag{14}$$

which provides a lower bound to the computationally intractable log marginal likelihood (log evidence), $\log p(X,Y)$, of the model [19]. The derived algorithm is in essence an expectation-maximization-like algorithm. Each iteration comprises an E-step, on which the variational posteriors over the model latent variables are computed, and an M-step, on which the variational posteriors over the model parameters are updated. Let us denote as $\langle . \rangle$ the posterior expectation of a quantity.

M-Step. This step comprises the updates of the Gamma-distributed variables $\Lambda_i(x_n)$

$$q(\Lambda_i(x_n)) = \mathcal{G}(\Lambda_i(x_n)|\beta_{ni}, \xi_{ni}) \tag{15}$$

where

$$\beta_{ni} = \alpha k_i(x_n; \hat{z}_{\partial_n}) + q(z_n = i) \tag{16}$$

$$\xi_{ni} = 1 + \frac{1}{\sum_{j=1}^{K} \langle \Lambda_j(x_n) \rangle} \tag{17}$$

and

$$\langle \Lambda_j(x_n) \rangle = \frac{\beta_{nj}}{\xi_{nj}} \tag{18}$$

as well as of the parameters θ_i, for which we obtain the general solution

$$\log q(\theta_i) \propto \log p(\theta_i) + \sum_{n=1}^{N} q(z_n = i) \log p(y_n|\theta_i) \tag{19}$$

which is similar to the corresponding solution for models imposing simple DP priors over their cluster assignment distributions.

E-Step. This step comprises the updates of the posteriors $q(z_n = i)$:

$$q(z_n = i) \propto \exp\left(\langle \log\Lambda_i(x_n) \rangle\right) \exp\left(\varphi_{ni}\right) \qquad (20)$$

where

$$\langle \log\Lambda_i(x_n) \rangle = \psi(\beta_{ni}) - \log\xi_{ni} \qquad (21)$$

and

$$\varphi_{nj} = \langle \log p(\boldsymbol{y}_n | \boldsymbol{\theta}_j) \rangle \qquad (22)$$

It also consists in updating the estimates of the assignment variables $\hat{\boldsymbol{z}} = (\hat{z}_n)_{n=1}^N$ which are used in computing the pointwise MRF priors employed in our model to regulate cluster discounting. For this purpose, we simply set

$$\hat{z}_n = \operatorname{argmax}_{i=1}^K q(z_n = i) \qquad (23)$$

Finally, regarding the model hyperparameters Ξ, in this work we obtain estimates of the hyperparameter γ by maximization of the lower bound $\mathcal{L}(q)$, and we heuristically select the values of the rest of the model hyperparameters.

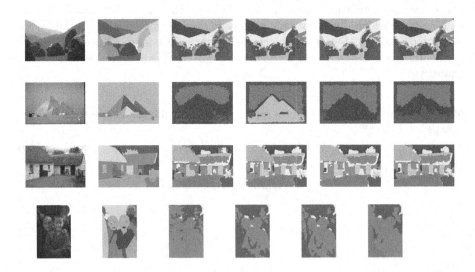

Fig. 1. Few selected 321x481 images from the Berkeley image segmentation dataset. **Left-to-right**: a) Original image, b) One human groundtruth, c) K-means initialization, d) iHMRF, e) MRF-NGP. **Top-to-bottom**: a) #241004, b) #161062, c) #385028, d) #246053.

Table 1. Obtained PRI results for the considered subset of the Berkeley benchmark

Image #	DPM	iHMRF	MRF-NGP	Image #	DPM	iHMRF	MRF-NGP
159029	0.7688	0.7727	0.7842	246053	0.5896	0.6006	0.6526
20008	0.8376	0.8514	0.8478	239096	0.7570	0.7957	0.7871
100075	0.7851	0.7795	0.7861	323016	0.8283	0.8436	0.8479
301007	0.8438	0.8432	0.8460	231015	0.8019	0.8185	0.8138
122048	0.7396	0.7520	0.7421	25098	0.8270	0.8231	0.8394
145053	0.6189	0.6315	0.7304	8143	0.6294	0.6605	0.7011
236017	0.5997	0.6035	0.6346	35010	0.7701	0.7854	0.8051
170054	0.6841	0.7453	0.7628	15004	0.7561	0.7865	0.7994
385028	0.8520	0.8393	0.8544	100080	0.8067	0.8093	0.8036
67079	0.7344	0.7347	0.7599	161062	0.6610	0.6280	0.6742
209070	0.6335	0.7006	0.7380	159045	0.6984	0.7315	0.7482
27059	0.8359	0.8470	0.8669	170057	0.6948	0.7243	0.7498
176019	0.6930	0.7470	0.7343	89072	0.7377	0.7590	0.7841
103070	0.7164	0.7307	0.7530	175032	0.5594	0.6783	0.6686
241004	0.8643	0.8642	0.8738	86016	0.7379	0.7664	0.7573

4 Experimental Evaluation

We investigate the efficacy of our approach considering an unsupervised image segmentation application. Specifically, we consider segmentation of real-world images, using a subset of the Berkeley image segmentation benchmark [20]. The Berkeley image segmentation benchmark comprises real-world color images along with their segmentation maps provided by different individuals. To obtain an objective performance evaluation of the proposed algorithm, we employ the probabilistic rand index (PR index or PRI) [21]. The PR index counts the fraction of pairs of pixels whose labelings are consistent between a computed segmentation and the given groundtruth, averaging across multiple groundtruth segmentations to account for scale variation in human perception. In all our experiments, we use Gaussian likelihoods, and choose to impose a Normal-Wishart prior over the likelihood parameters.

We compare the performance of our approach to iHMRF and DPM, both using the same likelihood function and prior over the likelihood function parameters as in the case of our model. All the evaluated algorithms are initialized by means of the k-means algorithm. Regarding the potential functions of the imposed pointwise MRFs for both the evaluated iHMRF and MRF-NGP models, we opt for a simple Potts model with a second order (8-neighbors) neighborhood system, yielding

$$p(z_n = c|\hat{\boldsymbol{z}}_{\partial_n}; \gamma) = \frac{\exp(\gamma \sum_{l \in \partial_n} \delta(c - \hat{z}_l))}{\sum_{h=1}^{K} \exp(\gamma \sum_{l \in \partial_n} \delta(h - \hat{z}_l))} \tag{24}$$

for the pointwise MRF priors, where K is the truncation threshold, and $\delta(.)$ stands for the Kronecker's delta function.

Feature extraction is effected as follows: First, each image is segmented into approximately $N = 1000$ superpixels using the method proposed in [22]. We then compute feature vectors at superpixel level, comprising RGB and HSV color information along with the values of the Maximum Response (MR) filter banks [23]. The truncation level of the variational Bayesian algorithm for all the treated models is set to $K = 10$.

To account for the effect of poor model initialization, which may lead model training to yield bad local optima as model estimators, we execute our experiments multiple times for each image, with different initializations each time, common for all the evaluated algorithms. The visual segmentation result is presented for 4 selected images in Fig. 1, along with the original image, one human groundtruth, and the initialization. The mean PRI results (over the executed repetitions) for the whole considered dataset are presented in Table 1. Total results across all images are presented in Table 2. Based on the obtained PRI metric results, we can conclude that the MRF-NGP performs considerably better than all the considered rival methods. Note also that small differences in the values of the PRI metric correspond to significant differences in the quality of the obtained segmentation results [24]. The illustrated segmentation results vouch for this assertion.

Table 2. Mean and median of the PRI metric across the considered subset of the Berkeley benchmark

PRI(%)	DPM	iHMRF	MRF-NGP
Mean	73.54	75.51	77.15
Median	73.88	76.27	77.34

5   Conclusions

In this paper, we proposed a method for nonparametric clustering of data with general spatial interdependencies. Our method, coined the MRF-NGP, consists in postulating a normalized Gamma-process, the cluster prior probabilities of which are discounted by means of a simplified pointwise Markov random field imposed over data point allocation into clusters. As a result of this construction, the MRF-NGP imposes the belief that spatially proximate data are more likely to cluster together. To examine the efficacy of our approach, we evaluated it in unsupervised image segmentation tasks using a real-life benchmark dataset, namely the Berkeley image segmentation benchmark. We showed that it yields a considerable improvement in the obtained performance of the clustering algorithm compared to both the DPM, and the recently proposed iHMRF model.

Acknowledgment. This work has been partially funded by the EU FP7 ALIZ-E project (contract #248116).

Appendix

Here, we prove the almost sure finiteness of the normalizing factor $\sum_{j=1}^{\infty} \Lambda_j(x_n)$ in (7). Let

$$S_T \triangleq \sum_{j=1}^{T} \Lambda_j(x_n) \tag{25}$$

It follows that $S_1 \leq S_2 \leq \cdots \leq S_T \leq \cdots \leq S$, where

$$S \triangleq \lim_{T \to \infty} S_T \tag{26}$$

since the random variables $\Lambda_j(x_n)$ are non-negative, as they follow a Beta distribution.

Then, to prove that S is finite almost surely, we only need to prove that $\mathbb{E}[S]$ is finite. From the monotone convergence theorem, we yield

$$\mathbb{E}[S] = \lim_{T \to \infty} \mathbb{E}[S_T] = \lim_{T \to \infty} \sum_{j=1}^{T} \mathbb{E}[\Lambda_j(x_n)] = \alpha \tag{27}$$

since $\lim_{T \to \infty} \sum_{j=1}^{T} k_j(x_n; \hat{z}_{\partial_n}) = 1$, as the $k_j(x_n; \hat{z}_{\partial_n})$ comprise prior MRF-derived probabilities of the observation at the nth site being assigned to any of the postulated model states. Hence, we have proven that S is finite almost surely.

References

1. Walker, S., Damien, P., Laud, P., Smith, A.: Bayesian nonparametric inference for random distributions and related functions. J. Roy. Statist. Soc. B 61(3), 485–527 (1999)
2. Neal, R.: Markov chain sampling methods for Dirichlet process mixture models. J. Comput. Graph. Statist. 9, 249–265 (2000)
3. Muller, P., Quintana, F.: Nonparametric Bayesian data analysis. Statist. Sci. 19(1), 95–110 (2004)
4. Antoniak, C.: Mixtures of Dirichlet processes with applications to Bayesian nonparametric problems. The Annals of Statistics 2(6), 1152–1174 (1974)
5. Blei, D., Jordan, M.: Variational methods for the Dirichlet process. In: 21st Int. Conf. Machine Learning, New York, NY, USA, pp. 12–19 (July 2004)
6. Orbanz, P., Buhmann, J.: Nonparametric Bayes image segmentation. International Journal of Computer Vision 77, 25–45 (2008)
7. Zhang, J.: The mean field theory in EM procedures for Markov random fields. IEEE Transactions on Image Processing 2(1), 27–40 (1993)
8. Celeux, G., Forbes, F., Peyrard, N.: EM procedures using mean field-like approximations for Markov model-based image segmentation. Pattern Recognition 36(1), 131–144 (2003)
9. Chatzis, S.P., Tsechpenakis, G.: The infinite hidden Markov random field model. In: Proc. 12th International IEEE Conference on Computer Vision (ICCV), Kyoto, Japan, pp. 654–661 (September 2009)

10. Chatzis, S.P., Tsechpenakis, G.: The infinite hidden Markov random field model. IEEE Transactions on Neural Networks 21(6), 1004–1014 (2010)
11. Ferguson, T.: A Bayesian analysis of some nonparametric problems. The Annals of Statistics 1, 209–230 (1973)
12. Blackwell, D., MacQueen, J.: Ferguson distributions via Pólya urn schemes. The Annals of Statistics 1(2), 353–355 (1973)
13. Maroquin, J., Mitte, S., Poggio, T.: Probabilistic solution of ill-posed problems in computational vision. Journal of the American Statistical Assocation 82, 76–89 (1987)
14. Geman, S., Geman, D.: Stochastic relaxation, Gibbs distributions and the Bayesian restoration of images. IEEE Transactions on Pattern Analysis and Machine Intelligence 6, 721–741 (1984)
15. Clifford, P.: Markov random fields in statistics. In: Grimmett, G., Welsh, D. (eds.) Disorder in Physical Systems. A volume in Honour of John M. Hammersley on the Occasion of His 70th Birthday. Oxford Science Publication, Clarendon Press, Oxford (1990)
16. Chatzis, S.P., Varvarigou, T.A.: A fuzzy clustering approach toward hidden Markov random field models for enhanced spatially constrained image segmentation. IEEE Transactions on Fuzzy Systems 16(5), 1351–1361 (2008)
17. Qian, W., Titterington, D.: Estimation of parameters in hidden Markov models. Philosophical Transactions of the Royal Society of London A 337, 407–428 (1991)
18. Blei, D.M., Jordan, M.I.: Variational inference for Dirichlet process mixtures. Bayesian Analysis 1(1), 121–144 (2006)
19. Jordan, M., Ghahramani, Z., Jaakkola, T., Saul, L.: An introduction to variational methods for graphical models. In: Jordan, M. (ed.) Learning in Graphical Models, pp. 105–162. Kluwer, Dordrecht (1998)
20. Martin, D., Fowlkes, C., Tal, D., Malik, J.: A database of human segmented natural images and its application to evaluating segmentation algorithms and measuring ecological statistics. In: Proc. 8th Int'l Conf. Computer Vision, Vancouver, Canada, pp. 416–423 (July 2001)
21. Unnikrishnan, R., Pantofaru, C., Hebert, M.: A measure for objective evaluation of image segmentation algorithms. In: Proc. IEEE Conf. Computer Vision and Pattern Recognition, San Diego, CA, USA, pp. 34–41 (June 2005)
22. Mori, G.: Guiding model search using segmentation. In: Proc. 10th IEEE Int. Conf. on Computer Vision, ICCV (2005)
23. Varma, M., Zisserman, A.: Classifying Images of Materials: Achieving Viewpoint and Illumination Independence. In: Heyden, A., Sparr, G., Nielsen, M., Johansen, P. (eds.) ECCV 2002, Part III. LNCS, vol. 2352, pp. 255–271. Springer, Heidelberg (2002)
24. Nikou, C., Galatsanos, N., Likas, A.: A class-adaptive spatially variant mixture model for image segmentation. IEEE Transactions on Image Processing 16(4), 1121–1130 (2007)

GamRec: A Clustering Method Using Geometrical Background Knowledge for GPR Data Preprocessing

Ruth Janning, Tomáš Horváth, Andre Busche, and Lars Schmidt-Thieme

University of Hildesheim, Information Systems and Machine Learning Lab,
Marienburger Platz 22, 31141 Hildesheim, Germany
{janning,horvath,busche,schmidt-thieme}@ismll.uni-hildesheim.de

Abstract. GPR is a nondestructive method to scan the subsurface. On the resulting radargrams, originally interpreted manually in a time consuming process, one can see hyperbolas corresponding to buried objects. For accelerating the interpretation a machine shall be enabled to recognize hyperbolas on radargrams autonomously. One possibility is the combination of clustering with an expectation maximization algorithm. However, there is no suitable clustering algorithm for hyperbola recognition. Hence, we propose a clustering method specialized for this problem. Our approach is a *directed* shape based clustering combined with a sweep line algorithm. In contrast to other approaches our algorithm finds hyperbola shaped clusters and is (1) able to recognize intersecting hyperbolas, (2) noise robust and (3) does not require to know the number of clusters in the beginning but it finds this number. This is an important step towards the goal to fully automatize the buried object detection.

Keywords: Ground penetrating radar (GPR), Object detection, Clustering, Sweep line algorithm, Preprocessing.

1 Introduction

Ground penetrating radar (GPR) is a method to scan the shallow subsurface without destroying the road surface. It can be used to detect buried objects like pipes, cables, ducts and sewers. To get a 2D GPR radargram image (fig. 1 (b), 2 (a)) of a cross-sectional area of the subsurface, electromagnetic waves are transmitted into the ground and the reflected signals are caught by an antenna. The vector of reflections (intensities on the radargram image) measured at one certain point for different answer times is called an A-Scan. The radargram image, the B-Scan, is a sequence of consecutive A-Scans. If the antenna is moved in a line perpendicular to a pipe, then the signals reflected from the pipe have the shape of a hyperbola branch in the radargram image, because (as fig. 1 (a) shows) the more the antenna nears the pipe the shorter is the time in which the corresponding signal is reflected. Originally, GPR radargram images were interpreted manually by human experts, but the pipe localization takes much time in this way (cp. [6]). Hence, an automatization of the radargram interpretation

L. Iliadis et al. (Eds.): AIAI 2012, IFIP AICT 381, pp. 347–356, 2012.

which can accelerate the whole process is required. However, in the beginning it is not known how much hyperbolas there are and where the hyperbola branches are located. Additionally, there might be noise in the radargram image and the different hyperbola branches might intersect (see fig. 2, 3).

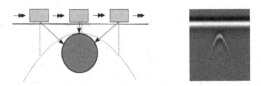

Fig. 1. The antenna (*rectangle*) is moved perpendicular to the buried object (*circle*). The downwards directed arrows represent transmitted signals, the dotted lines indicate the position of the received signal values in the corresponding A-Scans. On the radargram (right) one can see a hyperbola branch corresponding to the buried object.

2   General Problem

For $n \in \mathbb{N}$ let $[n] := \{x \in \mathbb{N} \mid 0 \le x \le (n-1)\}$. Given is a dense 2D image

$$\boldsymbol{B} \subseteq [c] \times [r] \times [\iota_{min}, \iota_{max}], \ c, r \in \mathbb{N}, \ \iota_{min}, \iota_{max} \in \mathbb{R}, \tag{1}$$

a B-Scan composed of c columns (A-scans) with intensities in a range from ι_{min} to ι_{max}. The r rows of $\boldsymbol{B}$ correspond to the signal travel times. In this B-scans upper branches (the origin is the upper left corner of the image) of North-South opening hyperbolas with extreme intensities are searched.

$\boldsymbol{B}^T := \{(j, i) \mid (j, i, \iota) \in \boldsymbol{B} \wedge \iota > \tau\}$ is a sparse representation of $\boldsymbol{B}$ containing only points with extreme intensities larger than a threshold τ.

A searched upper hyperbola branch can be defined as $f_{a,b,j_0,i_0} : \mathbb{R} \to \mathbb{R}, \ i \mapsto \sqrt{(1 + \frac{(j-j_0)^2}{b^2}) \cdot a^2} + i_0$ where (j_0, i_0) is the center of the hyperbola, a is the distance between the center and the apexes of the branches and b influences the curvature of the hyperbola. Let the distance between a point and a hyperbola branch be $D((i, j), f) := d((j, i), (j, f(j)))$ for any distance measure between points, e.g. the Euclidean distance $d((j, i), (j', i')) := \sqrt{(j - j')^2 + (i - i')^2}$. The hyperbola recognition problem can then be formalized as follows:

Definition 1 (Hyperbola recognition problem). *Given $\boldsymbol{B}^T$ and D, find a set of hyperbola branches $F := \{f_1, ..., f_K\}$ and a decomposition $\boldsymbol{B}^T = (\bigcup_{f_k \in F} h_{f_k}) \cup B_{noise}, \forall f_k \in F : B_{noise} \cap h_{f_k} = \emptyset$, of $\boldsymbol{B}^T$ into clusters h_{f_k} and a noise cluster B_{noise} with $|B_{noise}| \approx \nu^{|F|} \cdot |\boldsymbol{B}^T|$ for some $\nu \in [0, 1)$ such that the following sum of distances of all points $\in \boldsymbol{B}^T \setminus B_{noise}$ to their assigned hyperbola branch is minimal:*

$$err := \sum_{k=1}^{K} \sum_{(j,i) \in h_{f_k}} D((j, i), f_k) \tag{2}$$

3   Related Work

In the last few years different approaches were applied to the hyperbola recognition problem on GPR radargrams. The Hough Transform (HT) (e.g. used in [6], [7]) is an often used method. However, the HT is a brute force method and is hence computationally intensive (see e.g. [2]). A more efficient approach is presented in [2]. This method uses clustering combined with a classification expectation maximization algorithm. In an iterative process the algorithm fits hyperbola branches to clusters and then rearranges the clusters by assigning each point to one cluster according to the maximum posterior probability of the given point to be in this cluster. The hyperbola fitting and a rearrangement of the clusters is done alternately until convergence. However, for getting the initial clusters a K-means algorithm is used. The K-means clustering has some important drawbacks in relation to the hyperbola recognition problem:

1. It is not resistant against noise (even if a noise cluster is used which contains the $q\%$ most distant points of the other clusters, see fig. 2 (b), (d)).
2. It can recognize only convex clusters (see fig. 2 (c)).
3. The number K of clusters to detect has to be known beforehand.

In [2] a Bayesian information criterion (BIC) is proposed to estimate K, but also for this procedure a range for the possible number of hyperbolas has to be estimated beforehand and for each of the values in this range the whole approach has to be applied. Afterwards the model with the highest overall likelihood is selected as the final result.

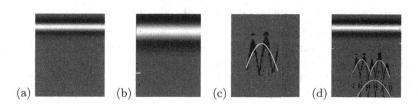

Fig. 2. (a) Radargram image of 5 buried objects. (b) K-means does not recognize the noise points as noise and assigns them to a cluster (*black*, fitted hyperbola: *white*). (c) K-means can only find convex clusters (*black*, fitted hyperbola: *white*). (d) final result of K-means with $K = 5, t = 20$ and $q = 10$ for the noise cluster.

A clustering method which avoids the above mentioned drawbacks of K-means is the shape based clustering method DBSCAN [3]. This method is able to treat noise, to find clusters of arbitrary shapes and the number of occurring clusters has not to be known beforehand. Nevertheless, the original DBSCAN

1. is not able to recognize intersecting hyperbola branches, which are caused by nearby buried objects and often occur in GPR radargrams (it finds one cluster for two or more intersecting hyperbolas (see fig. 3 (b))), and
2. does not differentiate between clusters with hyperbola shape and clusters with other shapes.

4   Our Contribution

The drawbacks of DBSCAN in relation to the hyperbola recognition problem can be treated by integrating geometrical background knowledge. For this purpose we propose *GamRec* (*Geometric approach for multi-hyperbola Recognition*), a kind of a *directed* DBSCAN combined with a sweep line algorithm. Sweep line algorithms (see e.g. [1]) originate from algorithmic geometry. The idea is to imagine that a line moves (*sweeps*) over the considered plane from one side to the other one and pauses if it touches a special point to apply some action to this point. Such points are visited just once, i.e. every point behind the sweep line is never visited again. The characteristic of sweep line algorithms is that they reduce the computational complexity by translating an n dimensional static problem into an $n - 1$ dimensional dynamic problem. A clustering algorithm for spatial data which uses a sweep line algorithm is presented in [5], but it is not able to separate overlapping clusters, i.e. it can also not recognize intersecting hyperbola branches. In GPR radargram images apexes are the highest points of hyperbola branches and if an apex is once found one has to search for the left and right hyperbola branch side below it in left-down and right-down direction. Hence, a sweep line can be moved top down over the image and collect hyperbola shaped clusters composed of an apex, a right and a left side. In doing so intersections are treated by considering that a point of an intersection must belong to the right side of one hyperbola branch and to the left side of another one. The usual sorting phase of a sweep line algorithm is not needed, as the considered points are already sorted by their occurrence in the input image. Our approach treats all the above mentioned problems of K-means and DBSCAN in relation to the hyperbola recognition problem, i.e. it (see also fig. 3 (c), (d))

1. ignores background noise,
2. finds hyperbola branch shaped clusters composed of apex, right and left side,
3. has not to know the number of occurring hyperbolas in the beginning,
4. recognizes intersecting hyperbola branches as different hyperbola branches,
5. distinguishes between hyperbola shaped clusters and non-hyperbola clusters.

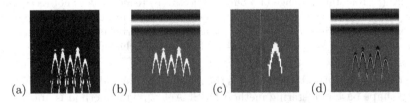

(a)   (b)   (c)   (d)

Fig. 3.   (a) Thresholded radargram image with $p = 18$. (b) One Single DBSCAN cluster (*white*) found for actually 5 hyperbolas. (c) Single GamRec cluster (*white*). (d) final resulting clusters (*black*) and fitted hyperbola branches (*white*) of GamRec applied to the thresholded radargram image of the radargram in fig. 2 with $e = 8$.

The shapes of the resulting clusters are very close to the shapes of the original hyperbola branches (see fig. 3 (c)). That indicates that a hyperbola fitting has to be applied just once after running GamRec instead of applying it several times in an iterative process to rearranged clusters until the clusters are close enough to the hyperbola branches, as done in the approach of [2]. To strengthen this assumption in our experiments in section 6 we compare the original initialization (K-means with K given) of the approach in [2] with GamRec as an initialization and show that if we use GamRec instead of K-means with K given just one hyperbola fitting step is needed for achieving already very good results. But first we will introduce GamRec in the following section 5.

5 The GamRec Algorithm

Our approach (see also fig. 4) consists of three consecutive steps: the preparation, the sweep step and the noise removing step.

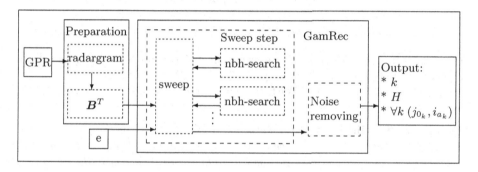

Fig. 4. Architecture of GamRec (nbh-search = *neighborhood search*)

5.1 Preparation

Before applying our algorithm (and any other clustering method) the B-scan B has to be translated into B^T. Hence, for every row i of the radargram image an intensity threshold τ_i is computed for $B^T := \{(j,i) \mid (j,i,\iota) \in B \wedge \iota > \tau_i\}$. Threshold τ_i corresponds to the lower limit of the upper $p\%$ quantile of the intensities ι of the points $(j,i,\iota) \in B$ in row i. Parameter p depends on the contrast of the image but can be considered as a constant for the same kind of images (for our experiments we used $p = 18$).

5.2 The Sweep Step

GamRec searches for a set of clusters $H := \{h_1, h_2, ..., h_K \mid h_k \subseteq B^T, k = 1, ..., K\}$ with $h_k := \{(j,i) \in B^T \mid \exists (j',i') \in h_k \neq (j,i) : (|j-j'| \leq e) \wedge (|i-i'| \leq e)\}$ where e is the maximal distance between two neighboring points in a cluster. Such clusters h_k shall possess the following properties:

- $h_k = h_{k_L} \cup h_{k_C} \cup h_{k_R}$ is a union of point sets.
  - h_{k_C} contains the apex (j_a, i_a).
  - $h_{k_L} := \{(j,i) \in h_k \mid j < j_a, i \geq i_a\}$ contains all points of the left side.
  - $h_{k_R} := \{(j,i) \in h_k \mid j > j_a, i \geq i_a\}$ contains all points of the right side.
- For every point $(j,i) \in h_k$ must hold $|(\sqrt{(1 + \frac{(j-j_0)^2}{b^2})} \cdot a^2 + i_0) - i| < \delta$ with appropriate a, b, i_0, j_0 and δ. That means the points in the cluster are positioned around and near to a hyperbola branch.

```
Sweep:
H := {};
for i from 1 to r do estimate all middle points in row i;
    for every found middle point poi = (j_mid, i) do
        if poi is not used right and not used left then do
            h_k := directedNeighborhoodSearch(poi);
            if |h_{k_L}| > e^2 ∧ |h_{k_R}| > e^2 then do H := H ∪ {h_k};
                for all points poi_L ∈ h_{k_L} do mark poi_L as used left;
                for all points poi_R ∈ h_{k_R} do mark poi_R as used right;
```

Fig. 5. GamRec algorithm: sweep step

Our algorithm (fig. 5, 6) applies a sweep line algorithm by moving top down over the thresholded image $T := \{(j,i,\iota) \mid (j,i,\iota') \in B \wedge ((j,i) \in B^T \Rightarrow \iota = 1) \wedge ((j,i) \notin B^T \Rightarrow \iota = 0)\}$. With the sweep line all *middle points* (see def. 2) are visited.

Definition 2 (Middle point). *Let* $s = ((j_1,i), ..., (j_n,i))$ *with* $j_1 < j_2 < ... < j_n$, $|j_{w+1} - j_w| \leq e$, $w = 1, ..., n$ *be a sequence of points* $\in B^T$ *in row* i. *For every such sequence* s *with length* n *in row* i *a middle point* (j_{mid}, i) *is the point in* s *with the index* $mid = div(\frac{n}{2})$.

For every *middle point* a directed neighborhood search is started (fig. 6), if the point was not *used left* or *used right* before (a point is assigned as *used left* (*used right*), if it belongs to the left (right) side of a hyperbola cluster already found). For the points $\in B^T$ of the left side h_{k_L} of the appendant cluster is searched in the neighborhood in left and down direction and for the points $\in B^T$ of the right side h_{k_R} in right and down direction. A neighboring point $\in B^T$ is inserted into h_{k_L}, (1) if the point was not *used left* and not *used right* before or (2) if it was yet *used right* for another cluster and is now located left from the current apex. h_{k_R} is treated in the same way but with *left* and *right* transposed. In this way GamRec is able to treat intersections of different hyperbola branches. It allows that points may belong to more than one cluster if they are located in the intersection of two hyperbola shaped clusters. GamRec uses one hyper parameter e for the maximal distance between neighboring cluster points (equivalent to ϵ in DBSCAN). Parameter e depends on the closeness of the points belonging to a hyperbola cluster. For our experiments we used a constant value of $e = 8$, estimated by a grid search. The second hyper parameter of DBSCAN, the minimal number of neighbors min_N, is not needed in GamRec, as noise is identified mainly by considering the hyperbola shape as described in the following section 5.3.

---

directedNeighborhoodSearch$((j_{mid}, i))$

1. $h_{k_C} := \{(j_{mid}, i)\};$
2. $h_{k_L}^{(0)} := \{poi_L = (j_{mid} - e_1, i + e_2) \in \boldsymbol{B}^T \mid e_1 = 1, ..., e, e_2 = 0, ..., e,$
   $\quad poi_L$ is not *used left* $\vee$ poi_L is *used right*$\};$
   until $h_{k_L}^{(t)} - h_{k_L}^{(t-1)}$ **do**
   $\quad h_{k_L}^{(t)} := h_{k_L}^{(t-1)} \cup \{poi_L = (j, i) \in \boldsymbol{B}^T \mid \exists (j_v, i_v) \in h_{k_L}^{(t-1)} : j = j_v - e_3,$
   $\quad i = i_v + e_4, e_3 = 0, ..., e, e_4 = 0, ..., e,$ poi_L is not *used left* $\vee$ poi_L is *used right*$\};$
   $\quad h_{k_L} := h_{k_L}^{(t)};$
3. $h_{k_R}^{(0)} := \{poi_R = (j_{mid} + e_1, i + e_2) \in \boldsymbol{B}^T \mid e_1 = 1, ..., e, e_2 = 0, ..., e,$
   $\quad poi_R$ is not *used right* $\vee$ poi_R is *used left*$\};$
   until $h_{k_R}^{(t)} = h_{k_R}^{(t-1)}$ **do**
   $\quad h_{k_R}^{(t)} := h_{k_R}^{(t-1)} \cup \{poi_R = (j, i) \in \boldsymbol{B}^T \mid \exists (j_v, i_v) \in h_{k_R}^{(t-1)} : j = j_v + e_3,$
   $\quad i = i_v + e_4, e_3 = 0, ..., e, e_4 = 0, ..., e,$ poi_R is not *used right* $\vee$ poi_R is *used left*$\};$
   $\quad h_{k_R} := h_{k_R}^{(t)};$
4. **return** $h_{k_L} \cup h_{k_C} \cup h_{k_R};$

Fig. 6. GamRec: directed neighborhood search

5.3 The Noise Removing Step

Finally, clusters which do not correspond to hyperbola branches are removed from H (see fig. 7). GamRec identifies such a *non-hyperbola* cluster by considering geometrical properties of hyperbolas. It investigates if

- the side length of one side of the rectangle around all of its points is less than $2e$ (to eliminate larger background noise),
- in this rectangle the points in the upper left corner, in the upper right corner or in the lower center belong to the cluster and
- the height of one side is less than half of the height of the other side.

In this cases the cluster cannot have the shape of a hyperbola branch.

Noise removing:
for all $h_k \in H$ **do**
$\quad$**if** $(j_{max} - j_{min}) \leq 2e \vee (i_{max} - i_{min}) \leq 2e$ **then do** $H := H \setminus h_k;$
$\quad$**else do**
$\quad\quad$**if** $\quad (\{(j_{min} + e_3, i_{min} + e_4) \mid e_3 = 0, ..., e, e_4 = 0, ..., e\} \subset h_{k_L})$
$\quad\quad\quad \vee (\{(j_{max} - e_3, i_{min} + e_4) \mid e_3 = 0, ..., e, e_4 = 0, ..., e\} \subset h_{k_R})$
$\quad\quad\quad \vee (\{(j_{mid} \pm e_3, i_{max} - e_4) \mid e_3 = 0, ..., e, e_4 = 0, ..., e\} \subset h_k)$
$\quad\quad$**then do** $H := H \setminus h_k;$[a]
$\quad\quad$**else do**
$\quad\quad\quad$**if** $\quad ((i_{max_L} - i_{min_L}) < 0.5(i_{max_R} - i_{min_R}))$
$\quad\quad\quad\quad \vee ((i_{max_R} - i_{min_R}) < 0.5(i_{max_L} - i_{min_L}))$ **then do** $H := H \setminus h_k;$

---

[a] j_{min}, i_{min} are the minimal j and i values of all points $\in h_k$, j_{max}, i_{max} the maximal j and i values (correspondingly i_{min_R}, i_{max_R} and i_{min_L}, i_{max_L} for the right and left side) and j_{mid} is the j value of the apex of h_k.

Fig. 7. GamRec algorithm: noise removing step

5.4   Complexity

Let n be the number of considered points for a clustering. In our experiments we compare the original initialization (K-means with K given) used in [2] with GamRec as an initialization. Hence, we consider the complexity of GamRec as well as of the K-means clustering. The complexity $O(n \cdot K \cdot t)$ of K-means clustering results from the t iterations multiplied by the n investigations for every point of its distances to all K cluster centers. The complexity of GamRec corresponds to the sum of the complexities of the sweep and the noise removing step. As GamRec is applied to points of an image, the sorting of points for the sweep can be skipped and an efficient neighborhood search is enabled. The sweep consists of two main steps. The first step is the search for the m *middle points*, which takes $O(n)$ time, as every point in every row is visited. The second step includes a directed neighborhood search with a complexity equal to the size of the corresponding cluster for every found *middle point* times $2e^2$, as for every point $2e^2$ points in the neighborhood are investigated. The complexity of the second step overall is $O((\sum_{i=1}^{m} size_i) \cdot e^2)$, where $size_i$ is the size of the cluster belonging to the ith *middle point*. Hence, the complexity of the sweep is $O(n + (\sum_{i=1}^{m} size_i) \cdot e^2)$. To specify $O((\sum_{i=1}^{m} size_i) \cdot e^2)$ we investigate three cases:

1. If all points would be *middle points* ($m = n$), then $O((\sum_{i=1}^{m} size_i) \cdot e^2) = O(n \cdot e^2)$, because then every point has either no neighbors (two *middle points* of the same row cannot have a distance less or equal than e, see def. 2) and $size_i = 1$ for every i, or a *middle point* possesses neighbors below it which are *middle points* and belong to its cluster but these points are then marked as *used left* or *used right*. That means that from these points no neighborhood search will be started in the following. Additionally, these points will either be visited never again or only once again if it belongs to an intersection.
2. If no point would be a *middle point* ($m = 0$), then $O((\sum_{i=1}^{m} size_i) \cdot e^2) = O(n \cdot e^2)$ because then holds $n = 0$.
3. If $0 < m < n$, then $O((\sum_{i=1}^{m} size_i) \cdot e^2) = O(n \cdot e^2)$, as every point may belong to at most two clusters (just intersection points belong to two clusters) so that the sum of all cluster sizes $size_i$ has to be less than $2n$.

In every case the complexity of the sweep is $O(n + n \cdot e^2) = O(n \cdot e^2)$. The noise removing step needs $O(m + m \cdot e^2 + m)$ time, as for every cluster h_i two constant time operations are applied and e^2 points at four places in the rectangle around h_i are investigated. Because of $m \leq n$ we assign a complexity of $O(n + n \cdot e^2 + n) = O(n \cdot e^2)$ time to the noise removing. Altogether GamRec has a complexity of $O(n \cdot e^2 + n \cdot e^2) = O(n \cdot e^2)$ time. As one can see, the complexities $O(n \cdot K \cdot t)$ and $O(n \cdot e^2)$ of K-means clustering and GamRec are similar efficient. In our experiments e.g. we have chosen $K = 5, t = 20, e = 8$, which results in $O(n \cdot 100)$ and $O(n \cdot 64)$.

6 Experiments

We simulated a cross-sectional area of 2.50 meter length and 0.60 meter depth. The radargrams (see 2 (a)) of this cross-sectional area were produced by GprMax [4], an electromagnetic wave simulator for modeling GPR data based on the Finite-Difference Time-Domain numerical method. GprMax is an often used tool in this field, as real data are rare and expensive to achieve. Additionally, with this data we can easily compare the results of GamRec and of the original initialization of the approach in [2], as we know exactly the ground truth, i.e. where the pipes are located. This means we can compare the positions found with the real positions of the buried objects. In each of 45 scenarios, 5 perfect conductors (e.g. wires) with a radius of 3 centimeters are buried in a 2.50×0.60 meter area of wet sand at different locations with a distance of 25 centimeters between neighboring pipes. We used a 600 MHz antenna frequency for the measurements. In this way we investigated 225 pipes at different horizontal positions and different depths. For the experiments we applied GamRec (with $e = 8$) as well as the original initialization (K-means with K given, $K = 5, t = 20$) of the approach in [2] to the thresholded radargram images (with $p = 18$, see fig. 3 (a)). To the clusters found a hyperbola fitting algorithm was applied (from the K-means clusters the 10% most distant points were removed into the noise cluster). Subsequently, the positions found (fitted apex for K-means, depth of apex of hyperbola cluster and horizontal position of fitted apex for GamRec) were compared to the upper positions of the corresponding buried objects by computing the Euclidean distance between them (see table 1). Table 1 shows that after one run of GamRec the real positions of the buried objects can be estimated already with an error of just a few millimeters and at most about one centimeter. K-means with K given in contrast delivers position estimations with an error from a few centimeters up to more than one meter. These results indicate that GamRec delivers not just the better initialization but also that there is no need for a rearrangement of the clusters found by GamRec. One hyperbola fitting step after the application of GamRec suffices to reach already very good position estimation results.

Table 1. Average distance errors in meter of the 5 pipes in each of the 45 scenarios ordered from minimal to maximal Euclidean distance. The overall average Euclidean distance error of all 225 pipes is presented in the last column.

	1.	2.	3.	4.	5.	average
GamRec	0.0016 m	0.0018 m	0.0027 m	0.0066 m	0.0092 m	0.0044 m
(standard deviation)	(0.0008 m)	(0.0008 m)	(0.0015 m)	(0.0031 m)	(0.0033 m)	(0.0037 m)
K-means with K given	0.0678 m	0.1292 m	0.2413 m	0.5078 m	1.0707 m	0.2365 m
(standard deviation)	(0.0512 m)	(0.0497 m)	(0.1185 m)	(0.3052 m)	(0.5791 m)	(0.4718 m)

7 Conclusions

We presented GamRec, a clustering method specialized for an improved GPR data preprocessing. Just one run of GamRec and one hyperbola fitting step applied to the hyperbola shaped clusters found is needed to reach already very good pipe position estimation results. Furthermore, GamRec finds the number of occurring hyperbolas by itself, it is able to recognize intersecting hyperbola branches and it is resistant against background noise. This work is an initial study in the field of pipe detection by GPR. Our final goal is a full automatization of the whole pipe detection process, including pipe position estimation and pipe course detection.

Acknowledgments. This work is co-funded by the European Regional Development Fund project AcoGPR under the grant agreement no. WA3 80122470.

References

1. Boissonnat, J.-D., Yvinec, M.: Algorithmic Geometry. Cambridge University Press (1998)
2. Chen, H., Cohn, A.G.: Probabilistic Robust Hyperbola Mixture Model for Interpreting Ground Penetrating Radar Data. In: Proceedings of the 2010 IEEE World Congress on Computational intelligence (WCCI 2010), Barcelona (2010)
3. Ester, M., Kriegel, H., Sander, J., Xu, X.: A density-based algorithm for discovering clusters in large spatial databases with noise. In: Proceedings of the Second International Conference on Knowledge Discovery and Data Mining, KDD 1996 (1996)
4. Giannopoulos, A.: Modelling ground penetrating radar by GprMax. Construction and Building Materials 19(10), 755–762 (2005)
5. Zalik, K.R., Zalik, B.: A sweep-line algorithm for spatial clustering. In: Advances in Engineering Software, vol. 40, pp. 445–451 (2009)
6. Simi, A., Bracciali, S., Manacorda, G.: Hough transform based automatic pipe detection for array GPR: algorithm development and on-site tests. In: Radar Conference RADAR 2008, Rome, pp. 1–6 (2008)
7. Windsor, C.G., Capineri, L., Falorni, P.: The Estimation of Buried Pipe Diameters by Generalized Hough Transform of Radar Data. In: Progress in Electromagnetics Research Symposium, Hangzhou, China, pp. 345–349 (2005)

Enhancing Clustering by Exploiting Complementary Data Modalities in the Medical Domain

Samah Jamal Fodeh[1], Ali Haddad[2], Cynthia Brandt[3],
Martin Schultz[4], and Michael Krauthammer[5]

[1,3] Yale University School of Medicine,
[2] Department of Mathematics,
[4] Department of Computer Science,
[5] Department of Pathology,
Yale University, New Haven, CT, USA
{samah.fodeh,ali.haddad,cynthia.brandt,
Schultz-martin,michael.krauthammer}@yale.edu

Abstract. Data Clustering has been an active area of research in many different application areas, with existing clustering algorithms mostly focusing on partitioning one modality or representation of the data. In this study, we delineate and demonstrate a new, enhanced data clustering approach whose innovation is its exploitation of multiple data modalities. We propose BI-NMF, a bi-modal clustering approach based on Non Negative Matrix Factorization (NMF) that clusters two differing data modalities simultaneously. The strength of our approach is its combining of multiple aspects of the data when forming the final clusters. To assess the utility of our approach, we performed several experiments on two distinct biomedical datasets with two modalities each. Comparing the clusters of BI-NMF with NMF clusters of single data modality, we observed consistent performance enhancement across both datasets. Our experimental results suggest that BI-NMF is advantageous for boosting data clustering.

Keywords: BI-NMF, clustering, biomedical, images, non negative matrix factorization.

1   Introduction

Clustering has been an active area of research in data mining and machine learning due to the rapidly growing data in different domains such as biology and clinical medicine. In biology, for instance, there is an avalanche of data from novel high throughput and imaging technologies. When applied to cancer images, clustering has been effective in identifying malignant and normal breast images [1]. Biomedical publications often present the results of biological experiments in figures and graphs that feature detailed, explanatory footnotes and captions. This annotation comprises a simple, textual representation of the images. In the clinical literature, a new semantic representation has evolved as a result of mapping the words in physicians' clinical notes to the corresponding semantic descriptors in the Unified Medical Language System (UMLS).

L. Iliadis et al. (Eds.): AIAI 2012, IFIP AICT 381, pp. 357–367, 2012.

Each representation of the data e.g. images, captions and semantic descriptors, is a unique data modality generated by a particular process wherein the objects have different features, structure and dimensionality. Differential encoding of the features of each modality causes variability in the obtained partitions when clustering around the individual data modality. In this discussion we explore alternative methods of building clusters around the complementary data modalities of a particular dataset to obtain more cohesive clusters. Unlike current algorithms which cluster on a single data modality, our proposed approach creates clusters by extracting information from completely different domains of information that describe the same data.

There have been recent efforts to perform multi-modal clustering. For example, Chen, Wang and Dong [10] proposed a co-clustering method using textual data that employs non negative matrix factorization (NMF) that draws from two data modalities: textual documents and their corresponding categories. Their method, however, is semi-supervised and requires user input to allow the algorithm to "learn" the distance metric. Comar, Tan and Jain [5] proposed the joint clustering of multiple social networks to identify cohesive communities characterized by reduced levels of noise. In this paper, we propose BI-NMF that combines information from two complementary data modalities to enhance clustering. NMF is a matrix factorization approach that has been shown to be effective for improving data clustering [6] as it produces meaningful clusters due to the non-negative nature of the solution. Specifically, NMF aims to factorize a data matrix into two non-negative matrices which are more compact (with lower dimensionality) and their product approximates the original matrix. One hopes that the new representation uncovers the hidden clusters in a given dataset. In this study, we cluster by drawing information from two different data matrices pertaining to complementary data modalities, thereby allowing us to exploit different aspects of the data while simultaneously reducing the distortion associated with clustering on a single modality. BI-NMF can be useful for any data described with multiple sources of information, i.e., modalities. We demonstrate our algorithm on two clinical datasets that each has information from two modalities. The first dataset contains images and their corresponding text captions and the second features textual notes reported by a clinical radiologist and their complementary semantic descriptors. The major contribution in this paper is the demonstration of a new method that simultaneously clusters two data modalities by jointly factorizing their corresponding matrices. The chief advantage of our method is enhanced clustering via the exploitation of information from complementary data modalities

The remainder of this paper is organized as follows. Section 2 presents the related work on clustering using NMF. Section 3 derives the proposed method along with the formal proofs. Section 4 presents the experimental results, followed by Section 5 featuring some concluding remarks.

2   Related Work

NMF has gained considerable attention recently in many domains such as pattern recognition and machine learning. Paatero and Tapper [6] proposed to use NMF algorithm to identify certain parts of objects like human faces. In a similar fashion, Xu, Liu and Gong used NMF to find clusters of documents [9]. They considered each dimension in the NMF space as one cluster and mapped a document d to the column *cluster* that has the maximum entry with *d*. As NMF performs learning in the Euclidean space, it fails to consider the intrinsic geometrical structure as suggested in [2], hence the authors extended NMF by imposing a new constraint that captures the geometrical representation of the data. Unlike previous methods which apply NMF to only one data modality, our proposed method aims to learn from two different modalities simultaneously. Reference [5] proposed to jointly cluster multiple networks using tri non-negative matrix factorization. Their updating rules, however, are different from ours since they minimized the KL-divergence metric in the cost function. In a similar context, the authors in [10] proposed a co-clustering method based on NMF that combines two modalities of the data. Their approach requires the user to provide input to learn a distance metric.

3   Methodology

BI-NMF is our proposed method for extracting information from two data modalities as a means of enhanced clustering. As our method is based on NMF, we describe NMF first and then discuss BI-NMF.

3.1   Non-Negative Matrix Factorization NMF

NMF [3] is a matrix factorization algorithm that deals with non-negative data matrices. Given a data matrix $X = [x_1, x_2, ...x_n] \in R^{(p \times n)}$. NMF produces two non-negative matrices $U \in R^{(p \times k)}$ and $V \in R^{(n \times k)}$ as a result of minimizing the following objective function:

$$O = \|X - UV^T\|_F^2 \tag{1}$$

where $\|.\|_F$ denotes the matrix Frobenius norm. Lee and Seung [3] proposed an iterative approach using multiplicative rules to solve for U and V.

$$u_{ij}^{t+1} = u_{ij}^t \frac{(XV)_{ij}}{(UV^TU)_{ij}} \tag{2}$$

$$v_{ij}^{t+1} = v_{ij}^t \frac{(X^TU)_{ij}}{(VU^TU)_{ij}} \tag{3}$$

Each column in the original matrix X is a linear combination of the columns of U weighted by the components of the corresponding column in V. Therefore U can be regarded as containing a basis that is optimized for the linear approximation of the data in X [3]. It is proven by Lee and Seung that the objective function O in (1) is nonincreasing under the update rules (2) and (3).

3.2   BI-NMF

Our algorithm extends NMF using two modalities of the data. We argue that each modality covers certain aspects of the data, therefore utilizing two modalities maximizes the gained benefit and potentially improves the clusters. The two data modalities are represented by the matrices A and B. Let $A \in R^{(mxn)}$, $B \in R^{(pxn)}$, $U_1 \in R^{(mxk)}$, $U_2 \in R^{(pxk)}$, $V \in R^{(nxk)}$, we seek to approximate the new compact representation of the data by simultaneously factorizing A and B. BI-NMF minimizes the following objective function:

$$J = \|A - U_1 V^T\|^2 + \|B - U_2 V^T\|^2 \tag{4}$$

where V is anticipated to capture the agreement between A and B about the clusters. The objective function above can be rewritten as follows:

$$J = (A - U_1 V^T)(A - U_1 V^T)^T + (B - U_2 V^T)$$
$$(B - U_2 V^T)^T$$
$$= tr(AA^T) - 2tr(AVU_1^T) + tr(U_1 V^T V U_1^T) + \tag{5}$$
$$tr(BB^T) - 2tr(BVU_2^T) + tr(U_2 V^T V U_2^T)$$

in the second step we used the matrix property $tr(XY)=tr(YX)$ and $tr(X)=tr(X^T)$. The objective function J needs to be solved under the constraints $u_1(i,j)>0$, $u_2(i,j)>0$ and $v(i,j)>0$. This is a typical constrained optimization problem that can be solved using Lagrange multiplier method. Let $\alpha = [\alpha_{ij}]_{mxk}$, $\beta = [\beta_{ij}]_{pxk}$ and $\varphi = [\varphi_{ij}]_{nxk}$ be the Lagrange multipliers for the constraints $u_1(i,j)>0$, $u_2(i,j)> 0$ and $v(i,j)>0$, respectively. For notational convenience, we are using the same indices i and j even though the dimensions of U_1, U_2 and V are not necessarily the same. The Lagrange L is:

$$L = tr(AA^T) - 2tr(AVU_1^T) + tr(U_1 V^T V U_1^T)$$
$$+ tr(BB^T) - 2tr(BVU_2^T) + tr(U_2 V^T V U_2^T) \tag{6}$$
$$+ tr(\alpha U_1^T) + tr(\beta U_2^T) + tr(\varphi V^T)$$

the partial derivatives of the Lagrange function L with respect to U_1, U_2 and V are:

$$\frac{\partial L}{\partial U_1} = -2AV + 2U_1 V^T V + \alpha \tag{7}$$

$$\frac{\partial L}{\partial U_2} = -2BV + 2U_2 V^T V + \beta \tag{8}$$

$$\frac{\partial L}{\partial V} = -2A^T U_1 + 2V U_1^T U_1 - $$
$$2B^T U_2 + 2V U_2^T U_2 + \varphi \tag{9}$$

Solving with respect to α, β, φ and utilizing the Kuhn-Tucker conditions $\alpha_{ij} u_{1(i,j)} = 0$, $\beta_{ij} u_{2(i,j)} = 0$, and $\varphi_{ij} v_{(i,j)} = 0$, we get the following equations:

$$-(AV)_{(i,j)} u_{1(i,j)} + (U_1 V^T V)_{(i,j)} u_{1(i,j)} = 0 \tag{10}$$

$$-(BV)_{(i,j)}u_{2(i,j)} + (U_2 V^T V)_{(i,j)}u_{2(i,j)} = 0 \tag{11}$$

$$-(A^T U_1)v_{(i,j)} - (B^T U_2)v_{(i,j)} + (VU_1^T U_1)v_{(i,j)} + (VU_2^T U_2)v_{(i,j)} = 0 \tag{12}$$

after rearranging the last 3 equations we obtain the following update rules:

$$u_{1(i,j)} = u_{1(i,j)} \frac{(AV)_{(i,j)}}{(U_1 V^T V)_{(i,j)}} \tag{13}$$

$$u_{2(i,j)} = u_{2(i,j)} \frac{(BV)_{(i,j)}}{(U_2 V^T V)_{(i,j)}} \tag{14}$$

$$v_{(i,j)} = v_{(i,j)} \frac{(A^T U_1 + B^T U_2)_{(i,j)}}{(VU_1^T U_1 + VU_2^T U_2)_{(i,j)}} \tag{15}$$

The objective function J in (4) is nonincreasing under the update rules in (13) (14) (15) (see appendix). The update rules of U_1, U_2 and V converge and the final solution is a local minimum. Lee and Seung [3] used an auxiliary function to prove the convergence of (1); which essentially minimizes a distance function. Equation (4), however, is the summation of two distance functions. Following the steps in [5] show that minimizing the auxiliary function of the summation is sufficient to decrease the objective function of the sum of distances. The matrix V computed in (15) is used to define the clusters as proposed by [9]. Each column $V_{.j}$ corresponds to one cluster and each row $V_{i.}$ to a point. A point is assigned to the cluster associated to the maximum value in its corresponding row. Formally, assign x_i to cluster c if $c = arg \max_j V_{ij}$. Note that the clusters in V are computed by joining information from two data modalities represented by the matrices A and B. It is important to mention that we normalized the matrices A and B using *TFIDF*. Further, we rescaled both matrices using the following formula:

$$X = X * [\text{diag}(X^T X e)]^{-1/2} \tag{16}$$

where X is a data matrix and e is a unit vector. Transforming the matrices using (16) before applying BI-NMF was proposed in [9]. We noticed that this transformation helped improve the clustering results. The pseudo code of our algorithm is summarized below.

Algorithm 1. BI-NMF

Input: data modality A, data modality B, maximum number of iterations I_{max}. Clusters C.
1. **Initialize U_1^t, U_2^t, V^t , normalize A, B**   using (16)
2. **for** $t = 1$ **to** I_{max} **do**
   - compute u_1^{t+1} using (13), u_2^{t+1} using (14), v^{t+1} using (15)
   - set $u_1^t = u_1^{t+1}$, $u_2^t = u_2^{t+1}$, $v^t = v^{t+1}$
3. **end**
4. **C = AssignClusters(V)**

---

4   Experimental Evaluation

We evaluated the proposed algorithm on two biomedical datasets. We demonstrate the effectiveness of BI-NMF by comparing its output clusters with the two NMF clustering solutions of each individual data modality, and with the NMF clusters of the two modalities merged. In the latter method, classic NMF [3] is applied to the merged matrices A and B after normalizing using *TFIDF*. We also compare BI-NMF with the two ensemble clusters computed for each individual data modality and with the combined ensemble clustering proposed in [8]. Combined ensemble clustering is fundamentally based on combining two data modalities using ensemble clustering. In this method, the co-association matrices are generated for each individual data modality and subsequently combined into one co-association matrix whereupon k-means is applied to obtain the consensus clustering. We also report the clusters of each data modality based on k-means.

4.1   Datasets

Pubmed Images Dataset. It consists of 3000 images extracted from articles of PubMed Central. Images with no captions were dropped and 2607 were retained. The images in the dataset were classified into 5 different categories by domain expert annotators. Discrepancies among the annotators were resolved by assigning the image to the category with the majority of votes. The list of annotations is: 564 images were assigned to the experimental category, 1131 images to the graph category, 645 images to the diagrams category, 86 images to the clinical category, and 181 images were assigned to the others category. We generated two modalities for the images. In one modality the images were represented using the pictorial and textural features computed using the Haralick method [7]. The other modality is a Bag of Words *BOW* representation generated using captions.

Radiology Reports Dataset. It consists of radiology reports collected from clinical records of patients for research purposes. The radiology reports were annotated by domain experts and classified into four categories. The categories and the counts of their content reports are: 35 abdominal MRI reports, 486 abdominal CT reports, 248 abdominal ultrasound reports and 500 non-abdominal radiology reports. For simplicity, we will call these MRI, CT, Ultrasound, and non-abdominal, respectively. The reports are represented using two data modalities: Textual features BOW and Bag of Concepts (*BOC*). In the BOW modality, the reports are represented using the original words that appear in the clinical narratives and weighted using their *TFIDF* score. In the *BOC* modality, the vectors are indexed by semantic concepts derived from cTAKES [4], a natural language processing tool that maps text to concepts from the *UMLS* ontology.

4.2   Evaluation Metrics

The clustering results are evaluated by comparing to gold standard annotations of images and radiology reports. We use three measures to evaluate the quality of the clusters: micro-averaged precision, purity and Normalized Mutual Information (*NMI*). Micro-averaged precision is an average over data points, which by default gives higher weight to those classes with many data points. *NMI* measures the amount of information by which our knowledge about the classes increases upon definition of the clusters.

$$micro\ averaged\ precision = \frac{\sum_{i=1}^{k} TP_i}{\sum_{j=1}^{k} TP_j + FP_j}$$

$$purity = \sum_i \frac{|C_i|}{n} max\ (precision(C_i, L_j))$$

$$NMI = \frac{I(X;Y)}{log\ k + log\ c}$$

(17)

where *TP* is true positive, *FP* is false positive, n is the number of data points, k is the number of clusters, c is the number of classes, C_i is the ith cluster, L_j is the jth class, $I(X;Y)$ is the mutual information between two random variables X (the cluster) and Y (the class).

4.3   Single Modality Clustering: BI-NMF vs. NMF, k-Means and Ensemble Clustering

We compare the clustering solutions produced by BI-NMF which draws information from different data modalities with the output clusters obtained using single data modality in order to demonstrate the benefit of leveraging multiple representations of the data. We show the performance of regular NMF on single modalities, along with comparable approaches such as k-means and ensemble clustering [8]. In ensemble clustering, a number of clustering solutions are aggregated in a co-association matrix that measures the number of times each pair of data points are placed into the same cluster. K-means is applied to the co-association matrix to get the final clusters. Table 1 shows a comparison between the performances of several clustering methods on single data modalities: K-means clusters of each data modality, the cluster ensembles of each data modality and NMF applied to each individual data modality. For the sake of clarity, the method descriptor has two parts: the applied method and the data modality used. For radiology reports, we observed that the ensemble clustering method applied to one data modality performed poorly when compared to NMF of single data modality, while outperforming single-modality k-means. With the exceptions discussed below, BI-NMF clusters were significantly better than single modality NMF, single modality ensemble clusters and k-means clusters as shown in Table 1 vs Table 2. It is important to mention that for the Pubmed images data, the clusters of k-means for the captions modality yield comparative clusters to BI-NMF based on purity as shown in Table 1. Nevertheless, *NMI* and micro averaged precision measures suggest that BI-NMF clusters are better than k-means clusters. To further assure this result, we computed the average of 100 BI-NMF runs and got consistent results. This result strongly emphasizes the benefit of our method that draws information from two data modalities.

Table 1. One data modality: Performance of different clustering methods of each data modality

Data	Method Descriptor	Micro Avg Precision	Purity	NMI
Radiology Reports	k-means_words	0.506	0.639	0.240
	Ensemble_words	0.506	0.640	0.238
	NMF_words	**0.676**	**0.791**	**0.599**
	k-means_concepts	0.555	0.758	0.490
	Ensemble_concepts	0.581	0.764	0.503
	NMF_concepts	**0.665**	**0.884**	**0.787**
Pubmed Images	k-means_Haralick	0.318	0.505	0.141
	Ensemble_Haralick	0.306	0.513	**0.150**
	NMF_Haralick	**0.331**	**0.516**	0.145
	k-means_captions	0.456	**0.558**	**0.180**
	Ensemble_captions	**0.479**	0.519	0.153
	NMF_captions	0.445	0.518	0.134

4.4 Two Modality Clusterting: BI-NMF vs. NMF_Merged and Combined Ensemble Clustering

To assess the effectiveness of BI-NMF, we compared its performance against another bi-modality clustering approach called combined ensemble clustering. In combined ensemble clustering, two co-association matrices are generated from two data modalities then linearly combined into one co-association matrix upon which k-means is applied to obtain the final clusters. We also compare the output clusters of our method with the clusters obtained when applying NMF to the merged data modalities. We implemented the combined ensemble clustering algorithm in [8] and applied it to our biomedical datasets. Table 2 shows a comparison in performance between NMF_merged, combined ensemble clustering and BI-NMF for radiology reports data and PubMed images data. Recall that in the NMF_merged method the matrices A and B pertaining to both data modalities are first combined and *NMF* is subsequently applied to the combined matrix after normalization. The performance of the two methods depends on their respective emphases on forming the BI-NMF clusters from various modalities versus combining different features of the data modalities prior to the formation of clusters.

Table 2. Two data modalities: Performance of different clustering methods for both modalities

Data	Method Descriptor	Micro Avg Precision	Purity	NMI
Radiology Reports	NMF_merged	0.584	0.793	0.599
	Combined Ensemble Clustering	0.582	0.761	0.513
	BI-NMF	**0.777**	**0.903**	**0.825**
Pubmed Images	NMF_merged	0.367	0.461	0.119
	Combined Ensemble Clustering	0.483	0.542	0.190
	BI-NMF	**0.551**	**0.558**	**0.200**

The quality of the clusters obtained by BI-NMF was superior compared to that of combined ensemble clustering and NMF_merged for both datasets in terms of all reported measures. On radiology reports, compared to combined ensemble clustering, BI-NMF achieved a relative improvement of the order of 33%, 18% and 60% in terms of micro averaged precision, purity and NMI, respectively. Similarly, it outperformed NMF_merged and yield a better clustering solution with a difference of 32%, 13% and 38% in terms of micro averaged precision, purity and NMI, respectively. BI-NMF also outperformed combined ensemble clustering and NMF_merged for Pubmed images as shown in Table 2. The micro averaged precision reported for BI-NMF was .551 compared to .483 for combined ensemble clustering. Likewise, purity and NMI showed a relative improvement of 3% and 5%, respectively. Superior performance is also observed for the proposed method compared to NMF_merged, it yield a relative improvement of 50%, 21% and 68% in terms of micro averaged precision, purity and NMI, respectively.

5   Conclusion

In this paper, we demonstrate an enhanced data clustering approach whose innovation is its exploitation of multiple data modalities called BI-NMF. Our proposed method is a bi-modal clustering algorithm based on non negative matrix factorization. It utilizes two modalities of the data to improve clustering. We applied the method on two biomedical datasets and demonstrated enhanced performance relative to ensemble clustering and NMF based on single and merged data modalities, on three standard metrics. Given our results, we conclude that BI-NMF is advantageous for enhanced biomedical data clustering and potentially useful for data from other domains.

Acknowledgements. This study was funded by NIH/NLM 5R01LM009956 (MK), and a VA grant HIR 08-374/HSR&D: Consortium for Healthcare Informatics (CB,SF,MK).

References

1. Chandra, B., Nath, S., Mlhothra, A.: Classification and Clustering of Cancer Images. In: The 6th International Joint Conference on Neural Networks, pp. 3843–3847 (2006)
2. Cai, D., He, X., Wang, X., Bao, H., Han, J.: Locality preserving non-negative matrix factorization. In: Proc. 27th Annual Inte'l ACM SIGIR, pp. 96–103 (2004)
3. Lee, D.D., Seung, H.S.: Algorithms for non-negative matrix factorization. Advances in Neural Information Processing Systems (13) (2001)
4. Savova, G.K., Masanz, J.J., Ogren, P.V., Zheng, J., Sohn, S., Kipper-Schuler, K.C., Chute, C.G.: Mayo clinical Text Analysis and Knowledge Extraction System (cTAKES): architecture, component evaluation and applications. Journal AMIA (17), 507 (2010)
5. Mandayam-Comar, P., Tan, P.N., Jain, A.K.: Identifying Cohesive Subgroups and Their Correspondences in Multiple Related Networks, vol. (1), pp. 476–483. WI-IAT (2010)
6. Paatero, P., Tapper, U.: Positive matrix factorization: A non-negative factor model with optimal utilization of error estimates of data values. Environmetrics 5(2), 111–126 (1994)

7. Haralick, R.M.: Statistical and structural approaches to texture. IEEE 67, 786–804 (1979)
8. Fodeh, S.J., Punch, W.F., Tan, P.N.: Combining statistics and semantics via ensemble model for document clustering. In: ACM symposium on Applied Computing, pp. 1446–1450 (2009)
9. Xu, W., Liu, X., Gong, Y.: Document clustering based on non-negative matrix factorization. In: Proc. 26th Annual Int'l ACM SIGIR, pp. 267–273 (2003)
10. Chen, Y., Wang, L., Dong, M.: Non-Negative Matrix Factorization for Semisupervised Heterogeneous Data Coclustering. TKDE, 1459–1474 (2009)

Appendix

Theorem 1. The objective function J is non-increasing under the rules (13) (14) (15).

The proof follows the one given by Lee and Sung [3] since update rules for U_1 and U_2 do not change. For the update rule (15), we use the auxiliary function trick.

Definition 1. $G(v,v')$ is an auxiliary function for $F(v)$ if the following are satisfied:

$$G(v, v^t) \geq F(v), \qquad G(v, v) = F(v) \tag{18}$$

Lemma 1. If $G_1(v, v^t)$ and $G_2(v, v^t)$ are auxiliary functions for $F_1(v)$ and $F_2(v)$ respectively, then: (a) $G(v, v^t) = G_1(v, v^t) + G_2(v, v^t)$ is the auxiliary function for $F(v) = F_1(v) + F_2$, (b) $F(v)$ is non-increasing under the update:

$$v^{t+1} = arg \min_v G(v, v^t) \tag{19}$$

The proof of (a) is trivial, for (b) we have:

$$
\begin{aligned}
F(v^{t+1}) \quad &= F_1(v^{t+1}) + F_2(v^{t+1}) \\
&\leq G_1(v^{t+1}, v^t) + G_2(v^{t+1}, v^t) \\
\leq G_1(v^t, v^t) + G_2(v^t, v^t) &= F_1(v^t) + F_2(v^t) = F(v^t)
\end{aligned}
\tag{20}
$$

Note that the third line is a result of the fact that v^{t+1} minimizes the auxiliary function G, then $G(v^{t+1}, v^t) \leq G(v^t, v^t)$ and $F(v^{t+1}) \leq F(v^t)$ as shown in [5]. To conclude the proof of *Theorem 1*, we show that the update rule (15) is the update given by (19), i.e.

$$v_{(i,j)}^{t+1} = arg \min_v G(v, v_{(i,j)}^t) \tag{21}$$

for a suitable auxiliary function $G(v, v^t)$. The objective function of eq.(5) can be written:

$$J = \Sigma_{i,j} F_{i,j}(v_{i,j}) \tag{22}$$

where $F_{i,j}$ is a quadratic function that depends only on $v_{i,j}$, the generic term of the matrix V. We need to show that the function $F_{i,j}$ is non-increasing under the update rule (15), or equivalently find an auxiliary function for $F_{i,j}$ such that the update rule (15) corresponds to (21). We compute the first and second order derivatives of $F_{i,j}$. One can easily check that:

$$F'_{i,j} = 2(-A^T U_1 - B^T U_2 + V(U_1^T U_1 + U_2^T U_2))_{i,j} \tag{23}$$

$$F''_{i,j} = 2(U_1^T U_1 + U_2^T U_2)_{j,j} \tag{24}$$

Then we consider:

$$G(v, v_{i,j}^t) = F_{i,j}(v_{i,j}^t) + F'_{i,j}(v_{i,j}^t)(v - v_{i,j}^t)$$
$$+ \frac{(V(U_1^T U_1 + U_2^T U_2))_{i,j}}{v_{i,j}^t}(v - v_{i,j}^t)^2 \tag{25}$$

now we need to show that $G(v, v_{i,j}^t)$ corresponds to an auxiliary function for $F_{i,j}$:

It is obvious that $G(v_{i,j}^t, v_{i,j}^t) = F_{i,j}(v_{i,j}^t)$. We only need to show $G(v_{i,j}, v_{i,j}^t) \geq F_{i,j}(v_{i,j}^t)$. Since $F_{i,j}$ is a quadratic form, consider the following Taylor series for $F_{i,j}$:

$$F_{i,j}(v) = F_{i,j}(v_{i,j}^t) + F'_{i,j}(v_{i,j}^t)(v - v_{i,j}^t) + \frac{1}{2}F''_{i,j}(v_{i,j}^t)(v - v_{i,j}^t)^2 \tag{26}$$

we need to show that:

$$\frac{(V(U_1^T U_1 + U_2^T U_2))_{i,j}}{v_{i,j}^t} \geq (U_1^T U_1 + U_2^T U_2)_{j,j} \tag{27}$$

the inequality (27) is obvious since

$$\frac{(V(U_1^T U_1))_{i,j}}{v_{i,j}^t} = \sum_k \frac{v_{i,k}^t}{v_{i,j}^t}(U_1^T U_1)_{k,j} \geq (U_1^T U_1)_{j,j} \tag{28}$$

and the same inequality holds for U_2:

$$\frac{(V(U_2^T U_2))_{i,j}}{v_{i,j}^t} = \sum_k \frac{v_{i,k}^t}{v_{i,j}^t}(U_2^T U_2)_{k,j} \geq (U_2^T U_2)_{j,j} \tag{29}$$

Thus $G(v_{i,j}, v_{i,j}^t) \geq F_{i,j}(v_{i,j}^t)$. We conclude the proof of *Theorem 1* by checking that (21) corresponds to (15). Indeed, given (22) and (26), we can get by solving $G'(v_{i,j}, v_{i,j}^t) = 0$.

$$v_{i,j}^{t+1} = v_{i,j}^t(1 - \frac{F'_{i,j}(v_{i,j}^t)}{2(V(U_1^T U_1 + U_2^T U_2))_{i,j}}) \tag{30}$$

After arranging the equation, one can easily show that (30) is equivalent to (15).

Extraction of Web Image Information:
Semantic or Visual Cues?

Georgina Tryfou and Nicolas Tsapatsoulis

Cyprus University of Technology,
Department of Communication and Internet Studies,
Limassol, Cyprus

Abstract. Text based approaches for web image information retrieval have been exploited for many years, however the noisy textual content of the web pages makes their task challenging. Moreover, text based systems that retrieve information from textual sources such as image file names, anchor texts, existing keywords and, of course, surrounding text often share the inability to correctly assign all relevant text to an image and discard the irrelevant. A novel method for indexing web images is discussed in the present paper. The main concern of the proposed system is to overcome the obstacle of correctly assigning textual information to web images, while disregarding text that is unrelated to them. The proposed system uses visual cues in order to cluster a web page into several regions and compares this method to the use of semantic information and the realization of a k-means clustering. The evaluation reveals the advantages and disadvantages of the different clustering techniques and confirms the validity of the proposed method for web image indexing.

1   Introduction and Related Work

Numerous web image search engines have been developed as the amount of digital image collections on the web constantly increases [1]. These systems share the objective to minimize the necessary human interaction while offering an intuitive image search. The two main approaches that exist in the literature for content extraction and representation of web images are: (i) the text-based and (ii) the visual feature-based methods. The text-based approaches use associate text (*i.e* image file names, anchor texts, surrounding paragraphs) to derive the content of images. The text blocks that are used as concept sources for images may be extracted with several methods, with the following four being the most popular: (i) fixed-size sequence of terms [2], (ii) DOM tree structure [3, 4], (iii) Web page segmentation [5] and (iv) hybrid versions of the above [6]. The first approach is time-efficient but yields poor results since the extracted text may be irrelevant to the image, or on the other hand, important parts of the relevant text may be discarded. Approaches that use the DOM tree structure of the web page are in general not adaptive and they are designed for specific design patterns. Web page segmentation is a more adequate solution to the problem since it is adaptable to different web page styles. Most of the proposed algorithms in this

L. Iliadis et al. (Eds.): AIAI 2012, IFIP AICT 381, pp. 368–373, 2012.

field though, are not designed specifically for the problem of image indexing and therefore often deliver poor results. The proposed system uses information obtained following the web page segmentation approach.

The paper is organized as follows: Section 2 presents the architecture of the proposed system. Section 3 presents the evaluation of the proposed system. Finally some conclusions and future perspectives are given in Sec. 4.

2 System Architecture

The general architecture of the proposed system is depicted in Fig. 1. As shown there the system consists of two main parts, which are described in the following sections.

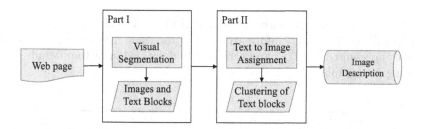

Fig. 1. The general architecture of the proposed system

2.1 Visual Segmentation

The content extraction of web images is based on textual information that exists in the same web document and refers to this image. In order to determine to which image the various text parts of a web page refer to, we use the visual cues which are connected to the outline and the presentation of the hosting web page. In order to obtain the set of visual segments that form a web page, we use the Visual Based Page Segmentation (VIPS) algorithm [7], which extracts the semantic structure of a web page based on its visual representation. It attempts to make full use of the page layout structure by extracting blocks from the DOM tree structure of the web page and locating separators among these blocks. Each web page is represented as a set of blocks that bare similar Degree of Coherence (DOC). With the permitted DOC (pDOC) set to its maximum value, we obtain a set of visual blocks that consist of visually indivisible contents. An example of a web page segmentation using pDOC = 10 (*i.e.* maximum allowed value) is illustrated in Fig. 2.

2.2 Text to Image Assignment

Each text block found in the web page has to be assigned to an image block. In other words, we attempt to determine to which image, each textual block refers

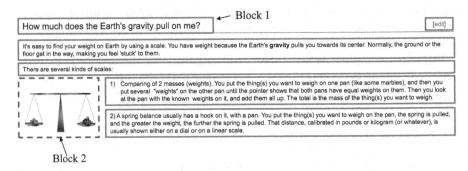

Fig. 2. The results of VIPS algorithm on a fragment of a web page. Each visual block is marked with a rectangle region: dashed-red when the block is an image and continuous-blue when it is text.

to. After this decision, the corresponding text blocks will be adequate to use for the extraction of the image information. The processing that takes place for this module is presented in Fig. 3 and each part is described in further details in the following paragraphs.

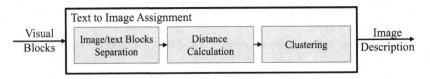

Fig. 3. The processing that takes place in the second part of the system: Text to Image Assignment

Image/text Blocks Separation. The first task towards the assignment of each text block to the image block it refers to, is to determine whether a block contains image content or text. The use of the maximum pDOC during VIPS execution certifies that for a well formed HTML document, each block will either be an image or a text block indicating that no blocks with mixed content will be returned. The HTML source code that corresponds to each one of these blocks, is returned by the VIPS algorithm and it is used in order to classify them into two categories: (i) image blocks and (ii) text blocks.

Distance Calculation. Once the blocks are separated into these categories, the Euclidean distance between every image/text block pair is calculated making use of the Cartesian coordinates returned by the VIPS algorithm. The distance calculation in this case is not a trivial problem since the goal is to quantify the intuitive understanding of how close two visual blocks are. This understanding depends not only on the distance of the centres of the two visual blocks, but also on their shape, size and relative position. In order to solve the distance calculation problem several approaches were taken into consideration. The calculation of the distance between the closest edges of the blocks was found to offer the better representation of the block distance.

Clustering. After the distance calculation the web page has to be clustered into regions. Each region or cluster is defined by an image in its center and contains all the text blocks that have been found to refer to this image. For the clustering we took into consideration two different approaches. The first one is based only in visual cues and the location of the various blocks while the second approach mines semantic information and implements a k-means clustering.

Clustering Based on Visual Cues. In this approach, the text blocks are assigned to the corresponding cluster making use of the visual information that is available for them: the Euclidean coordinates which are returned be the VIPS algorithm are used for distance calculation and each text block is assigned to the cluster, whose center it is closest to. However, it is possible that one or more blocks of text do not refer to a certain image of the web page. For this reason, it is necessary to discard one or more text blocks from the calculated clusters. In order to determine which blocks of text are irrelevant to the image they are connected to, the blocks whose distance to the cluster center (*i. e.* corresponding web image) is longer than a defined threshold t, are discarded. In order to calculate this threshold, the distances d_i^c that appear in a cluster c are normalized to the maximum distance. Using the normalized values $\tilde{d}_i^c$ the threshold t is calculated as $t_{ed} = t' + m_{ed} - s_{ed}$, where t' is a static, predefined threshold (in our experiment is empirically set to 0.1), m_{ed} is the mean value of the euclidean distances found in the cluster c and s_{ed} is their standard deviation.

Clustering Based on Semantic Information. Semantic information, as obtained from the application of the Vector Space Model [8], is used in the second approach in order to create a clustering which is based on the content of each block rather than its location on the web page. The web page is considered the corpus from which the vocabulary is extracted. Each text block tb_i is one of the corpus' documents and it is expressed as a vector of term weights as $v_i = [w_{1i} w_{2i} \ldots w_{Ni}]^T$. The term w_{ti} indicates the weight of text t in text block tb_i and is calculated using the **tf*idf** [8] statistic which expresses how important is each word for the representation of a text block. The k-means algorithm is then used on the vectors of term weights in order to cluster the text blocks into M regions. Since each text block may refer to any image but it may also be irrelevant to every image, the number of clusters M is equal to the total number of images found in the web page plus one for text blocks irrelevant to every image. Once the k-means algorithm is executed each text block is assigned to a specific cluster. However, it is not yet determined to which image each cluster refers to. To find the solution in this problem we considered the average distances among images and clusters as well as an initialization to the k-means algorithm based on the output of the first approach to the clustering procedure. The results presented in Sec. 3 are obtained using this initialization.

3   Evaluation

A set of manually labelled web images has been collected in order to create a corpus, based on which, the clustering of text blocks is evaluated. This corpus was obtained using an annotation tool that we designed in order to facilitate web image labelling. The annotator had to assign relevant text to each one of the images that exist in web pages which were rendered in the default browser. A dataset that consists of 40 web pages and their annotations was created in order to evaluate our method. Using the above described corpus and the evaluation measures *Precision, Recall* and *F-score* as defined in [9], we obtained two different sets of results using the clustering methods described earlier. In the first set of results, the whole processing is based on visual information as obtained from the application of the VIPS algorithm. In the second set, the results are based on semantic information, as it is obtained from the Vector Space Model applied on the content of the web page. As shown in Table 1 there is a significant variation on the efficiency of the two approaches with the first approach having the highest average F-score.

Table 1. The results for different Text Block Discarding Methods

Clustering Method	Precision	Recall	F-score
Visual	0.8610	0.8836	0.8962
Semantic	0.3576	0.4528	0.4533

The execution of the proposed method yields an average *F-score* equal to **0.8610** for the total of the 131 annotated images. As shown in Fig. 4, more than 80% of the text blocks are identified with *Recall* and *Precision* values higher than 0.8, indicating that the system succeeds in retrieving most of the blocks that according to the annotation refer to a certain image.

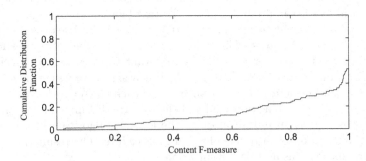

Fig. 4. Cumulative distribution function of F-measure

4   Conclusions

In this paper we presented an image indexing system that uses textual information in order to extract the concept of the images that are found in a web page. The method uses visual cues in order to identify the segments of the web page and calculates euclidean distances among these segments. It delivers a semantic or euclidean clustering of the contents of a web page in order to assign textual information to the existing images. During the experimenting and the evaluation it was found that a clustering based on semantic information of the contextual information delivers poor results compared to the clustering that is based on visual information. This stresses out the importance of understanding and processing the web pages as a structured visual document when it comes to web image indexing rather than an unstructured bag of words.

The weight vectors that were used for the k-means clustering are in general sparse vectors, since the length of the vocabulary is not proportional to the size of each text block. Moreover, when two neighbouring text blocks refer to the same image and contain similar semantic content, the authors usually select synonyms to express the same meaning. The Vector Space Model does not account such connections among different terms. It is therefore expected that the use of an ontology that describes the semantic distances and relations among different words and phrases will improve the results of the k-means clustering and this is the direction of our future study.

References

[1] Sclaroff, S., La Cascia, M., Sethi, S., Taycher, L.: Unifying textual and visual cues for content-based image retrieval on the world wide web. Computer Vision and Image Understanding 75(1-2), 86–98 (1999)
[2] Feng, H., Shi, R., Chua, T.S.: A bootstrapping framework for annotating and retrieving www images. In: Proc. of the 12th ACM Int. Conf. on Multimedia (2004)
[3] Hua, Z., Wang, X.J., Liu, Q., Lu, H.: Semantic knowledge extraction and annotation for web images. In: Proc. of the 13th ACM Int. Conf. on Multimedia, pp. 467–470 (2005)
[4] Fauzi, F., Hong, J.L., Belkhatir, M.: Webpage segmentation for extracting images and their surrounding contextual information. In: Proc. of the 17th ACM Int. Conf. on Multimedia (2009)
[5] He, X., Cai, D., Wen, J.R., Ma, W.Y., Zhang, H.J.: Clustering and searching www images using link and page layout analysis. ACM Trans. on Multimedia Computing, Communications and Applications 3(2) (June 2007)
[6] Alcic, S., Conrad, S.: A clustering-based approach to web image context extraction. In: Proc. of the 19th ACM Int. Conf. on Multimedia, pp. 74–79 (2011)
[7] Cai, D., Yu, S., Wen, J.R., Ma, W.Y.: Vips: a vision based page segmentation algorithm. Technical report, Microsoft Research (2003)
[8] Salton, G., Wong, A., Yang, C.S.: A vector space model for automatic indexing. Commun. ACM 18(11), 613–620 (1975)
[9] Alcic, S., Conrad, S.: Measuring performance of web image context extraction. In: Proc. of the 10th Int. Workshop on Multimedia Data Mining, pp. 1–8 (July 2010)

Trust-Aware Clustering Collaborative Filtering: Identification of Relevant Items

Cosimo Birtolo, Davide Ronca, and Gianluca Aurilio

Poste Italiane – Information Technology
Research and Development - R&D Center
Piazza Matteotti 3 – 80133 Naples, Italy
{birtoloc,roncadav,aurilio5}@posteitaliane.it

Abstract. Identifying a customer profile of interest is a challenging task for sellers. Preferences and profile features can range during the time in accordance with current trends. In this paper we investigate the application of different model-based Collaborative Filtering (CF) techniques and in particular propose a trusted approach to user-based clustering CF. We propose a Trust-aware Clustering Collaborative Filtering and we compare several approaches by means of Epinions, which contains explicit trust statements, and MovieLens dataset, where we have implicitly defined a trust information. Experimental results show an increased value of coverage of the recommendations provided by our approach without affecting recommendation quality. To conclude, we introduce a tool, based on recommender systems, able to assist merchants in delivering special offers or in discovering potential interests of their customers. This tool allows each merchant to identify the products to suggest to the target customer in order to best fit his profile of interests.

Keywords: Trust, Recommender System, Collaborative Filtering, Clustering, Web Intelligence.

1 Introduction

In the last decade, the concept of trust has become very important with the increase of Social Network platforms and with the use of social community in e-Commerce platforms. Trust can be considered as a remarkable element influencing a user's decision-making process. Furthermore, the concept of trust includes both the cognitive and the emotional dimensions.

Webs of trust are networks through which a trust-aware system can ask a user to evaluate other users already known. For example, Epinions suggests to put in a user's web of trust "those reviewers whose reviews and ratings resulted to be extremely useful". Online interpersonal relations are becoming one of the major characteristics of the Web 2.0, and are also useful for social aspects (MySpace, Msn, Facebook), working connections (LinkedIn) and information (Slashdot.org, Epinions.com) besides, obviously, commercial purposes (eBay.com, Amazon.Com).

L. Iliadis et al. (Eds.): AIAI 2012, IFIP AICT 381, pp. 374–384, 2012.

In general, trust is a directional relationship between two parties that can be called trustor and trustee. In the e-Commerce domain, trust is applied to a specific purpose, such as mutual trust between customers and sellers. According to Liu [1] trust and reputation management research is highly interdisciplinary, involving researchers from networking and communication, data management and information systems, e-commerce and service computing.

In this paper we associate the concept of trust with similarity between users based on shopping lists. We propose a model-based approach able to combine trust and similarity in order to improve both the quality and the coverage of the suggestions. Furthermore, we present a tool able to assist the merchant in discovering item potentially interesting for customers.

The remainder of this work is organized as follows: Section 2 provides a brief overview of recommendation and trust-aware systems; Section 3 describes how the problem has been modeled; Section 4 presents experimental results; and Section 5 outlines conclusions and future directions.

2 Recommendation System

Recommendation systems are aimed at helping users in the search of interesting items among a large set of items within a specific domain by using knowledge about user's preferences in the domain. In other words, Recommendation System is a way for improving personalization by giving personalized suggestions and have largely been adopted in different domains. Almost every e-Commerce site (e.g., Amazon) has its own recommendation engine; different Web sites are focusing on suggesting a personalized content such as a movie (e.g., *MovieLens* and *Netflix*) or a song (*Yahoo!Music*).

Recommendation systems were introduced in 1992 by means of the Tapestry project [2]. Several different approaches have recently been proposed in order to increase the accuracy of the predicted values, thus minimizing the prediction error and improving the quality of the recommendation while taking into account performance issues. However, some issues related to the quality of recommendations and to computational aspects still arise because collaborative systems rely solely on users preferences to make recommendations. Among recommendation techniques, Collaborative Filtering has gained great success in the practical application of e-Commerce [3] and has been proven to be one of the most successful techniques. Recently model-based CF, such as Clustering CF, which uses the user-item database to infer a model which is then applied for predictions, are investigated and discussed in order to outperform traditional memory-based CF. Different works [4,5] prove that these algorithms improve the quality of predictions, in particular when sparse data is considered.

Moreover, recent works have proved some benefits in terms of an increased quality of suggestions, by including social factors [6] (e.g., user's past behaviors and reputation) or trust information [7] in recommendation systems. These arising systems are called *Trust-based Recommendation Systems* [8] and combine the potentialities of a traditional recommendation system with a trust-aware system.

They adopted the social ties established among online users and extended traditional recommender systems by considering a trust matrix in addition to the ratings matrix.

On the other hand, J. Sun et al. [9] proposed a *Trust-Aware Recommender Model* which utilizes trustworthy experts and their search experiences to recommend their search histories to the common user according to the profile similarity between common users and experts. Furthermore, the same authors extended the model [10] by replacing similarity weight by trust information which is evaluated in a direct and indirect way and decreases along its propagation in the network.

Burke [11] investigated some techniques of hybrid recommendation systems. He compared four different recommendation techniques and several different hybridization strategies. According to his findings, it is possible to distinguish four different hybridization techniques: (i) *Weighted*, (ii) *Mixed*, (iii) *Switching*, and (iv) *Cascade*. In the first approach the score is a weighted linear combination of two recommendation techniques; the second one is a merge based on predicted rating or on recommender confidence; in the third one, the score is equal to one of the two engines according to an external selection criteria that establishes the best confidence; and in the last one, the idea is to create a strictly hierarchical hybrid, one in which a weak recommender cannot overturn decisions made by a stronger one, but can merely refine them.

Although different hybrid algorithms there exist, the major problems are the time necessary to explicitly define the online relations among users and, first and foremost, the small number of social links defined by users themselves, aspect that leads to a scarce quality of recommendations. The second problem, instead, is related to the definition of users' trust that seems to be a challenging task.

3   TRACCF: Trust-Aware Clustering CF

In Collaborative Filtering [12], once the items to consider are defined, a prediction $p_i(u)$ for the active user u is generally evaluated by:

$$p_i(u) = \bar{r}(u) + \frac{\sum\limits_{v \in N_u} sim(u,v) \cdot (r_i(v) - \bar{r}(v))}{\sum\limits_{v \in N_u} |sim(u,v)|} \tag{1}$$

where $r_i(u)$ is the rating given to item i by the user u, $\bar{r}(u)$ is the average rating given by the user u, N_u is the set of neighbors of user u among users, and $sim(u,v)$ is the similarity function between user u and user v.

In this paper we propose a model-based collaborative filtering algorithm which takes into account at the same time clustering and trustworthy information. We introduce $trust(u,v)$ between user u and user v in order to emphasize trustworthy information which comes from a subset of trusted users. This index is evaluated starting from the common viewed or bought items and it is calculated according to the Eq. 2.

$$trust(u,v) = \frac{card(A_{u,v})}{card(A_u)} \tag{2}$$

where $A_{u,v}$ is the set of common viewed or bought items by user u and user v, and A_u is the set of items viewed or bought by user u.

The proposed approach, named Trust-aware Clustering CF (TRACCF), evaluates the prediction of a rating p of item i for the user u as expressed by Eq. 3.

$$p_i(u) = \bar{r}(u) + \frac{\sum\limits_{v \in V_u} (\sigma \cdot trust(u,v) + (1-\sigma) \cdot sim(u,v)) \cdot (r_i(v) - \bar{r}(v))}{\sum\limits_{v \in V_u} |\sigma \cdot trust(u,v) + (1-\sigma) \cdot sim(u,v)|} \qquad (3)$$

where $r_i(u)$ is the rating given to item i by the user u, $\bar{r}(u)$ is the average rating given by the user u, $trust(u,v)$ and $sim(u,v)$ are respectively trust index and similarity function between user u and user v; σ is a weight ranging in the $[0,1]$ interval, and V_u is the set of users belonging to the same cluster of user u. The adopted hybridization techniques take into account the weighted model proposed by Burke [11], where the score is a weighted linear combination of the trust and of the similarity between two users.

Depending on the σ value, the proposed TRACCF becomes a traditional Clustering Collaborative Filtering (CCF) when $sigma$ is equal to 0. While, when $sigma$ is equal to 1, it becomes a Clustering-based version of the Trust-aware Recommender System (TaRS) proposed by Massa [8]. Consequently, the proposed approach arises from the hybridization of CCF and Clustering-based TaRS (CTaRS).

The proposed recommendation system is based on four steps: (i) User Clustering, (ii) Elicitation of trust information, (iii) Prediction Engine , and (iv) Top-N CF. In *User Clustering* step, we select a clustering method in order to group together users in clusters, minimizing the dissimilarity between users assigned to the same cluster. The second step elicits the trust information, while the *Prediction Engine* predicts ratings for items similar to rated ones and by taking into account trust information, and finally, *Top-N CF* selects the Top-N items to recommend.

4   Experimentation

4.1   Dataset

In the experimentation we consider MovieLens and Epinions datasets. The MovieLens dataset was collected by the GroupLens Research Group [13] and consists of 1,000,204 ratings for 3,900 movies by 6,040 users. Each user in the dataset has rated at least 20 movies.

The Epinions dataset has been extracted by [8] from *Epinions.com* which is a consumer opinion website where people can rate products and can express their Web of Trust, i.e. users whose ratings they have consistently found to be valuable. The extracted dataset provides 13,668,319 ratings by 132,000 users on 1,560,140 articles. Furthermore, Epinions dataset collects 841,372 statements divided in 717,667 trusts and 123,705 distrusts. In both datasets, the ratings are integer numbers on a 1("bad")-to-5("excellent") scale.

In order to assess the goodness of a dataset we evaluate the sparsity level[1]. The overall sparsity of MovieLens and Epinions is quite high and as the majority of ratings datasets (such as BookCrossing and Netflix), it presents a distribution that follows Zipf's power law distribution [14] (a lot of ratings for few items and few ratings for the remaining items). The value of sparsity is 0.9575 for MovieLens dataset and 0.9993 for Epinions one, entailing respectively 95.75% and 99.93% of unrated items.

The datasets are randomly divided into a training set, which ranges from 80% to 95% of the ratings per user and a testing set (the remaining ratings). Starting from the training set recommendation algorithms predict unknown ratings, while the testing set is used to evaluate the accuracy of the predictions.

4.2   Evaluating the Quality of a Recommendation System

The quality of a recommendation system can be decided on the basis of evaluation. The type of metric adopted depends on the type of applications [12]. Indeed, Herlocker [12] highlighted the problem of comparing different recommendation algorithms and defined some metrics that had been used to evaluate them.

The commonly-used metrics are accuracy metrics as Root Mean Squared Error (RMSE) which is becoming popular partly because it is commonly used for the evaluation of movie recommendation performance. It amplifies the contributions of the absolute errors between the predictions and actual ratings and is defined as:

$$RMSE = \sqrt{\frac{1}{N} \cdot \sum_{u,i} (p_i(u) - r_i(u))^2} \qquad (4)$$

where N is the total number of ratings over all users, $p_i(u)$ is the predicted rating for user u on item i, and $r_i(u)$ is the actual rating. The lower values of $RMSE$ entail better predictions.

Another kind of metrics in recommendation system is the coverage that can be defined as "the percentage of a dataset that the recommender system is able to provide prediction for" [12].

4.3   Results

Analysis of performance is aimed at comparing the results offered by Trust-based Recommendation System (TRS), and Clustering Collaborative Filtering (CCF). TRS takes into account the different formulations proposed in the Section 3, while CCF is implemented by k-means algorithm.

---

[1] Sparsity level of a dataset is evaluated by means of $1 - \frac{N}{m \cdot n}$ where n is the number of items, m is the number of users and N is the total number of ratings over all users.

According to [5], we measure the similarity with Pearson correlation coefficient between users u and v as:

$$sim(u,v) = \frac{\sum\limits_{i \in A_{u,v}} (r_i(u) - \bar{r}(u))(r_i(v) - \bar{r}(v))}{\sqrt{\sum\limits_{i \in A_{u,v}} (r_i(u) - \bar{r}(u))^2} \sqrt{\sum\limits_{i \in A_{u,v}} (r_i(v) - \bar{r}(v))^2}} \tag{5}$$

where $\bar{r}(u)$ is the average rating of user u.

We adopt two trust indexes, the first one is explicitly provided by Epinion dataset, while the other is evaluated according to Eq.2.

We consider Epinions (dataset ID ranging from 1 to 3) and MovieLens (dataset ID ranging from 4 to 6) dataset with increased size of training set (i.e., 80%, 90%, and 95% of the entire dataset) and setup the algorithms with the following parameters: (i) number of clusters $k = 20$, (ii) trust evaluated according to Eq.2, and (iii) similarity function as in (5)). We experimented 5 different approaches: Clustering-based TaRS with explicitly defined values of trust (eCTaRS), Clustering-based TaRS with implicitly defined values of trust (iCTaRS), Clustering CF (CCF) and two versions of the proposed hybrid algorithm, i.e., eTRACCF and iTRACCF which consider at the same time trust and similarity information, but differ for the criteria adopted to evaluate the trust information: in the first approach the trust is explicitly provided by the users (only for the Epinion dataset), while in the second one, it is evaluated starting from the common viewed or bought items without taking into account the expressed rating.

In Fig. 1 and in Fig. 2, we compared the different algorithms at varying training size. Charts respectively refer to the average of RMSE and the average of coverage of 5 different runs (in each run, ratings per user in training set are randomly selected).

Fig. 3 shows a bar chart, where a comparison between the trust approach and pure clustering approach is depicted, while Fig. 4 compares explicit and implicit trust within the proposed algorithm.

On the one hand, iCTaRS and iTRACCF seems to outperform in term of accuracy (a lower value of RMSE). On the other hand, iCTaRS and iTRACCF adopt the implicit evaluation of trust which guarantees a higher coverage of predictions as proved in Fig. 2 where the Epinions dataset is taken into account and in Fig. 5 where the MovieLens dataset is also investigated. These findings can be useful for the adoption of trust information even if they are not explicitly provided.

Investigating these results more deeply, we consider a pairwise (two-sided) Wilcoxon signed-rank test, reporting results in Table 1, where the average value of RMSE of 15 different runs per technique (5 repetitions per training size) is shown. Assuming 0.05 as upper limit to reject the null hypothesis, we can affirm that there is statistical difference between iTRACCF and other RS approaches (i.e., CCF and iCTaRS). We prove that eTRACCF outperforms CCF, but we cannot reject the null hypothesis when eTRACCF and eCTaRS is compared.

380   C. Birtolo, D. Ronca, and G. Aurilio

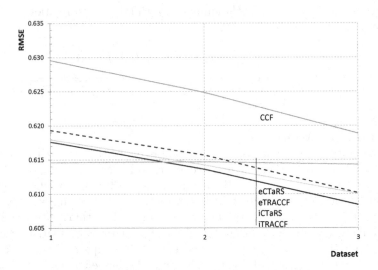

Fig. 1. Comparison of eCTaRS, iCTaRS, CCF, eTRACCF, and iTRACCF by means of average RMSE at varying the training size (from 80% to 95% of the Epinions dataset)

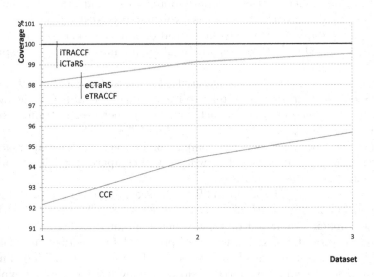

Fig. 2. Comparison of eCTaRS, iCTaRS, CCF, eTRACCF, and iTRACCF by means of average coverage at varying the training size (from 80% to 95% of the Epinions dataset)

To sum up, the proposed approach provides three main advantages: first of all, in every experiment conducted on the same dataset iTRACCF outperforms iCTaRS and CCF, as shown in Fig. 1, then even if there is no explicit trust

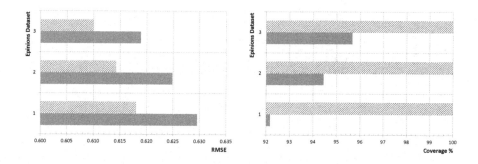

Fig. 3. Comparison of iCTaRS (light-grey bar) and CCF (dark-grey bar) by means of average RMSE and average coverage of 5 different runs

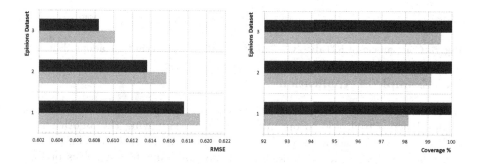

Fig. 4. Comparison of eTRACCF (grey bar) and iTRACCF (black bar) by means of average RMSE and average coverage of 5 different runs

Table 1. Wilcoxon signed-rank test on the Epinions dataset: Average RMSE and p-values

| | Average | | p-value | | |
Algorithm	RMSE	CCF	eCTaRS	iCTaRS	eTRACCF	iTRACCF
CCF	0.6244	-	1.526e-03	6.104e-05	6.104e-05	6.104e-05
eCTaRS	0.6145	1.526e-03	-	6.387e-01	8.469e-01	5.614e-01
iCTaRS	0.6141	6.104e-05	6.387e-01	-	1.245e-02	3.052e-04
eTRACCF	0.6150	6.104e-05	8.469e-01	1.245e-02	-	6.104e-05
iTRACCF	0.6132	6.104e-05	5.614e-01	3.052e-04	6.104e-05	-

information, it is possible to evaluate it by means of user web usage mining, as depicted in Fig. 3 and in Fig. 4, and finally, trust information can increase the coverage of recommendation, as proved by Fig. 2 and Fig. 5.

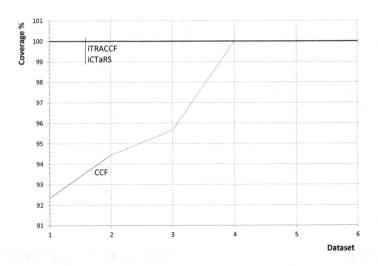

Fig. 5. Comparison of iCTaRS, CCF, and iTRACCF at varying the adopted dataset

4.4 Example of Application

In order to assist the merchant in suggesting interesting items to his customers, we developed a tool able to optimize web pages employing the algorithm described above. The tool interface (see Fig. 6) is organized in four main panels.

The first panel on the top allows to specify the input files which is preliminarily divided into training and testing. The input file contains all the ratings expressed by all the users for the items in the catalogue. The second panel, named Parameter, allows to specify the recommendation algorithm and related parameters. Selecting a model-based algorithm it is possible to set the required parameters, otherwise a default value is taken into account. Once a cluster-based algorithm is selected, the tool allows the visualization of users in the different clusters. The third panel, named Recommendation, is divided into two sections: (i) on the left, the list of users is shown, where each user is colored by green if recommendations are available for that user, otherwise he is colored by red; (ii) on the right, once a user is selected, the recommended items and predicted rating are shown. Double-clicking on user opens a pop-up dialog that shows details on the selected user such as his expressed ratings (History section), a graphical representations of the user preferences, and the items recommended by means of the selected algorithm. Fig. 6 shows the results for the users 1641 of MovieLens dataset, his rated movies and the top suggestions.

The proposed tool gives the merchant different features in e-Commerce domain. First of all, by selecting a target user it is possible to discover customer's profile of interest, his interaction history and a set of candidate items that he can potentially consider very interesting. In other words, this tool allows knowledge extraction regarding customers and suggests implicitly which products add in some personalized marketing campaigns enhancing customer loyalty.

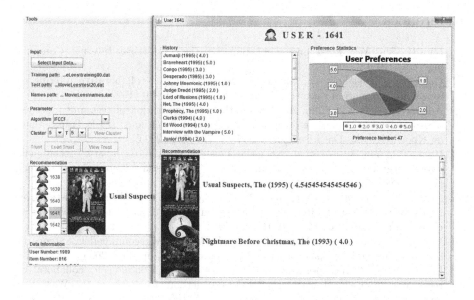

Fig. 6. Tool for selecting contents

5   Conclusions and Future Work

In this paper we investigate the application of different collaborative filtering techniques as a viable approach to identify interesting products for customers.

We evaluated the trust by means of shopping chart of the customers and proved the benefits of this information in recommending items. We proposed a hybrid model-based CF which takes into account estimated trust and user similarity. Similarity and trustworthy information is adopted in order to elaborate the rating predictions.

Experimental results showed that the proposed approach slightly increases the accuracy of the predicted values when the same dataset is considered. Moreover, we proved that trust information increase the coverage of predictions without affecting recommendation quality.

However, we aim at extending the proposed approach in the future, taking into account other information arising from social network analysis such as the relationship level (e.g.,friend, colleague), the number of common friends, and the number of common social interaction (e.g., thematic groups, content sharing). These factors could lead to a more accurate measurement of trust entailing further improvement of suggestions.

Acknowledgments. This work was partially supported by MSE under the Intelligent Virtual Mall (InViMall) Project MI01-00123.

References

1. Liu, L., Shi, W.: Trust and reputation management. IEEE Internet Computing 14(5), 10–13 (2010)
2. Goldberg, D., Nichols, D., Oki, B.M., Terry, D.: Using collaborative filtering to weave an information tapestry. Commun. ACM 35, 61–70 (1992)
3. Adomavicius, G., Tuzhilin, A.: Toward the next generation of recommender systems: a survey of the state-of-the-art and possible extensions. IEEE Transactions on Knowledge and Data Engineering 17(6), 734–749 (2005)
4. Huang, C., Yin, J.: Effective association clusters filtering to cold-start recommendations. In: 2010 Seventh Int. Conf. on Fuzzy Systems and Knowledge Discovery (FSKD), vol. 5, pp. 2461–2464 (August 2010)
5. Birtolo, C., Ronca, D., Armenise, R., Ascione, M.: Personalized suggestions by means of collaborative filtering: A comparison of two different model-based techniques. In: NaBIC, pp. 444–450. IEEE (2011)
6. Kitisin, S., Neuman, C.: Reputation-based trust-aware recommender system. In: Securecomm. and Workshops, August 28-September 1, pp. 1–7 (2006)
7. Liu, B., Yuan, Z.: Incorporating social networks and user opinions for collaborative recommendation: local trust network based method. In: Proc. of the Workshop on Context-Aware Movie Recommendation, CAMRa 2010, pp. 53–56. ACM, New York (2010)
8. Massa, P., Avesani, P.: Trust-aware recommender systems. In: Proceedings of the 2007 ACM Conference on Recommender Systems, RecSys 2007, pp. 17–24. ACM, New York (2007)
9. Sun, J., Yu, X., Li, X., Wu, Z.: Research on trust-aware recommender model based on profile similarity. In: International Symposium on Computational Intelligence and Design, ISCID 2008, vol. 1, pp. 154–157 (October 2008)
10. Wu, Z., Yu, X., Sun, J.: An improved trust metric for trust-aware recommender systems. In: First International Workshop on Education Technology and Computer Science, ETCS 2009, vol. 1, pp. 947–951 (March 2009)
11. Burke, R.: Hybrid Web Recommender Systems. In: Brusilovsky, P., Kobsa, A., Nejdl, W. (eds.) Adaptive Web 2007. LNCS, vol. 4321, pp. 377–408. Springer, Heidelberg (2007)
12. Herlocker, J.L., Konstan, J.A., Terveen, L.G., Riedl, J.T.: Evaluating collaborative filtering recommender systems. ACM Transactions on Information Systems 22(1), 5–53 (2004)
13. Resnick, P., Iacovou, N., Suchak, M., Bergstorm, P., Riedl, J.: Grouplens: An open architecture for collaborative filtering of netnews. In: Proc. of ACM 1994 Conference on Computer Supported Cooperative Work, pp. 175–186. ACM (1994)
14. Zaier, Z., Godin, R., Faucher, L.: Evaluating recommender systems. In: Proceedings of the 2008 International Conference on Automated solutions for Cross Media Content and Multi-channel Distribution, pp. 211–217. IEEE Computer Society, Washington, DC (2008)

Unsupervised Detection of Fibrosis in Microscopy Images Using Fractals and Fuzzy c-Means Clustering

S.K. Tasoulis, Ilias Maglogiannis, and V.P. Plagianakos

Department of Computer Science and Biomedical Informatics,
University of Central Greece, Papassiopoulou 2–4, Lamia, 35100, Greece
{stas,imaglo,vpp}@ucg.gr

Abstract. The advances in improved fluorescent probes and better cameras in collaboration with the advent of computers in imaging and image analysis, assist the task of diagnosis in many fields of biologic and medical research. In this paper, we introduce a computer-assisted image characterization tool based on a Fuzzy clustering method for the quantification of degree of Idiopathic Pulmonary Fibrosis (IPF) in medical images. The implementation of this algorithmic strategy is very promising concerning the issue of the automated assessment of microscopic images of lung fibrotic regions.

Keywords: Image Analysis, Cluster Analysis, Fuzzy Clustering, Fractal Dimension.

1 Introduction

Idiopathic pulmonary fibrosis (IPF), also referred to as cryptogenic fibrosing alveolitis, is a chronic, progressive and usually lethal lung disorder of unknown aetiology. Median survival of newly diagnosed patients with IPF is about 3 years, similar to that of clinical non-small cell lung cancer. At present there are no proven therapies for IPF [18]. IPF usually affects patients aged 50–70 years, with a male preponderance, evidenced by a male-to-female ratio of 2:1. The incidence of IPF has been estimated at 10.7 cases per 100,000 men and 7.4 cases per 100,000 women and the incidence appears to be rising [6].

Advances in improved fluorescent probes [5] and better cameras [11, 21] have expanded the capabilities of the light microscope and its usefulness in biologic and medical research. The advent of computers in imaging and image analysis permits quantitative computer analysis. Thus, automatic classification and assessment of microscopic images is possible in an attempt to assist the task of diagnosis.

The field of microscopy image analysis has occupied several research teams and significant research work may be found in the literature of this field. In [14] a tool that classifies biological microscopic images of lung tissue sections with idiopathic pulmonary fibrosis was presented. Similar tools have also been proposed for the assessment of liver fibrosis [2, 4, 16, 22], the study of micro vascular

L. Iliadis et al. (Eds.): AIAI 2012, IFIP AICT 381, pp. 385–394, 2012.
© IFIP International Federation for Information Processing 2012

circulating leukocytes [9], the assessment of testicular interstitial fibrosis [19, 20], or that of lung fibrosis [12]. The use of pattern recognition or classification methods like Support Vector Machines or Neural Networks has enabled the design of decision-making algorithms, appropriate to microscopic data. Within this context, a method for evaluation of electron microscopic images of serial sections based on the Gabor wavelets and the construction of a mapping between the model and the target image has been proposed in [13].

In the present study, we employ one of the most widely used fuzzy clustering algorithms; namely the Fuzzy c-means [3]. The goal is to identify the presence and the degree of lung pathology caused by idiopathic pulmonary fibrosis (IPF), which is a chronic, progressive and usually lethal lung disorder of unknown etiology [1], whose variability concerning the severity of the lesions it incurs in the lung is great when assessed by microscopic histological images [12]. Fuzzy c-means clustering (and more over an improved version of fuzzy c-means) has also been applied in environmental cases [10]

The Fuzzy c-means clustering method was applied to digital images of sections, captured using a Nikon ECLIPSE E800 microscope and a Nikon digital camera DXM1200 at a magnification of 4x. Each image was partitioned into windows of specified size and features were extracted for each window. The main contribution of the this study is the calculation of the fractal dimension of the microscopic images and its incorporation to the clustering algorithm.

The rest of the paper is structured as follows. In Section 2 we briefly review the Fuzzy c-means clustering algorithm and we outline the fractal dimension and the box-counting method for its computation. Next in Section 3 we introduce the proposed methodology and finally in Section 4 we present the experimental analysis and results. The paper ends with concluding results and pointers for future work.

2   Background Information

2.1   Fuzzy Clustering

Data clustering is the process of partitioning a set of data vectors into disjoint groups (clusters), so that objects of the same clusters are more similar to each other than objects in different clusters. Different measures of similarity may be used in the clustering process, where the similarity measure controls how the clusters are formed. Some examples of measures that can be used in clustering include distance, connectivity, and intensity. In traditional (hard) clustering, data are divided into distinct clusters, where each data vector belongs to exactly one cluster. On the other hand, in fuzzy (soft) clustering, data vectors can belong to more than one cluster, and associated with each element is a set of membership levels. These indicate the strength of the association between that data element and a particular cluster. Fuzzy clustering is a process of assigning these membership levels and then using them to assign data elements to one or more clusters.

One of the most widely used fuzzy clustering algorithms is the Fuzzy C-Means (FCM) algorithm. FCM is a clustering method that assigns each each data point to a cluster to some degree that is specified by a membership grade. This technique was originally introduced by Jim Bezdek in [3] as an improvement on earlier clustering methods and attempts to partition a finite collection of data vectors into a collection of fuzzy clusters with respect to some given criterion. A theoretical discussion of FCM can be found in [7].

Given a finite set of data, the algorithm returns a list of cluster centers and a partition matrix indicating the degree to which each element belongs to a given cluster. Like the k-means algorithm, the FCM aims to minimize an objective function, like the following:

$$J_m = \sum_{i=1}^{N} \sum_{j=1}^{C} u_{ij}^m \|x_i - c_j\|^2, \quad 1 \le m \le \infty, \tag{1}$$

where m is any real number greater than 1, u_{ij} is the degree of membership of x_i in the cluster j, x_i is the i-th of d-dimensional measured data, c_j is the d-dimension center of the cluster, and $\| \cdot \|$ is any norm expressing the similarity between any measured data and the center. Fuzzy partitioning is carried out through an iterative optimization of the objective function shown above, with the update of membership u_{ij} and the cluster centers c_j by:

$$u_{ij} = \frac{1}{\sum_{k=1}^{C} \left(\frac{\|x_i - c_j\|}{\|x_i - c_k\|} \right)^{\frac{2}{m-1}}}, \tag{2}$$

where

$$c_j = \frac{\sum_{i=1}^{N} u_{ij}^m \cdot x_i}{\sum_{i=1}^{N} u_{ij}^m}.$$

This iteration stops when

$$max_{ij}\{|u_{ij}^{k+1} - |u_{ij}^k|\} \le \epsilon, \tag{3}$$

where k is the iteration number and ϵ is a constant between 0 and 1 that controls the termination of algorithm. This procedure converges to a local minimum or a saddle point of J_m.

2.2   Fractals and Fractal Dimension

It is known that a fractal designates a rough or fragmented geometric shape that can be subdivided into parts, each of which is a reduced-size copy of the whole [8]. Fractals are generally self-similar and independent of scale. In general nature conforms to fractals much more than it does to classical shapes and hence fractals can serve as models for many natural phenomena.

Contrary to classical geometry, fractals are not regular and may have a non-integer dimension. Fractal concepts have provided a new approach for quantifying the geometry of complex or noisy shapes and objects. Fractal geometry has been proven capable of quantifying irregular patterns, such as tortuous lines, crumpled surfaces and intricate shapes, and estimating the ruggedness of systems [15].

In this study, we incorporate the fractal dimension in an attempt to guide the FCM algorithm to construct meaningful clusters. The box-counting approach is one of the frequently used techniques to estimate the FD of an image. Among a number of techniques [15], it was found to be the most appropriate method of fractal dimension estimation.

The box-counting method is easy, automatically computable, and applicable for patterns with or without self-similarity [8, 17]. In the box counting procedure, each input image is covered by a sequence of grids of descending sizes and for each of the grids, two values are recorded. These are the number of square boxes intersected by the image, $N(s)$, and the side length of the squares, s.

Then the fractal dimension is calculated by the regression slope D ($1 \leq D \leq 2$) of the straight line formed by plotting $\log(N(s))$ against $\log(1/s)$ [15]. An image having a fractal dimension of 1, or 2, is considered as completely differentiable, or very rough and irregular, respectively. The linear regression equation used to estimate the fractal dimension is

$$\log(N(s)) = \log(K) + D\log(1/s), \tag{4}$$

where K is a constant and $N(s)$ is proportional to $(1/s)D$.

3   Proposed Methodology

For this study, Age- and sex- matched, 6-8 week-old mice were used for the induction of pulmonary fibrosis by a single intravenous injection with a dose of 100mg/Kg of body weight (100 mg/kg body weight; 1/3 LD50; Nippon Kayaku Co. Ltd., Tokyo). Bleomycin administration initially induces lung inflammation that is followed by a progressive destruction of the normal lung architecture. To monitor disease initiation and progression, mice were sacrificed at 7, 15 and 23 days after bleomycin injection. Mice injected with saline alone and sacrificed 23 days post injection, served as the control group. For pathology assessment, at each time point, bronchoalveolar lavage (BAL) has been performed (3x 1ml Saline) for the estimation of total and differential cell populations. Finally after perfusion of lungs via the heart ventricle with 10ml Phosphate Buffer Saline, lungs were then removed, weighed dissected and collected, for histology. Sagittal sections from the right lung were used for Hematoxylin and Eosin staining and histopathologic analysis.

When we retrieve the microscopic images the procudure continues as follows. Initially, the input microscopic images are being binirized using the Otsu's method. This method is one of many binarization algorithms, i.e. it automatically performs reduction of a graylevel image to a binary one. The algorithm assumes that the image contains two classes of pixels (e.g. foreground and background) and calculates the optimum threshold separating those two classes so that their combined spread (intra-class variance) is minimal. At the next step we segment the binary image to $n \times n$ windows and then we calculate Fractal Dimension (FD) of each window. Fractal dimension is a useful feature proposed to characterize roughness and self-similarity in a picture, which has been used for texture segmentation, shape classification, and graphic analysis in many fields. It can be defined as a ratio providing a statistical index of complexity comparing how detail in a pattern changes with the scale at which it is measured. Finally based on the Fractal Dimension and other image properties we perform clustering using Fuzzy c-means (FCM) clustering algorithm. The fuzzification parameter m that we use in this work is set to the default value 1.25. The complete scheme of the proposed method is shown at Figure 1. Next at Figure 2 an example of an image is illustrated in various steps of the procedure. From left to right we can see the input microscopic image, the greyscale image, the binary image, and the image that shows the clustering result.

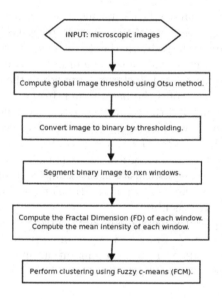

Fig. 1. Diagram of the proposed methodology

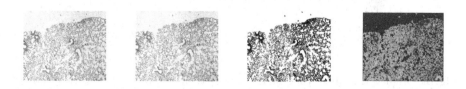

Fig. 2. An example of an image in various steps of the proposed methodology

4   Experiment Setup and Results

The results that are presented in this section are based on images that correspond to 7, 15 and 23 days after bleomycin administration in mice (see Figure 3). The lung images captured using a Nikon ECLIPSE E800 microscope and a Nikon DXM1200 digital camera, were initially separated to windows of size 100 × 100 pixels and the box-counting method was applied at each window to calculate the corresponding fractal dimension. Next, the images were separated into windows of smaller sizes (5, 10, and 20, respectively). Each of this windows corresponds to a data vector, with attributes the fractal dimension of the initial 100 × 100 window that contains it and the mean value of the intensity of the grayscale window image. Finally, to cluster each of the datasets constituted by such data points, we employed the Fuzzy c-Means clustering algorithm, for 3 and 4 clusters, respectively.

At Figures 4 and 5 we can see the clustering results with respect to the window size for all images where windows of the same cluster are colored equally for the 4 and 3 clusters case respectively. In the 4 cluster case the cluster colored in red corresponds to the severe pathology, the orange cluster corresponds to the mild pathology and green colored cluster corresponds to the normal lung. Finally the blue colored cluster denotes the background. In the 3 cluster case the mild pathology class is missing and the colors red, green and blue corresponds to the severe pathology, normal lung and background class respectively. The results are quite encouraging with respect to their medical content. The clustering algorithm seems to work pretty fine in distinguishing pathologic from normal. Thus the majority of the blocks that are clustered according to their fractal dimension into pathological clusters (i.e. severe and mild) belong into the regions indicated by the expert pathologists as fibrotic.

Tables 1 and 2 report the percentage, with respect to the window size for all images (retrieved after 7, 15 and 23 days), of each cluster found. The computed percentages coincide with manual annotation and scoring performed by our collaborating experts pathologists.

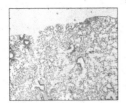 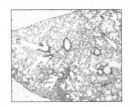 

Fig. 3. Original IPF Microscopy Images retrieved after 7, 15 and 23 days respectively

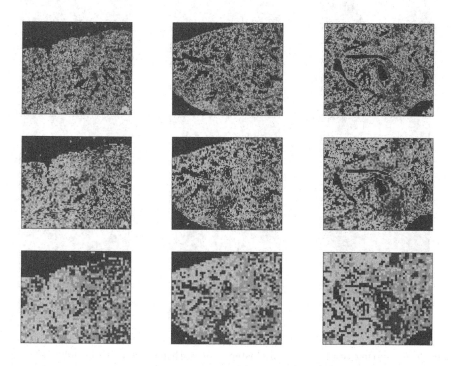

Fig. 4. Clustering result for the 4 clusters case with respect to the window sizes for the images retrieved after 7, 15 and 23 days respectively

Table 1. Percentages of the 4 clusters found by the clustering algorithm

Window Size	Severe pathology	Mild pathology	Normal lung	Backround
	Image retrieved after 7 days			
5	10.89	21.59	23.19	44.31
10	12.59	26.86	26.72	33.81
20	16.55	31.55	28.48	23.40
	Image retrieved after 15 days			
5	11.49	21.99	22.00	44.50
10	14.17	27.01	25.16	33.64
20	14.52	33.33	30.56	21.58
	Image retrieved after 23 days			
5	15.01	24.74	25.50	34.74
10	16.83	27.19	29.33	26.64
20	16.25	29.95	35.82	17.96

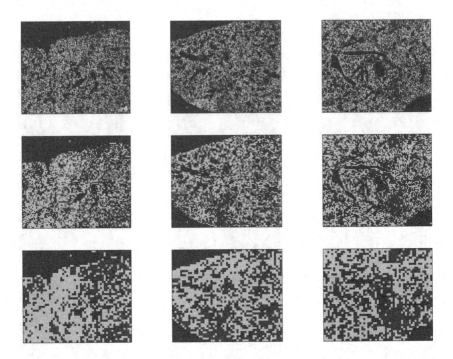

Fig. 5. Clustering result for the 3 clusters case with respect to the window sizes for the images retrieved after 7, 15 and 23 days respectively

Table 2. Percentages of the 3 clusters found by the clustering algorithm

Window Size	Pathology	Normal lung	Backround
Image retrieved after 7 days			
5	18.68	31.30	50.00
10	23.07	38.37	38.55
20	28.76	45.19	26.03
Image retrieved after 15 days			
5	20.52	30.56	48.91
10	23.79	37.40	38.79
20	29.64	44.98	25.37
Image retrieved after 23 days			
5	23.61	33.58	42.80
10	24.68	39.87	35.44
20	26.72	44.43	28.83

5   Conclusions

In the present paper, a computer-assisted image characterization tool based on a Fuzzy clustering method was introduced for the quantification of degree of

IPF in medical images. The implementation of this algorithmic strategy is very promising concerning the issue of the automated assessment of microscopic images of lung fibrotic regions. The results obtained so far show that the proposed strategy addresses the vital biological issues concerning their imaging part as this is contained in the specific type of microscopic images.

In a future study, we intent to use the degree of membership to postprocess the clustering results in an attempt to eliminate small isolated regions and increase the accuracy of the proposed approach.

Acknowledgments. The authors would like to thank the European Union (European Social Fund ESF) and Greek national funds through the Operational Program "Education and Lifelong Learning" of the National Strategic Reference Framework (NSRF) - Research Funding Program: "Heracleitus II. Investing in knowledge society through the European Social Fund." for financially supporting this work.

References

1. Antoniou, K.M., Pataka, A., Bouros, D., Siafakas, N.M.: Pathogenetic pathways and novel pharmacotherapeutic targets in idiopathic pulmonary fibrosis. Pulmonary Pharmacology and Therapeutics 20(5), 453–461 (2007)
2. Bedossa, P., Dargére, D., Paradis, V.: Sampling variability of liver fibrosis in chronic hepatitis c. Hepatology 38(6), 1449–1457 (2003)
3. Bezdek, J.C.: Pattern Recognition with Fuzzy Objective Function Algorithms. Kluwer Academic Publishers, Norwell (1981)
4. Caballero, T., Pérez-Milena, A., Masseroli, M., O'Valle, F., Salmerón, F., Del Moral, R., Sánchez-Salgado, G.: Liver fibrosis assessment with semiquantitative indexes and image analysis quantification in sustained-responder and non-responder interferon-treated patients with chronic hepatitis c. Journal of Hepatology 34(5), 740–747 (2001)
5. Chalfie, M., Tu, Y., Euskirchen, G., Ward, W.W., Prasher, D.C.: Green fluorescent protein as a marker for gene expression. Science 263(5148), 802–805 (1994)
6. Coultas, D.B., Zumwalt, R.E., Black, W.C., Sobonya, R.E.: The epidemiology of interstitial lung diseases. Am. J. Respir. Crit. Care Med. 150(4), 967–972 (1994)
7. Cox, E.: Fuzzy Modeling and Genetic Algorithms for Data Mining and Exploration. Elsevier Inc., USA (2005)
8. Foroutan-pour, K., Dutilleul, P., Smith, D.: Advances in the implementation of the box-counting method of fractal dimension estimation. Applied Mathematics and Computation 105(23), 195–210 (1999),
   http://www.sciencedirect.com/science/article/pii/S0096300398100966
9. Hussain, M., Merchant, S., Mombasawala, L., Puniyani, R.: A decrease in effective diameter of rat mesenteric venules due to leukocyte margination after a bolus injection of pentoxifylline digital image analysis of an intravital microscopic observation. Microvascular Research 67(3), 237–244 (2004),
   http://www.sciencedirect.com/science/article/pii/S0026286204000081

10. Iliadis, L., Vangeloudh, M., Spartalis, S.: An intelligent system employing an enhanced fuzzy c-means clustering model: Application in the case of forest fires. Computers and Electronics in Agriculture 70(2), 276–284 (2010), http://www.sciencedirect.com/science/article/pii/S0168169909001367 special issue on Information and Communication Technologies in Bio and Earth Sciences

11. Inoué, S., Spring, K.: Video Microscopy. Plenum, New York (1997)

12. Izbicki, G., Segel, M., Christensen, T., Conner, M., Breuer, R.: Time course of bleomycin-induced lung fibrosis. Int J. Exp. Path 83(3), 111–119 (2002)

13. König, P., Kayser, C., Bonin, V., Würtz, R.: Efficient evaluation of serial sections by iterative gabor matching. Journal of Neuroscience Methods 111(2), 141–150 (2001), http://www.sciencedirect.com/science/article/pii/S0165027001004393

14. Maglogiannis, I., Sarimveis, H., Kiranoudis, C.T., Chatziioannou, A.A., Oikonomou, N., Aidinis, V.: Radial basis function neural networks classification for the recognition of idiopathic pulmonary fibrosis in microscopic images. IEEE Transactions on Information Technology in Biomedicine 12(1), 42–54 (2008)

15. Mandelbrot, B.: The Fractal Geometry of Nature. W.H. Freeman and Co., New York (1983)

16. Masseroli, M., Caballero, T., O'Valle, F., Del-Moral, R., Pérez-Milena, A., Del-Moral, R.: Automatic quantification of liver fibrosis: design and validation of a new image analysis method: comparison with semi-quantitative indexes of fibrosis. Journal of Hepatology 32(3), 453–464 (2000)

17. Peitgen, H.O., Jürgens, H., Saupe, D.: Chaos and Fractals: New Frontiers of Science. Springer (February 1993), http://www.amazon.com/exec/obidos/redirect?tag=citeulike07-2&path=ASIN/0387979034

18. Selman, M., Pardo, A.: Idiopathic pulmonary fibrosis: an epithelial/fibroblastic cross-talk disorder. Respiratory Research 3(1), 3 (2002), http://www.pubmedcentral.nih.gov/articlerender.fcgi?artid=64814&tool=pmcentrez&rendertype=abstract

19. Shiraishi, K., Takihara, H., Naito, K.: Quantitative analysis of testicular interstitial fibrosis after vasectomy in humans. Aktuelle Urologie 34(4), 262–264 (2003)

20. Shiraishi, K., Takihara, H., Naito, K.: Influence of interstitial fibrosis on spermatogenesis after vasectomy and vasovasostomy. Contraception 65(3), 245–249 (2002)

21. Shotton, D.: Image resolution and digital image processing in electronic light microscopy. Cell Biology, a Laboratory Handbook 3, 85–98 (1998)

22. Yagura, M., Murai, S., Kojima, H., Tokita, H., Kamitsukasa, H., Harada, H.: Changes of liver fibrosis in chronic hepatitis c patients with no response to interferon-alpha therapy: including quantitative assessment by a morphometric method. J. Gastroenterol 35(2), 105–111 (2000), http://www.biomedsearch.com/nih/Changes-liver-fibrosis-in-chronic/10680665.html

Image Threshold Selection Exploiting Empirical Mode Decomposition

Stelios Krinidis and Michail Krinidis

Information Management Department
Technological Institute of Kavala,
Ag. Loukas, 65404 Kavala, Greece
stelios.krinidis@mycosmos.gr, mkrinidi@gmail.com

Abstract. Thresholding process is a fundamental image processing method. Typical thresholding methods are based on partitioning pixels in an image into two clusters. A new thresholding method is presented, in this paper. The main contribution of the proposed approach is the detection of an optimal image threshold exploiting the empirical mode decomposition (EMD) algorithm. The EMD algorithm can decompose any nonlinear and non-stationary data into a number of intrinsic mode functions (IMFs). When the image is decomposed by empirical mode decomposition (EMD), the intermediate IMFs of the image histogram have very good characteristics on image thresholding. The experimental results are provided to show the effectiveness of the proposed threshold selection method.

Keywords: Threshold selection, clustering, empirical mode decomposition, ensemble empirical mode decomposition, intrinsic mode.

1 Introduction

Image thresholding is one of the main and most important tasks in image analysis and computer vision. Thresholding principle is based on distinguishing an object from the background in order to extract useful information from the image.

A large number of image thresholding techniques have been proposed in the literature [1]. However, the design of a robust and an efficient thresholding algorithm is far from being a simple process, due to the existence of images depicting complex scenes at low resolution, uneven illumination, scale changes, etc.

Thresholding techniques could be categorized in six groups [1] according to information being exploited. These categories are:

- Shape-based methods, which analyze the shape of the image histogram (i.e., the peaks, valleys and curvature). Each method uses different forms of shape properties, such as distance from the histogram convex hull [2], autoregressive modelling [3], overlapping peaks, etc.
- Clustering-based methods, which label the gray-level samples as background or foreground (object), or alternatively they model them as a mixture of two Gaussians [4–8]. In this category, the gray-level data undergoes a clustering analysis, with the number of clusters being always equal to two.

L. Iliadis et al. (Eds.): AIAI 2012, IFIP AICT 381, pp. 395–403, 2012.

- Entropy-based methods, which use the entropy of the foreground and the background regions, the cross-entropy between the original and the binarized image etc. [9, 10]. These algorithms exploit the entropy of the distribution of the gray levels in an image, by maximizing the entropy of the thresholded image or by minimizing the cross entropy.
- Attribute-based methods, which seek a measure between the gray-level and the binarized images, such as fuzzy shape similarity, edge coincidence, etc. [11, 12]. These algorithms evaluate the threshold value by using attributes quality or similarity measures between the initial gray-level image and the output binary image, such as edge matching, shape compactness, gray level moments, connectivity, texture or stability of segments objects [12], or fuzzy measures [11], etc.
- Spatial methods, which exploit higher-order probability distribution and/or correlation between the image pixels [13, 14]. The algorithms in this category utilize not only the gray level distribution, but also the dependency of pixels in a neighborhood, for example, the probabilities, correlation functions, cooccurrence probabilities, local linear dependence models of image pixels, 2-D entropy, etc.
- Local methods, which adapt the threshold value on each image pixel to the local image characteristics, such as range, variance [15–17], contrast, surface-fitting parameters of the pixel neighborhoods, etc.

This paper presents a novel, fast and robust image thresholding method. The method is based on the decomposition of the image histogram by the *Empirical Mode Decomposition* (EMD) [18] to its *Intrinsic Mode Functions* (IMFs). More specific, the decomposition is performed by the *Ensemble Empirical Mode Decomposition* (EEMD) [19], which provides noise resistance and assistance to data analysis. The properties of the desired IMFs [18, 19] will be shown that provide an efficient threshold for the image under examination.

The remainder of the paper is organized as follows. The thresholding method is introduced in Section 2. Experimental results are shown in Section 3 and conclusions are drawn in Section 4.

2   Threshold Selection Based-On EEMD

In this Section, the image threshold selection method is introduced. This method is fully automated and is based on the IMFs extracted by the EEMD [19] algorithm applied on the histogram of the image under examination. More details regarding the Ensemble Empirical Mode Decomposition (EEMD) and the derived Intrinsic Mode Functions (IMFs), their properties and all the adopted assumptions are presented in [18, 19].

The histogram $h(k)$ is computed for an input image I with $k = 0 \ldots G$ and G being the maximum luminance value in the image I, typically equal to 255 when 8-bit quantization is assumed. Then, the probability mass function of the image histogram is defined as the normalized histogram by the total pixel number:

$$p(k) = \frac{h(k)}{N}, \tag{1}$$

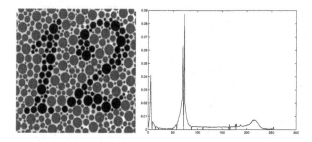

Fig. 1. An image for color blindness test and its normalized histogram

where N is the total number of image pixels. An example of an image and its normalized histogram is depicted in Figure 1.

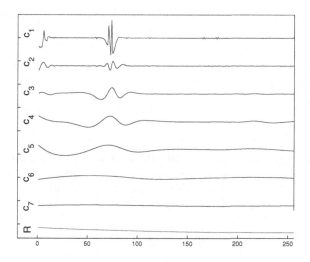

Fig. 2. The IMFs of the histogram for the image depicted in Figure 1 with a Gaussian noise of amplitude 0.2 and 1000 trials are performed

In the sequence, the proposed method analyzes the normalized histogram $p(k)$ of an image into its intrinsic mode functions (IMFs) exploiting the ensemble empirical mode decomposition (EEMD) algorithm. The IMFs of the histogram of the image shown in Figure 1, are presented in Figure 2. The IMFs are produced using the EEMD algorithm with a Gaussian noise of amplitude equal to 0.2 and 1000 trials. The number of the extracted IMFs (including the residue function) for a 8-bit quantized image is $log_2(256) = 8$.

One can easily notice in Figure 2 that the first IMF c_1 mainly carries the histogram "noise", irregularities and the sharp details of the histogram, while IMFs c_6, c_7 and the residue R mostly describe the trend of the histogram. On the other hand, IMFs c_2 to c_5 describe the initial histogram with simple and

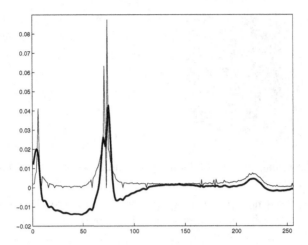

Fig. 3. The histogram of the image shown in Figure 1 (thin line) and the summation of the c_2 to c_5 IMFs (fat line)

uniform pulses. This is the main reason that the proposed method is focused on c_2 to c_5 IMFs. Let us now define the summation c_m of these IMFs as follows:

$$c_m = \sum_{i=2}^{5} c_i. \tag{2}$$

Figure 3 depicts the summation c_m (fat line) in contrast to the initial normalized histogram (thin line). One can notice that this summation c_m describes the main part of the histogram leaving out all its meaningless details.

The minimum of summation c_m is given by:

$$T^* = \arg\left\{\min_{0 \leq T \leq G} c_m(T)\right\}, \tag{3}$$

where T^* is the desired image threshold. Since the summation c_m provides a better, more clear and uniform formation of the image histogram, its minimum can be considered as an optimal threshold for the input image and its efficiency will be experimentally shown in the next Section.

Finally, the overall algorithm could be summarized in Figure 4.

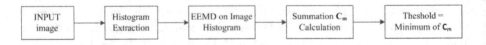

Fig. 4. The histogram of the image shown in Figure 1 (thin line) and the summation of the c_2 to c_5 IMFs (fat line)

3   Experimental Results

In this Section, the performance of the proposed method is examined by presenting numerical results using the introduced thresholding approach on various synthetic and real images, with different types of histogram. The obtained results are compared with the corresponding results of four well-known thresholding methods [4–6, 11]. In all the experiments, the EEMD algorithm was used with a noise of amplitude equal to 0.2 and 1000 trials are performed.

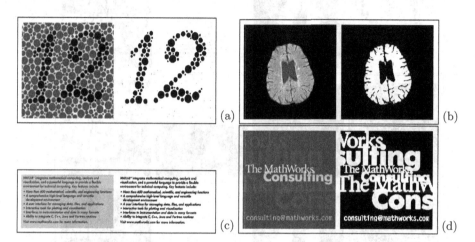

(a)   (b)   (c)   (d)

Fig. 5. Various images (left column) and their corresponding thresholded images produced by the proposed method (right column)

Figure 5 presents various real and synthetic images and their corresponding thresholded images obtained by the proposed approach. The left column shows the initial images, while the right column depicts the corresponding thresholded images produced by the proposed algorithm. One can clearly see that the proposed method can efficiently threshold the images under examination. Table 1 confirms the results in terms of the well known Tanimoto/Jaccard error [20] $E(\cdot)$ defined here as:

$$E(o, m) = 1 - \frac{\displaystyle\int_{I_o \cap I_m} dxdy}{\displaystyle\int_{I_o \cup I_m} dxdy}, \qquad (4)$$

where I_m and I_o are the extracted and the desired thresholded images respectively. In Table 1, the desired thresholded images have been extracted manually and then, compared (4) with the acquired thresholded images produced by the proposed method and four well known thresholding methods [4–6, 11]. The errors of the proposed methods are small enough to enforce one to claim that they

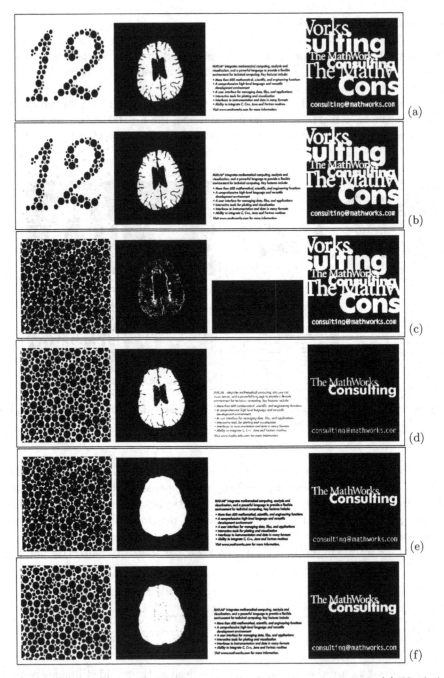

Fig. 6. Thresholded images: **(a)** ground truth, **(b)** proposed method, **(c)** Kittler's method, **(d)** Otsu's method, **(e)** Huang's method and **(f)** Kwon's method

are insignificant. On the contrary, the other four methods produce larger errors, a fact that is also depicted in Figure 6. The thresholded images produced by the proposed algorithm are more efficient. In Table 1, is also shown, the thresholds produced by the corresponding algorithms (the value in the parenthesis).

Table 1. Threshold values determined by five threshold selection methods with the corresponding area difference measure results

Method	Color blindsness (Fig. 5(a))	MRI image (Fig. 5(b))	Doc. image 1 (Fig. 5(c))	Doc. image 2 (Fig. 5(d))
Proposed	0.006 (46)	0.006 (135)	0.017 (126)	0.136 (75)
Kittler's [4]	0.738 (179)	0.834 (193)	0.998 (195)	0.136 (75)
Otsu's [6]	0.615 (136)	0.100 (86)	0.091 (45)	0.781 (140)
Huang's [11]	0.711 (169)	0.230 (1)	0.085 (170)	0.760 (122)
Kwon's [5]	0.577 (110)	0.228 (48)	0.063 (152)	0.730 (101)

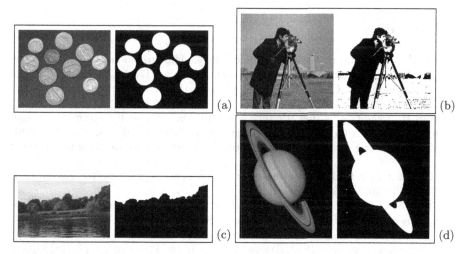

Fig. 7. Various images (left column) and their corresponding thresholded images produced by the proposed method (right column)

Figure 6 presents the thresholded images extracted by the proposed and the rest algorithms [4–6, 11]. Figure 6a shows the ground truth which manually extracted in order to calculate the numerical results depicted in Table 1. Figure 6b depicts the thresholded images produced by the proposed algorithm, while Figures 6c-6f show the thresholded images acquired by the Kittler's method [4] (Fig. 6c), Otsu's method [6] (Fig. 6d), Huang's method [11] (Fig. 6e) and Kwon's method [5] (Fig. 6f).

Finally, Figure 7 shows various images (left column) and their corresponding thresholded images acquired by the proposed method. All images in Figure 7 depict complex scenes, especially Figure 7(d). However, the proposed algorithm thresholds these images in a very reasonable way, a fact that is also proved in Figures 5 and 6 and in parallel provides better performance than the other four methods.

However, the algorithm fails to threshold an image correctly when the desired cluster is too small.

4   Conclusion

In this paper, a novel image thresholding method is introduced. The proposed approach exploits ensemble empirical mode decomposition (EEMD) to analyze the normalized histogram of the image under examination into a number of intrinsic mode functions (IMFs). The proposed algorithm uses only specific components, the intermediate IMFs of the EEMD decomposition, in order to evaluate an optimal image threshold. The effectiveness of the proposed threshold selection method is proved in the experimental results Section where the proposed thresholding algorithm is applied to various images with simple and complex scenes.

References

1. Sezgin, M., Sankur, B.: Survey over image thresholding techniques and quantitative performance evaluation. Journal of Electronic Imaging 13(1), 146–165 (2004)
2. Sahasrabudhe, S., Gupta, K.: A valley-seeking threshold selection technique. In: Computer Vision and Image Understanding, pp. 55–65 (1992)
3. Guo, R., Pandit, S.: Automatic threshold selection based on histogram modes and a discriminant criterion. Machine Vision and Applications 10(5-6), 331–338 (1998)
4. Kittler, J., Illingworth, J.: Minimum error thresholding. Pattern Recognition 19(1), 41–47 (1986)
5. Kwon, S.: Threshold selection based on cluster analysis. Pattern Recognition Letters 25(9), 1045–1050 (2004)
6. Otsu, N.: A threshold selection method from gray level histograms. IEEE Transactions on Systems, Man, and Cybernetics 9(1), 62–66 (1979)
7. Krinidis, M., Pitas, I.: Color texture segmentation based on the modal energy of deformable surfaces. IEEE Transactions on Image Processing 18(7), 1613–1622 (2009)
8. Krinidis, S., Chatzis, V.: Fuzzy energy-based active contours. IEEE Transactions on Image Processing 18(12), 2747–2755 (2009)
9. Sahoo, P., Wilkins, C., Yeaget, J.: Threshold selection using renyi's entropy. Pattern Recognition 30(1), 71–84 (1997)
10. Yen, J., Chang, F., Chang, S.: A new criterion for automatic multilevel thresholding. IEEE Transactions on Image Processing 4(3), 370–378 (1995)
11. Huang, L., Wang, M.: Image thresholding by minimizing the measures of fuzziness. Pattern Recognition 28(1), 41–51 (1995)
12. Pikaz, A., Averbuch, A.: Digital image thresholding based on topological stable state. Pattern Recognition 29(5), 829–843 (1996)
13. Cheng, H., Chen, Y.Y.: Fuzzy partition of two-dimensional histogram and its application to thresholding. Pattern Recognition 32(5), 825–843 (1999)

14. Krinidis, S., Chatzis, V.: A robust fuzzy local information c-means clustering algorithm. IEEE Transactions on Image Processing 19(5), 1328–1337 (2010)
15. Sauvola, J., Pietikainen, M.: Adaptive document image binarization. Pattern Recognition 33(2), 225–236 (2000)
16. Aghagolzadeh, M., Soltanian-Zadeh, H., Araabi, B., Aghagolzadeh, A.: A hierarchical clustering based on mutual information maximization. In: IEEE International Conference on Image Processing, vol. I, pp. 277–280 (September 2007)
17. Yu, F.X., Lei, Y.Q., Wang, Y.G., Lu, Z.M.: Robust image hashing based on statistical invariance of dct coefficients. Journal of Information Hiding and Multimedia Signal Processing 1(4), 286–291 (2010)
18. Huang, N., Shen, Z., Long, S., Wu, M., Shih, E., Zheng, Q., Tung, C., Liu, H.: The empirical mode decomposition method and the Hilbert spectrum for non-stationary time series analysis. Proceedings of the Royal Society of London 454, 903–995 (1998)
19. Wu, Z., Huang, N.: Ensemble empirical mode decomposition: A noise-assisted data analysis method. Advances in Adaptive Data Analysis 1(1), 1–41 (2009)
20. Tohka, J.: Surface Extraction from Volumetric Images Using Deformable Meshes: A Comparative Study. In: Heyden, A., Sparr, G., Nielsen, M., Johansen, P. (eds.) ECCV 2002, Part III. LNCS, vol. 2352, pp. 350–364. Springer, Heidelberg (2002)

Modelling Crowdsourcing Originated Keywords within the Athletics Domain

Zenonas Theodosiou and Nicolas Tsapatsoulis

Dept. of Communication and Internet Studies,
Cyprus University of Technology,
31 Archbishop Kyprianos Str., CY-3036, Limassol
{zenonas.theodosiou,nicolas.tsapatsoulis}@cut.ac.cy

Abstract. Image classification arises as an important phase in the overall process of automatic image annotation and image retrieval. Usually, a set of manually annotated images is used to train supervised systems and classify images into classes. The act of crowdsourcing has largely focused on investigating strategies for reducing the time, cost and effort required for the creation of the annotated data. In this paper we experiment with the efficiency of various classifiers in building visual models for keywords through crowdsourcing with the aid of Weka tool and a variety of low-level features. A total number of 500 manually annotated images related to athletics domain are used to build and test 8 visual models. The experimental results have been examined using the classification accuracy and are very promising showing the ability of the visual models to classify the images into the corresponding classes with the highest average classification accuracy of 74.38% in the purpose of SMO data classifier.

Keywords: Crowdsourcing Annotation, Keyword Modelling, Image Classification.

1 Introduction

Image tagging helps search engines to better retrieve desired images in response to text queries. Automatic image annotation has concentrated on the difficulty of relating high-level human interpretations with low-level visual features. The interpretation inconsistency between image descriptors and high-level semantics is known as the semantic gap [1] or the perceptual gap [2]. This is due to the fact that the visual image features extracted from an image cannot be automatically translated reliably into high-level semantics [3]. A manually annotated set of multimedia data is used to train a system for the identification of joint or conditional probability of an annotation occurring together with a certain distribution of multimedia content feature vectors [4]. Different models and machine learning techniques are developed to learn the correlation between image features and textual words from the examples of annotated images and then apply the learned correlation to predict words for unseen images [5].

L. Iliadis et al. (Eds.): AIAI 2012, IFIP AICT 381, pp. 404–413, 2012.

On the other hand, manual image tagging annotation is an extremely difficult and elaborate task and cannot always be considered as correct due to visual information that always lets the possibility for more individual interpretation and ambiguity [6]. Crowdsourcing [7] has magnetized the interest of several researchers, since it is a very attractive solution to the problem of cheaply and quickly acquiring annotations and has a potential to improve evaluation of information retrieval systems by scaling up relevance assessments and creating test collections with more complete judgments [8]. Amazon Mechanical Turk [9] opened a new way of satisfying the need for large collections of human-annotated data as presented in the recent past [10] by extending the interactivity of crowdsourcing tasks using more comprehensive user interfaces and micro-payment mechanisms.

In this paper we get the advantage of the availability of a large dataset related to the athletics domain created during the FP6 Boemie project [11] and deal with the experimental evaluation of the efficiency of various low-level features and data classifiers in modelling crowdsourcing originated keywords. A set of images was annotated by 15 users using a predefined set of keywords. Images sharing a common keyword are grouped together and used to create the visual model which corresponds to this keyword. Eight different keyword models are created using low-level features and tested with the aid of well known data classifiers. We have used publicly available tools for the computation of the low level features [12], [13] and the model creation (the Weka tool [14]) and classified the images into 8 keyword classes.

This paper is organized as follows: Section 2 gives a detailed description of the method we have followed to create the dataset and build the visual models while Section 3 presents and discusses the experimental results. Finally, conclusions are drawn and further work hints are given in Section 4.

2   Method Overview

This section presents the method we have followed to model the crowdsourced keywords. It consists of 3 main steps: the dataset creation, the feature extraction, and the keyword modelling. The overall procedure is illustrated in Fig. 1.

2.1   Dataset Creation

The crowdsourcing annotation was based on a randomly selected set of 500 images taken from a large dataset created during the BOEMIE project [11]. The dataset was manually annotated by 15 users using the MuLVAT annotation tool [15] with the aid of a structured xml dictionary consists of 33 different keywords. For our experiments we have selected 8 representative keywords and for each keyword, 50 images that were annotated from more than 5 annotators with this keyword were grouped together to create a set of 8 different groups of images (Fig. 2).

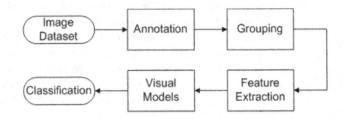

Fig. 1. The flowchart of overall method

Fig. 2. The set of image groups

2.2 Feature Extraction

Among the possible low-level features that can be extracted from the image groups, we have chosen the following popular and widely used features:

Histogram of Gradients Features. The HOG features exploit the idea that local object appearance can be described by the distribution of intensity gradients or edge directions. The image is divided into small connected regions, called cells. For each cell, a histogram of gradient directions or edge orientations within this cell is compiled. For the implementation of HOG, each pixel within the cell casts a weighted vote for an orientation-based histogram channel. For the current study we have used the implementation proposed in [13] with the aid of 25 rectangular cells and 9 bins histogram per cell. The 16 histograms with 9 bins were then concatenated to make a 225-dimensional feature vector.

Scale-Invariant Feature Transform Features. SIFT transforms image data into scale-invariant coordinates relative to local features and performs a set of features that are not affected by object scaling and rotation. Key points are detected as the maxima of an image pyramid built using difference-of-Gaussians. The multi-scale approach results in features that are detected across different scales of images. For each detected key-point, a 128 dimensional feature vector is computed describing the gradient orientations around the key-point. The strongest gradient orientation is selected as reference, thus giving rotation invariance to SIFT features. For our experiments each SIFT vector is quantized into a 100-dimensional feature vector using k-means clustering.

MPEG-7 Features. MPEG-7 visual descriptors include the color, texture and shape descriptor. A total of 22 different features are included, nine for color, eight for texture and five for shape. The dominant color features include color value, percentage and variance and require especially designed metrics for similarity matching. Furthermore, their length is not known a priori since they are image dependent (for example an image may be composed from a single color whereas others vary in color distribution). The previously mentioned difficulties cannot be easily handled in machine learning schemes, therefore we decided to exclude these features for the current experimentation. The texture browsing features (regularity, direction, scale) have not been included in the description vectors since in the current implementation of the MPEG-7 experimentation model [12] the corresponding descriptor cannot be reliably computed (it is a known bug of the implementation software). The scalable color and shape descriptor features have been also excluded because vary depending on the form of an input object and can not be used for the holistic image description. Among all MPEG-7 descriptors only the Color Layout (CL), Color Structure (CS), Edge Histogram (EH) and Homogenous Texture (HT) descriptors are used in our experiments. The combination of the selected descriptors creates a 186-dimensional feature vector.

2.3   Keyword Modelling

To overcome the multiclass classification problem and facilitate effective and efficient learning, each keyword is treated as a separate binary classification problem. We have followed the one-against-rest approach [16] and we have built a total number of 8 models, one for each keyword. The feature vectors of each keyword class were split into two groups, called the training (80%) and testing (20%) set. Each model is trained and tested between one class and the 7 other classes. The training and testing set for each model contain the feature vectors of the corresponding keyword class and the same number of randomly selected feature vectors of the the rest 7 classes. Keywords models were created using Weka tool [14]. Since statistical methods have their limitations, particularly in relation to distributional assumptions and to the restrictions on data input, we have decided to use artificial intelligence classifiers such as Support Vector Machines (SVM) and Decision Trees (DT).

2.4   Support Vector Machines

SVM separates the classes with a decision surface that maximizes the margin between the classes. The surface is often called the optimal hyperlane and the data points closer to the hyperplane are called support vectors. For our experiments we decided to use two of the state of the art implementations of the Support Vector Machines (SVMs), the SMO [17] and the LibSVM [18]. They have been reported in several publications as the best performing machine learning algorithms for a variety of classification tasks. The performance of SVM classifiers can vary significantly with variation in parameters of the models. During training we experimented with different parameters and kernels and for each kernel we built models for several combinations of the parameters, with the Pearson VII universal and polynomial kernel performing better than the others for the SMO and LibSVM classifier respectively.

2.5   Decision Trees

Unlike other classification approaches that use a set of features jointly to perform classification in a single decision step, the decision tree is based on a multistage or hierarchical decision scheme or a tree like structure. The tree is composed of a root node (containing all data), a set of internal nodes (splits) and a set of terminal nodes (leaves). Each node of the decision tree structure makes a binary decision that separates either one class or some of the classes from the remaining classes. The processing is generally carried out by moving down the tree until the leaf node is reached. Turning to the classifiers, Random Forest [19] and Logistics Model Tree (LMT) [20] have been employed to model the keywords.

3   Experimental Results

We used the dataset and keyword modelling process described in the previous Section to examine the performance and effectiveness of the created models: "Discus", "Hammer", "High Jump", "Hurdles", "Javelin", "Long Jump", "Running", and "Triple Jump". Fig. 3, 4, 5, 6 show the accuracy of correctly classified instances for all classes using the various data classifiers. The results shown in these figures were examined under three perspectives: First, in terms of the efficiency of the various classifiers in modelling crowdsourced keywords, second in terms of the efficiency of the low-level features to create accurate visual models and third, in terms of the ability of the created models to classify the images into the corresponding classes.

The efficiency of the training algorithms is examined through the effectiveness of the created models, the time required to train the models and the robustness to the variation of learning parameters. The SMO and Random Forest classifiers require by far the lower time and effort to create an effective model, while LMT is the slowest classifier among all. In the case of SMO and Random Forest the learning takes no more than a few seconds for the majority of the keyword models. Furthermore, the fluctuation in classification performance during parameters tuning is significantly lower than that of the LibSVM and LMT. There is a significant difference on the performance of the models created using the individual classifiers. It is evident from Table 1 that SMO is the most reliable classifier with a total average classification accuracy equal to 74.38%. The LibSVM and Random Forest give the same average classification accuracy with the difference that LibSVM performs better for the HOG while the Random Forest performs better for the MPEG-7 features. The most disappointing classifier is the LMT which has the worst average classification accuracy values. The best classification accuracy was occurred in the case of SMO classifier using MPEG-7 features and the worst in the case of LMT classifier using SIFT features.

Concerning the efficiency of the various low-level features, the experimental results indicate that the MPEG-7 features perform better than HOG and SIFT. The classification accuracy obtained using these features is quite good in comparison with the other two and has average classification accuracy values in the range of 71.25%- 81.25%, with the lowest value given by LMT and the highest by SMO classifier respectively. The second more reliable low-level features for modeling keywords are the HOG that can obtain the maximum average classification accuracy value of 72.5% in the purpose of the SMO classifier. The most disappointing classification performance is achieved by the SIFT which can obtain the maximum value of 75% only for "Discus", "Hammer" and "Hurdles" classes.

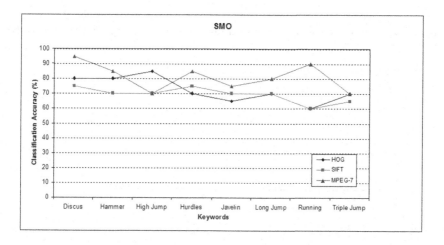

Fig. 3. Accuracy of the correctly classified instances using the SMO data classifier

Nearly all models are able to classify the images into the corresponding classes with classification accuracy in the range of 55%-95%. The best efficiency is perceived when testing models created by images having objects with a well defined shape such as "Discus" and "Hurdles". As a consequence, the worst results are occurred when testing the "Running" and "Triple Jump" because the content of images belong to these keywords has many similarities with the content of images belong to other keywords.

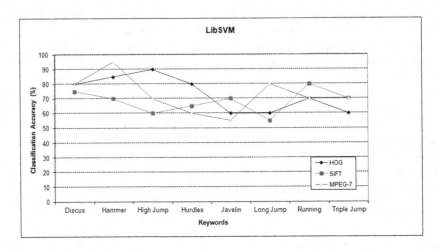

Fig. 4. Accuracy of the correctly classified instances using the LibSVM data classifier

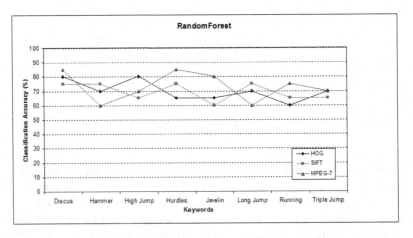

Fig. 5. Accuracy of the correctly classified instances using the Random Forest data classifier

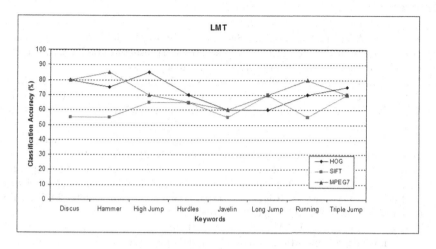

Fig. 6. Accuracy of the correctly classified instances using the LMT data classifier

Table 1. Average classification accuracy values (%) for the different classifiers

Classifier	HOG	SIFT	MPEG-7	Overall
SMO	72.5	69.38	81.25	74.38
LibSVM	73.13	68.13	71.25	70.83
Random Forest	70.0	69.38	73.13	70.83
LMT	71.88	61.25	72.5	68.54

4   Conclusions and Future Work

We present an experimental evaluation of modelling crowdsourcing originated keywords within the athletics domain. Specifically, 8 different keywords were modeled using various low-level features and data classifiers. According to our experimental results, nearly all created models can classify the images into the 8 classes with medium to high classification accuracy. Although there is a significant variation on the efficiency of the various classifiers with the SMO having the highest performance, a great improvement can achieved when the MPEG-7 features are used. Our future perspectives involve the evaluation of the proposed method on larger and different datasets as well as the experimentation of additional training algorithms and other classifications schemes. In addition, the efficiency of more low-level features in creation of visual models will be investigated.

Acknowledgements. This work falls under the Cyprus Research Promotion Foundation's Framework Programme for Research, Technological Development and Innovation 2009-2010 (DESMI 2009-2010), co-funded by the Republic of Cyprus and the European Regional Development Fund, and specifically under Grant $\Pi ENEK/0609/95$.

References

1. Smeulders, A.W.M., Worring, M., Santini, S., Gupta, A., Jain, R.: Content-based image retrieval at the end of the early years. IEEE Transactions on Pattern Analysis and Machine Intelligence 22(12), 1349–1380 (2000)
2. Jaimes, A., Christel, M., Gilles, S., Sarukkai, R., Ma, W.-Y.: Multimedia information retrieval: what is it, and why isn't anyone using it? In: Proc. of the 7th ACM SIGMM International Workshop on Multimedia Information Retrieval (MIR 2005), pp. 3–8 (2005)
3. Datta, R., Joshi, D., Li, J., Wang, J.Z.: Image retrieval: Ideas, influences, and trends of the new age. ACM Computing Surveys 40(2), 1–60 (2008)
4. Athanasakos, K., Stathopoulos, V., Jose, J.: A Framework for Evaluating Automatic Image Annotation Algorithms. In: Gurrin, C., He, Y., Kazai, G., Kruschwitz, U., Little, S., Roelleke, T., Rüger, S., van Rijsbergen, K. (eds.) ECIR 2010. LNCS, vol. 5993, pp. 217–228. Springer, Heidelberg (2010)
5. Zhang, P., Zhang, Z., Li, M., Ma, W.Y., Zhang, H.J.: A probabilistic semantic model for image annotation and multi-modal image retrieval. Multimedia Systems 12(1), 27–33 (2006)
6. Volker, T., Thom, A., Tahaghoghi, S.M.M.: Modelling human judgment of digital imagery for multimedia retrieval. IEEE Transactions on Multimedia 9(5), 967–974 (2007)
7. Howe, J.: Crowdsourcing: Why the Power of the Crowd Is Driving the Future of Business. Crown Publishing Group (2008)
8. Kazai, G., Kamps, J., Koolen, M., Milic-Frayling, N.: Crowdsourcing for book search evaluation: Impact of quality on comparative system ranking. In: Proc. of the 34th Annual International ACM SIGIR Conference on Research and Development in Information Retrieval. ACM Press, New York NY (2011)
9. Amazon Mechanical Turk - Artificial Artificial Intelligence, http://www.mturk.com
10. Eickhoff, C., de Vries, A.P.: How Crowdsourcable is Your Task? In: Proc. of the Workshop on Crowdsourcing for Search and Data Mining, Hong Kong, China (2011)
11. BOEMIE - Bootstrapping Ontology Evolution with Multimedia Information Extraction, http://www.boemie.org
12. MPEG-7 Visual Experimentation Model (XM), Version 10.0, ISO/IEC/JTC1/SC29/WG11, Doc. N4063 (2001)
13. Ludwig, O., Delgado, D., Goncalves, V., Nunes, U.: Trainable Classifier-Fusion Schemes: An Application To Pedestrian Detection. In: Proc. of 12th International IEEE Conference on Intelligent Transportation Systems, vol. 1, pp. 432–437 (2009)
14. Witten, I.H., Frank, E.: Data Mining: Practical machine learning tools and techniques, 2nd edn. Morgan Kaufmann, San Francisco (2005)
15. Theodosiou, Z., Kounoudes, A., Tsapatsoulis, N., Milis, M.: MuLVAT: A Video Annotation Tool Based on XML-Dictionaries and Shot Clustering. In: Alippi, C., Polycarpou, M., Panayiotou, C., Ellinas, G. (eds.) ICANN 2009, Part II. LNCS, vol. 5769, pp. 913–922. Springer, Heidelberg (2009)
16. Tax, D.M.J., Duin, R.P.W.: Using two-class classifiers for multiclass classification. In: Proc. of 16th International Conference of Pattern Recognition, pp. 124–127 (2002)

17. Platt, J.: Fast Training of Support Vector Machines using Sequential Minimal Optimization. In: Schoelkopf, B., Burges, C., Smola, A. (eds.) Advances in Kernel Methods-Support Vector Learning. MIT Press (1998)

18. Fan, R.-E., Chen, P.-H., Lin, C.-J.: Working set selection using the second order information for training SVM. Journal of Machine Learning Research 6, 1889–1918 (2005)

19. Breiman, L.: Random Forests. Machine Learning 45(1), 5–32 (2001)

20. Landwehr, N., Hall, M., Frank, E.: Logistic Model Trees. Machine Learning 59(1), 161–205 (2005)

Scalable Object Encoding
Using Multiplicative Multilinear Inter-camera
Prediction in the Context of Free View 3D Video

Ioannis M. Stephanakis[1] and George C. Anastassopoulos[2,3]

[1] Hellenic Telecommunication Organization S.A. (OTE),
99 Kifissias Avenue, GR-151 24, Athens, Greece
stephan@ote.gr
[2] Democritus University of Thrace, Medical Informatics Laboratory,
GR-681 00, Alexandroupolis, Greece
anasta@med.duth.gr
[3] Hellenic Open University, Parodos Aristotelous 18, GR-262 22, Patras, Greece

Abstract. Recent advancements in 3D television allow for the capture of scene depth from multiple cameras and the interactive selection of view point and direction within a certain range, the so-called Free Viewpoint Video (FVV). State-of-the-art video codecs such as H.264/AVC exploit the large amount of inter-view statistical dependencies by combined temporal and inter-view prediction, i.e. prediction from temporally neighboring images as well as from images in adjacent views. This is known as Multi-view Video Coding (MVC). We propose herein an alternative object oriented video coding scheme for multi-view video with associated multiple depth data (N-video plus N-depth). A structure that we call a *Multi-view Video Plane* (MVP) is introduced. Object planes associated with a certain view are approximated as multilinear components of an image that are projected onto other views in a tensor-like fashion. The order of the tensor equals the number of multiple views. The coefficients of the tensor subspace projections as well as the updates of the multi-linear components (object-planes) are quantized and transmitted in the MPEG stream. Motion-compensated prediction is carried out in order to transmit the residual object planes (P-frames) using conventional MPEG algorithms.

Keywords: object oriented coding; multiview video; 3D television; multilinear principal component analysis; Generalized Singular Value Decomposition; H264/AVC.

1 Introduction

1.1 Approaches to 3-D Capture, Multi-view And Free View Video

Recent advances in stereoscopic display and capture technologies have led to the enhancement of existing coding standards as well as relevant processing algorithms [1].

L. Iliadis et al. (Eds.): AIAI 2012, IFIP AICT 381, pp. 414–424, 2012.
© IFIP International Federation for Information Processing 2012

A straightforward way to encode stereoscopic video sequences is for example MPEG-2 multi-view profile (MVP) [2]. Multi-view video (MVV) support is intended for 3D video applications, where 3D depth perception of a visual scene is provided by a 3D display system [3]. Such 3D display systems include classic stereo systems that require special-purpose glasses as well as more sophisticated multi-view auto-stereoscopic displays that do not require glasses. Multi-view video enables free-viewpoint video, i.e. it allows the interactive selection of viewpoint and view direction within a specified range [4]. Each output view can either be one of the input views or a virtual view that was generated from a smaller set of multi-view inputs and other data that assists in view generation process. With such a system, viewers can freely navigate through the different viewpoints of the scene – within a range covered by the acquisition cameras. Next-generation 3D video services have already appeared into the entertainment market. The *Society of Motion Picture and Television Engineers* (SMPTE) formed a task force to investigate the production requirements in order to realize 3D video to the home [5]. The final report of the task force recommends standardization of a *3D Home Master* which would essentially be an uncompressed and high-definition stereo image format.

A simple compression method that may be used is to encode all video signals independently using a state-of-the-art video codec such as H.264/AVC [6,7]. This solution features low complexity and keeps computation and processing delay to a minimum. It is the so-called simulcast coding (Fig. 1.a.). Nevertheless multi-view video contains a large amount of inter-view statistical dependencies that can be exploited for combined temporal and inter-view prediction [8]. Frames are not only predicted from temporal neighboring frames but also from corresponding frames in adjacent views. Multi-view encoding based on temporal and inter-view prediction is illustrated Fig. 1.b. Several techniques for inter-view prediction have been proposed [9,10]. Such predictions are key features of the MVC design and are enabled through flexible reference picture management of AVC, where decoded pictures from other views are essentially made available in the reference picture list. A reference picture list is maintained for each picture to be decoded in a given view according to the state-of-the-art encoding standards [11]. Prediction of a picture in the current view may be based upon the disparity of references generated from neighboring views (*Disparity-Compensated Prediction* - DCP) or from synthesized references generated from neighboring views (*View Synthesis Prediction* - VSP). Multi-view video with associated multiple depth data is standardized as ISO/IEC 23002-3 (also referred to as MPEG-C Part 3). It specifies the representation of auxiliary video and supplemental information and enables signaling for depth map streams to support 3D video applications. View synthesis prediction (VSP) is possible from depth data.

The proposed encoding scheme is a free viewpoint approach that employs a novel method for synthesis prediction from *Video Object Planes* (VOPs) that are obtained from different views. It is assumed that a video frame associated with the k-th view (denoted as $\mathbf{V}^{(k)}$) is composed by I_k VOPs that are considered to be orthogonal for the sake of simplicity, i.e.

$$\mathbf{V}^{(k)} = \begin{bmatrix} VOP_1^k & VOP_2^k & \cdots & VOP_{I(k)}^k \end{bmatrix} \text{ and} \qquad (1)$$

$$\left[\mathbf{V}^{(k)}\right]^T \mathbf{V}^{(k)} = diag\{(\sigma_1^{(k)})^2,(\sigma_2^{(k)})^2\cdots(\sigma_{I(k)}^{(k)})^2\}. \tag{2}$$

Let us define the *Multi-view Video Plane* of an N-view system at t as an N-th order tensor according to the following equation,

$$MVP(t) = S(t)\times_1 \mathbf{V}^{(1)}\times_2 \mathbf{V}^{(2)}\times_3 \mathbf{V}^{(3)}\times\cdots\times_N \mathbf{V}^{(N)}. \tag{3}$$

It follows that,

$$MVP(t) = \sum_{i_1=1}^{I_1}\sum_{i_2=1}^{I_2}\cdots\sum_{i_N=1}^{I_N} s(i_1,i_2,\ldots;t)\ VOP_{i_1}^1 \circ VOP_{i_2}^2 \cdots VOP_{i_N}^N. \tag{4}$$

Should one keep P_k objects for view k, the following approximations are possible regarding the *Multiview Video Plane* and the separate frames per view,

$$MVP(t) = \sum_{i_1=1}^{P_1}\sum_{i_2=1}^{P_2}\cdots\sum_{i_N=1}^{P_N} s(i_1,i_2,\ldots;t)\ VOP_{i_1}^1 \circ VOP_{i_2}^2 \cdots VOP_{i_N}^N + \Delta(t)\ \text{and} \tag{5}$$

$$\mathbf{V}^{(k)}(t) = \left[VOP_1^k \cdots VOP_{P(k)}^k\ \mathbf{0}\ \cdots\right]+\Delta^{(k)}(t) = \tilde{\mathbf{V}}^{(k)}(t)+\Delta^{(k)}(t). \tag{6}$$

Herein the concept of transmitting predictions of the structural elements of the *Multi-view Video Planes* at the start of each *Group of Pictures* (GOP) is investigated according to the scheme illustrated in Figs. 2.a and 2.b. It is a scalable approach to multi-view video coding that features base and higher layers as well as inter-camera predictions. It may well be compared against other scalable methods proposed in the literature [12].

1.2   Multi-linear Principal Component Representation of Multi-view Video Plane (MVP) and Multiplicative View Predictions

An Nth-order tensor A resides in the tensor multi-linear space $R^{I_1} \otimes R^{I_2} \otimes \cdots \otimes R^{I_N}$

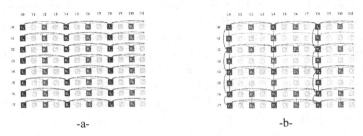

-a-   -b-

Fig. 1. a- Simulcast coding structure with hierarchical B pictures for temporal prediction and - b- Multi-view coding structure with hierarchical B pictures for temporal and inter-view prediction.

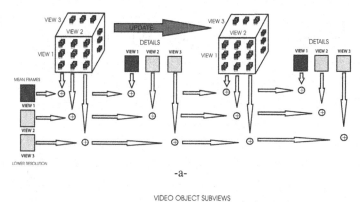

-a-

VIDEO OBJECT SUBVIEWS

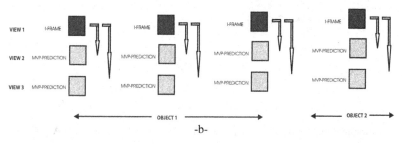

-b-

Fig. 2. a- Encoding sequence according to the proposed approach (lower resolution and MVP encoding) and -b- MVP inter view prediction according to *Step* II-*iii* of the algorithm

where $R^{I_1}, R^{I_2} \cdots, R^{I_N}$ are the N vector linear spaces. The "k-mode vectors" of A are defined as the I_k-dimensional vectors obtained from A by varying its index in k-mode while keeping all the other indices fixed [13,14]. Unfolding A along the k-mode is denoted as

$$\mathbf{A}_{(k)} \in R^{I_k \times (I_1 \times \cdots \times I_{k-1} I_{k+1} \times \cdots \times I_N)}, \quad (7)$$

where the column vectors of $A_{(k)}$ are the k-mode vectors of A (see Fig. 3). Unfolding the *Multi-view Video Plane* of an N-view system along the k-mode view results into the following matrix representation

$$\mathbf{MVP}_{(k)}(t) = \mathbf{V}^{(k)} . S_{(k)}(t) . \left(\mathbf{V}^{(k+1)} \otimes \mathbf{V}^{(k+2)} \otimes \cdots \otimes \mathbf{V}^{(N)} \otimes \mathbf{V}^{(1)} \otimes \cdots \otimes \mathbf{V}^{(k-1)} \right)^{\mathsf{T}}, \quad (8)$$

where $\otimes$ denotes the Kronecker product. The core tensor S (in a representation similar to the one described in Eq. 3) is analogous to the diagonal singular value matrix of the traditional SVD and coordinates the interaction of matrices to produce the original tensor. Matrices $\mathbf{V}^{(k)}$ are orthonormal and their columns span the space of the corresponding flattened tensor denoted as $\mathbf{MVP}_{(k)}$. The objective of MPC analysis for predetermined dimensionality reduction is the estimation the N projection matrices $\{ \tilde{\mathbf{V}}^{(k)}(t) \in R^{I_k \times P_k}, k = 1,...,N \}$ that maximize the total tensor scatter [15],

$$\left\{\tilde{\mathbf{V}}^{(k)}(t),k=1,...,N\right\}= \arg \max_{\tilde{\mathbf{V}}^{(1)},\tilde{\mathbf{V}}^{(2)}...\tilde{\mathbf{V}}^{(N)}} \left\|MVP(t)-\Delta(t)\right\|^2 , \tag{9}$$

where $\sum_{k=1}^{N} P_k \leq c$.

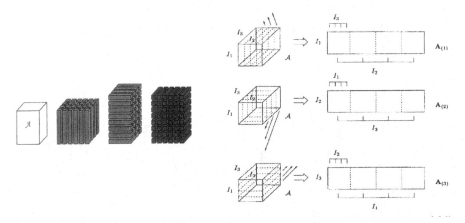

Fig. 3. Unfolding the 3-mode tensor $A \in R^{I_1 \times I_2 \times I_3}$ to the matrices $\mathbf{A}_{(1)} \in R^{I_1 \times (I_2 \times I_3)}$, $\mathbf{A}_{(2)} \in R^{I_2 \times (I_3 \times I_1)}$ and $\mathbf{A}_{(3)} \in R^{I_3 \times (I_1 \times I_2)}$ using 1-mode, 2-mode and 3-mode vectors of A respectively (from Lathauwer et al. 2000)

By solving Eq. 9 one determines the N projections to the N vector subspaces of the underlying multi-view system. It simplifies calculations to solve the maximization problem iteratively by successively estimating the set of vectors (VOPs) that maximize the scatter in each view mode vector space. It is straightforward to show that the approximation matrix $\tilde{\mathbf{V}}^{(k)} \in R^{I_k \times P_k}$ - where $k=1,...,N$ - that maximizes the scatter in the k-mode vector space is estimated by the following relationship for the expectation values,

$$\text{maximize}\,\mathscr{G}\left(\tilde{\mathbf{V}}^{(k)}\right)=\frac{1}{2}\mathscr{E}\left\{\left\|(\tilde{\mathbf{V}}^{(k)})^\mathsf{T}(\mathbf{MVP}_{(k)}-\overline{\mathbf{MVP}}_{(k)})\tilde{\mathbf{V}}_{\Phi^{(k)}}\right\|^2\right\}, \tag{10}$$

where $\tilde{\mathbf{V}}_{\Phi^{(k)}} = \mathbf{V}^{(k+1)} \otimes \mathbf{V}^{(k+2)} \otimes \cdots \otimes \mathbf{V}^{(N)} \otimes \mathbf{V}^{(1)} \otimes \cdots \otimes \mathbf{V}^{(k-1)}$ and $\mathbf{MVP}_{(k)}$ stands for the unfolded form of the *Multi-view Video Plane* along the k-mode view (see Eq. 8). The gradient of Eq. 10 is given as

$$\frac{\partial \mathscr{G}\left(\tilde{\mathbf{V}}^{(k)}\right)}{\partial \tilde{\mathbf{V}}^{(k)}}=\frac{\partial \mathscr{G}\left(\tilde{\mathbf{V}}^{(k)}\right)}{2\,\partial \tilde{\mathbf{V}}^{(k)}}\text{Tr}\{(\tilde{\mathbf{V}}^{(k)})^\mathsf{T}\mathbf{\Phi}^{(k)}\tilde{\mathbf{V}}^{(k)}\}=\mathbf{\Phi}^{(k)}\tilde{\mathbf{V}}^{(k)} , \tag{11}$$

where $\mathbf{\Phi}^{(k)} = \mathscr{E}\{(\mathbf{MVP}_k-\overline{\mathbf{MVP}}_k)\tilde{\mathbf{V}}_{\Phi^{(k)}}(\tilde{\mathbf{V}}_{\Phi^{(k)}})^\mathsf{T}(\mathbf{MVP}_k-\overline{\mathbf{MVP}}_k)^\mathsf{T}\}$. By decomposing into non-negative parts and employing a multiplicative update rule that maintains orthonormality [16,17], one gets

$$[\mathbf{V}^{(k)}]_{ij}^{new} = [\mathbf{V}^{(k)}]_{ij} \frac{[\mathbf{\Phi}_{+}^{(k)}\mathbf{V}^{(k)}]_{ij} + [\mathbf{V}^{(k)}(\mathbf{V}^{(k)})^{\mathsf{T}}\mathbf{\Phi}_{-}^{(k)}\mathbf{V}^{(k)}]_{ij}}{[\mathbf{\Phi}_{-}^{(k)}\mathbf{V}^{(k)}]_{ij} + [\mathbf{V}^{(k)}(\mathbf{V}^{(k)})^{\mathsf{T}}\mathbf{\Phi}_{+}^{(k)}\mathbf{V}^{(k)}]_{ij}}, \tag{12}$$

where

$$[\mathbf{\Phi}_{+}^{(k)}]_{ij} = \begin{cases} [\mathbf{\Phi}^{(k)}]_{ij} & if\ [\mathbf{\Phi}^{(k)}]_{ij} > 0 \\ 0 & otherwise \end{cases} \ and\ [\mathbf{\Phi}_{-}^{(k)}]_{ij} = \begin{cases} -[\mathbf{\Phi}^{(k)}]_{ij} & if\ [\mathbf{\Phi}^{(k)}]_{ij} < 0 \\ 0 & otherwise. \end{cases} \tag{13}$$

2 A Scalable Model For Multi-linear Principal Object Encoding For Multi-view Systems

2.1 Hierarchies of Multi-view Encoding According to the Proposed Approach

An MVC coder consists of N parallelized single view coders. Each of them uses temporal prediction structures, were a sequence of successive pictures is coded as intra (I), predictive (P) or bi-predictive (B) frames. Further improvement of coding efficiency may be achieved by using hierarchical B pictures, where a B picture hierarchy is created by a cascade of B pictures that are references for other B pictures. A current picture in the coding process can have temporal as well as inter-view references pictures for prediction. Advanced formats for 3D video coding require geometry data [9,10,11]. Depth data are estimated based on the acquired pictures. They are obtained by the application of depth estimation algorithms and should not be regarded as ground truth. The proposed approach assumes three distinct frame hierarchies (see Fig. 2.a), namely lower resolution mean value frames that are transmitted as intra (I) frames at the beginning of each GOP, structural encoding of multiple views with reference to a scalable multi-view object plane (MVP) and MVP-based predicted frames per view (Fig. 2.b) and, finally, residual frames per view.

2.2 Outline of the Encoding Algorithm

The following algorithm outlines in detail the steps of the proposed encoding approach for multi-view systems:

```
Encode Multi-view GOPs in three distinct hierarchies
I - Lower resolution encoding: Transmit mean values of
blocks as I-frames
II - MVP encoding: Determine the number of objects and
the number of subviews per object.
   i- Encode initial object subviews or updates
      - Initial object subviews are transmitted as
        I- frames whereas updates are transmitted as
        P-frames.
```

- Transmit updates according to Eqs. 12 and 13.
- Set an upper limit of the multiplicative factor and multiply the most significant elements of $[\mathbf{V}^{(k)}]_i$ (ignore near-zero elements). Normalize $[\mathbf{V}^{(k)}]_i$ in Eq. 12.

ii- Scalable encoding of master view
- Transmit actual transform coefficients of the differences of master subviews.

iii- Estimation of secondary views
- Estimate secondary views using Eq.(8). Assume that k denotes master view and that $\mathbf{MVP}_{master\,view}$ is the muliview video plane for the entire GOP. $\mathbf{V}^{master\,view}$ stands for the actual transmitted VOPs of the master view within the GOP whereas $\hat{\mathbf{V}}^{secondary\,views}$ stands for the estimation of cross secondary views within the GOP, hence

$$\hat{\mathbf{V}}^{secndary\,views} = (\mathbf{V}^{master\,view})^{\top} diag(\frac{1}{(\sigma_1^{master\,view})^2}, \quad \frac{1}{(\sigma_2^{master\,view})^2} \quad \cdots \quad \frac{1}{(\sigma_p^{master\,view})^2})\mathbf{MVP}_{master\,view} \qquad (14)$$

Estimate cross secondary views within the GOP by averaging the rows of $\hat{\mathbf{V}}^{secondary\,views}$ (see Fig. 2.b).

III - Encode residual images for each view (using motion estimation and rate control algorithms)
IV - Repeat until end of GOP (go to *Step* II)

a – Initial right eye frame (View 1)

b – Initial left eye frame (View 2)

c – Right eye initial depth image (View 1)

d – Left eye initial depth image (View 2)

Fig. 4. Image frames used in numerical simulations

3   Numerical Simulations

Numerical simulations for the proposed encoding method have been carried out for the image sequences used for video view interpolation as described in [18]. Each sequence is 100 frames long. The camera resolution is 1024x768 and the capture rate is 15fps. The initial uncompressed frames for the ballet sequence are depicted in   Fig. 4. The multi-view GOP for the numerical simulations consists of sixteen (16) frames. It is assumed that one stereo object analyzed into six (6) orthogonal video subviews is encoded according to the proposed algorithm. Luminance as well as chrominance and depth frames are decomposed as described in *Section* 2.2. Their interdependencies determine the MVP structure of the GOP. They are depicted in Fig. 5. Depth images are more correlated than the images of the luminance and the chrominance components. The VOPs corresponding to the luminance components are illustrated in Fig. 6. Only the elements of the VOPs featuring an absolute magnitude higher than 0.5% of the maximum value are updated multiplicatively each time. The higher multiplication factor is limited to 3. Convergence is slow as indicated by Fig. 7 for the luminance VOPs. We reorder VOPs by sorting their eigenvalues at each multiplicative step. The average estimated rate for the first two encoding steps is estimated to about 0.1 bit per pixel. We assume that each VOP transmitted as I-frame at the beginning of the GOP requires 0.25 bit per pixel. The estimated residual frames for luminance are presented in Figs. 8. PSNR values (Peak Signal-to-Noise-Ratio) of the transmitted frames (without residual frame encoding as described in *Step* III of the proposed algorithm) are given in Figs. 9.a and 9.b. Estimated residual frames according to Eq. 14 feature slightly lower PSNR values as compared against true corresponding values.

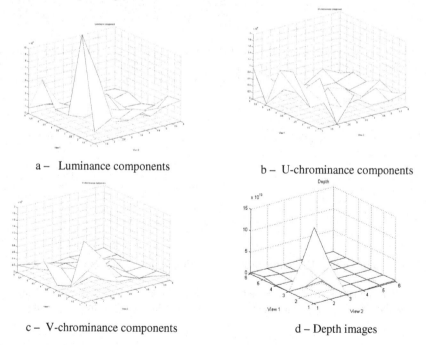

a – Luminance components   b – U-chrominance components

c – V-chrominance components   d – Depth images

Fig. 5. Interdependencies between object subviews in Views 1 and 2

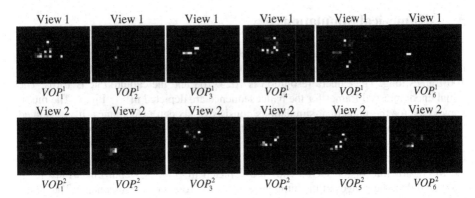

View 1	View 1	View 1	View 1	View 1	View 1
VOP_1^1	VOP_2^1	VOP_3^1	VOP_4^1	VOP_5^1	VOP_6^1
View 2	View 2	View 2	View 2	View 2	View 2
VOP_1^2	VOP_2^2	VOP_3^2	VOP_4^2	VOP_5^2	VOP_6^2

Fig. 6. Planes of video subviews for luminance View 1 and View 2
(six orthogonal subviews for one stereo object)

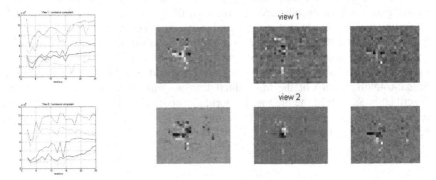

Fig. 7. Convergence of luminance eigenvalues for view 1 and view 2

Fig. 8. Residual frames for luminance and chrominance (six object subviews - residual frames for view 2 are obtained by estimation from view 1 according to Eq. 14)

View 1 - Luminance & U/V-chrominance

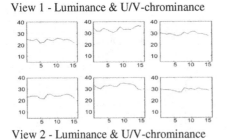

View 2 - Luminance & U/V-chrominance

Fig. 9.a- Actual PSNR [dB vs frame #]
(without residual frame encoding)

View 1 - Luminance & U/V-chrominance

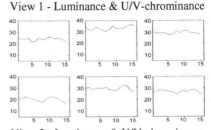

View 2 - Luminance & U/V-chrominance

Fig. 9.b- Estimated PSNR [dB vs frame #]
(without residual frame encoding)

4   Conclusion

A novel scalable approach to multi-view video encoding based on the so-called MVP structure is proposed. Views are defined as lower order projections of tensorial objects. VOPs that constitue the MVP are transmitted at the beginning of each GOP and are multiplicatively updated within the GOP at the beginning of each subgroup of frames. Transmission rates are comparable to the rates reported in the literature for state-of-the-art multi-view encoders. Scalability is determined by the number of VOP for luminance, chrominance and depth frames. The number of VOPs is determined by the stereo-objects in the scene, their relative angles and their velocities with respect to shooting cameras. The proposed method may be combined with other morphing and fusion techniques described in the literature for view synthesis prediction.

References

1. Smolic, A., Kauff, P.: Interactive 3-D Video Representation And Coding Technologies. Proceedings of the IEEE 93(1), 98–110 (2005)
2. Ohm, J.-R.: Stereo/Multiview Video Encoding Using the MPEG Family of Standards. In: Proceedings SPIE, Stereoscopic Displays and Virtual Reality Systems VI, vol. 3639, pp. 242–253 (1999)
3. Vetro, A., Wiegand, T., Sullivan, G.J.: Overview of the Stereo Multiview Video Coding Extensions of the H.264/MPEG-4 AVC Standard. Proceedings of the IEEE 99(4), 626–642 (2011)
4. Kimono, T.: Free Viewpoint Television (FTV), http://www.tanimoto.nuee.nagoya-u.ac.jp/study/FTV (last accessed in April 2012)
5. SMPTE: Report of SMPTE Task Force on 3D to the Home (2009)
6. Wiegand, T., Sullivan, G.J., Bjøntegaard, G., Luthra, A.: Overview of the H.264/AVC Video Coding Standard. IEEE Transactions on Circuits and Systems for Video Technology 13(7), 560–576 (2003)
7. Richardson, I.E.G.: H.264 and MPEG-4 Video Compression – Video Coding for Next-generation Multimedia. Wiley (2003)
8. Girod, B., Aaron, A.M., Rane, S., Rebollo-Monederato, D.: Distributed Video Coding. Proceedings of the IEEE 93(1), 71–83 (2005)
9. Ouaret, M., Dufaux, F., Ebrahimi, T.: Iterative Multiview Side Information for Enhanced Reconstruction in Distributed Video Coding. EURASIP Journal on Image and Video Processing 2009, article ID 591915 (2009)
10. Fujii, T., Kimono, T., Tanimoto, M.: Free-viewpoint TV System Based on Ray-Space Representation. In: Proc. SPIE ITCom, vol. 4864-22, pp. 175–189 (2002)
11. ITU-T and ISO/IEC JTC 1: Advanced Video Coding for Generic Audiovisual Services. ITU-T Recommendation H.264 and ISO/IEC 14496-10 (MPEG-4 AVC), Version 1: May 2003, Version 2: May 2004, Version 3: March 2005 (including FRExt extension), Version 4: September 2005, Version 5 and Version 6: June 2006, Version 7: April 2007, Version 8: July 2007 (including SVC extension), Version 9: July 2009 (including MVC extension)
12. Ouaret, M., Dufaux, F., Ebrahimi, T.: Error-resilient Scalable Compression Based on Distributed Video Coding. Signal Processing: Image Communication 24, 437–451 (2009)

13. de Lathauwer, L., de Moor, B., Vandewalle, J.: A Multilinear Singular Value Decomposition. SIAM Journal of Matrix Analysis and Applications 21(4), 1253–1278 (2000)
14. Lu, H., Plataniotis, K.N., Venetsanopoulos, A.N.: MPCA: Multilinear Principal Component Analysis of Tensor Objects. IEEE Trans. on Neural Networks 19(1), 18–39 (2008)
15. Lu, H., Plataniotis, K.N., Venetsanopoulos, A.N.: A Survey of Multilinear Subspace Learning for Tensor Data. Pattern Recognition 44(7), 1540–1551 (2011)
16. Yang, Z., Laaksonen, J.: Multiplicative Updates for Non-negative Projections. Neurocomputing 71, 363–373 (2007)
17. Zhang, Z., Jiang, M., Ye, N.: Effective Multiplicative Updates for Non-negative Discriminative Learning in Multimodal Dimensionality Reduction. Artificial Intelligence Review 34, 235–260 (2010)
18. Zitnick, C.L., Kang, S.B., Uyttendaele, M., Winder, S., Szeliski, R.: High-quality video view interpolation using a layered representation. ACM SIGGRAPH and ACM Trans. on Graphics, 600–608 (2004)

Correlation between Seismic Intensity Parameters of HHT-Based Synthetic Seismic Accelerograms and Damage Indices of Buildings

Eleni Vrochidou[1], Petros Alvanitopoulos[1],
Ioannis Andreadis[1], and Anaxagoras Elenas[2]

[1] Department of Electrical and Computer Engineering,
Democritus University of Thrace, GR-67100 Xanthi, Greece
[2] Department of Civil Engineering, Democritus University of Thrace,
GR-67100 Xanthi, Greece

Abstract. In this work, a correlation study between well-known seismic parameters and damage indices is presented. The correlation analysis is first performed on a set of natural seismic signals and afterwards on a set of artificial accelerograms generated from natural records. The artificial seismic signals are generated by a combination of techniques; the Hilbert-Huang transform and an optimization algorithm. In addition, they are compatible with the design spectra of a chosen seismic area. Results reveal that the seismic parameters of the synthetic earthquake accelerograms provide the same degree of correlation with the used damage indices as the natural earthquakes. Thus, the proposed synthetic accelerograms technique appropriately represents a seismic event and, therefore, it is a useful tool in earthquake engineering.

Keywords: Artificial seismic signals, Hilbert-Huang transform, correlation, seismic parameters, damage indices.

1   Introduction

In order to examine the resistance of a building, a dynamic analysis is required. Disaster scenarios need seismic accelerograms that could lead the building to the limits of its strength. Accelerograms are signal records of the acceleration versus time measured during an earthquake ground motion. Unfortunately, there is lack of natural strong motion earthquake records for some areas. Thus, there is need of artificial seismic records compatible with the design spectra of the area under observation. Methods for generating artificial earthquakes have been reported in the literature [1-4]. The presented methodology uses the Hilbert-Huang transform (HHT) which appears to be an effective technique for analyzing non-stationary and non-linear signals such as seismic signals [5].

The HHT decomposes the initial signal into a number of intrinsic mode functions (IMFs) and presents the results to the energy-frequency-time field, designing the so-called Hilbert spectrum. The IMFs extraction is based on the local characteristics of the signals and not on some predetermined functions. This is the key feature characterizing the HHT providing a more physical meaning to the analysis of the

L. Iliadis et al. (Eds.): AIAI 2012, IFIP AICT 381, pp. 425–434, 2012.

examined signals [5]. Moreover, the finite number of the extracted IMFs, decreases the computational complexity of the algorithm. An additional advantage of the method is that the initial seismic signal used for the synthesis of the artificial signal, could be of any intensity and could belong to any region. This provides flexibility, since there can be selected signals with finite number of points and decrease even more the complexity of the algorithm.

The main challenge is to generate realistic simulated earthquake records with the characteristics of natural signals. The proposed methodology is applied to an eight-storey reinforced concrete frame structure. Both seismic parameters and damage indices have been calculated from natural and artificial signals. A comparison between them is made and correlation analysis demonstrates that the generated earthquake signals have the same behavior as the natural ones, and have approximately the same distribution of values among the seismic parameters and the damage indices. For example, the spectrum displacement and spectrum velocity are correlated with the global damage index of Park/Ang by 0.905 and 0.929, respectively for the natural accelerograms, while for the artificial accelerograms the values are 0.905 and 0.905, respectively. Additionally, the correlation study reveals that the seismic parameters that are known to be most correlated with the damage indices [6] have the same degree of correlation for both natural and artificial signals.

2   Earthquake Engineering

The intensity of a seismic accelerogram is expressed with the help of several seismic parameters that are connected to the damages caused to a specific structure [6]. Damages are described by the damage indices (DIs). The maximum inter-storey drift ratio (MISDR) and the global damage index of Park/Ang ($DI_{P/A}$) are used to characterize the structural damage caused to buildings.

2.1   Damage Indices

The level of the post-seismic corruption of a structure can be evaluated by the MISDR [7] and is given by the following equation:

$$MISDR = \frac{|u|_{max}}{h} 100 [\%] \tag{1}$$

where $|u|_{max}$ is the maximum absolute inter-storey drift and h the inter-storey height. In addition, the global DI after Park/Ang is also used as a damage index [7]:

$$DI_{P/A} = \frac{\sum_{i=0}^{n} DI_L E_i}{\sum_{i=0}^{n} E_i} \tag{2}$$

where E_i is the energy at the location i, n is the number of location where the energy is estimated and DI_L is the local DI given by the equation:

$$DI_L = \frac{\theta_m - \theta_r}{\theta_u - \theta_r} + \frac{\beta}{M_y \theta_u} E_T \tag{3}$$

where θ_m is the maximum rotation during the load history, θ_r is the recoverable rotation at unloading, θ_u is the ultimate rotation capacity, β is a strength degrading parameter (0.1-0.15), M_y is the yield moment of the section and E_T is the dissipated hysteric energy.

2.2 Seismic Parameters

Accelerograms can be described by some well-known intensity parameters. In this work a set of 8 parameters has been utilized and is presented in Table 1 along with their literature references. The selection is based on the correlation degree of the parameters with the damage indices [6, 7].

Table 1. Seismic intensity parameters strong correlated with the structural damage indices

No.	Seismic Intensity Parameters	Ref.
1	Spectrum Intensity after Housner (SI_H)	[8, 9]
2	Spectrum Intensity after Kappos (SI_K)	[8, 9]
3	Spectral Displacement (SD)	[8, 9]
4	Spectral Velocity (SV)	[8, 9]
5	Spectral Acceleration (SA)	[8, 9]
6	Seismic Intensity after Fajfar/Vidic/Fischinger (I_{FVF})	[8, 9]
7	Peak Ground Velocity (PGV)	[8, 9]
8	Seismic Energy Input (E_{inp})	[8]

2.3 Concrete Frame Structure

The examined reinforced concrete structure is depicted in Figure 1. The eigenfrequency of the 8-storey building is 0.85 Hz and its design is based on the recent Eurocode rules EC2 and EC8. The cross-sections of the beams are T-beams with 40 cm width, 20 cm slab thickness, 60 cm total beam height and 1.45 m effective slab width. The distance between the frames of the structure is 6 m. The structure has been characterized as an "importance class II and ductility class medium"- structure according to the EC8 Eurocode. The subsoil is of type C and the region seismicity of category 2. External loads are taken under consideration and are incorporated into load combinations due to the rules of EC2 and EC8.

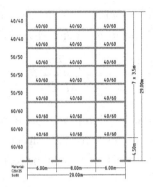

Fig. 1. Reinforced concrete frame structure

3   Synthesis of Artificial Accelerograms

The proposed synthesis of artificial seismic signals is based on the HHT. According to HHT, each natural earthquake record can be decomposed into finite frequency and amplitude components. These components are appropriately modified to compose the artificial signal. This is achieved through an optimization algorithm which is used to minimize the mean square error between the response spectrum of the synthetic earthquake and the design spectrum of the selected seismic area.

3.1   The Hilbert-Huang Transform

The HHT [5] includes the Empirical Mode Decomposition (EMD) and the Hilbert Spectral Analysis (HSA).

The EMD decomposed the signal into a finite number of IMFs. The decomposition is based upon the local characteristics of the signal, thus it is an appropriate method for non-linear and non-stationary data analysis. Two conditions must be fulfilled for the generated IMFs:

1. the number of extrema and zero-crossings must be equal or differ the most by one and
2. at any point, the mean value of the envelop that is defined by the local maxima and minima must be equal to zero.

For a seismic signal $X(t)$, all the local extrema must be connected by a cubic spline so as to form the upper envelope of the signal $y_{max}(t)$ and the same process is followed for the lower envelope $y_{min}(t)$. The signal must be enclosed between these two envelopes. The mean value of the envelopes is:

$$m_1(t) = \frac{(y_{max}(t) - y_{min}(t))}{2}$$ (4)

and the difference between the signal and the $m_1(t)$ is the first component h_1 :

$$h_1(t) = X(t) - m_1(t) \tag{5}$$

The procedure continues and now $h_1(t)$ is considered to be the signal, so:

$$h_{11}(t) = h_1(t) - m_{11}(t) \tag{6}$$

where $m_{11}(t)$ is the new mean of the two envelopes of $h_1(t)$. The same process is repeated for k times and finally the $h_{1k}(t)$ is:

$$h_{1k}(t) = h_{1(k-1)}(t) - m_{1k}(t) \tag{7}$$

where the $h_{1k}(t) = c_1(t)$ is the first IMF. The process finishes when the standard deviation $SD \leq 0.2$:

$$SD = \sum_{t=0}^{T} \frac{[h_{1(k-1)}(t) - h_{1k}(t)]^2}{h_{1(k-1)}^2(t)} \tag{8}$$

where T is the duration of the seismic signal. The residue $r_1(t)$ is derived by subtracting the first IMF from the initial signal:

$$r_1(t) = X(t) - c_1(t) \tag{9}$$

and it is considered as the new data. The new data is subjected to the same process until all $r_j(t)$ functions are obtained:

$$r_j(t) = r_{j-1}(t) - c_j(t), j = 2,3,...,n. \tag{10}$$

The iteration process stops when one of the two criteria is fulfilled:

1.   either the residue $r_n(t)$ or the component $c_n(t)$ is less than a predetermined value or
2.   the residue $r_n(t)$ is a monotonic function and no other IMF can be derived.

Finally, the initial signal is the sum of all IMFs plus the residue:

$$X(t) = \sum_{j=1}^{n} c_j(t) + r_n(t) \tag{11}$$

The Hilbert Spectral Analysis derives the instantaneous frequency data and forms the energy-frequency-time distribution by applying the HHT to each IMF. For all components $c_j(t)$, the Hilbert transform is:

$$y_j(t) = \frac{1}{\pi} P \int_{-\infty}^{+\infty} \frac{c_j(\tau)}{t - \tau} d\tau \tag{12}$$

where P is the Cauchy Principal Value. The combination of the IMF $c_j(t)$ and its Hilbert transform $y_j(t)$ form an analytic signal $z_j(t)$:

$$z_j(t) = c_j(t) + iy_j(t) = a_j(t)e^{i\theta_j(t)} \tag{13}$$

where the amplitude $a_j(t)$ and the phase $\theta(t)$ are:

$$a_j(t) = \sqrt{c_j^2(t) + y_j^2(t)} \tag{14}$$

$$\theta_j(t) = \tan^{-1}\frac{y_j(t)}{c_j(t)} \tag{15}$$

The instantaneous frequency is given by the equation:

$$\omega_j(t) = 2\pi f_j(t) = \frac{d\theta(t)}{dt} \tag{16}$$

The IMF components are defined as follows:

$$c_j(t) = \text{Re}(a_j(t)e^{i\theta_j(t)}) = a_j(t)\cos\theta_j(t) \tag{17}$$

where Re() is the real part. The initial signal can be written:

$$X(t) = \text{Re}(\sum_{j=1}^{n} a_j(t)\cos(\int 2\pi f_j(t)dt)) \tag{18}$$

In the previous equation, the residue is not invoked as it is a constant or a monotonic function. According to this equation both amplitude and frequency can be expressed as functions of time. The time-frequency distribution of the amplitude is known as the Hilbert spectrum $H(\omega, t)$.

3.2   Method for Generating Artificial Seismic Accelerograms

The objective of the proposed method is to supply earthquake engineering analysts with artificial seismic accelerograms compatible with the design spectrum of the current Greek antiseismic code [10] and it is a two step procedure.

The first step applies the HHT to the initial seismic signal which is decomposed into n IMFs. The HHT amplitude and frequency components of the initial signal are then properly modified to form an artificial seismic signal $AE(x,t)$ which is given by the equation:

$$AE(x,t) = \text{Re}\sum_{j=1}^{n}(x_j a_j(t))e^{i\int(\omega_j(t) + x_{n+j})dt} \tag{19}$$

where $x_1, x_2,..., x_n$ are the amplitude scaling parameters and $x_{n+1}, x_{n+2},..., x_{2n}$ are the frequency modifying parameters. The following equations describe the amplitude scaling and the frequency modification:

$$a_i(t) = x_j a_i(t) \quad , \quad j = 1,2,...,n \tag{20}$$

$$\omega_i(t) = \omega_i(t) + x_k \quad , \quad k = n+1,...,2n \tag{21}$$

The second step solves the optimization problem so as the response spectrum of the synthetic signal to be compatible with the design spectrum of the area under-study. The optimization problem is actually finding the appropriate parameters $x = \{x_1, x_2,..., x_{2n}\}$ of the equations (20) and (21) to match the aforementioned spectrum. In particular, the genetic algorithm used, tries to minimize a fitness function which in our case is the Mean Square Error (MSE) between the two spectra:

$$MSE = \frac{1}{n} \sum_{j=1}^{n} \left| SA_{generated} - SA_{t\,\arg et} \right|^2 \tag{22}$$

where n is the number of sample cases.

In this study, twenty natural seismic signals where used to generate synthetic spectrum-compatible seismic signals. The generated signals were afterwards submitted into a correlation study, to examine the relation between the seismic parameters and the damage indices, compared with the same correlation percentages of the initial seismic signals.

4   Correlation Coefficients

Correlation analysis proves the statistical relationship between two sets of data. There are many correlation coefficients; among them the most widely used are the Pearson correlation coefficient and the Spearman rank correlation coefficient.

The Pearson correlation shows how well the data fit a linear relationship, while the Spearman correlation shows how close the examined data are to monotone ranking. The latter coefficient is more important in the present study. For a set of n measurements of X and Y, where $i = 1,2,...,n$, the sample correlation coefficient is given by the equation:

$$P_{Spearman} = 1 - \frac{6 \sum_{i=1}^{n} D^2}{n(n^2 - 1)} \tag{23}$$

where D is the difference between the ranking degree of X and Y, respectively.

In case the correlation coefficient has a value greater than 0.8 means that there is a strong connection between the parameters. On the other hand, values less than 0.45 demonstrate a weak connection between the parameters. All other cases between 0.45 and 0.8 reveal medium connection [6].

5   Numerical Results

In this study, twenty natural earthquake records were used as initial signals in order to generate synthetic accelerograms. Information about these seismic events is included in Table 2. Twenty artificial seismic accelerograms have been generated from these seismic signals. The objective was that the response acceleration spectrum of every generated signal to be compatible with the Greek design spectrum. Figure 2 displays the results obtained for the Friuli (Feltre station) earthquake record.

Table 2. Data for the 20 natural seismic signals

Seismic Event	Station	Date
Round Valley	Mcgee Creek Surface (USGS 1661)	11/23/84
Coyote Lake	Coyote Lake Dam-San Martin, (CDMG 57217)	08/06/79
Friuli	Conegliano	05/06/76
Loma Prieta	Coyote Lake Dam Downst, (CDMG 57504)	10/18/89
Coalinga	Chp, (CDMG 46T04)	07/22/83
Alaska	Sitka Observatory	07/30/72
Mt Lewis	Halls Valley, (CDMG 57191)	03/31/86
San Fernando	Fort Tejon, (USGS 998)	02/09/71
West Moreland	Brawley Airport, (USGS 5060)	04/26/81
Oroville	Medical Center, (CIT 1544)	08/02/75
Morgan Hill	Apeel 1E Hayward, (USGS/CDMG 1180)	04/24/84
Dinar	Dinar	01/10/95
Lytle Creek	Devils Canyon, (CDWR 620)	09/12/70
Friuli	Barcis	05/06/76
Borah Peak	Cem	10/29/83
Morgan Hill	Capitola, (CDMG 47125)	04/24/84
Mammoth Lakes	Fis	05/27/80
Friuli	Feltre	05/06/76
Kocaelli	Eregli, (ERD)	08/17/99
Imperial Valley	Ec Co Center FF, (CDMG 5154)	10/15/79

Figure 2(a) shows the original signal used to generate the synthetic earthquake signal that is shown in Figure 2(b). Figure 2(c) presents the response spectrum of the original signal, the response spectrum of the synthetic accelerogram and the Greek design spectrum of the considered seismic area. It is obvious from Figure 2(c) that the response spectrum of the generated signal is compatible with the desired design spectrum. This is also justified numerically, as the MSE for this experiment was 1.73×10^{-3} which is significantly small. For this event, the response spectrum of the original accelerogram was lower than the design spectrum considered.

Results are also promising for original signals that their response spectra are higher than the design spectrum. In this category belongs the seismic event of Dinar. The seismic parameters were calculated for all 20 original and artificial seismic accelerograms. Afterwards, the considered global damage indices were evaluated for each accelerogram, considered as seismic load on the examined frame structure. Finally, the Spearman rank correlation coefficient between the seismic parameters and the damage indices is calculated and presented in Table 3.

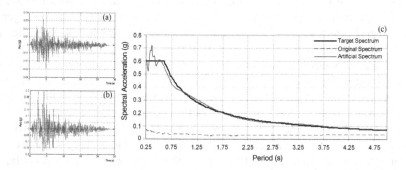

Fig. 2. (a) Friuli (Feltre station) seismic signal (b) synthetic generated signal and (c) response spectra and design spectrum

As it can be seen from the numerical results of Table 3, the values of correlation are approximately the same for the natural acceleration records and artificial generated accelerograms. For example, the parameter SA for the natural signal is correlated with the MISDR and the $DI_{P/A}$ by 0.833 and 0.881, respectively. The same parameter for the artificial signals is correlated with the same damage indices by 0.762 and 0.833 respectively, which is almost the same degree as the natural records.

The degrees of interdependences for all 8 parameters are in the same range of values for natural and artificial signals. This indicates that the proposed method produces realistic simulations of earthquake signals.

Table 3. Spearman correlation coefficients between the seismic parameters and the damage indices

Seismic Intensity Parameters	Original Signal		Artificial Signal	
	MISDR	$DI_{P/A}$	MISDR	$DI_{P/A}$
SI_H	0.810	0.786	0.929	0.857
SI_K	0.786	0.762	0.810	0.881
SD	0.857	0.905	0.690	0.905
SV	0.857	0.929	0.690	0.905
SA	0.833	0.881	0.762	0.833
I_{FVF}	0.857	0.929	0.690	0.905
PGV	0.905	0.952	0.690	0.905
E_{inp}	0.714	0.738	0.929	0.857

6   Conclusions

The need of large sets of accelerograms for the dynamic analysis of structures and the shortage of natural records on many regions globally, justifies the search for generation of artificial spectrum compatible seismic signals. The proposed method generates spectrum compatible seismic signals based on the HHT and on a genetic algorithm that helps solve an optimization problem. One random seismic signal is used every time. This makes the method flexible as there can be chosen signals with a finite number of points and thus reduce the computational complexity of the method.

All generated signals were analyzed to prove that there are reliable simulations of natural signals. For this purpose, the values of the 8 well-known seismic parameters were calculated along with the damage indices for a certain frame structure, for both natural and artificial signals. The values extracted were then compared with the respective values of the initial seismic signals, and a correlation study was carried out. The numerical results indicate that the synthetic signals behave like real earthquake signals. The degree of interdependence was very similar for all damage indices, for natural and artificial earthquake signals, respectively. For example, the parameter SA is correlated with the $DI_{P/A}$ for the natural and artificial earthquake signals by 0.881 and 0.833, respectively.

References

1. Suarez, L.E., Montejo, L.A.: Generation of artificial earthquakes via the wavelet transform. International Journal of Solids and Structures 45, 5905–5919 (2005)
2. Mavroeidis, G.P., Papageorgiou, A.S.: A Mathematical Representation of Near-Fault Ground Motions. Bulletin of the Seismological Society of America 93, 1099–1131 (2003)
3. Das, S., Gupta, V.K.: Wavelet-based simulation of spectrum-compatible aftershock accelerograms. Earthquake Engineering & Structural Dynamics 37, 1333–1348 (2008)
4. Spanos, P.D., Giaralis, A., Li, J.: Synthesis of accelerograms compatible with the Chinese GB 50011-2001 design spectrum via harmonic wavelets: artificial and historic records. Earthquake Engineering and Engineering Vibration 8, 189–206 (2009)
5. Huang, N.E., Shen, Z., Long, S.R., et al.: The empirical mode decomposition and the Hilbert spectrum for nonlinear and non-stationary time series analysis. Proceedings of the Royal Society A 454, 903–995 (1998)
6. Elenas, A., Meskouris, K.: Correlation study between seismic acceleration parameters and damage indices of structures. Engineering Structures 23, 698–704 (2001)
7. Elenas, A.: Correlation between seismic acceleration parameters and overall structural damage indices of buildings. Soil Dynamics and Earthquake Engineering 20(1-4), 93–100 (2000)
8. Alvanitopoulos, P., Andreadis, I., Elenas, A.: Interdependence Between Damage Indices and Ground Motion Parameters Based on Hilbert Huang Transform. Journal of Measurement Science & Technology 21, 1–14 (2010)
9. Alvanitopoulos, P., Papavasiliou, M., Andreadis, I., Elenas, A.: Seismic Feature Construction Based on the Hilbert Huang Transform. IEEE Transactions on Instrumentation & Measurement 61, 326–337 (2012)
10. EAK: National Greek Antiseismic Code. Earthquake Planning and Protection Organization (OASP) Publication, Athens (2003)

Improving Current and Voltage Transformers Accuracy Using Artificial Neural Network

Haidar Samet[1], Farshid Nasrfard Jahromi[1], Arash Dehghani[1], and Afsaneh Narimani[2]

[1] Shiraz University
[2] Foolad Technic International Engineering Company
samet@shirazu.ac.ir,
{farshid.nasr2010,dehghani_arash,afsaneh_narimani}@yahoo.com

Abstract. Capacitive Voltage Transformers (CVTs) and Current Transformers (CTs) are commonly used in high voltage (HV) and extra high voltage (EHV) systems to provide signals for protecting and measuring devices. Transient response of CTs and CVTs could lead to relay mal-operation. To avoid these phenomena, this paper proposes an artificial neural network (ANN) method to correct CTs and CVTs secondary waveform distortions caused by the transients. PSCAD/EMTDC software is employed to produce the required voltage and current signals which are used for the training process and finally the results show that the proposed method is accurate and reliable in estimation of the CT primary current and the CVT primary voltage.

Keywords: Artificial Neural Network (ANN), Capacitive Voltage Transformers (CVTs), Current Transformers (CTs), Transient.

1 Introduction

The operation of the protection relays is dependent on the measured signals such as current and voltage signals which are measured by CTs and CVTs. Thus an error in these signal measurements could lead to mal-operation or substantial delay in tripping of the protection relays. The CT and CVT output signals may not follow their input signals due to the transients mainly caused by current transformers saturation and discharging of the capacitive voltage transformers internal energy during faults.

Several techniques on the compensation of the distorted secondary current and voltage signals have been published. An algorithm to estimate the magnetizing current at each time step by using the magnetization curve of a CT was reported in [1], the other algorithm detects the saturation using the Discrete Wavelet Transform (DWT) and compensates the distorted section of a secondary current with features extracted from the healthy section using a least mean square fitting method [2]. Another method is to Compensate the error of CT by using hysteresis and Eddy characteristic [3]. Some digital methods to correct the secondary waveforms of the CT are proposed in Ref.[4]. One method is to detect high Source Impedance Ratio (SIR) conditions that could lead to severe CVT transients and add a time delay to the tripping decision of distance relay [5].

L. Iliadis et al. (Eds.): AIAI 2012, IFIP AICT 381, pp. 435–442, 2012.
© IFIP International Federation for Information Processing 2012

The method discussed in Ref. [6] proposes the use of artificial neural networks (ANNs) to correct CVT secondary voltage distortions due to CVT transient. Another method is to compensate the distorted secondary waveforms of the CVT in the time domain by considering the hysteresis characteristics of the core [7]. The method introduced in Ref. [8] proposes the use of compensation algorithms based on of the CVT transfer function. Some least squares phasor estimation techniques for estimation of voltage and current phasors are presented in Refs. [9] and [10].

This paper begins with analysis of the contributing factors in CT and CVT transients. The details of the simulated power system to produce the different cases which are used in the training and testing process are presented in the later section. An ANN method for CT and CVT error compensation is presented in section 5. Results clearly show that the proposed method could estimate the primary signals accurately.

2   CT Transients

CT model used in simulations is shown in Figure1 [11]. Real iron-cored CTs are not ideal and since the reluctance of the core is not infinite, the current which is obtained by dividing the primary current (I_p) to the turns ratio (N), is different from the secondary current (I_s) in amplitude and phase due to existence of exciting current (I_e).

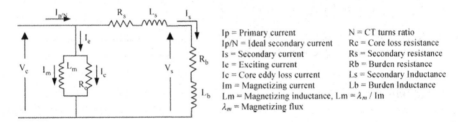

Ip = Primary current   N = CT turns ratio
Ip/N = Ideal secondary current   Rc = Core loss resistance
Is = Secondary current   Rs = Secondary resistance
Ie = Exciting current   Rb = Burden resistance
Ic = Core eddy loss current   Ls = Secondary Inductance
Im = Magnetizing current   Lb = Burden Inductance
Lm = Magnetizing inductance, Lm = λ_m / Im
λ_m = Magnetizing flux

Fig. 1. Equivalent circuit of a CT [11]

3   CVT Transients

CVT model used in simulations is shown in Figure 2 [12]. C_1 and C_2 are stack capacitances. L, R, L_{T1} and R_{T1} are inductance and resistance, respectively, of the tuning reactor and the step down transformer. L_0 and R_0 are burden inductance and resistance and f is a subscript for parameters of the anti-resonance circuit. The CVT Transients are basically controlled by the following factors:

A. Fault inception angle (point on wave)
The CVT output voltage does not follow the primary voltage for several cycles after the fault occurs. The timing of the faults on the voltage waveform (fault inception) affects the CVT transients. The transients of the faults occurring at voltage peaks and voltage zeros are the least severe and the most severe, respectively [3].

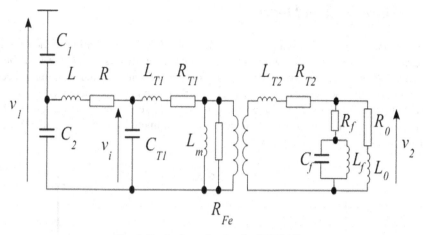

Fig. 2. Equivalent circuit of a CVT [12]

B. Coupling capacitors

The higher the sum of the capacitances $(C=C_1+C_2)$, the lower the magnitude of the transients. High-C CVTs decrease the magnitude of the CVT transients but are more expensive. So there should be a balance between CVT performance and CVT cost [12].

C. Ferroresonance suppression circuit

There are two models of a ferroresonance suppression circuit which are shown in Figure 3. The transients of the CVT with the passive model are less severe so the output voltage is less distorted [12], [6].

D. CVT burden

The energy accumulated in the CVT storage elements could be dissipated in the CVT burden. Therefore, the different CVT burden could change the shape and duration of the CVT transients. As a rule, if the CVT is fully loaded, the transients would be less severe [12].

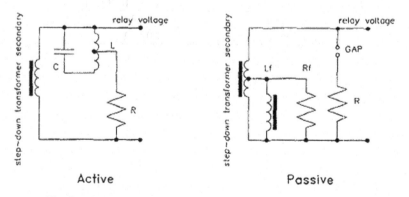

Fig. 3. Active and Passive Ferroresonance suppression circuit [3]

4   Power System Model

A three-phase 230 kV, radial power system is used in the simulations .The one-line diagram of this system is shown in Figure 4 and its parameters are shown in Table 1. Since ANN methods require an adequate amount of data for the training process, samples of the secondary voltage and current signals are achieved by using PSCAD/EMTDC software and then the primary voltage and current signals could be estimated by using the proposed method. The parameters of the CT and CVT are shown in tables 2 and 3, respectively.

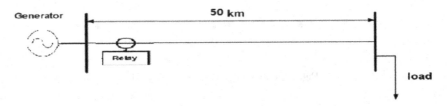

Fig. 4. Structure of the power system model

Table 1. Simulated power System parameters

Generator impedance(Ω)	1	
Line impedance(Ω/km)	0.072 + j 0.416	
Load	Active power(MW)	100
	Reactive power(MVAR)	2

Table 2. CT circuit parameters

Primary turns	20
Secondary turns	200
Secondary resistance	0.5Ω
Secondary Inductance	0.8e3H
Area	2.601e-3m*m
Path length	0.6377m
Burden resistance	0.5 Ω

Table 3. CVT circuit parameters

Capacitor-1	2920 pF
Capacitor-2	134952 pF
Compensating inductance	42 H
VT ratio	43.48
Primary inductance	0.47e-3 H
Primary resistance	0.05 Ω
Secondary inductance	0.47e-3 H
Secondary inductance	0.18 Ω
Eddy current loss at normal conditions	2.5 W
Secondary operation voltage	115 V
Operating flux density	0.8 T
Hysteresis loss at normal conditions	5 W
Burden resistance	301 Ω
Burden inductance	2.4 H

5   Proposed Neural Network Compensating Method

5.1   Training Patterns

To generate training patterns, the following conditions are changed from the base case: fault inception angle, fault type and fault resistance. The training pattern data generation is shown in Table 4.

Table 4. Training pattern data generation

Fault inception angle(°)	Different values between (0-360)with a step of 45 degrees
Fault type	Single-phase-to-ground, phase-to-phase-to-ground, three-phase -to- ground
Fault resistance(Ω)	0.01, 1, 100

5.2   Network Structure and Training

An ANN has the ability to learn from data. Such a situation is shown in Figure 5. Typically many of such input/output pairs are needed to train a network. A very important feature of these networks is their adaptive nature, where "learning by example" replace "programming" in solving problems. ANNs can be used to perform different tasks in power system relaying for signal processing and decision making [13]. A major problem with ANNs is that there is no exact rule to choose the number of hidden layers and neurons per hidden layer[14].The most widely used learning algorithm in an ANN is the Back-propagation algorithm [13],[14].

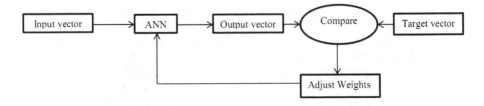

Fig. 5. A learning cycle in the ANN model

In this paper, for each CT and CVT a feedforward neural network with two hidden layers is used to process the training input data. The sampling rate is 40 samples per one 50-Hz cycle. The neural network includes 20 inputs, 20 outputs, two hidden layers (the first one with 4 neurons and the second one with 3 neurons). The inputs are the instantaneous values of the secondary current and voltage of the CT and CVT, respectively and the outputs are the instantaneous values of the primary sides of the CT and CVT, respectively. A log-sigmoid function as the activation function of the hidden layer neurons and a tan-sigmoid function for the output layer are used. The neural network is trained by Back-Propagation (BP) algorithms.

5.3   Test Results

The accuracy of the proposed method is tested with patterns which are different from the training patterns. Figures 6 and 7 show the test results of the CT. Furthermore, Figures 8 and 9 show the test results of the CVT. The left side figures show the ideal (primary side), actual (secondary side) and the output of ANN while the right side

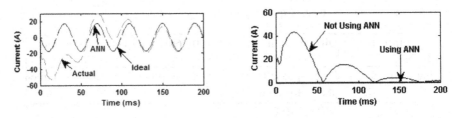

Fig. 6. a) CT waveforms for B-G fault with inception angle=300° and fault resistance=0. 1Ω b) Estimation error

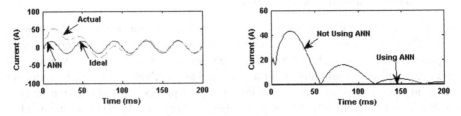

Fig. 7. a) CT waveforms for B-G fault with inception angle=120° and fault resistance=0. 1Ω b) Estimation error

figures show the estimation errors. The results show that the ANN output could exactly follow the ideal CT and CVT outputs ("ideal output" here means the primary value of CT and CVT and the "actual output" means the secondary side value of CT and CVT).

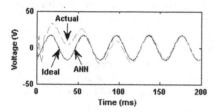 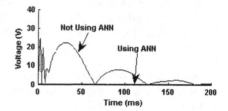

Fig. 8. a) CVT waveforms for A-B-G fault with inception angle=60° and fault resistance=0.1Ω b) Estimation error

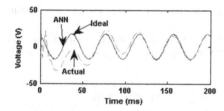 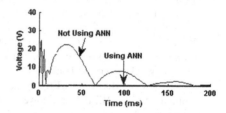

Fig. 9. a) CVT waveforms for A-B-C fault with inception angle=240° and fault resistance=0.1Ω b) Estimation error

6   Conclusion

The operation of the protection relays is dependent on the measured signals such as current and voltage signals which are measured by CTs and CVTs. The CT and CVT output signals may not follow their input signals due to their inherent transients. An ANN method for CT and CVT error compensation is presented in this paper which is an approximation of the inverse transfer function. Test results clearly show that the proposed method could estimate the primary signals accurately under different conditions.

References

1. Kang, Y.C., Kang, S.H., Jones, A.T., Aggarwal, R.K.: An algorithm for compensating secondary current of current transformers. IEEE Trans. Power Delivery 12(1), 116–124 (1997)
2. Li, F.: ombined wavelet transform and regression technique for secondary current compensation of current transformer. IEEE Proc.-Gener. Transm. Distrib. 149, 497–503 (2002)

3. Locci, N., Muscas, C.: Hysteresis and eddy currents enompsation in CT. IEEE Trans. Power Delivery 16(2), 147, 154–159 (2001)
4. Locci, N., Muscas, C.: A Digital Compensation Method for Improving Current Transformer Accuracy. IEEE Trans. Power Delivery 15(4), 1104–1109 (2000)
5. Hou, D., Roberts, J.: Capacitive Voltage Transformers: Transient Overreach Concerns and Solutions for Distance Relaying. In: Proc. Canadian Conference on Electrical and Computer Engineering, vol. 147, pp. 26–29 (1996)
6. Zadeh, K.H., Li, Z.: A compensation scheme for CVT transient effects using artificial neural network. SienceDirect, pp. 30–38. Elsevier (2006)
7. Kang, Y.-C., Zheng, T.-Y., Choi, S.-W., Kim, Y.-H., Kim, Y.-G., Jang, S.-I., Kang, S.-H.: Design and evaluation of a compensating algorithm for the secondary voltage of a coupling capacitor voltage transformer in the time domain. IET Gener. Transm. Distrib. 3(9), 793–800 (2009)
8. Izykowski, J., Kasztenny, B., Rosolowski, E., Saha, M.M., Hillstrom, H.: Dynamic Compensation of Capacitive Voltage Transformers. IEEE Trans. Power Delivery 13(1), 116–122 (1998)
9. Ajaei, F.B., Sanaye-Pasand, M.: Minimizing the Impact of Transients of Capacitive Voltage Transformers on Distance Relay. In: Proc. Indian Conference on Power System Technology, pp. 1–6 (2008)
10. Pajuelo, E., Ramakrishna, G., Sachdev, M.S.: An Improved Voltage Phasor Estimation Technique to Minimize the Impact of CCVT Transients in Distance Protection. IEEE Trans. Power Delivery, 454–457 (2005)
11. Pan, J., Vu, K., Hu, Y.: An Efficient Compensation Algorithm for Current Transformer Saturation Effects. IEEE Trans. Power Delivery 19(4), 1623–1628 (2004)
12. Kasztenny, B., Sharples, D., Asaro, V., Pozzuoli, M.: Digital Relays and Capacitive Voltage Transformers: Balancing Speed and Transient Overreach. In: 53rd Annual Conference for Protective Relay Engineers, College Station, pp. 1–22 (2000)
13. Jhr, G.K.: Artificial Neural Network. Indian Agricultural Research Institute PUSA, 1–10 (1981), http://www.iasri.res.in/ebook/EB_SMAR/ebook_pdf%20files/Manual%20IV/3-ANN.pdf
14. Saha, M.M., Rosolowski, E., Izykowski, J.: Artificial Intelligent Application to Power System Protection, pp. 195–197, http://zas.ie.pwr.wroc.pl/er_india00_ai.pdf

Modeling of Syllogisms in Analog Hardware

Darko Kovacevic[1], Nikica Pribacic[1], Radovan Antonic[1],
Asja Kovacevic[1], and Mate Jovic[2]

[1]Faculty of Maritime Studies Split, University of Split,
Zrinsko-Frankopanska 38, 21000 Split, Croatia
{nikica.pribacic,darko.kovacevicl}@st.t-com.hr
[2]Technical School Imotski, B. Busica b.b., 21260 Imotski
jovic.mate@gmail.com

Abstract. Syllogistic reasoning is modeled in analog hardware and some hardware models, i.e. syllogisms *Baroco* and *Darii* are presented. Chaining of syllogisms is modeled by using original *min-max entities* (circuits), "to see" whether the two rules, modeled in dedicated hardware, i.e., IF *A* THEN *B* and IF *B* THEN *C* imply the "hardware" rule IF *A* THEN *C*. The preliminaries include original *min-max circuits* based on operational amplifiers (Op-Amp), straight lines Op-Amp generators and different test circuits designed in Electronics WorkBench simulation environment and in real hardware. The "stage" to perform modeling is the phase plane.

Keywords: modeling syllogisms, analog hardware, min-max entities.

1 Introduction

In AI community, the *thinking rational approach* [1] deals with searching for ideal thinking and reasoning methods to provide means for indisputable and efficient problem solving. The thinking rational approach can be expressed by using syllogisms. The aim of this approach is to build systems that are able to solve problems correctly and ideally in the most satisfactory manner regardless of whether humans follow the same reasoning processes. Logicians amongst the AI researchers use the thinking rational approach to design and develop intelligent systems. In this paper, we will use our original min-max-Op-Amp technique approach to model syllogisms in hardware environment. Min-max-Op-Amp technique is based on an idea that *min* and *max* circuits can be designed by using Op-Amp. On the other hand, if we accept an inverting or non-inverting Op-Amp circuit as a straight-line generator in the phase plane (PP) then any kind of transfer function can be synthesized by using *min*, *max*, and Op-Amp circuits. All presented models are realized in Electronics Workbench (commercial software) environment as well as in a real hardware (not shown in the paper). Syllogistic reasoning with min-max-Op-amp technique approach, where the consequence of a rule in one reasoning stage is passed to the next stage as a fact, is essential in building up a large-scale system with high-level intelligence. Already the very first computable AI systems were based on logical reasoning in this sense and many recent developments were initiated or established/advanced by AI scientists (Fuzzy Logic, Default Logic, BDI Logics, etc.).

L. Iliadis et al. (Eds.): AIAI 2012, IFIP AICT 381, pp. 443–452, 2012.
© IFIP International Federation for Information Processing 2012

The thinking rational approach has its origin in philosophy and draws mainly on philosophy and mathematics. The roots of the thinking rational approach are strongly related to the fundamental philosophical problem of "right thinking" which deals with determining how humans could and should reason in order to obtain optimal results. The Greek philosopher Aristotle tried to identify what he called "laws of thought" as foundations of rational thinking. He formulated an informal system of *syllogisms* for the representation of and the deductive reasoning with knowledge [2]. The most famous example of such a syllogism is the argument: "Socrates is a man. All men are mortal. Therefore, Socrates is mortal". It is an example of the most fundamental syllogism; the "syllogism of the first mood of the first figure" to which the other syllogisms are reducible:

All *S* are *M*.
All *M* are *P*.
Therefore, all *S* are *P*.

Such syllogisms as "laws of thought" were designed to deduce correct conclusions from given correct premises (i.e. any conclusions drawn from reasoning with these laws should be indisputable, which means not capable of being proven wrong). The laws of thought were assumed to describe the logical reasoning processes of the rational mind and therefore, the laws of thought are considered as origin of research in logic and logical reasoning. These philosophical roots of this approach are the reason why this approach is also called the *laws of thought approach*.

In the two valued propositional calculus, the chaining syllogism

$$(p \rightarrow q) \wedge (q \rightarrow r) \rightarrow (p \rightarrow r)$$

where p, q, and r are propositional variables, is a tautology, i.e. a formula that is true in every possible interpretation. The chaining rule is one of the most important deduction schemes from a theoretical and practical point of view. It allows combining min-max (fuzzy) IF-THEN rules and because of that reduces the complexity of expert systems [3].

The presentation concept of this paper is figure rich one; "a figure says more than a 1000 words".

Modeling syllogism in analog hardware environment is based on some preliminaries that are described in the second section. In that section, straight-line generators and original min-max circuits based on Op-Amp-technique as well as set test circuits are briefly described. Among logic figures and their valid moods we have chosen Figure I (*Darii*) and Figure II (*Baroco*) to be modeled in analog hardware environment. These models are presented in the third section. In the same section, the simplest way to interpret "hardware deduction", i.e. conclusion is shown.

2   Preliminaries: Min-Max Hardware Circuits (Op-Amp Technique)

Hardware circuit modeling of syllogisms is based on a special class of hardware circuits that have their theoretical stand in mathematical environment of max and min algebra [4]. Min and max circuits "go" with straight-line generators to generate set environment.

Straight-Line Generators. The straight lines can be generated in two ways:

- by using standard operational amplifiers that support the modeling of the expression $l \equiv y = kx + l$ (*Op-Amp-technique*),
- by using EWB component called „Three-Way Voltage Summer" (TWVS straight lines modeling)

as illustrated with schematics shown in Fig. 1 and 2 respectively.

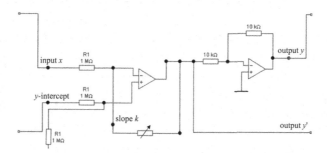

Fig. 1. Hardware model of a straight-line generator based on Op-Amp-technique. Transfer characteristic of a two-stage amplifier represents a straight line in PP.

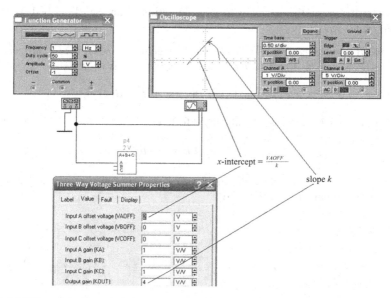

Fig. 2. TWVS straight-line modeling: non-inverting Op-Amp generates a straight line in the PP

Min Circuit. Properties of *min-circuit* operations can be described with the following rule [5]:

If A and B are two subsets of X then their intersection is subset D of X marked as
$D = A \cap B$ *such as that for every* $x \in X$ *is valid:*

$$D(x) = \min[A(x), B(x)] = A(x) \cap B(x) \tag{1}$$

Min operation can be expressed in algebraic form as:

$$\min(a,b) = \left((a+b) - |a-b|\right)/2 \tag{2}$$

Equation (2) defines the (obvious) design of the min circuit; two instrumentation amplifiers are used along with an absolute values circuit (see Fig. 3) [6].

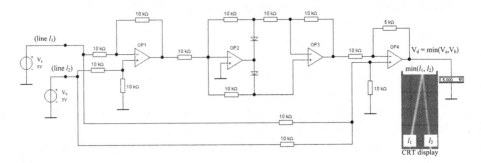

Fig. 3. Min circuit in Op-Amp-technique and its "DC" and "AC" response; Op-Amps are used as generators of straight lines l_1 and l_2

Max Circuit. Properties of *max-circuit* operations can be described with the following rule [5]:

If A and B are two subsets of X then their intersection is subset C of X marked as C = A ∩ B such as that for every x∈ X is valid:

$$C(x) = \max[A(x), B(x)] = A(x) \cup B(x) \tag{3}$$

Max operation can be expressed in algebraic form as:

$$\max(a,b) = \left[(a+b) + |a-b|\right]/2 \tag{4}$$

By using equation (4), we have designed *max-circuit* in Op-Amp-technique as it is shown with the schematics in Fig. 4 [6].

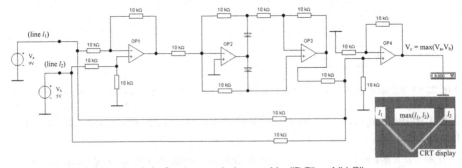

Fig. 4. Max circuit in Op-Amp-technique and its "DC" and "AC" response

Set Test Circuits. There are three different electronic set test circuits to check whether the test point P_T is an element of the solution space (a part of the phase plane).

In the first case, Op-Amp (LM741) is used as a sensing comparator. When the test point lies inside the solution area both comparators will respond (output voltage) with + voltage level turning the AND-gate on. Indicator glows [7].

Action algorithms for all three test circuits are shown along corresponding schematics (see Fig. 5, 6, and 7 respectively).

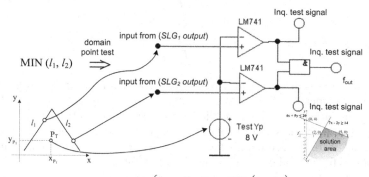

$$\text{Fig. 5. } f_{out} = \begin{cases} 1 & if \quad P_T \in MIN\ (L_1, L_2) \\ 0 & else \end{cases}$$

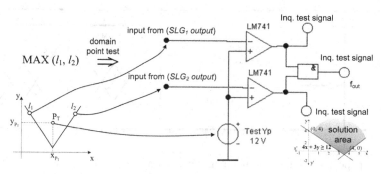

$$\text{Fig. 6. } f_{out} = \begin{cases} 1 & if \quad P_T \in MAX\ (L_1, L_2) \\ 0 & else \end{cases}$$

$$\text{Fig. 7. } f_{out} = \begin{cases} 1 & if \quad P_T \in MIN\ (L_1, y = const\ ) \lor MAX\ (y = const\ , L_2) \\ 0 & else \end{cases}$$

All of the mentioned set test circuits can be used for testing hardware models of syllogisms, i.e. whether the certain point P_T belongs to a chosen set, or not.

3   Logic Figures Modeling

Indirect deductive rule often appears in the form of categorical syllogism that can be presented with this scheme [8]:

$$M \quad P$$

$$\underline{S \quad M}$$

$$S \quad P$$

where minor term is denoted with letter S, major term with letter P, and the middle term with the letter M. There are four different figures of categorical syllogism, depending on the position of the middle term in the premises ($M \quad P$, respectively $S \quad M$) [8]. Each of the figures consists of several moods; in the first figure, there are four, in the second, there are also four, in the third, there are six, and in the fourth, there are five. First two figures and their moods are shown in Table 1.

Table 1. Logic figures and their min-max interpretation diagrams

	Figure I	Figure II	I Darii	II Baroco
Major premise	$M \; P$	$P \; M$	M \ min(M,S)	max (S, M, P)
Minor premise	$S \; M$	$S \; M$		
Valid moods	*Barbara Darii* *Darii Ferio*	*Baroco Cesare* *Camestres Festino*		

In this paper, valid moods of categorical syllogisms *Baroco* and *Darii* will be modeled with analog hardware circuits (min-max entities) in order to show that there is a possibility to process indirect deductive rules in analog hardware environment.

Specification of the algorithm for solving syllogisms with min-max entities requires strategies for:

- choosing premises representations,
- unifying premises representations,
- deciding whether there are valid conclusions,
- formulating valid conclusions.

The following example will illustrate the method. Fig. 8 illustrates the method applied to the syllogism *All P are M. Some S are not M.* Each premise is first represented by the diagram of its maximal model, i.e. the one containing the most types of individuals consistent with the premise. Within these diagrams, sub-regions representing the minimal models of the premise are marked with an "*x*". The two premise diagrams then have to be "registered" which means their *M* entity superimposed and their *P* and *S* entities arranged to form the maximal model consistent with each premise. If their region is bisected by the third entity when the premise diagrams are registered, then the *x* is removed. This ensures that the remaining *x* still mark the minimal models of the conjunction of the two premises.

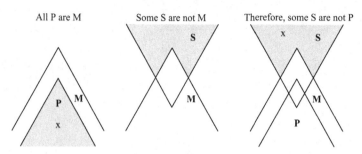

Fig. 8. Min-max entities graphical algorithm applied to the syllogism *All P are M. Some S are not M.*

Figure *Baroco* can be modeled in analog hardware (as it is shown in Fig. 9):

Baroco

All P are M
Some S are not M

Some S are not P

All informative things are useful.
Some websites are not useful.

Some websites are not informative.

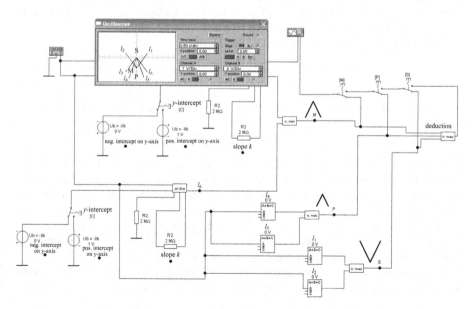

Fig. 9. Hardware model of figure *Baroco*

Graphic representation (an oscillogram) of relations between terms in figure *Baroco* is shown in Fig. 10.

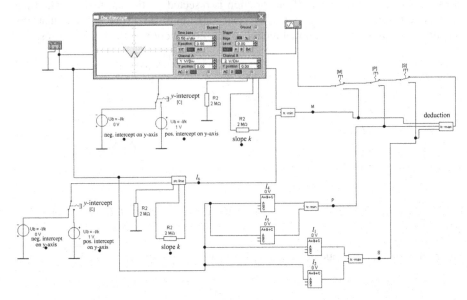

Fig. 10. "Hardware deduction" for figure *Baroco*: max $(S, P, M) \equiv$ some S are not P

Function generator plays the role of a universe of discourse and can be related with an output limiter (not shown in the model) which limits output voltage to an arbitrary output expressed as $\min(U_A, \max(U_S, U_P, U_M))$ where voltages U_S, U_P, U_M represents sets S, P and M respectively, and U_A is an arbitrary limiting voltage.

Valid mood of categorical syllogism called *Darii*

Darii

All M are P   *All cypress trees are evergreens.*
Some S are M   *Some trees are cypress trees.*
----------   ----------
Some S are P   *Some trees are evergreens.*

can also be modeled in the analog hardware circuit environment (see Fig. 11 and 12). Note that a different hardware approach, based on semicircle generators and square root circuits along with dedicated *min-max circuits*, is used in this *Darii*-hardware model [9].

Test circuits showed before (see Fig. 5, 6 and 7) can be used to check the validity of "hardware deduction", i.e. conclusion in both models. The simplest way to interpret "hardware deduction" is to use simple passive integrator as shown in Fig. 13.

Hardware circuit models of other figures can be a part of *future work*, together with the interpretation of their effect in a *hardware model of system control*.

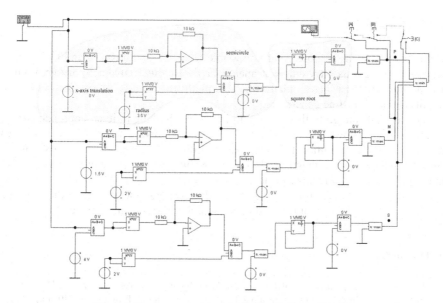

Fig. 11. Hardware model of figure Darii: Some S are P

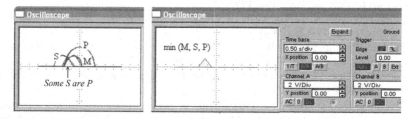

Fig. 12. "Hardware deduction" (conclusion) for figure Darii: some S are $P \equiv$ min (M, S, P) where M, S and P stand for related voltages U_S, U_P and U_M

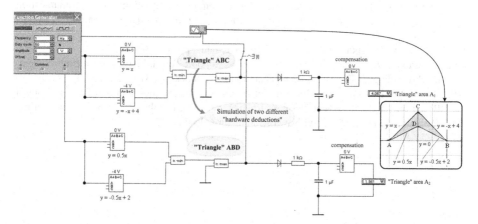

Fig. 13. Simple interpretation of "hardware deduction" There are more S_1s in P_1s then S_2s in P_2s, therefore $A_1 : A_2 = 2 : 1$

4   Conclusion

In this paper, we have shown that is possible to model syllogism(s) and to interpret deduction process as an voltage in analog hardware environment by using original operational amplifier based platform. The platform consists of original min-max circuits and straight-line generators.

The basic advantage of this method in modeling thinking rational approach can be summarized as follows: simple circuitry, no special hardware design (devices) is (are) necessary, simple definition terms are easy to obtain, comprehensible, inexpensive, student (education) oriented and many useful laboratory experiments at an undergraduate level can be performed.

Hardware models of syllogisms can be used as response modulators of the adaptive process control machine that is based on a learning transfer characteristic [10].

References

1. Kalenka, S.: Modelling Social Interaction Attitudes in Multi-Agent Systems, Department of Electronic Engineering Queen Mary, University of London, England, pp. 40–111 (2001)
2. Zadeh, L.: Knowledge Representation in Fuzzy Logic. IEEE Transactions on Knowledge and Data Engineering 1, 89–100 (1989)
3. Igel, C., Temme, K.H.: The chaining syllogism in fuzzy logic. IEEE Transactions on Fuzzy Systems, 849–853 (2004)
4. Butkovic, P.: Max-algebra: the linear algebra of combinatorics, Linear algebra and its applications, vol. 367, pp. 313–335. Elsevier, Holland (2003)
5. Yager, R., Filev, D.: Essentials of Fuzzy Modeling and Control, pp. 7–14. J. Wiley & Sons, New York (1994)
6. Kovacevic, D., Kovacevic, A.: Fuzzy Logic Controller Based on Standard Operational Amplifiers in Fuzzy Logic. In: Baldwin, J.F. (ed.), pp. 231–244. John Wiley & Sons, Chichester (1996)
7. Kovacevic, D., Pavic, I., Pribacic, N.: Solving Inequalities in Electronic MIN/MAX environment. In: 33rd International Convention on Information and Communication Technology, Electronics and Microelectronics, Mipro, Opatija, Croatia, pp. 170–175 (2010)
8. Petrovic, G.: Logika, Skolska knjiga Zagreb, Croatia, pp. 42–163 (1975)
9. Kovacevic, D.: Circuit Simulation Program as Support Toll in Mathematical Environment. In: Proceedings IEEE Africon 2007, Namibia, Windhoek, pp. 319–328 (2007)
10. Kovacevic, D.: Hardware Modelling of Newspaper Distribution and Sell Process Control Based on Intelligent Agent Network Realization, PhD, Faculty of Electrical Engineering, Mechanical Engineering, and Naval Architecture, Split (2012)

A New Approach to High Impedance Fault Detection Based on Correlation Functions

Najmeh Faridnia[1], Haidar Samet[2], and Babak Doostani Dezfuli[3]

[1]Research and Designing Department, Fars Regional Electric Company, Shiraz, Iran
[2] School of Electrical and Computer Engineering, Shiraz University, Shiraz, Iran
[3] Control and Instrumentation Engineeing Department,
Iranian Offshore Oil Company, Lavan, Iran
faridnia@frec.co.ir, samet@shirazu.ac.ir, bdoostani@gmail.com

Abstract. Detection of High Impedance Fault (HIF) that is mainly occurred in MV power systems has been a challenge due to its low current magnitude. A new HIF detection approach based on correlation functions is applied on the voltage and current signals in this paper. Twelve indices based on correlation functions are implemented and tried on a lot of cases. Some of them which have clearly different values for no-fault and HIF conditions are chosen. Simulation results prove the efficiency of proposed method for HIF detection.

Keywords: high impedance fault detection (HIF), auto correlation function (ACF), partial auto correlation function (PACF), SACF, SPACF.

1   Introduction

The high impedance fault (HIF) is an abnormal electrical condition which may affect human lives wherever the feeders and the living areas have any overlaps. The HIF is usually occurred in the distribution feeders because of their low height in comparison with high voltage transmission lines. This fault is happened when an unexpected connection appears between a feeder and the ground by a high impedance object. Most of the time, it is generated by falling a broken conductor of an energized line on the ground or connection between tree branches to the line. In electrical point of view if a high impedance object produces a parallel loop with the load in a circuit while the line is loaded as much as nominal current, then the over-current which is imposed by fault would be low in contrast to load current. So, this phenomenon is called HIF. Since the basis of operation for conventional relays is current magnitude, HIF flow cannot be detected by mentioned relays.

A review of many approaches on HIF detection from classical to heuristically algorithms is done in [3] that is a useful guideline.

Operating principle of HIF detection is reviewed and possible future technologies is discussed in [5]. Discrete wavelet transformer technique in [2], [6] and [7], artificial neural network in [4] is used to detection of HIF. Besides, some others such as [1] and [8] used low frequency harmonic analysis to detection.

L. Iliadis et al. (Eds.): AIAI 2012, IFIP AICT 381, pp. 453–462, 2012.

In this paper, a new approach for HIF detection based on correlation functions is proposed. It is found that the values of correlation functions applied on voltage, current and also their derivations during the system healthy conditions are different from their values when high impedance fault occurs. To show the benefits of correlation functions in HIF detection, the paper will be continued with five sections. A brief summary of the correlation function theory and proposed indices for this purpose will be discussed in section 2. Then a case study system with using of two different models of high impedance fault is used for data collection in section 3. Simulation results of the case study system using PSCAD/EMTDC software are presented in section 4. The conclusion is given in section 5.

2   A Brief Theory of Correlation Functions

Auto correlation function (*ACF*) and partial auto correlation function (*PACF*) that are defined by equations 1 and 2 respectively are applied to detect any abnormal changes in a signal.

$$r_k = \frac{\sum_{t=1}^{n-k}(Z_t - \mu)(Z_{t+k} - \mu)}{\sum_{t=1}^{n}(Z_t - \mu)^2} \qquad k = 1, 2, 3, \dots \qquad (1)$$

$$r_{kk} = \begin{cases} r_1 & \text{if } k = 1 \\ \dfrac{r_k - \sum_{j=1}^{k-1} r_{k-1,j} r_{k-j}}{1 - \sum_{j=1}^{k-1} r_{k-1,j} r_j} & \text{if } k = 2, 3, \dots \end{cases} \qquad (2)$$

where

$$r_{k,j} = r_{k-1,j} - r_{kk} r_{k-1,k-j} \qquad j = 1, 2, \dots, k-1$$

r_k and r_{kk} symbolize ACF_k and $PACF_k$ respectively. n is time series window length and μ is the mean of time series. To get a good performance it's better to use summation of *ACF* and *PACF* components. So, in the present approach two more functions are defined to distinguish the normal case from HIF.

SACF: Summation of *ACF* components of a window including k number of samples that is called *SACF*. Related equation is as follows:

$$SUM_{ACF} = \sum_{k=1}^{k_{max}} r_k \qquad (3)$$

SPACF: This function is defined similar to *SACF* but the difference is that the *PACF* components have been used to make it. Equation (4) belongs to *SPACF*.

$$SUM_{PACF} = \sum_{k=1}^{k_{max}} r_{kk} \tag{4}$$

Here six signals are used instead of Z in equations (1) and (2) which are as follows:

$$di_n = i_n - i_{n-1} \tag{5}$$

$$dv_n = v_n - v_{n-1} \tag{6}$$

$$d^2 i_n = di_n - di_{n-1} \tag{7}$$

$$d^2 v_n = dv_n - dv_{n-1} \tag{8}$$

where i and v are the current and voltage at relay placement. d and d^2 denote the first and second differences. Both $SACF$ and $SPACF$ functions are applied to all six above signals to obtain 12 indices as follow:

$$sacfi = \sum_{k=1}^{k_{max}} r_k(i) \tag{9} \qquad sacfv = \sum_{k=1}^{k_{max}} r_k(v) \tag{10}$$

$$spacfi = \sum_{k=1}^{k_{max}} r_{kk}(i) \tag{11} \qquad spacfv = \sum_{k=1}^{k_{max}} r_{kk}(v) \tag{12}$$

$$sacfdi = \sum_{k=1}^{k_{max}} r_k(di) \tag{13} \qquad sacfdv = \sum_{k=1}^{k_{max}} r_k(dv) \tag{14}$$

$$spacfdi = \sum_{k=1}^{k_{max}} r_{kk}(di) \tag{15} \qquad spacfdv = \sum_{k=1}^{k_{max}} r_{kk}(dv) \tag{16}$$

$$sacfd^2 i = \sum_{k=1}^{k_{max}} r_k(d^2 i) \tag{17} \qquad sacfd^2 v = \sum_{k=1}^{k_{max}} r_k(d^2 v) \tag{18}$$

$$spacfd^2 i = \sum_{k=1}^{k_{max}} r_{kk}(d^2 i) \tag{19} \qquad spacfd^2 v = \sum_{k=1}^{k_{max}} r_{kk}(d^2 v) \tag{20}$$

where $r_k(v)$ denotes kth component of voltage's ACF. Other equations are similarly defined.

3   Case Study System

One case study system with two different HIF models is implemented in figures 1 and 2. Case study system belongs to a radial cable feeder of a medium voltage (25 kv) system. Single line diagram of the case study system with the HIF model 1 is given in figure.1. V_s and I_{st} are the relay voltage and current which are used for sensing the HIF. The connected load to the feeder is assumed to be constant at the end of the line. HIF model 1 is explained through equation (21) and figure.1.

$$L_f = 3mH, R_f = R_{f0}(1 + \alpha(\frac{i_f}{i_{f0}})^\beta) \tag{21}$$

Where $R_{f0}= 20$, $i_{f0}= 70$, $\alpha= 0.6$, $\beta =2$, $V_{L-L}= 25\ kV$, $f=60\ Hz$

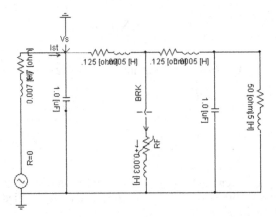

Fig. 1. The case study system with HIF model 1

The model circuit with the description of second HIF model is given in figure.2. Figure.3 shows the case study system used in figure.1 which applied HIF model 2 in figure.2. Where $i(t)$, u_p, R and l_p represent arc current, constant voltage parameter per arc length, resistive component per arc length and arc length respectively. It is considered that R ($R=9\Omega/cm$) and l_p ($l_p=10$ cm) are constant. The fault is occurred in the middle of the feeder.

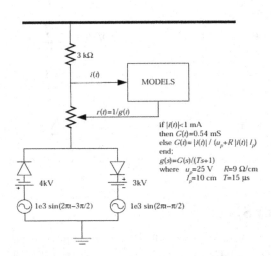

Fig. 2. Circuit and description of HIF model 2 [2]

The voltage and current curves of the case study system with both of HIF models which are simulated in PSCAD/EMTDC software are shown in figures 4 and 5. In these simulations, the time duration is 500 ms, the fault occurs in 202 ms and the sampling time is 50 μs.

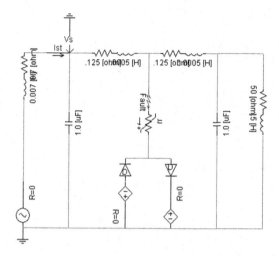

Fig. 3. The case study system with HIF model 2

As can be seen in figure.5, no sign of any events for model 2 is visible in both current and voltage of relay.

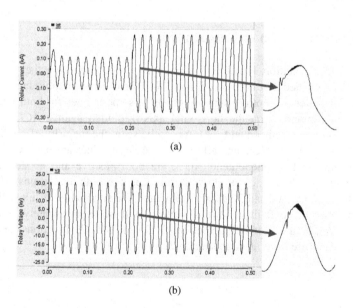

(a)

(b)

Fig. 4. (a) current and (b) voltage results of case study system with HIF model 1

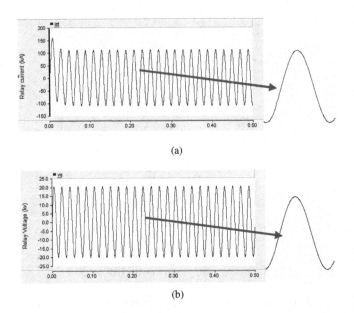

(a)

(b)

Fig. 5. (a) current and (b) voltage results of case study system with HIF model 2

4   Implementation of New Approach

After utilizing correlation functions on voltage, current and also their derivations samples, it is specified that both *SACF* and *PSACF* can be acceptable factors for HIF detection. For better perception, stem diagrams of some indices for both HIF models are shown in figure.6. k on the horizontal axis is plotted against *ACF* and *PACF* values. In this case, k_{max} is equal to 20. As can be seen, the majority of *pacfi (PACF$_i$)* and *pacfddv (PACF$_{d^2v}$)* values and all of *acfdv (ACF$_{dv}$)* values are positive in no fault condition while in HIF phenomenon there are both positive and negative. Briefly, feature pattern of three indices show differences for both conditions. If in each diagram the summation of all 20 data is calculated, the summation result will be a big positive number in no fault condition while it will be approximately zero in HIF event. So, it can be used for HIF detection. *SACF* and *SPACF* functions for all six of relay-side signals are introduced in section.2.

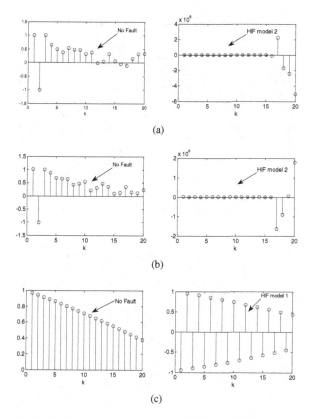

Fig. 6. Comparing feature pattern of (a) pacfi (b) acfd2v (c) acfdv for both no-fault and HIF conditions (n=100, k=1,…,20)

On the first step, the time window (including N samples with sampling time equal to 50 μs) is assumed. Two different windows are considered.

The two windows are assumed to have the same length.

The indices values are experimented for $n=40, 80, 100, 200$ and $k_{max}=10, 20, 40$ in both HIF and no fault states. The values of all 12 mentioned indices for $n=40$ and $k_{max}=10$ are given in tables I to IV.

Table 1. Current's ACF results for static window in no fault condition ($n=40$ and $kmax=10$)

Fault time (s)	Current					
	sacfv	sacfdv	sacfd²v	spacfv	spacfdv	spacfd²v
0.200	6.5900	6.5085	6.3878	4.0011	4.5190	4.6588
0.201	6.6192	6.4370	6.4153	4.0940	4.4195	4.6484
0.202	6.6264	5.3030	6.4222	4.5049	4.9448	5.1292
0.203	6.6248	6.2117	6.4206	4.2429	4.7193	5.0486
0.204	6.6116	6.4814	6.4083	4.4522	4.6284	4.7855
0.205	6.5568	6.5176	6.3564	4.4555	4.8640	5.0938
0.206	5.9560	6.5267	5.7754	4.3321	4.6874	4.8726
0.207	6.1057	6.5263	5.9225	4.1080	4.4797	4.7678
0.208	6.5651	6.5162	6.3643	4.5088	4.6646	4.8468
0.209	6.6134	6.4754	6.4099	3.8506	4.3387	4.6272
0.210	6.6252	6.1363	6.4210	3.5469	4.3965	4.6285

Table 2. voltage's acf results for static window in no fault condition (n=40 and kmax=10)

Fault time (s)	Voltage					
	sacfv	sacfdv	sacfd2v	spacfv	spacfdv	spacfd2v
0.200	6.6228	6.3595	6.4187	5.0753	5.9513	7.0459
0.201	6.6040	6.4967	6.4011	4.2844	4.8190	5.2745
0.202	6.5153	6.5213	6.3171	4.0917	4.5285	4.8304
0.203	4.8080	6.5273	4.5936	4.3407	4.6798	5.0479
0.204	6.3758	6.5249	6.1840	4.3514	4.7660	5.1581
0.205	6.5861	6.5103	6.3841	4.3868	4.9144	5.4635
0.206	6.6182	6.4469	6.4144	4.0035	4.2751	4.5497
0.207	6.6263	5.5800	6.4220	4.6931	5.3362	5.6569
0.208	6.6251	6.1491	6.4210	9.7443	9.7535	10.0218
0.209	6.6131	6.4764	6.4097	4.3266	4.7334	5.4211
0.210	6.5639	6.5164	6.3632	4.2424	4.8392	5.1792

Table 3. Current's acf results for static window in hif condition (n=40 and kmax=10)

Fault time (s)	Voltage					
	sacfv	sacfdv	sacfd2v	spacfv	spacfdv	spacfd2v
0.200	6.3466	3.0206	1.6699	0.3590	0.2429	0.6375
0.201	2.6460	0.8672	-1.0995	-0.0993	1.0627	0.1453
0.202	0.1420	0.1680	-0.4644	-0.8723	2.4235	0.1064
0.203	-0.9982	0.0624	-0.1323	0.2219	1.8844	-0.3442
0.204	-1.4295	-0.0277	-0.1417	0.2232	1.8580	-0.2611
0.205	-1.5050	0.0226	-0.3410	-0.6238	1.7768	-0.1060
0.206	-0.1160	0.2922	-1.0224	-0.3133	1.2205	0.5994
0.207	4.8707	1.6456	-0.9479	-0.0176	0.6933	0.2696
0.208	7.2593	2.4569	2.1355	-0.1606	0.4349	0.0973
0.209	3.8737	1.6491	-1.0497	0.0396	0.8800	0.2729
0.210	0.8931	0.2576	-1.0511	0.1691	0.8014	0.5908

Table 4. voltage's acf results for static window in hif condition (n=40 and kmax=10)

Fault time (s)	Voltage					
	sacfv	sacfdv	sacfd2v	spacfv	spacfdv	spacfd2v
0.200	6.9339	1.6671	-0.4695	0.2669	0.1739	0.0934
0.201	5.4783	-0.9503	-0.2817	0.5827	0.0167	0.7039
0.202	3.4873	-0.4204	-0.2167	0.6335	0.0104	0.2035
0.203	0.9900	-0.0886	-0.0216	0.7565	-0.4900	-0.1591
0.204	0.5949	-0.0708	-0.0063	0.6957	-0.5060	-0.8871
0.205	3.2958	-0.1868	-0.0500	0.5217	-0.3864	-1.9170
0.206	5.4064	-0.8062	-0.5091	-0.1666	0.6426	-0.8450
0.207	6.5456	-1.2097	-0.8134	0.5620	0.1342	0.6332
0.208	7.0033	0.8362	-0.0565	0.4719	0.1510	0.6231
0.209	6.0259	-0.7936	0.2544	0.5499	0.0080	0.5214
0.210	4.2548	-0.8951	-0.3937	-0.0565	0.5887	0.1744

These tables clearly show that all indices values are positive and almost large for both current and voltage in no-fault case, while most of them are around zero in HIF phenomenon. In this stage, *sacfv*, *sacfi* and *sacfdi* indices are put aside because their values in HIF are far from zero and they are not suitable for fault detection.

Now, it's time to go one step ahead and analyze the results of moving time window for different values of *n* and *k*. Finally, feature patterns of two indices for HIF with clear changes are shown in figure.7. These features belong to *spacfdv* and *spacfd²v* for HIF model 1.

Using a low-pass filter would be useful to make the curve flat to get a good and clear feature pattern for HIF detection. So, it is suggested that use a filter of degree 4 with equation number (22). Filtering of *SPACF* curve can be got similarly.

$$f_n(SACF) = \sum_{k=(n-3)}^{n} SACF_k \tag{22}$$

Some indices curves like *sacfd²i*, *sacfd²v* and *spacfi* for HIF model 2 before and after filtering are brought in figure.8.

Although voltage and current of relay don't show any problem in appearance, HIF can be detected easily by using of SACF and SPACF functions.

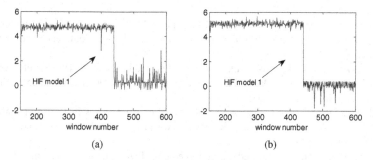

(a)   (b)

Fig. 7. Feature pattern of (a) spacfdv, (b) spacfd2v for HIF condition (time duration=3 s)

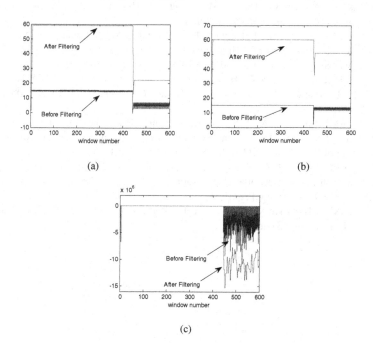

(a)   (b)

(c)

Fig. 8. Feature pattern of (a) sacfd2i, (b) sacfd2v and (c) spacfi for HIF (model 2) before and after using filter (time duration=3 s)

5   Conclusion

A new approach for HIF detection in distribution systems based on correlation functions is proposed in this paper. After simulating a sample system, *SACF and SPACF* functions were experimented on *i, di, d^2i, v, dv* and d^2v signals that resulted in twelve indices to HIF detection. After getting the results of both normal and HIF conditions for different values of variables, it was founded that *SACF* and *SPACF* operators can detect HIF phenomenon from voltage and current samples of relay.

References

1. Sharaf, A.M., Wang, G.: High Impedance Fault Detection using Feature-Pattern Based Relaying. In: 2003 IEEE PES Transmission and Distribution Conference and Exposition, vol. 1, pp. 222–226 (2003)
2. Michalik, M., Rebizant, W., Lukowicz, M., Lee, S.-J., Kang, S.-H.: Wavelet Transform Approach to High Impedance Fault Detection in MV Networks. In: 2005 IEEE Power Tech., Russia, pp. 1–7 (2005)
3. Sedighizadeh, M., Rezazadeh, A., Elkalashy, N.I.: Approaches in High Impedance Fault Detection A Chronological Review. Advances in Electrical and Computer Engineering 10(3) (2010)
4. Zadeh, H.K.: An ANN-Based High Impedance Fault Detection Scheme: Design and Implementation. International Journal of Emerging Electric Power Systems 4(2) (2005)
5. Adamiak, M., Wester, C., Thakur, M., Jensen, C.: High Impedance Fault Detection On Distribution Feeders. GE Industrial solutions
6. Elkalashy, N.I., Lehtonen, M., Darwish, H.A., Izzularab, M.A., Taalab, A.I.: Modeling and Experimental Verification of a High Impedance Arcing Fault in MV Networks. In: IEEE 2006 Power System Conference and Exposition, pp. 1950–1956 (2006)
7. Ibrahim, D.K., Eldin, E.S.T.: High-Impedance Fault Detection in EHV Transmission Lines. In: International Middle-East Power System Conference, MEPCON 2008, pp. 192–199 (2008)
8. Sharaf, A.M., Snider, L.A., Debnath, K.: Residual Third Harmonic Detection of High Impedance Fault in Distribution System Using perception Neural Networks. In: Proceeding of the ISEDEM 1993, Singapore (October 1993)
9. Sharaf, A.M., Snider, L.A., Debnath, K.: Harmonic based Detection of High Impedance Faults in Distribution Networks Using Neural Networks. In: Proceeding of the IASTED Conference 1993, Pittsburg, PA (1993)
10. Sharaf, A.M., EI-Sharkawy, R.M., Jalaat, H., Badr, M.A.: Fault Detection on Radial and Meshed Transmission Systems

Network Selection in a Virtual Network Operator Environment

Ioannis Chamodrakas and Drakoulis Martakos

National and Kapodistrian University of Athens,
Department of Informatics and Telecommunications, Athens, Greece
{ihamod,martakos}@di.uoa.gr

Abstract. Service offerings of virtual network operators should focus on the needs of distinct groups of users in order to face challenges posed in the saturated telecommunications mass market. Since a virtual operator may hire capacity from a number of host networks with varied characteristics, the selection of an optimal host network according to network attributes, user preferences and QoS requirements results to the provision of services with desirable characteristics adding value for the user. Network selection in such an environment has not been studied as all previous works have focused exclusively on handover decisions in heterogeneous wireless networks. This article describes a network selection method in a virtual operator environment that uses a modified version of TOPSIS to rank alternative networks. The proposed modification enables the inclusion of QoS requirements in the ranking results. The uncertainty of user preferences is handled through the use of linguistic assessments and triangular fuzzy numbers. Simulations conducted show that the proposed approach is effective in deciding the optimum network according to all decision factors.

Keywords: network selection, virtual network operators, TOPSIS, Always Best Connected, fuzzy numbers, QoS.

1   Introduction

The deregulation process in the telecommunications sector has led to a vertical disintegration of the network industry [1]. Traditional network operators owned the physical network infrastructure and exercised strict control over network management and service provision. The "liberalization" of the telecommunications market led to the entry of virtual operators that carry out a subset of the functions and activities of traditional operators without possessing their own physical network and fulfill one or several roles in the value chain by hiring network capacity and basic network facilities from traditional operators under exclusive contracts [1], [2]. Virtual operators can be distinguished into three types according to their technical dependence on their "network suppliers": service providers (SPs) resell network services and products to the end-users under a different brand name, provide tariffing, billing and customer support, and undertake distribution and marketing activities; enhanced service providers (ESPs) differentiate their service offerings through the development of extra and

L. Iliadis et al. (Eds.): AIAI 2012, IFIP AICT 381, pp. 463–473, 2012.

innovative services; virtual network operators (VNOs) pertain to the most technically developed type of virtual operators since apart from differentiated services they also maintain their own routing infrastructure.

It has been argued that three main areas are key to VNO success: efficient and effective distribution channels to reach the target audience, retention/loyalty programs that minimize churn cost, and design of a technology road map that enables the virtual operator to quickly adopt new technologies and launch innovative products [3]. Various studies have identified as the main factor of the success of a new virtual operator entering a saturated mass market, its ability to provide differentiated service offerings in order to target specific market segments by offering value-added services that focus on the satisfaction of specific needs of distinct groups of users [2].

In this context, since a virtual operator may have established contracts with a number of traditional operators providing network capacity with different characteristics such as monetary cost, Quality of Service (QoS) attributes (bandwidth, packet loss, latency, jitter), bit/frame error rate, etc, we propose the application of the Always Best Connected concept in order to devise an intelligent network selection mechanism. The Always Best Connected concept [4], of a user being connected to a network in the best possible way based on a number of different criteria such as user preferences, size and capabilities of the service, application requirements, security, operator or corporate preferences, has been introduced in the context of heterogeneous wireless network environments, and guided the development of various vertical handover management systems [5], [6]. Vertical handover decision algorithms have been categorized into 5 groups by Kassar et al. [5], i.e. decision-function based strategies, user-centric strategies, multiple attribute decision strategies, fuzzy logic and neural network based strategies and context aware strategies, and into 4 groups by Yan et al. [6], i.e. received signal strength (RSS) based algorithms, bandwidth based algorithms, cost function based algorithms and combination based algorithms (fuzzy logic and neural network).

Inspired by the research mentioned above, this paper proposes a network selection method for the users of a virtual operator taking into account both network conditions and user preferences. This is a relatively unexplored field of research since all previous studies focused on the problem of network selection for vertical handover in the context of heterogeneous wireless environments. Apart from being an original direction for research, the proposed network selection method may be utilized by virtual operators in order to provide differentiated services according to the specific preferences and needs of the users regarding both QoS and cost of the provided services, contributing, thus, to their success in the market. The motivation of this research stems from the analysis of the requirements of an active virtual network operator in Greece and aims to be put into use after a period of simulation and real experiments.

In this article we consider QoS, monetary cost and user preferences as decision factors, Since QoS preferences are usually expressed as upper and lower limits, the structure of the selection problem fits well with the assumptions of the TOPSIS decision method. The TOPSIS method was introduced by Hwang and Yoon [7] and ranks the alternative solutions according to their distances from the zenith (the ideal alternative) and the nadir (the negative ideal alternative) points. In this case, we propose a modification of the TOPSIS method by defining the zenith point according to the QoS upper or lower limit defined for each type of service and the respective monetary costs of the network suppliers. In this way, the modified TOPSIS method produces results that take into account QoS requirements, enabling the treatment of different QoS profiles. The QoS profiles can be related either to specific service types (service QoS profiles) or to user-defined QoS requirements (user QoS profiles). TOPSIS has been applied with success in a number of selection/evaluation problems with a finite number of alternatives both in and out of the context of network selection [8], [9]. Moreover, the TOPSIS method is one of the best methods in addressing the problem of rank reversal [10] and its logic is consistent with the rationale of human choice. Since human judgments are uncertain with regard to the weighting of different decision criteria, many researchers have proposed fuzzy extensions of the TOPSIS method in order to grasp the vagueness of user preferences [8], [11]. For that, our method employs the use of linguistic assessments and their expression as fuzzy numbers in order to calculate the weights of different criteria. The proposed method is illustrated and validated in the article by applying it in different network selection scenarios. The simulated scenarios correspond to different service/user QoS profiles in order to observe and discuss how the proposed method would work.

2   Technique for Order Preference by Similarity to the Ideal Solution

The TOPSIS method ranks alternatives according to their distance from the ideal solution (zenith) and the negative ideal solution (nadir). Each alternative is represented as a point in a n-dimensional Euclidean space, and a basic assumption is that the utility of all attributes is monotonically increasing or decreasing. Thus, the zenith and nadir points are modeled as hypothetical alternatives that have respectively the best or the worst values for each attribute, from the set of attribute values of all the alternatives. The ranking principle of TOPSIS is that the best alternative is simultaneously farthest from the nadir point and closest to the zenith point in terms of Euclidean distance. In this article, we apply a modified version of TOPSIS since the zenith point is defined according to the QoS upper or lower limit of each type of service for relevant attributes according to the monotonicity of their utility.

The steps of the TOPSIS method in order to apply it to the network selection problem are the following:

1. Identify all attributes impacting the decision process as well as the alternatives under consideration. Since QoS and monetary cost preferences are taken into account as decision factors for network selection, the list of attributes may include cost per byte or cost per second of call duration (*monetary cost*), total bandwidth and allowed bandwidth (*throughput*), packet delay, packet jitter and response time (*timeliness*), bit error rate, packet loss, burst error and average number of retransmissions per packet (*reliability*), utilization, etc.
2. Construct the normalized decision matrix representing the alternatives under consideration. This step entails the collection of QoS data from the alternative networks in order to determine the respective value of each attribute under consideration.
3. Determine the weights representing the relative importance of each attribute and construct the weighted normalized decision matrix.

4. Determine the nadir and zenith points for each type of service: the best value for each attribute is defined according to the QoS upper or lower limit and the worst value for each attribute is either the maximum or the minimum value, depending on the monotonicity of the attribute's utility. As far as the attribute of cost is concerned, the best value is the minimum cost and the worst value is the maximum cost.
5. Measure the separation S of alternative networks from the nadir and zenith points using Euclidean distances. It must be noted that when an attribute value of an alternative is better than the respective value of the zenith point for a specific type of service, its superiority is ignored and the separation of the specific attribute from the zenith point is assumed to be zero.
6. Calculate the level of preference C of each alternative network according to the TOPSIS measure of the "relative closeness to the ideal solution"

$$C = \frac{S_{nadir}}{S_{nadir} + S_{zenith}} \tag{1}$$

7. Finally, select the network with the highest level of preference C.

Apart from steps 3 and 4 the procedure described above is relatively straightforward. However, step 3 entails the representation of uncertain human judgments for the determination of the attribute weights and thus crisp values are inadequate. Various researchers have proposed different fuzzy versions of TOPSIS. The fuzzification and defuzzification procedures involve the use of linguistic variables and assessments and their representation as (usually triangular) fuzzy numbers. In some cases the calculation of the distances of the fuzzified alternatives from the zenith and nadir points has been handled through the use of fuzzy distance norms, fuzzy ranking approaches or grey related analysis methods [11], [12]. In other cases, the linguistic assessments which are represented as triangular fuzzy numbers are transformed directly into crisp numbers [8], [13]. This approach is followed in order to avoid complicated fuzzy arithmetic operations and dubious fuzzy ranking approaches since it has been observed that excessive fuzzification entailing the use of complex algorithms can be considered a fallacy [13]. In this article we follow the second approach described in [8], [13].

Furthermore, step 4 entails the configuration of the QoS profiles with regard to the upper and lower limits of the QoS attribute values. QoS profiles of individual services are defined by the virtual operator, are applicable to all users and may refer to e.g. Voice service profile, Web browsing profile, Multimedia streaming profile, Messaging service profile, Dedicated data transfer profile (database backup, facsimile, etc). For example, Voice service profile upper and lower limits for packet loss are 1% to 0% and for latency are 0 to 150 milliseconds. On the other hand, Video streaming application profile upper and lower limits for packet loss and latency are 2% to 0% and 0 to 4000 milliseconds respectively. Moreover, the user may define his/her own QoS user profile applicable to all services, and thus override individual service profiles by explicitly defining upper and lower limits of the QoS attribute values.

3   Network Selection Scheme

The proposed model for the network selection process in a virtual network operator environment is presented in Fig. 1. This model includes a pre-configuration stage where the virtual operator determines QoS service profiles and users identify selection attributes, provide input for the determination of their relative importance and optionally determine QoS user profiles.

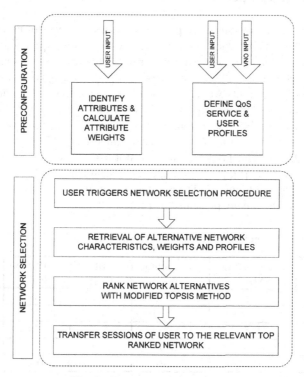

Fig. 1. Network Selection Scheme

The actual selection process is initiated by the user. Another option would be to automate the initiation step through a rule-based system that would incorporate a monitoring mechanism periodically probing the QoS network characteristics and a set of initiation rules provided by the virtual operator for each type of service. Nevertheless, we present a simplified model for the sake of the clarity of presentation since the details of such a mechanism are rather trivial.

- In the next step of the selection process, network characteristics relevant to the user-identified QoS attributes are retrieved along with attribute weights and QoS profiles.
- Then alternative networks are ranked for each type of service.
- In the last step of the selection process the sessions of the users are transferred to the top ranked network for the corresponding services. It must be noted that the

virtual operator routing infrastructure handles the management of sessions and their transfer to the appropriate network and, therefore, it is possible to select different networks for different services without facing problems relevant to network selection in a heterogeneous wireless environment (excessive power consumption, authentication of the user terminal with multiple networks, etc.).

4   Simulation Experiment

In order to illustrate and validate the proposed decision method for network selection in a virtual operator environment we performed a simulation experiment. In the experiment we consider three users consuming services provided by a virtual operator that has established contracts with five host network operators supplying network capacity. Without loss of generality we have chosen five criteria to be taken into account for the network selection decision represented as attributes of the alternative networks: allowed bandwidth per user, latency, jitter, packet loss and cost per byte.

The preferences of the users for the relative importance of the criteria are given through linguistic assessments which are transformed into crisp values according to the method described in [8], [13].The linguistic assessments and the attribute weights per user are depicted in Figure. 2. The attribute values of the five alternative networks at the moment of retrieval are presented in Table 1.

Table 1. Attribute values of alternative networks

Networks	Bandwidth (Mbs)	Latency (ms)	Jitter (ms)	Packet loss	Cost per byte
1	10	200	15	0,01	40
2	5	100	10	0,005	50
3	0,512	200	25	0,01	45
4	1	400	50	0,02	30
5	0,256	50	5	0,001	40

Furthermore, we consider seven network selection scenarios corresponding to 5 service QoS profiles: voice service, video streaming, text messaging, video conferencing, web browsing and 2 user QoS profiles: Low QoS levels profile, High QoS levels profile. The modified TOPSIS method previously described is used to get the ratings and rankings of the alternative networks for each service profile. In order to compare the proposed approach with the standard TOPSIS method that does not take into account different service or user profiles we also include the ratings and rankings which standard TOPSIS produces. The upper and lower QoS limits for the chosen attributes are depicted in Fig. 3 and the results of simulations for each service profile and for the standard TOPSIS method are shown in Fig. 4.

The results depicted in Fig. 4 indicate that high discrepancies in the relative importance of the attributes influence greatly the relative rating and ranking of alternative networks, i.e. network ratings for User 2 are much different than network ratings for

User 1 and User 3. For example, it can be seen that since User 2 weights the relative importance of the cost attribute much higher than the other attributes, it follows that Network 2 which is characterized by the highest cost per byte is rated much lower than in the case of Users 1 and 3 for all QoS profiles, whereas Network 4, which is characterized by the lowest cost, is rated much higher than in the case of Users 1 and 3 for whom it gets the lowest rating for all QoS profiles, reaching even a first ranking for User High and User Low QoS profiles. On the contrary, smaller discrepancies play a less significant role as the almost identical ratings for User 1 and User 3 show.

On the other hand, for the same relative importance of attributes, network ratings vary according to the QoS profiles due to the different service requirements pertaining to each service or user profile. For example, if User 1 is considered, it can be seen that Network 5 is optimal for Voice, Messaging and User Low QoS profiles followed by Network 2 whereas the rank order is reversed, i.e Network 2 is optimal followed by Network 5 for Video streaming, Video conferencing and User High QoS profiles.

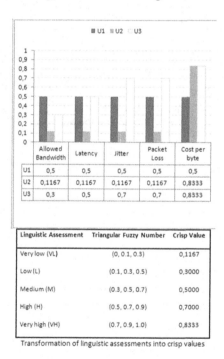

	Allowed Bandwidth	Latency	Jitter	Packet Loss	Cost per byte
U1	0,5	0,5	0,5	0,5	0,5
U2	0,1167	0,1167	0,1167	0,1167	0,8333
U3	0,3	0,5	0,7	0,7	0,8333

Linguistic Assessment	Triangular Fuzzy Number	Crisp Value
Very low (VL)	(0, 0.1, 0.3)	0,1167
Low (L)	(0.1, 0.3, 0.5)	0,3000
Medium (M)	(0.3, 0.5, 0.7)	0,5000
High (H)	(0.5, 0.7, 0.9)	0,7000
Very high (VH)	(0.7, 0.9, 1.0)	0,8333

Transformation of linguistic assessments into crisp values

Fig. 2. Attribute weights per user / Linguistic assessment transformation

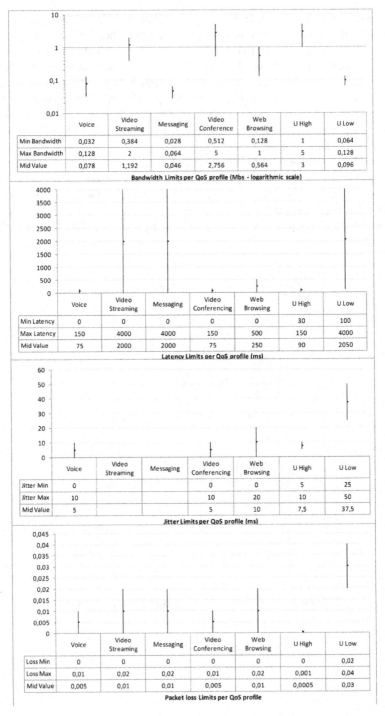

Fig. 3. QoS Profiles

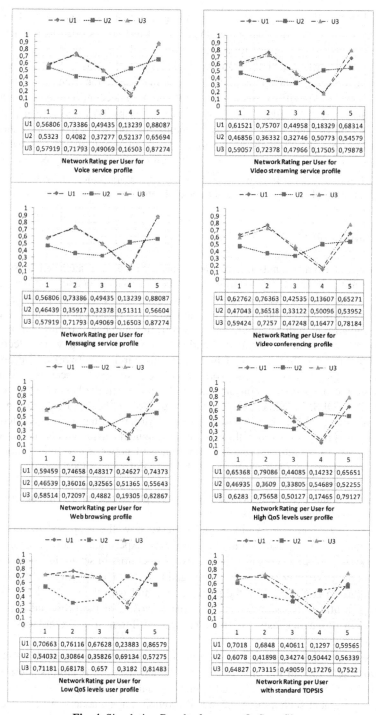

Fig. 4. Simulation Results for seven QoS profiles

Nevertheless, if User 3 is considered it can be seen that discrepancies in the relative importance of attributes may neutralize the effect of different service requirements, i.e. for User 3, Network 5 is optimal for all QoS profiles followed by Network 2.

As far as comparison with the standard TOPSIS method is concerned, the results of the experiments show that it is not adequate to handle selection scenarios corresponding to different QoS profiles, since it does not take into account requirements pertaining to different QoS profiles. The approach followed by Bari and Leung [9] in the context of heterogeneous wireless environments that proposes the representation of different QoS profiles through the tuning of the relative importance of attributes by the network operator is inflexible since, in the context of the same QoS profile, users cannot express their own preferences for the relative importance of attributes as they are fixed and predefined by the operator for each profile.

5   Conclusions

The success of a virtual network operator participating in a saturated mass market depends mainly on its ability to offer value-added services focusing on the satisfaction of specific needs of distinct groups of users. In this article we propose a novel intelligent network selection method for the users of a virtual operator that consumes network capacity with different characteristics from a number of traditional operators. Our approach constitutes an original direction for research since network selection has been studied only in the context of heterogeneous wireless networks. The proposed method takes into account both network criteria and user preferences as well as QoS profiles so that the user will consume network services in the best possible way. To this aim, the TOPSIS method is modified in order to produce results according to the QoS requirements of different types of services. Moreover, the user is situated in the center of the process not only through using linguistic assessments and triangular fuzzy numbers to capture the relative importance of network attributes according to the user preferences, but also through the ability to define user QoS profiles according to which networks are rated and ranked. This feature enhances the flexibility of the proposed mechanism in comparison with works published earlier in the context of heterogeneous wireless networks.

In order to demonstrate, validate and observe how the proposed method would work we have designed and executed a simulation experiment where three users of a virtual operator select among five host network operators in seven different scenarios corresponding to five service QoS profiles and two user QoS profiles. The simulation also facilitates a comparison of the proposed method with the standard TOPSIS method. The results of the simulation experiment show that the proposed approach is successful in producing results corresponding to different user preferences and QoS requirements.

The results of this article provide a basis for the initiation of research into the area of network selection in the environment of virtual network operators. Furthermore, the proposed method may be utilized from a virtual network operator in order to design and implement such a mechanism and provide it as an added-value service to its users.

The main limitation of this study is that the proposed method models QoS attributes as precise, crisp data. The method should be modified in order to model these parameters as random variables in time and space which are also dependent on routes and then extract weighted averages on specific time intervals. This modification, which constitutes one of the main future directions for our research, entails the design of appropriate mechanisms for QoS measurements, which should be also implemented in user equipment. A further future direction is the evaluation of the proposed method through the consideration of other decision methods and the comparison with our approach.

References

1. Cricelli, L., Grimaldi, M., Ghiron, N.L.: The competition among mobile network operators in the telecommunication supply chain. Int. J. Prod. Econ. 131, 22–29 (2011)
2. Jaspers, F., Hulsink, W., Theeuwes, J.: Entry and Innovation in Maturing Markets: Virtual Operators in Mobile Telecommunications. Tech. Anal. Strategic Manage. 19, 205–225 (2007)
3. Bluestein, W.: Three keys to MVNO success. Telephony 247, 20 (2006)
4. Gustaffson, E., Jonsson, A.: Always Best Connected. IEEE Wirel. Commun. 10, 49–55 (2003)
5. Kassar, M., Kervella, B., Pujolle, G.: An overview of vertical handover decision strategies in heterogeneous wireless networks. Comput. Commun. 31, 2607–2620 (2008)
6. Yan, X., Sekercioglu, A., Narayanan, S.: A survey of vertical handover decision algorithms in Fourth Generation heterogeneous wireless networks. Comput. Netw. 54, 1848–1863 (2010)
7. Hwang, C.L., Yoon, K.: Multiple Attribute Decision Making: Methods and Applications. Springer (1981)
8. Yong, D.: Plant location selection based on fuzzy TOPSIS. Int. J. Adv. Manuf. Tech. 28, 839–844 (2006)
9. Bari, F., Leung, V.C.M.: Automated Network Selection in a Heterogeneous Wireless Network Environment. IEEE Network 21, 34–40 (2007)
10. Zanakis, S.H., Solomon, A., Wishart, N., Dublish, S.: Multi-attribute decision making: A simulation comparison of select methods. Eur. J. Oper. Res. 107, 507–529 (1998)
11. Chen, C.T.: A fuzzy approach to select location of the distribution center. Fuzzy Set. Syst. 18, 65–73 (2001)
12. Desheng Wu, J.Z., Olson, D.L.: The Method of Grey Related Analysis to Multiple Attribute Decision Making Problems with Interval Numbers. Math. Comput. Model. 42, 991–998 (2006)
13. Chamodrakas, I., Alexopoulou, N., Martakos, D.: Customer evaluation for order acceptance using a novel class of fuzzy methods based on TOPSIS. Expert Syst. Appl. 36, 7409–7415 (2009)
14. Ribeiro, R.A.: Fuzzy multiple attribute decision making: A review and new preference elicitation techniques. Fuzzy Set. Syst. 78, 155–181 (1996)

Position and Velocity Predictions of the Piston in a Wet Clutch System during Engagement by Using a Neural Network Modeling

Yu Zhong[1], Bart Wyns[1], Abhishek Dutta[1], Clara-Mihaela Ionescu[1], Gregory Pinte[2], Wim Symens[2], Julian Stoev[2], and Robin De Keyser[1]

[1] Department of Electrical energy, Systems and Automation (EeSA), Ghent University, 913 Technologiepark, Zwijnaarde, 9052, Belgium
{Yu.Zhong,Bart.Wyns,Dutta.Abhishek,Claramihaela.Ionescu, Robain.DeKeyser}@ugent.be
[2] Flanders' Mechatronics Technology Center (FMTC), Celestijnenlaan 300D, Heverlee, 3001, Belgium
{Gregory.Pinte,Wim.Symens,Julian.Stoev}@fmtc.be

Abstract. In a wet clutch system, a piston is used to compress the friction disks to close the clutch. The position and the velocity of the piston are the key effectors for achieving a good engagement performance. In a real setup, it is impossible to measure these variables. In this paper, we use transmission torque and slip to approximate the piston velocity and position information. By using this information, a process neural network is trained. This neural predictor shows good forecasting results on the piston position and velocity. It is helpful in designing a pressure profile which can result in a smooth and fast engagement in the future. This neural predictor can also be used in other model-based control techniques.

Keywords: Neural predictor, wet clutch system, nonlinear system, process neuron model.

1 Introduction

A clutch is a mechanical device which provides for the transmission power from input device to output device. Clutches are used whenever the transmission of power or motion needs to be controlled either in amount or over time. Normally inside of a clutch system, there are hydraulic, electrical and mechanical components. It is very difficult to identify and model the wet clutch system since this is a typical nonlinear system [1].

Recently, some advanced learning control methods are used to control and optimize the engagement performance of the wet clutch system. Zhong, *et al.* [2], used a genetic algorithm (GA) to optimize the parametric signal to improve the engagement performance. Depraetere, *et al.* [3, 4], introduced a two-level iterative learning controller (ILC) to generate and track an optimal pressure profile, so that a good

L. Iliadis et al. (Eds.): AIAI 2012, IFIP AICT 381, pp. 474–482, 2012.

engagement performance can be achieved. Dutta, *et al.* [5], proposed to solve this problem by using robust model predictive control (MPC) technique. Gagliolo, *et al.* [6], validated the potential of implementing reinforcement learning techniques on the wet clutch system. This paper focuses on the use of neural network to solve this complex problem.

An artificial neural network has many useful properties and capabilities such as nonlinearity; input-output mapping; adaptivity; and so on [7]. With such advantages, the applications of the neural network can be found in every engineering field such as vehicle system [8], target recognition [9], and so on. In this paper, we will use the neural network modeling algorithm to get the prediction of the position and velocity of the piston in a wet clutch system during its engagement process. The difficulties in training this neural network are:

1. we cannot measure the position and the velocity of the piston in the real test, so we need to find out the information based on the available measurements;
2. the inputs and the outputs of the neural network are time-varying.

To solve these problems, we will first model the wet clutch based on fundamental physical laws and trying to find out the relations between the position/velocity of the piston and the available measurements. Then a process neural network will be trained.

2   Principle of a Wet Clutch and Theoretical Modeling

2.1   Wet Clutch System

A wet clutch is a clutch that is immersed in a cooling lubricating fluid which keeps the surfaces clean and gives smoother performance and longer life [10]. However, the wet clutch could lose some energy to the liquid since the surfaces can be slippery. One way to overcome this drawback is to stack multiple clutch disks. Fig. 1 shows the design of a wet clutch. An electro-hydraulic pressure-regulated proportional valve regulates the pressure inside the clutch, such that the position of a piston which presses the multiple clutch disks together can be controlled, thus the torque is transmitted. Once the command to engage the clutch is received, the left chamber of the clutch is filled with oil, so that the piston is pushed forward. A good engagement is then defined as decreasing the distance between the piston and the disks as fast as possible to zero, without the piston and the disks making brutal contact.

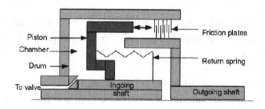

Fig. 1. Wet clutch design

The test bench used in this paper contains an electromotor and a flywheel. The electromotor drives a flywheel via two mechanical transmissions: one transmission is controlled in this project; the other transmission is used to vary the load and to adjust the braking torque. The sampling frequency on this setup is 1000Hz.

2.2   Feed Forward Control Strategy and Parameterized Signal

Wet clutches used in industry are filled with a feed-forward parametric signal of the current to the electro-hydraulic valve. Fig. 2 shows a typical parameterized, feed-forward current signal, which is sent to the valve [11].

The shape of this signal perfectly illustrates the underlying idea behind the actual industrial control design. First, a step signal with height a and width b is sent to the valve to generate a high pressure level in the clutch. With this pressure, the piston will overcome the force from the return spring, and start to get closer to the clutch disks. After this pulse, the signal will give some lower current with height c and width d to decelerate the piston and trying to position it close to the clutch disks. Once the piston is close to the clutch disks and with very low velocity, a force is needed to push the piston forward so that the clutch disks are compressed together. This force will be generated by the pressure which is caused by the step current with height e and width f. Then a ramp current signal with slope alpha and the end height g is sent to the valve so that the pressure inside the clutch will increase again gradually. In order to secure the full closing of the clutch, the signal will be kept as a high current level afterwards. Many research efforts can be found in tuning such kind of signals in order to achieve a good engagement [2, 3, 12].

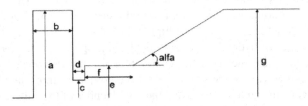

Fig. 2. Typical parameterized signal for controlling the wet clutch system

2.3   Theoretical Modelling

In Fig. 3(a), an example of the output torque and slip of a wet clutch during the whole engagement process are illustrated. The unit for the torque is $N \cdot m$, while the slip is calculated by (Fig. 3(b)):

$$Slip = (\omega_1 - \omega_2)/\omega_1 \tag{1}$$

where ω_1 is the input speed, and ω_2 is the output speed, and the resulted slip has no unit. Since the existence of the breaking torque, the final transmission torque does not go to zero. For confidential reasons, all results in this paper will be scaled.

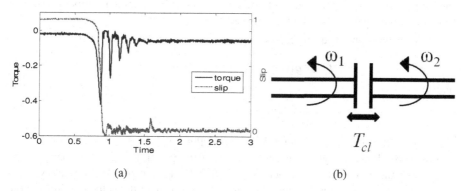

(a)   (b)

Fig. 3. (a) Illustrative example of the performance of a wet clutch; (b) Torque transmission in the wet clutch

When closing a wet clutch, the inside of the system can be explained by some fundamental physical laws. An illustrative figure is given to show the closing of the wet clutch (Fig. 4).

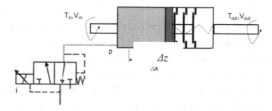

Fig. 4. The closing of a wet clutch, i is the triggle signal for the valve, p is the flow pressure, T_{in} and V_{in} are the input torque and velocity, T_{out} and V_{out} are the output torque and velocity, and Δz is the displacement of the piston.

The pressure of the fluid in the left side of the chamber is denoted as p_c, and if the area of the piston is known as A, then the normal force applied on the piston F can be expressed as:

$$F = p_c \cdot A \tag{2}$$

and by using Newton's law, we can have:

$$F = m \cdot \ddot{z} + b \cdot \dot{z} + k(z + z_0) \tag{3}$$

where m is the mass of the piston, b is the total damping efficient in the chamber, k is the spring constant of the return spring, and z_0 is the position of the piston when the clutch is fully open.

In Fig. 3(b), we schematically illustrate the torque transmission in a wet clutch. The rotational motion of the input shaft is on the left side with an angular velocity ω_1, and the output shaft is on the other side of the figure with an angular velocity ω_2. The transmission torque on the clutch is denoted as T_{cl}.

The transmission torque T_{cl} can be expressed as a function of slip [13]:

$$T_{cl} = g[(\omega_1 - \omega_2)] \tag{4}$$

and:

$$T_{cl} = \mu \cdot (\omega_1 - \omega_2) \tag{5}$$

where μ is a generalized coefficient which is proportional to the velocity of the piston.

From the above analysis, we can notice that the position and the velocity of the piston are the key factors for achieving a good engagement. Unfortunately, it is impossible to use any sensor to measure either the position or the velocity of the piston. To overcome this drawback, a neural predictor can be used to predict the position and velocity of the piston.

3   A Process Neural Network and Its Input and Outputs

3.1   Inputs and Outputs

From the analysis in section 2.3, we can understand that the purpose for training a neural predictor is to predict the position and the velocity of the piston with regards to the input pressure. Since there is no measurement for either position or velocity, we have to obtain the information based on the measureable outputs.

On the test bench, the available measurable variables are input rotational speed, output rotational speed, and transmission torque. Based on the input and output rotational speed, slip value can be calculated by (1). It can be assumed that when the piston has moved 90% of the complete distance to the friction plates, the output shaft starts to accelerate. From this position, the speed of the piston is critical since it will directly affect the transmission torque. Thus, we can approximate the position and velocity based on the slip and torque respectively. And for the position, it is very hard to define in accurate, so we set the position as a binary variable, and defined as [14]:

$$z(t) = \begin{cases} 0, & Slip \geq 0.95 \\ 1, & Slip < 0.95 \end{cases} \tag{6}$$

And the velocity:

$$v(t) = K_v[T(t) - T(t-1)] \tag{7}$$

where $T(t)$ is the torque measurement, and K_v is the gain. The value we get is not the real velocity, but this can give an indication of the speed changes according to the different input pressure profiles. The input of the system is the pressure $p(t)$. In the training process, since K_v and A are constants, the values can be arbitrary. For simplification, these values are set as 1.

3.2   Process Neuron Model

Since the inputs and the outputs are all time-varying functions, a conventional neuron model is not suitable for this case study. A process neuron and process neuron network [15], similar to additive model [16], will be implemented here to obtain a neural predictor.

The major difference between a process neuron and a traditional artificial neuron is in the process neuron, i.e., the inputs and the synaptic weights are time-varying functions. Additionally, besides the linear combiner to sum up the weighted inputs, there is also a time aggregation operator, which can integrate the input time-varying signal. Fig. 5(a) shows a process neuron model.

Therefore, we can write the output y as:

$$y = \varphi\left(\left(\int_0^T (\textstyle\sum_{i=1}^n w_i(t)x_i(t))\right) dt - B_k(t)\right) \tag{8}$$

If considering a neural network which is composed by neurons presented in (8), we can have the output of this neural network y(t) as:

$$y(t) = \textstyle\sum_{j=1}^m v_j(t) f_j\left(\int_0^T (\textstyle\sum_{i=1}^n w_{ij}(t)x_{ij}(t))dt - B_j(t)\right) \tag{9}$$

where $v_j(t)$ is the connection weight from the hidden layer to the output layer; n is the number of inputs; and m is the number of neurons in the hidden layer (see Fig. 5(b)).

Taking the pressure $p(t)$ as the input and the position $z(t)$ as the output, and $z'(t)$ as the desired output for the given input, the error of the neural network can be written as:

$$e = \|z(t) - z'(t)\| \tag{10}$$

where $z(t)$ is:

$$z(t) = \textstyle\sum_{j=1}^m v_j(t) f\left(\int_0^T w_j p(t)dt - \varphi_j(t)\right) \tag{11}$$

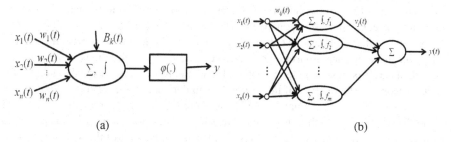

(a)   (b)

Fig. 5. (a)Process neuron model, $x_1(t)$, $x_2(t)$,...$x_n(t)$ are the time –varying inputs; $w_1(t)$, $w_2(t)$,...$w_n(t)$ are the time –varying connection weights; B_k is the bias of the neuron, also a time-varying function; $\sum$ is the adding operator; $\int$ is the time aggregation operator; $\varphi(.)$ is the activation function; and y is the output; (b) Process neural network

The learning rules for the connection weights and bias are:

$$w_{ij}^k(t) = w_{ij}^{k-1}(t) + \alpha \Delta w_{ij}^{k-1}(t) \tag{12}$$

$$v_j^k(t) = v_j^{k-1}(t) + \beta \Delta v_j^{k-1}(t) \tag{13}$$

$$\varphi_j^k(t) = \varphi_j^{k-1}(t) + \gamma \Delta \varphi_j^{k-1}(t) \tag{14}$$

where α, β, and γ are the learning rates, k is the number of iteration.

Finally, we have that:

$$\Delta w_{ij}^k(t) = -\partial e / \partial w_{ij}^{k-1}(t) \tag{15}$$

$$\Delta v_j^k(t) = -\partial e / \partial v_j^{k-1}(t) \tag{16}$$

$$\Delta \varphi_j^k(t) = -\partial e / \partial \varphi_j^{k-1}(t) \tag{17}$$

And the same method is used to build the neural model for the velocity prediction.

4   Experiments and Results

The data used to train and validate the neural network is from [2]. In this paper, engagement time and the torque loss are treated as two objectives to be optimized simultaneously. The tuning target is the parametric signal in Fig. 2. A nondominated sorting genetic algorithm (nsGA) [17] is used in order to obtain the Pareto solutions. In this method, the nondominated solutions constituting a nondominated front are assigned the same dummy fitness value. Nondomination here is defined as: in a minimization problem, if a vector X1 is partially less than X2, we say X1 dominates X2, and any member of such vectors that is not dominated by any other member is said to be nondominated [17]. These solutions are ignored in the further classification process. Then the front is extracted. This procedure is repeated until all individuals in the population are classified.

The genetic algorithm has a population size equal to 50, and stops at 8th generation. So in that paper [2], totally there are 400 test runs. Some of the individuals didn't pass the safety check, and were penalized. The safety check is defined as "For safety reasons, the individuals are tested under low external load and low breaking inertial condition in laboratory environments. Before sending the individuals to test, all the individuals are first tested under no external load and no breaking inertial condition for safety reasons. If the reading of the torque loss under such working condition is larger than 200 N·m for any individual, then it is considered as "unsafe", and will not be tested further." Individuals which failed to pass the safety check are not used in this paper, so the total number of remaining test runs is 369. In this study, 70% of the data are used for training and the remaining data is used for validating the network. The neural network has 1 input and 2 outputs, and 5 neurons in the hidden layer, so the structure is 1-5-2.

Some validation results are shown in Fig. 6. We can find that the obtained neural network can provide a very good forecast on the position and velocity of the piston.

In Fig. 6(a), the slip drops below 0.95 at 0.327s, while the neural network gives the position indicator switch from 0 to 1 at 0.324s. In Fig. 6(b), we can notice that from 0.278s to 0.309s, the transmission torque experiences a sudden drop, and the neural network at this time gives the prediction of the velocity at this time will be very large.

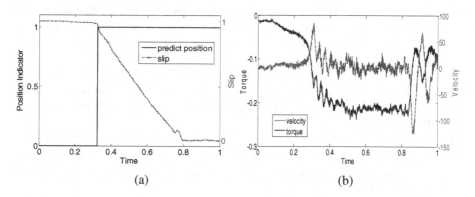

(a)   (b)

Fig. 6. (a) Predictive position indicator and slip; (b) Predictive velocity and transmission torque

5   Concluding Remarks and Future Work

In this paper, a process neural network is developed for predicting the position and velocity of the piston in a wet clutch system during its engagement. The input and outputs of the neural network are selected based on the fundamental analysis of the physical behavior of the wet clutch. The resulted neural network shows a good prediction on the position and velocity for a given input pressure profile.

Next, this neural network can be used for designing a pressure profile which can have the output as: switch on the position indicator as fast as possible while keep the velocity at this moment as low as possible. We expect that such pressure profiles can ensure a smooth and fast engagement on the test bench.

Acknowledgements. This work has been carried out within the framework of the LeCoPro project (grant nr. 80032). C.M. Ionescu is financially supported by the Flanders Research Center (FWO).

References

[1] Widanage, W.D., Stoev, J., Van Mulders, A., Schoukens, J., Pinte, G.: Nonlinear system-identification of the filling phase of a wet-clutch system. Control Engineering Practice 19(12), 1506–1516 (2011)
[2] Zhong, Y., Wyns, B., De Keyser, R., Pinte, G., Stoev, J.: An implementation of genetic-based learning classifier system on a wet clutch system. In: 14th Conference of the ASMDA International Society, Rome, Italy, pp. 1924–1931 (2011)

[3] Depraetere, B., Pinte, G., Swevers, J.: Iterative optimization of the filling phase of wet clutches. In: The 11th International Workshop on Advanced Motion Control, Nagaoka, Japan, pp. 94–99 (2010)

[4] Depraetere, B., Pinte, G., Swevers, J.: A reference free iterative learning strategy. In: 30th Benelux Meeting on Systems and Control, Lommel, Belgium (2011)

[5] Dutta, A., De Keyser, R., Zhong, Y., Pinte, G., Stoev, J., Wyns, B.: Robust predictive control of a wet-clutch using evolutionary algorithm optimized engagement profile. In: 15th International Conference on System Theory, Control, and Computing, Sinaia, Romania (2011)

[6] Gagliolo, M., Van Vaerenbergh, K., Rodriguez, A., Nowe, A., Goossens, S., Pinte, G., Symens, W.: Policy search reinforcement learning for automatic wet clutch engagement. In: 15th International Conference on System Theory, Control, and Computing, Sinaia, Romania (2011)

[7] Haykin, S.: Neural networks: a comprehensive foundation. Printice Hall, USA (1999)

[8] Ghazi Zadeh, A., Fahim, A., El-Gindy, M.: Neural network and fuzzy logic applications to vehicle system: literature survey. International Journal of Vehicle Design 18(2), 93–132 (1997)

[9] Roth, M.W.: Survey of neural network technology for automatic target recognition. IEEE Transactions on Neural Networks 1(1), 28–43 (1990)

[10] Lake, J., Romas, B.: Self-contained wet clutch system. US Patent No. 3,314,513 (1967)

[11] Hebbale, K.V., Kao, C.-K., McCulloch, D.E.: Adaptive electronic control of power-on upshifting in an autamatic transmission. US Patent No. 5,282,401 (1994)

[12] Lazar, C., Caruntu, C.-F., Balau, A.-E.: Modelling and predictive control of an electro-hydraulic actuated wet clutch for automatic transmission. In: IEEE International Symposium on Industrial Electronics, Bari, Indonesia, pp. 256–261 (2010)

[13] Dassen, M.: Modelling and control of automotive clutch systems. Report number 2003.73, Eindhoven (2003),
http://alexandria.tue.nl/repository/books/615474.pdf

[14] Pinte, G., Stoev, J., Symens, W., Dutta, A., Zhong, Y., Wyns, B., De Keyser, R., Depraetere, B., Swevers, J., Gagliolom, M., Nowe, A.: Learning Strategies for wet clutch control. In: 15th International Conference on System Theory, Control, and Computing, Sinaia, Romania (2011)

[15] He, X.G., Xu, S.H.: Process neural networks and its application in time-varying information processing. CAAI Transactions on Intelligent Systems 1(1), 1–6 (2006)

[16] Rall, W.: Cable theory for dendritic neurons. In: Methods in Neuronal Modeling, pp. 9–62. MIT Press, Cambridge (1989)

[17] Srinivas, N., Deb, K.: Multiobjective Function Optimization Using Nondominated Sorting Genetic Algorithms. Evolutionary Computation 3, 221–248 (1995)

Uniform Asymptotic Stability and Global Asymptotic Stability for Time-Delay Hopfield Neural Networks

Adnene Arbi, Chaouki Aouiti, and Abderrahmane Touati

University of Carthage, Faculty of sciences of Bizerta, Department of Mathematics,
7021 Jarzouna Bizerta, Tunisia
adnen.arbi@enseignant.edunet.tn, {chaouki.aouiti,Abder.Touati}@fsb.rnu.tn

Abstract. In this paper, we consider the uniform asymptotic stability and global asymptotic stability of the equilibrium point for time-delays Hopfield neural networks. Some new criteria of the system are derived by using the Lyapunov functional method and the linear matrix inequality approach for estimating the upper bound of the derivative of Lyapunov functional. Finally, we illustrate a numerical example showing the effectiveness of our theoretical results.

1 Introduction

Hopfield ([1-2]) has proposed Hopfield neural networks (HNN) which have found applications in a broad range of discipline where the targeted problems can reduce to optimization problems. It has been extensively studied and developed in recent years, and it has attracted much attention in the literature on Hopfield neural networks with time delays, . They are now recognized as candidates for information processing systems and have been successfully applied to associative memory, pattern recognition, automatic control, model identification, optimization problems, etc. (we refer to reader [3-10]). Therefore, the study of stability of HNN has caught many researchers' attention. HNN with time delays has been extensively investigated over the years, and various sufficient conditions for the stability of the equilibrium point of such neural networks have been presented via different approaches. In [5], [13], some sufficient conditions of stability by utilizing the Lyapunov functional method, and linear matrix inequality approach for delayed continuous HNN are derived. In [14], G.Zong and J.Liu established a novel delay-dependent condition to guarantee the existence of HNN and its global asymptotic stability by resorting to the integral inequality and constructing a Lyapunov-Krasovskii functional. In [16], S.Long and D.Xu got the sufficient conditions for global exponential stability and global asymptotic stability by using Lyapunov-Krasovskii-type functionally of negative definite matrix and Cauchy criterion.

This paper is organized as follows: In section 2, a model of time-delay Hopfield neural network is described. In addition, we present some basic definitions and lemmas. New stability criteria for Hopfield neural network are derived in section

L. Iliadis et al. (Eds.): AIAI 2012, IFIP AICT 381, pp. 483–492, 2012.

3. An example is given in section 4, to illustrate the advantage of the results obtained. Finally, some conclusions are drawn in section 5.

2   Preliminaries

Let $\mathbb{R}$ denote the set of real numbers, $\mathbb{Z}_+$ denote the positive integers and $\mathbb{R}^n$ denote the n-dimensional real space equipped with the Euclidean norm $\|.\|$. The identity matrix, with appropriate dimensions, is denoted by Id and $\operatorname{diag}(...)$ denotes the block diagonal matrix.

Consider the following delayed HNN model with impulses

$$
\begin{cases}
\dot{x}_i(t) = -c_i x_i(t) + \sum_{j=1}^{n} a_{ij} f_j(x_j(t)) \\
\quad + \sum_{j=1}^{n} b_{ij} g_j(x_j(t - \tau(t))) + I_i \ if \ t \neq t_k \\
\triangle x_i \backslash_{t=t_k} = d_k^{(i)}(x_i(t_k^-) - \bar{x}_i) \quad i = 1, ..., n, \quad n, k \in \mathbb{Z}_+,
\end{cases}
\tag{1}
$$

where $n \geq 2$ corresponds to the number of units in a neural network; the impulsive times t_k satisfy:

$0 \leq t_0 < t_1 < ... < t_k < ..., \lim_{k \longrightarrow +\infty} t_k = +\infty$; x_i corresponds to the state of the unit i at time t; c_i is positive constant; f_j, g_j, denote respectively, the measures of response or activation to their incoming potentials of the unit j at time t and $t - \tau(t)$; constant a_{ij} denotes the synaptic connection weight of the unit j on the unit i at time t; constant b_{ij} denotes the synaptic connection weight of the unit j on the unit i at time $t - \tau(t)$; I_i is the input of the unit i; $\tau(t)$ is the transmission delay such as $0 < \tau(t) \leq \tau$ and $\dot{\tau}(t) \leq \rho < 1$; $t \geq t_0$; $\tau, \rho, d_k^{(i)}$ are constants; $\bar{x}$ is the equilibrium point of the first equation in the system (1). The evolution of the neuron state i at time t_k is described by the equation:

$$
\triangle x_i \backslash_{t=t_k} = x_i(t_k) - x_i(t_k^-) = d_k^{(i)}(x_i(t_k^-) - \bar{x}_i), \ i = 1, 2, ..., n, \ k = 1, 2, ...
$$

The initial conditions associated with system (1) are of the form:

$$
x(s) = \phi(s), \ s \in [t_0 - \tau, t_0]
\tag{2}
$$

where

$$
x(s) = (x_1(s), x_2(s), ..., x_n(s))^T, \phi(s) = (\phi_1(s), \phi_2(s), ..., \phi_n(s))^T \in PC([-\tau, 0], \mathbb{R}^n)
$$

$PC([-\tau, 0], \mathbb{R}^n) = \{\psi : [-\tau, 0] \longrightarrow \mathbb{R}^n$, is continuous everywhere except at finite number of points t_k, at which $\psi(t_k^+)$ and $\psi(t_k^-)$ exist and $\psi(t_k^+) = \psi(t_k)\}$. For $\psi \in PC([-\tau, 0], \mathbb{R}^n)$, the norm of ψ is defined by

$$
\|\psi\|_\tau = \sup_{-\tau \leq \theta \leq 0} \|\psi(\theta)\|.
$$

For any $t_0 \geq 0$, let: $PC_\delta(t_0) = \{\psi \in PC([-\tau, 0], \mathbb{R}^n) : \|\psi\|_\tau < \delta\}$.

In this paper, we assume that some conditions are satisfied so that the equilibrium point of system (1) does exist, see ([5], [11]). Impulsive operator is viewed as perturbation of the equilibrium point $\bar{x}$ of such system without impulsive effects.

Since $\bar{x}$ is an equilibrium point of system (1), one can derive from system (1) that the transformation $y_i = x_i - \bar{x}_i$, $i = 1, 2, ..n$ transforms such system into the following system:

$$\begin{cases} \dot{y}_i(t) = -c_i y_i(t) + \sum_{j=1}^{n} a_{ij} F_j(y_j(t)) + \sum_{j=1}^{n} b_{ij} G_j(y_j(t - \tau(t))) \; si \; t \neq t_k \\ y_i(t_k) = (1 + d_k^{(i)}) y_i(t_k^-) \quad i = 1, ..., n, \quad n, k \in \mathbb{Z}_+, \end{cases} \quad (3)$$

where $F_j(y_j(t)) = f_j(\bar{x}_j + y_j(t)) - f_j(\bar{x}_j)$ and
$G_j(y_j(t - \tau(t))) = g_j(\bar{x}_j + y_j(t - \tau(t))) - g_j(\bar{x}_j)$.

To prove the stability of $\bar{x}$ of system (1), it is sufficient to prove the stability of the zero solution of system (3).

In this paper, we assume that there exist constants L_i, $M_i \geq 0$ such as

$$\mid F_i(y) \mid \leq L_i \mid y \mid, \; \mid G_i(y) \mid \leq M_i \mid y \mid, i \in \Lambda = \{1, 2, ...n\},$$

and we set

$$L_{max} = \max_{i \in \Lambda} L_i, M_{max} = \max_{i \in \Lambda} M_i, c_{max} = \max_{i \in \Lambda} c_i,$$

$$c_{min} = \min_{i \in \Lambda} c_i, D_k = diag(1 + d_k^{(1)}, 1 + d_k^{(2)}, ..., 1 + d_k^{(n)}).$$

So, the system (3) can be written as follows:

$$\begin{cases} \dot{y}(t) = -Cy(t) + A.F(y(t)) + B.G(y(t - \tau(t))) \; si \; t \neq t_k \\ y(t_k) = D_k y(t_k^-) \quad i = 1, ..., n, \quad n, k \in \mathbb{Z}_+, \\ y(t_0 + \theta) = \varphi(\theta), \end{cases} \quad (4)$$

where

$y(t) = (y_1(t), ..., y_n(t))^T; y(t - \tau(t)) = (y_1(t - \tau(t)), ..., y_n(t - \tau(t)))^T;$
$C = diag[c_1, ..., c_n]; A = (a_{ij})_{n \times n}; B = (b_{ij})_{n \times n}; F(y) = (F_1(y_1), F_2(y_2), ..., F_n(y_n))^T;$
$G(y) = (G_1(y_1), G_2(y_2), ..., G_n(y_n))^T.$

Some definitions and lemma of stability for system (1) at its equilibrium point are introduced as follows:

Definitions 21. *Assume $y(t) = y(t_0, \varphi)(t)$ be the solution of (3) through (t_0, φ), then the zero solution of (3) is said to be [12]*

P1 *stable, if for any $\epsilon > 0$ and $t_0 \geq 0$, there exists some $\delta(\epsilon, t_0) > 0$ such as $\varphi \in PC_\delta(t_0)$ implies $\|y(t_0, \varphi)(t)\| < \epsilon, \ t \geq t_0$.*

P2 *uniformly stable, if the δ in (P1) is independent of t_0.*

P3 *uniformly attractive, if there exists some $\delta > 0$ such as for any $\epsilon > 0$, there exists some $T = T(\epsilon, \delta) > 0$ such as $t_0 \geq 0$ and $\varphi \in PC_\delta(t_0)$ implies $\|y(t_0, \varphi)(t)\| < \epsilon, \ t \geq t_0 + T$.*

P4 *uniformly asymptotically stable, if (P2) and (P3) are held.*

P5 *globally asymptotically stable, if (P1) holds and for any given initial value $y_0 = \varphi$, $\|y(t_0, \varphi)(t)\| \longrightarrow 0$ as $t \longrightarrow +\infty$.*

Now, we need the following basic lemmas used in our work.

Lemma 21. *[15] For any $a, b \in \mathbb{R}^n$, the inequality*

$$\pm 2a^T b \leq a^T X a + b^T X^{-1} b$$

holds, where X is any $n \times n$ matrix with $X > 0$.

3   Robust Stability Criteria

In this section, we shall establish some theorems which provide sufficient conditions for uniform asymptotic stability and global asymptotic stability of system (1).

Theorem 31. *The system (1) is uniformly stable if there is $\epsilon^* \in [0, 1]$, $\sigma > 0$ and positive $n \times n$ definite matrix Q such as:*

(i) $\sigma \cdot \dfrac{\max\limits_{i,j} c_i \cdot q_{ij}}{\min\limits_{i,j} |b_{ij}|} \leq \dfrac{\tau^2 + 4 - \tau\sqrt{\tau^2 + 4}}{\tau^2 + 4 + \tau\sqrt{\tau^2 + 4}}$

(ii) $\max\limits_{1 \leq i \leq n} \{ \frac{1}{c_i} \sum\limits_{j=1}^{n} a_{ij} \} + \max\limits_{1 \leq j \leq n} \{ L_j^2 \sum\limits_{i=1}^{n} \frac{a_{ij}}{c_i} \} + \max\limits_{1 \leq j \leq n} \{ M_j^2 \sum\limits_{i=1}^{n} \frac{|b_{ij}|}{c_i} \} + \frac{\epsilon^*}{c_{min}}$
$+ \frac{1}{\sigma} \lambda_{max}(C^{-1} B Q^{-1} B^T C^{-1}) < 2$

(iii) $\dfrac{\prod\limits_{t_0 < t_k < t} \max\{c_{max} \xi_k, 1\}}{1 + \epsilon^*(t - t_0)^2} < \infty$, *where ξ_k is the largest eigenvalue of $D_k C^{-1} D_k$, and $k \in \mathbb{Z}_+$.*

In addition, if we have:

(iv) $\dfrac{\prod\limits_{t_0 < t_k < t} \max\{c_{max} \xi_k, 1\}}{1 + \epsilon^*(t - t_0)^2} \longrightarrow 0$, *if $t \longrightarrow +\infty$, then the system (1) is uniformly asymptotically stable and globally asymptotically stable.*

Proof. First, we prove the equilibrium point of system (1) is uniformly stable. We consider this Lyapunov function:

$$V(y)(t) = [1 + \epsilon^*(t - t_0)^2] \sum_{i=1}^{n} \frac{1}{c_i} y_i^2(t) + \sum_{i=1}^{n} \sum_{j=1}^{n} \frac{|b_{ij}|}{c_i} \int_{t-\tau(t)}^{t} (1 + (s - t_0)^2) G_j^2(y_j(s)) ds.$$

$$(5)$$

For condition (iii), there is a constant $M^* > 0$, such as:

$$\frac{\prod_{t_0 < t_k < t} \max\{c_{max}\xi_k, 1\}}{1 + \epsilon^*(t - t_0)^2} < M^*, t \geq t_0.$$

For any $t_0 \geq 0$, let $y(t_0, \varphi)(t)$ be a solution of system (1). So, $\forall \epsilon > 0$, we choose δ of the following manner:

$$\delta = \sqrt{\frac{1}{c_{max}[\frac{1}{c_{min}} + \max_{1 \leq i \leq n}\{\sum_{j=1}^{n} M_j^2 \frac{|b_{ij}|}{c_i}\}(\tau + \frac{\tau^3}{3})]M^*}} \cdot \epsilon$$

By simple calculation:

$$\frac{\partial V(y)(t)}{\partial t} = [1 + \epsilon^*(t - t_0)^2]\{-2\sum_{i=1}^{n} y_i^2(t) + 2\sum_{i=1}^{n}\sum_{j=1}^{n}\frac{a_{ij}}{c_i}y_i(t)F_j(y_j(t))$$

$$+ 2\sum_{i=1}^{n}\sum_{j=1}^{n}\frac{b_{ij}}{c_i}y_i(t)G_j(y_j(t - \tau(t)))\} + 2\epsilon^*(t - t_0)\sum_{i=1}^{n}\frac{1}{c_i}y_i^2(t)$$

$$+ \sum_{i=1}^{n}\sum_{j=1}^{n}\frac{|b_{ij}|}{c_i}(1 + (t - t_0)^2)G_j^2(y_j(t)) - \sum_{i=1}^{n}\sum_{j=1}^{n}\frac{|b_{ij}|}{c_i}(1 + (t - \tau(t) - t_0)^2)G_j^2(y_j(t - \tau(t))).$$

Therefore,

$$\frac{\partial V(y)(t)}{\partial t} = -2(1 + \epsilon^*(t - t_0)^2)\sum_{i=1}^{n} y_i^2(t) + (1 + \epsilon^*(t - t_0)^2)\{\sum_{i=1}^{n}\sum_{j=1}^{n}\frac{a_{ij}}{c_i}y_i^2(t)$$

$$+ \sum_{i=1}^{n}\sum_{j=1}^{n}\frac{a_{ij}}{c_i}F_j^2(y_j(t)) + \frac{1}{\sigma}\lambda_{max}(C^{-1}BQ^{-1}B^TC^{-1})\sum_{i=1}^{n} y_i^2(t)\}$$

$$+ 2\epsilon^*(t - t_0)\sum_{i=1}^{n}\frac{1}{c_i}y_i^2(t) + \sigma(1 + \epsilon^*(t - t_0)^2)\sum_{i=1}^{n}\sum_{j=1}^{n} q_{ij}G_j^2(y_j(t - \tau(t)))$$

$$- (1 + (t - \tau(t) - t_0)^2)\sum_{i=1}^{n}\sum_{j=1}^{n}\frac{|b_{ij}|}{c_i}G_j^2(y_j(t - \tau(t))) + (1 + (t - t_0)^2)\sum_{i=1}^{n}\sum_{j=1}^{n}\frac{|b_{ij}|}{c_i}G_j^2(y_j(t)).$$

It follows that,

$$\frac{\partial V(y)(t)}{\partial t} \leq (1 + (t - t_0)^2)\{-2\sum_{i=1}^{n} y_i^2(t) + \sum_{i=1}^{n}\sum_{j=1}^{n}\frac{a_{ij}}{c_i}y_i^2(t)$$

$$+ \sum_{i=1}^{n}\sum_{j=1}^{n}\frac{a_{ij}}{c_i}F_j^2(y_j(t)) + \sum_{i=1}^{n}\sum_{j=1}^{n}\frac{|b_{ij}|}{c_i}G_j^2(y_j(t)) + \frac{1}{\sigma}\lambda_{max}(C^{-1}BQ^{-1}B^TC^{-1})\sum_{i=1}^{n} y_i^2(t)$$

$$+ \frac{2\epsilon^*(t - t_0)}{1 + (t - t_0)^2}\sum_{i=1}^{n}\frac{1}{c_i}y_i^2(t)\} + \sum_{i=1}^{n}\sum_{j=1}^{n}[(1 + (t - t_0)^2)\sigma \cdot q_{ij}$$

$$- (1 + (t - \tau(t) - t_0)^2)\frac{|b_{ij}|}{c_i}]G_j^2(y_j(t - \tau(t))). \tag{6}$$

We have:

$$2\sum_{i=1}^{n}\sum_{j=1}^{n}\frac{1}{c_i}b_{ij}y_i(t)G_j(y_j(t-\tau(t)))=2y^T(t)C^{-1}BG(y(t-\tau(t)))$$

$$=2G^T(y(t-\tau(t)))B^TC^{-1}y(t)=2[G(y(t-\tau(t)))\sqrt{\sigma}]^T(B^TC^{-1}y(t)\frac{1}{\sqrt{\sigma}})$$

$$\leq\sigma G^T(y(t-\tau(t)))QG(y(t-\tau(t)))+\frac{1}{\sigma}y^T(t)C^{-1}BQ^{-1}B^TC^{-1}y(t)$$

$$\leq\sigma\sum_{i=1}^{n}\sum_{j=1}^{n}q_{ij}G_j^2(y_j(t-\tau(t)))+\frac{1}{\sigma}\lambda_{max}(C^{-1}BQ^{-1}B^TC^{-1})\sum_{i=1}^{n}y_i^2(t). \quad (7)$$

So, from (6) and (7):

$$\frac{\partial V(y)(t)}{\partial t}\leq(1+(t-t_0)^2)[-2+\max_{1\leq i\leq n}\{\frac{1}{c_i}\sum_{j=1}^{n}a_{ij}\}$$

$$+\max_{1\leq j\leq n}\{L_j^2\sum_{i=1}^{n}\frac{a_{ij}}{c_i}\}+\max_{1\leq j\leq n}\{M_j^2\sum_{i=1}^{n}\frac{|b_{ij}|}{c_i}\}+\frac{1}{\sigma}\lambda_{max}(C^{-1}BQ^{-1}B^TC^{-1})+\frac{\epsilon^*}{c_{min}}]\|y(t)\|^2$$

$$+\sum_{i=1}^{n}\sum_{j=1}^{n}[(1+(t-t_0)^2)\sigma\cdot q_{ij}-(1+(t-\tau(t)-t_0)^2)\frac{|b_{ij}|}{c_i}]G_j^2(y_j(t-\tau(t))).$$

Then we obtain,

$$\frac{\partial V(y)(t)}{\partial t}<0, \quad (8)$$

if

$$(1+(t-t_0)^2)\sigma\cdot q_{ij}\leq(1+(t-\tau(t)-t_0)^2)\frac{|b_{ij}|}{c_i}.$$

Therefore, it is sufficient that: $\frac{\sigma\cdot q_{ij}c_i}{|b_{ij}|}\leq\frac{(1+(t-\tau(t)-t_0)^2)}{(1+(t-t_0)^2)}$.

Let $u(t)=\frac{1+(t-\tau(t))^2}{1+t^2}$, next we show for $t\geq0$

$$u(t)\geq\frac{\tau^2+4-\tau\sqrt{\tau^2+4}}{\tau^2+4+\tau\sqrt{\tau^2+4}}. \quad (9)$$

First, for $t\in[\tau,+\infty[$, we have: $u(t)\geq\frac{1+(t-\tau)^2}{1+t^2}=v(t)$,
it is easy to compute that for $t\geq0$

$$v_{min}=v(\frac{\tau+\sqrt{\tau^2+4}}{2})=\frac{\tau^2+4-\tau\sqrt{\tau^2+4}}{\tau^2+4+\tau\sqrt{\tau^2+4}},$$

also we obtain $v(\tau)>v_{min}$, that is

$$\frac{1}{1+\tau^2}\geq\frac{\tau^2+4-\tau\sqrt{\tau^2+4}}{\tau^2+4+\tau\sqrt{\tau^2+4}}. \quad (10)$$

Second, for $t \in [0, \tau[$, we have:

$$u(t) = \frac{1 + (t - \tau(t))^2}{1 + t^2} \geq \frac{1}{1 + \tau^2}. \tag{11}$$

In view of (10) and (11), we obtain that (9) also holds for $t \in [0, \tau[$. Then we obtain that we have proved (9) holds for all $t \in [0, +\infty[$.

We have $\forall k \geq 1$:

$$V(y)(t_k) = [1 + (t_k - t_0)^2] \sum_{i=1}^{n} \frac{1}{c_i} y_i^2(t_k) + \sum_{i=1}^{n} \sum_{j=1}^{n} \frac{|b_{ij}|}{c_i} \int_{t_k - \tau(t_k)}^{t_k} (1 + (s - t_0)^2) G_j^2(y_j(s)) ds$$

$$= [1 + (t_k - t_0)^2] y^T(t_k^-) D_k C^{-1} D_k y(t_k^-) + \sum_{i=1}^{n} \sum_{j=1}^{n} \frac{|b_{ij}|}{c_i} \int_{t_k^- - \tau(t_k^-)}^{t_k^-} (1 + (s - t_0)^2) G_j^2(y_j(s)) ds$$

$$\leq [1 + (t_k - t_0)^2] \xi_k y^T(t_k^-) y(t_k^-) + \sum_{i=1}^{n} \sum_{j=1}^{n} \frac{|b_{ij}|}{c_i} \int_{t_k^- - \tau(t_k^-)}^{t_k^-} (1 + (s - t_0)^2) G_j^2(y_j(s)) ds$$

$$\leq [1 + (t_k - t_0)^2] \frac{\xi_k}{\lambda_{min}(C^{-1})} y^T(t_k^-) C^{-1} y(t_k^-) + \sum_{i=1}^{n} \sum_{j=1}^{n} \frac{|b_{ij}|}{c_i} \int_{t_k^- - \tau(t_k^-)}^{t_k^-} (1 + (s - t_0)^2) G_j^2(y_j(s)) ds$$

$$\leq \max\{\frac{\xi_k}{\lambda_{min}(C^{-1})}, 1\} V(t_k^-).$$

Therefore,

$$V(y)(t_k) \leq \max\{\xi_k c_{max}, 1\} V(t_k^-). \tag{12}$$

Then, we have:

$$\frac{1}{c_{max}} (1 + \epsilon^*(t - t_0)^2) \|y(t)\|^2 \leq V(t) \leq V(t_0) \times \prod_{t_0 < t_k \leq t} \max\{\xi_k c_{max}, 1\}. \tag{13}$$

From (5), we have:

$$\frac{1}{c_{max}} (1 + \epsilon^*(t - t_0)^2) \|y(t)\|^2 \leq V(t) \leq \frac{1}{c_{min}} (1 + \epsilon^*(t - t_0)^2) \|y(t)\|^2$$

$$+ \max_{1 \leq i \leq n} \{\sum_{j=1}^{n} M_j^2 \frac{|b_{ij}|}{c_i}\} (\tau(t) + \frac{(t - t_0)^3 - (t - t_0 - \tau(t))^3}{3}) \|y(t)\|^2.$$

Therefore,

$$V(t) \leq [\frac{1}{c_{min}} (1 + \epsilon^*(t - t_0)^2) + \max_{1 \leq i \leq n} \{\sum_{j=1}^{n} M_j^2 \frac{|b_{ij}|}{c_i}\} (\tau(t)$$

$$+ \frac{(t - t_0)^3 - (t - t_0 - \tau(t))^3}{3})] \|y(t)\|^2. \tag{14}$$

For $t = t_0$, we have:

$$V(t_0) \leq [\frac{1}{c_{min}} + \max_{1 \leq i \leq n} \{\sum_{j=1}^{n} M_j^2 \frac{|b_{ij}|}{c_i}\} (\tau + \frac{\tau^3}{3})] \|\varphi\|^2. \tag{15}$$

Then,

$$\frac{1}{c_{max}}(1+\epsilon^*(t-t_0)^2)\|y(t)\|^2 \leq [\frac{1}{c_{min}}+\max_{1\leq i\leq n}\{\sum_{j=1}^{n}M_j^2\frac{|b_{ij}|}{c_i}\}(\tau+\frac{\tau^3}{3})]\|\varphi\|^2\times \prod_{t_0<t_k<t}\{\xi_k c_{max}, 1\}.$$

This implies that:

$$\|y(t)\|^2 \leq [\frac{1}{c_{min}}+\max_{1\leq i\leq n}\{\sum_{j=1}^{n}M_j^2\frac{|b_{ij}|}{c_i}\}(\tau+\frac{\tau^3}{3})]c_{max}\|\varphi\|^2 \times \frac{\prod_{t_0<t_k<t}\{\xi_k c_{max}, 1\}}{(1+\epsilon^*(t-t_0)^2)} \leq \epsilon^2.$$

Hence, the zero solution of system (1) is uniformly stable.

In view of condition (iv), it is obvious that :

$\limsup_{t\longrightarrow+\infty}\|y(t)\|^2 = 0$, so the equilibrium point of system (1) is also uniformly asymptotically stable and globally asymptotically stable.

Which completes the proof.

If $\prod_{t_0<t_k\leq t}\max\{c_{max}\cdot\xi_k, 1\} < \infty$, then we can get the following criterion for stability with $\epsilon^* = 0$.

Corollary 32. *Assume that there is a constant $\sigma > 0$ and $n\times n$ positive definite matrix Q such as:*

(i) $\sigma \cdot \dfrac{\max\limits_{i,j} c_i\cdot q_{ij}}{\min\limits_{i,j}|b_{ij}|} \leq \dfrac{\tau^2+4-\tau\sqrt{\tau^2+4}}{\tau^2+4+\tau\sqrt{\tau^2+4}}$

(ii) $\max\limits_{1\leq i\leq n}\{\frac{1}{c_i}\sum\limits_{j=1}^{n}a_{ij}\} + \max\limits_{1\leq j\leq n}\{L_j^2\sum\limits_{i=1}^{n}\frac{a_{ij}}{c_i}\} + \max\limits_{1\leq j\leq n}\{M_j^2\sum\limits_{i=1}^{n}\frac{|b_{ij}|}{c_i}\}$
$+ \frac{1}{\sigma}\lambda_{max}(C^{-1}BQ^{-1}B^TC^{-1}) < 2.$

Then, the equilibrium point of system (1) is uniformly asymptotically stable and globally asymptotically stable.

If $\sigma = \frac{\tau^2+4-\tau\sqrt{\tau^2+4}}{\tau^2+4+\tau\sqrt{\tau^2+4}}$ and $Q = \frac{1}{60}Id$ in Corollary 32, then we can get the following criterion for stability.

Corollary 33. *Assume that the following conditions are satisfied:*

(i) $\dfrac{\max\limits_{i} c_i}{60\min\limits_{i,j}|b_{ij}|} \leq 1$

(ii) $\max\limits_{1\leq i\leq n}\{\frac{1}{c_i}\sum\limits_{j=1}^{n}a_{ij}\} + \max\limits_{1\leq j\leq n}\{L_j^2\sum\limits_{i=1}^{n}\frac{a_{ij}}{c_i}\} + \max\limits_{1\leq j\leq n}\{M_j^2\sum\limits_{i=1}^{n}\frac{|b_{ij}|}{c_i}\}$
$+ \frac{60}{\sigma}\lambda_{max}(C^{-1}BB^TC^{-1}) < 2.$
Then, the equilibrium point of system (1) is uniformly asymptotically stable and globally asymptotically stable.

4   Application

In this section, we present a numerical example to illustrate that our conditions are more feasible than those given in some earlier references [5].

Example 41. *Consider the two-neuron delayed neural network with impulses as follows:*

$$
\begin{cases}
\dot{x}_1(t) = -2.5x_1(t) - 0.5f_1(x_1(t)) + 0.1f_2(x_2(t)) \\
\quad -0.1g_1(x_1(t-\tau)) + 0.2g_2(x_2(t-\tau)) - 1 \\
\dot{x}_2(t) = -2x_2(t) + 0.2f_1(x_1(t)) - 0.1f_2(x_2(t)) \\
\quad +0.2g_1(x_1(t-\tau)) + 0.1g_2(x_2(t-\tau)) + 4 \\
\triangle x_i|_{t=t_k} = x_i(t_k) - x_i(t_k^-) = d_k^{(i)}(x_i(t_k^-) - \bar{x}_i) \\
k \in \mathbb{Z}_+,
\end{cases}
\tag{16}
$$

where $\tau = 0.87$, the activation functions are the following:

$$
f_1(x) = f_2(x) = g_1(x) = g_2(x) = 0.5(|x+1| - |x-1|)
$$

and $d_k^{(1)} = \sqrt{1 + \frac{1}{5k^2}} - 1$, $d_k^{(2)} = \sqrt{1 + \frac{1}{6k^2}} - 1$, $t_k = k$, $k \in \mathbb{Z}_+$.
So, the matrix A, B and C are:
$$
A = \begin{pmatrix} -0.5 & 0.1 \\ 0.2 & 0.1 \end{pmatrix}, \ B = \begin{pmatrix} -0.1 & 0.2 \\ 0.2 & 0.1 \end{pmatrix}, \ C = \begin{pmatrix} 2.5 & 0 \\ 0 & 2 \end{pmatrix}.
$$

By Matlab, we note
$$
C^{-1}BB^T C^{-1} = \begin{pmatrix} 0.008 & 0 \\ 0 & 0.0125 \end{pmatrix} \ so, \ \lambda_{max}(C^{-1}BB^T C^{-1}) = 0.0125.
$$
Considering the activation functions f_1, f_2, g_1 and g_2, we can choose $L_i = 1$, $M_i = 1$, $i = 1, 2$.
On the other hand, by Mathematica software, we note

$$
\prod_{s=1}^{\infty} \max_{i=1,2}(1 + d_s^{(i)})^2 = \prod_{s=1}^{\infty} \max_{i=1,2}(1 + \frac{1}{5s^2})^2 < 1.4
$$

We show the equilibrium point of (16) is uniformly stable and globally asymptotically stable.
From Corollary 33: $\sigma = \frac{\tau^2 + 4 - \tau\sqrt{\tau^2 + 4}}{\tau^2 + 4 + \tau\sqrt{\tau^2 + 4}} \simeq 0.4297$, $\frac{1}{\sigma} \simeq 2.3272$. Thus,

$$
\max_{1 \le i \le n} \{\frac{1}{c_i} \sum_{j=1}^{n} a_{ij}\} + \max_{1 \le j \le n} \{L_j^2 \sum_{i=1}^{n} \frac{a_{ij}}{c_i}\} + \max_{1 \le j \le n} \{M_j^2 \sum_{i=1}^{n} \frac{|b_{ij}|}{c_i}\} + \frac{60}{\sigma}\lambda_{max}(C^{-1}BB^T C^{-1}) < 2.
$$

Hence, by Corollary 33, the equilibrium point $(-0.2258, 1.9548)^T$ of system (16) is uniformly asymptotically stable and globally asymptotically stable.

Remark 41. *In [5], the authors proved that system (16) is globally asymptotically stable with upperbound of delays $\tau = 0.17$. For this example, we additionally get that the equilibrium point of system (16) is uniformly asymptotically stable and globally asymptotically stable with upperbound of delays $\tau = 0.87 > 0.17$. However, the criteria given in [5] are invalid for ($\tau \ge 0.87$). Therefore, our results are less conservative and more efficient than those given in [5].*

5 Conclusion

In this paper, a class of HNN with delays is considered. We obtain some new criteria ensuring the global and uniform asymptotic stability of the equilibrium point for such system by using the Lyapunov method and linear matrix inequality. Our results show the effects of delay on the stability of HNN. The results here are compared to earlier results. Our criteria are more simpler to verify. An example is given to illustrate the efficiency of the results.

References

1. Hopfield, J.J.: Neural networks and physical systems with emergent collective computational abilities. Proc. Natl. Acad. Sci. 79, 2554–2558 (1982)
2. Hopfield, J.J.: Neurons with graded response have collective computational properties like those of two-state neurons. Proc. Natl. Acad. Sci. 81, 3088–3092 (1984)
3. Singh, V.: On global robust stability of interval Hopfield neural networks with delays. Chaos Solitons and Fractals 33, 1183–1188 (2007)
4. Huang, H., Cao, J.: On global asymptotic stability of recurrent neural networks with time-varying delays. Applied Mathematics and Computation 142, 143–154 (2003)
5. Zhang, Q., Xu, X.W.J.: Delay-dependent global stability results for delayed Hopfield neural networks. Chaos Solitons and Fractals 34, 662–668 (2007)
6. Liu, B.: Almost periodic solutions for Hopfield neural networks with continuously distributed delays. Mathematics and Computers in Simulation 73, 327–335 (2007)
7. Zhou, J., Xiang, L., Liu, Z.: Synchronization in complex delayed dynamical networks with impulsive effects. Physica A 384, 684–692 (2007)
8. Zhang, Y., Sun, J.: Stability of impulsive neural networks with time delays. Physica A 384, 44–50 (2005)
9. Chen, Z., Ruan, J.: Global stability analysis of impulsive Cohen-Grossberg neural networks with delay. Physica A 345, 101–111 (2005)
10. Xiang, H., Yan, K.M., Wang, B.Y.: Existence and global exponential stability of periodic solution for delayed high-order Hopfield-type neural networks. Physica A 352, 341–349 (2006)
11. Zhang, Q., Wei, X., Xu, J.: Delay-dependent global stability condition for delayed Hopfield neural networks. Nonlinear Analysis 8, 997 (2007)
12. Fu, X.L., Yan, B.Q., Liu, Y.S.: Introduction of Impulsive Differential Systems. Science Press, Beijing (2005)
13. Li, X., Chen, Z.: Stability properties for Hopfield Neural Networks with delays and impulsive perturbations. Nonlinear Analysis: Real World Applications 10, 3253–3265 (2009)
14. Zong, G., Liu, J.: New Delay-dependent Global Asymptotic Stability Condition for Hopfield Neural Networks with Time-varying Delays. International Journal of Automation and Computing, 415–419 (2009)
15. Liao, X., Chen, G., Sanchez, E.: LMI approach for global periodicity of neural networks with time-varying delays. IEEE Transactions on Circuits Syst. I 49, 1033 (2002)
16. Long, S., Xu, D.: Delay-dependent stability analysis for impulsive neural networks with time varying delays. Neurocomputing 71, 1705–1713 (2008)
17. Chen, A., Cao, J., Huang, L.: An estimation of upperbound of delays for global asymptotic stability of delayed Hopfield neural networks. IEEE Trans Circuits. Syst. I 49, 1028–1032 (2002)

Author Index